Dead Reckoning

Dead Reckoning

Air Traffic Control, System Effects, and Risk

Diane Vaughan

The University of Chicago Press
Chicago and London

The University of Chicago Press, Chicago 60637
The University of Chicago Press, Ltd., London

Published 2021
Paperback edition 2023
Printed in the United States of America

32 31 30 29 28 27 26 25 24 23   1 2 3 4 5

ISBN-13: 978-0-226-79640-6 (cloth)
ISBN-13: 978-0-226-82657-8 (paper)
ISBN-13: 978-0-226-79654-3 (e-book)
DOI: https://doi.org/10.7208/chicago/9780226796543.001.0001

Library of Congress Cataloging-in-Publication Data

Names: Vaughan, Diane, author.
Title: Dead reckoning : air traffic control, system effects, and risk / Diane Vaughan.
Description: Chicago : University of Chicago Press, 2021. | Includes bibliographical references and index.
Identifiers: LCCN 2020056565 | ISBN 9780226796406 (cloth) | ISBN 9780226796543 (ebook)
Subjects: LCSH: Air traffic control. | Air traffic controllers.
Classification: LCC TL725.3.T7 V384 2021 | DDC 387.7/404260973—dc23
LC record available at https://lccn.loc.gov/2020056565

♾ This paper meets the requirements of ANSI/NISO Z39.48-1992 (Permanence of Paper).

For my support team that has endured
through all the books,
all the events, over the years,
my dearest loves,
my very best work,
my children

Contents

Figures and Tables

Figures

Tables

PART I

Beginnings

Figure 1: The Wright brothers' first flight, on December 17, 1903, at Kitty Hawk, North Carolina
Smithsonian National Air and Space Museum (NASM 2002-16646)

Figure 2: Archie League, first air traffic controller, with signal flags and wheelbarrow, Lambert Municipal Field, St. Louis, Missouri, 1929
FAA History Office/National Archives (237-G-183-13)

1

Dead Reckoning

Reckoning, according to the dictionary, is a cognitive activity: an act or instance of taking into account, calculating, estimating. *Dead reckoning* is a navigational term that has these cognitive processes at its core. It refers to a procedure that attempts to locate something in space or time by deduction—that is, unaided by direct observation or direct evidence—and thus the original term, *ded reckoning*. The historical origin of dead reckoning is in early marine navigation. Unable to identify their location by direct observation or in relation to familiar landmarks, early mariners developed methods of observing and recording their position, distances and directions traveled, and currents of wind and water. The purpose was to calculate where their vessel was, to compare progress with a predetermined route, and to correct for any deviations. For many centuries, navigators relied on the positions and motions of sun and stars and direction of winds for their direction finding. The calculative, intuitive, and cognitive aspects of dead reckoning dominated navigational practice; material technologies were absent.

But gradually, the cognitive and the technological began to merge in navigation. Marine charts, maps, the compass, and devices for measuring speed and distance were among the earliest material technologies for sea navigation. They enabled navigators not only to plot where a craft was but also to predict where it would be at given time. But technological advance notwithstanding, mistake was endemic, due to calculation errors based on using these early devices. With the invention of the airplane in the early twentieth century, human cognition and material technologies have merged in both air and sea navigation, the changes driven by the continuing assumption that increasing the sophistication of the technology will improve the accuracy of measurement and prediction, reduce mistakes, and therefore increase safety.

Now, in the twenty-first century, dead reckoning has even broader meaning. The amount of traffic, the amount and complexity of the technology, and the institutional and organizational contexts of navigation have changed dramatically. So have the goals: dead reckoning includes not only selection of the course, staying on it, and avoiding collision but also mandates to achieve cost efficiency by minimizing fuel consumption and adhering to a predetermined schedule. Moreover, dead reckoning occurs at the organizational and system levels, as administrators estimate and calculate in order to track, predict, and be responsive to changing demands and resources. Counting and measurement dominate, as science and technology are deployed in the interest of accuracy, safety, and efficiency, as well as the survival of the organizational system itself.

This book explores dead reckoning in air traffic control in the early twenty-first century. The central puzzle is, what makes air traffic control so safe? Although the Federal Aviation Administration (FAA), the agency responsible for regulating air traffic in the United States, has been repeatedly castigated by Congress and the press for inefficiencies, costly technologies, congestion, and delays, the FAA's air traffic control system, responsible for the management of airplane movement on the ground and in the sky, nonetheless has a surprisingly positive safety record. Failures, in the form of accidents and collisions for commercial airlines, are a rarity. When these occur, most often they are due to pilot error or technical failure. In contrast to commercial airlines, accidents are frequent in general aviation, where pilots are not typically trained to be professional pilots, are less experienced, and fly in uncontrolled airspace. However, the safety record of air traffic control for commercial aviation is impressive.

In light of this safety record, the continuing cries from critics to increase safety by increasing reliance on automation and decreasing the number of air traffic controllers doesn't make sense. Historically, arguments for reducing the number of controllers have rested on the notion that with better technology, safety can be preserved and efficiency (read: cutting costs, meeting schedules) increased. However, insufficient numbers of controllers mean tired controllers, and tired controllers means errors. Automation, yes—the high volume of traffic calls for the best technology possible. But to reduce controllers, or, as some have even argued, replace them with technology? We had powerful evidence of controllers' importance during the September 11 terrorist attacks when, in an unprecedented situation, one unimagined in their training, technologies, or system design, air traffic controllers nationwide cleared the sky of over four thousand airplanes in a little over two hours. Without them, it would have been an even greater tragedy.

Although the failures of the FAA have been publicly derided, the contribution of the FAA's air traffic control system and its controllers have never been isolated and identified. To discover why air traffic control is so safe, this book narrows in on controllers, and on the cognitive, technical, and material practices that they acquire during their training and deploy in everyday air traffic and emergencies. It takes into account the relationship between controllers and their technologies, how controllers give them meaning, repurpose them, and change them to fit the local situation, and also the reverse, or how the technologies, architecture, and socially organized arrangements of the control room affect controllers' work.[1] Equally important, answering the question of what makes air traffic control so safe also demands a focus on the large, complex socio-technical system in which the work is done. The sociologist Robert Merton observed that all systems of social action produce unanticipated consequences: they can be positive or negative.[2] Robert Jervis, writing about political systems, warned that the characteristics of a system are different from—not greater than—the sum of its parts, so that looking at only the individual parts and their relations with one another misses the essence of the system and its effects.[3] Merton and Jervis both stressed that despite the variation in the interconnectedness of system parts, they will always react to one another, producing unintended consequences.

Pursuing this line of thinking, Charles Perrow, in his 1984 *Normal Accidents,* identified the error-inducing characteristics of high-risk technical systems, arguing that the complexity and tight coupling of the technical system's parts produce unavoidable, unanticipated negative consequences: hence, the *normal* accident.[4] His emphasis is on the interaction of complex structures and the inevitability of failure. In Perrow's schema, the air traffic control system, although complex and tightly coupled, ranks as a low risk technical system. Other scholars have gone further, identifying air traffic control as an *error-reducing* system. In recognition of its safety achievements, they have described the air traffic control system as an exemplar of a high-reliability organization.[5] To understand what makes high-reliability organizations so safe, these scholars have primarily examined the social psychology and interactions of small groups engaged in risky work: airline cockpit crews, workers on aircraft carrier flight decks, wildland firefighting crews, to name a few.[6]

These social psychological studies have yielded many important lessons about how small work groups make sense of situations and coordinate activities that have been relevant to improving safety in many other kinds of organizations. However, for the most part, studies of high-reliability organizations have purposely isolated workers from the larger socio-technical system and its institutional environment in order to better explore the dynamics of individual interactions and collective understanding in a group.[7] For example, one study focused on teamwork on an aircraft carrier flight deck when a flight comes in but left air force budgets and resources unquestioned.[8] How political conditions and social actors in the institutional environment affected the resources available for training and recruitment practices affected the work and work conditions on the flight deck were not part of the study.

Research on how large-scale socio-technical systems affect the interpretations and meanings in small-group interactions of people doing risky technical work has rarely been done.[9] Moreover—and surprisingly—although technology is central to research on normal accidents and high reliability, neither scholarship from science and technology studies nor workplace studies that locate human-technology interactions in the socially organized activities and physical settings of their use have been incorporated into the research of either specialty.[10] The rule that a change in one part of a complex system will affect other parts in unanticipated ways generally holds true. How has air traffic control, operating under similar political and economic pressures as the other large-scale socio-technical systems that Perrow defined as more risky, managed to avoid these same deleterious effects—or has it?

My approach deviates from both previous approaches by studying controllers, their technology, and work practices within the larger social context in which they are located. I use *system effects* to mean the dynamic relationship between conditions, events, and social actors in the institutional environment as they impact the air traffic control system, its organization and technology, and so change it, and how, in turn, the air traffic control system impacts the work and experiences of the people who do the technical work. This includes their reactions, as they change, confirm, or contest system effects. Therefore, to understand the inner workings of air traffic control, it is also necessary to explore the system through its history, politics, and the problem-solving social actors both externally and internally that have formed, re-formed, and constrained it.[11]

Consequently, this book goes beyond previous work by focusing on the ongoing relationship between history, institutions, organizations, and the social, technological, and material arrangements that constitute controllers' everyday practices in work settings. Necessarily, I combine historical ethnography with interviews, archival research, and surveys in order to capture system dynamics over time and social space: the past, the time of the study, and now. The substantive contribution of this book is to identify the essential characteristics of this *error-reducing system* of air traffic control. In a challenge to advocates for the cost-efficiency and safety gains of maximum automation, this book reveals the liabilities of technological innovation and argues for the importance of people. The theoretical and practical implications of these findings are considerable.

By embracing history, the book captures the changing nature of organizations, technologies, and work. The idea that an organization's fate is tied to its institutional environment—its origin, evolution, persistence or demise, capacities and vulnerabilities—is well studied and accepted. We also know that both institutions and organizations are created, changed, and constrained by heterogeneous social actors, which has consequences for an organization's structure, technology, performance, its people, and their work. Therefore, the focus throughout this book is on the air traffic control system, standardized and rule-embedded, within its historical shifting political, cultural, technological, and economic environment, exposing the impact on both everyday work and the workplace, as well as the responses of problem-solving individuals to contingency and the unanticipated consequences—both positive and negative—that result. As a result, the book illuminates our understanding of institutions of all kinds: their emergence, transformation, and technologies, and the effects of those things on the people who work there.[12]

To a great extent, we can think of all organizational systems as engaged in dead reckoning: internally preoccupied with predicting their own future positions in social space and time in relation to the positions of other organizations in their environment by deduction—unaided by direct observation or direct evidence. The analysis elaborates theories of boundaries and boundary work, showing how systems and their boundaries are created, how they expand over time, their permeability and stubborn resistance, and the difficulty of crossing those boundaries.[13] In the workplace, the book opens to full view the effects of system changes on intraorganizational structure, culture, cognition, meaning making, and everyday work practice. Thus, it builds upon workplace studies that examine technologies to support cooperative work that requires coordination between multiple users across time and social space.[14] In addition, it reveals the role of organizational systems in the production of professional expertise, showing how the problem-solving and material practices in the workplace are affected by institutional, organizational actors and factors outside the control room.

The case demonstrates the complexities of modernizing: the ramifications of advancing from simple to complex—here, from flags to "shrimp boats" to radar to automation—for designing and implementing technological infrastructures for large information spaces,[15] as well as for small spaces to carry out coordinated, technologically mediated or assisted work.[16] As more complex specialized technologies were developed for air traffic control, they had to be adapted and embedded in an aging socio-technical system, fitting not only into the workspace but also locating the necessary technological infrastructure into the existing organization structure. Repeatedly, the design and implementation of technological innovations created tensions between the standardization of the system and the need to customize to local situations. This same problem plagues many organizations that are currently challenged to keep up by patching the new onto the existing organization and technologies.

Finally, the relevance of this case extends to concerns about technology as the medium of transnational connection in a global society and the future of work in an age when competition drives a need for greater speed, accuracy, and efficiency through automation. Complex organizational systems are dynamic, processual, and unpredictable, so in spite of planning, outcomes are fraught with unanticipated consequences, both positive and negative.[17] This book shows that the old and the new do not readily mesh, causing lag in responses to changing external conditions and unanticipated consequences for the socio-technical system and for the people who work in it. For the complex systems of today, Arthur Stinchcombe's writing about the liabilities of technological and organizational innovation rings true.[18] Moreover, this book reveals how, in the short run, an organization can reproduce its flaws even when trying its utmost to change in order to survive a major crisis. At the same time that it conveys warnings, however, the book demonstrates the agency of the workforce in maintaining the viability of the systems that they inhabit. Incrementally, problem-solving people and organizations inside the air traffic control system have developed strategies of resilience, reliability, and redundancy that provided perennial dynamic flexibility to the parts of the system structure, and they have improvised tools of repair to adjust innovations to local conditions, contributing to system persistence.

Why Air Traffic Control?

I came to this project after studying how and why things went wrong in organizations. I had completed three books on the topic. The first involved a computer crime in which one organization defrauded another, the second looked at how intimate relationships come apart, and the third explored the causes of the National Aeronautics and Space Administration's flawed decision to launch the space shuttle *Challenger*.[19] During a long-term project of developing explanations by analogical comparison—looking for similarities and differences across cases—I was struck by the analogies across three projects so obviously different.[20] All three were organizations—an intimate relationship being the smallest organization we create—and also in common they had all publicly failed in some way. Moreover, and unsuspected by me at the outset of each project, the explanation of how things went wrong had a common pattern across the three cases: an unanticipated outcome was preceded by a long incubation period during which early warning signs were plentiful but were missed, ignored, or misinterpreted. Only in retrospect, when the negative consequences were known, did the meaning of these early warning signs become clear.

Equally surprising, the causes of organization failure in each case were common and ordinary: the very aspects of organizations designed to promote positive outcomes—structure, division of labor, culture, technology, socialization, rules and procedures—had the unintended consequences of producing mistakes, misconduct, disaster, and other failures that fit no category. The results shifted attention away from the usual tendency to attribute responsibility for failure and negative outcomes to human factors alone. Instead, each case exposed the subtle but powerful impact of the organizational systems in which we live and work on what we think, say, and do. Post-*Challenger*, I realized that I wasn't going to learn anything more by going in after the fact, when all the early warning signs were clear. As an ethnographer, I wanted to avoid the problem of retrospection by locating myself in a research setting where I could watch decisions being made, where technology and risk were integral to everyday work, and where people were trained to identify anomalies and deviations early, correcting them so that small mistakes didn't turn into personal, organizational, and/or international catastrophes. Air traffic control met all these criteria. Following my cross-case comparison of how things go wrong, air traffic control would be my negative case—the counterfactual example that shows how it might have been otherwise—providing some insight into how a complex socio-technical system gets things (mostly) right.[21]

A frame is a set of tools—theories, research design, methods—that help sort out analogies and differences between what is expected and what is discovered; it suggests the directions to look but does not predict what we will find. I wanted to know how controllers identified anomalies—early warning signs—and corrected them so little mistakes didn't turn into tragic errors, and second, how they coordinated activities with pilots and other controllers physically distant from them to move aircraft across the sky in a time-critical way.[22] Both called for dead reckoning: predicting the position of objects in space and time by deduction, unaided by direct observation or direct evidence. Central to my questions was the nature of the work itself and human-technology interaction in the workplace. My approach differs from previous work in several ways.

First, and in contrast to research in computer science and cognitive psychology that isolates human-machine or human-computer interaction from its social context, I explore human-technology interaction within the work setting as small groups of controllers coordinate a range of tasks interacting with one another and deploying multiple devices and socially organized skills and practices in cooperative work. Also novel, I define technology broadly as *technologies of coordination and control.* I include not only the obvious technologies—computers, radar, radios, binoculars, automation—but also material objects, less familiar to the public but crucial for safety and coordination on a daily basis. Some examples serve as sensitizers for the chapters to follow: checklists, the glass in tower windows, runway markings and lighting, workplace architecture, rules and procedures, documents and charts, pad management systems, cartographic maps of the sky, signaling systems on the ground, and perhaps most important, the formal training of air traffic controllers. These also act, affecting their work.[23] The safety of the system is not solely in its structures but also in its processes: the interaction between air traffic controllers and the multiple technologies of coordination and control that are the material objects on which dead reckoning depends.[24]

Second, what happens in the workspace cannot be separated from the context of the organization in which controllers work and its environment.

How do events, conditions, and individual and organizational actors in the external environment impact the air traffic control system, changing it, and how, in turn, does the response of the system affect the workplace, the work, and the people who do it? To capture these layered interactive effects, I decided to take a *situated action* approach, investigating the dynamic between the system's institutional environment, the organization as a socio-technical system, and controllers' material practices, interpretive work, and the meanings the work has for them.[25] Situating action in its larger social context opens a window into situated change: how controllers themselves enact change as they incorporate cognitive, organizational, and technological innovations into their sense making, adjusting plans to fit the local situation.[26]

Framing the study as situated action helped expand my study of controllers and their work in three ways. First, it expanded my study "up" to the institutional environment to examine how institutional actors and events in the political, economic, technological, and cultural realm affected the air traffic control organization, and as a consequence, dead reckoning. Second, it expanded "down" and "across" the system boundaries at the organization level to show how controllers experience, enact, and give meaning to material practices and technologies, and coordinate with pilots, other controllers in the same air traffic facility, across facilities, and in other parts of the system. Third, it expanded my study "back" in history, to embrace events, conditions, and actions of institutional actors and problem-solving heterogeneous actors of the past on the system and its structure, culture, architecture, and technologies as they manifested in local actions, improvisation, change, and persistence in the present.

Particularly relevant to me for understanding dead reckoning was the relationship between culture and cognition—most certainly, the production of cultural understandings in the workplace, but in addition, framing the project to include research on institutionalized cultural belief systems and what would be known subsequently as the institutional logics approach allowed me to explore how past events and actors external to the system influenced the way controllers thought, acted, and worked in the present.[27] In addition, the rich research in science and technology studies and history of technology was essential to understanding the effect of the social context on the production of scientific and technical knowledge and the social construction of technology itself.[28] Notably, as a socio-technical system, technologies mattered at the institutional, organizational, and individual levels.

If these connections were to be found in a workplace, they were most likely to be observable in a complex socio-technical system in which the work was tightly rule-bound, a setting in which written institutional rules and procedures are extensively scripted into the organization structures and processes at every level. The Federal Aviation Administration's National Airspace System, standardized for global, national, and local coordination, represents an extreme case of formalization and coordination where possible links between the institutional environment, the organizational system, and cognition might be tracked in the thoughts and actions of air traffic controllers, who are subjected to rigorous training and retraining throughout their careers. The National Airspace System includes both civilian and military airspace, as well as navigational facilities and airports of the United States, and is responsible for establishing national programs, policies, regulations, and standards; for managing airspace; for operating air navigation and communications systems and air traffic facilities; for separating and controlling aircraft; and for providing flight assistance.

Within the National Airspace System, my interest was in the Air Traffic Organization, and within it, the air traffic control system and its airspace, facilities, devices, rules and procedures, and the managers, supervisors, and controllers responsible for the movement of aircraft across the sky and ground.[29] When I began this project, the US airspace was divided into nine sky regions, each with a corresponding region on the ground consisting of regional offices and air traffic control facilities. Distributed across the regions according to traffic needs were 21 regional Air Route Traffic Control Centers (ARTCCs, or centers), the large radar facilities responsible for high-altitude aircraft; 185 Terminal Radar Approach Control Facilities (TRACONs, or "Approach Control"), the intermediate-altitude radar facilities that guide arriving and departing aircraft between towers and high-altitude ARTCCs, and 352 Airport Traffic Control Towers. Regulating the flow of traffic throughout the parts of the system to minimize congestion and expedite delivery of foreign and domestic traffic was the Air Traffic Control System Command Center in Herndon, Virginia. Staffed with experienced controllers drawn from large facilities from all over the country, it was still known to controllers throughout the system by its original name, "Central Flow." All controllers throughout the system were and are civil service employees.

Rules, procedures, and other forms of standardization are central among the system's technologies of coordination and control. To make dead reckoning and coordination across the system predictable and safe, the connections between the air traffic control facilities are spelled out by letters of agreement and memoranda of understanding: documents that articulate the connection between the parts and the larger system in the United States through multiple rules and procedures designed to create common material practices to facilitate coordinated activity across physical and social space. Overarching these local connections are international standardized rules and procedures and standardized phraseology and language for communication between pilots and controllers in a globally coordinated system. Thus, the US system is one part of a complex international system that comprises member-country systems, all connected to one another in a grand overall design and regulated by the International Civil Aviation Organization. It is systems within systems within systems: in a tower, each controller interacts with the others in their tower, TRACON, or center, operating as a system of interdependent parts: a tower is one among many different facilities operating within a regional system; the region is one among nine others in the country; and the country is a part of the global air traffic system.

Always, but especially for an ethnographer, having a sense of a place, its people, and the routine interactions in a setting are essential to the framing of a research project. Because the inner workings of the air traffic control system are not readily accessible to outsiders, next I explain the further evolution of the research as I entered the system for the first time, taking readers along in order to introduce the basics about controllers, their work, technology, and the system as they were when the research began.

Introduction to the System: "A Monkey Could Do This Job"

Living and teaching in Boston in 1998, I hoped to do my fieldwork in four air traffic control facilities in the New England Region, one of the nine regions in the system. I selected these four because they varied in size, technology, architecture, type of aircraft, air traffic volume, complexity and density, and airspace characteristics. Together, they represented the spectrum of work that air traffic controllers do. The cross-case comparison of the four facilities would allow me to explore analogies and differences in dead reckoning . Moreover, located in the same region, the four had to coordinate with one another in order to exchange airplanes, so their differences and the relationships between them would give me a sense of how the larger system operated.

Two were at Boston Logan International Airport, then ranked nineteenth in the United States in the number of traffic operations annually: Boston Tower and Boston Terminal Radar Approach Control (the TRACON), the latter of which handles intermediate-altitude traffic descending into or ascending from the tower airspace. The third facility was Boston Air Route Traffic Control Center (the Boston ARTCC, or informally, Boston Center), the large radar facility in Nashua, New Hampshire, that handled all high-altitude traffic for the entire New England Region. The fourth was Bedford Tower, a small but high-traffic-count facility with a traffic mix including pilots in training, corporate jets, military, and commercial airlines at Hanscom Field, in Bedford, Massachusetts, near the picturesque communities of Lexington and Concord, where the initial battles of the American Revolution were fought.

Apart from the research skills, theoretical tools, and background in organizations, technology, and systems that I brought to the project, I was woefully unprepared for the world of air traffic controllers that I hoped to enter. In 1998, the available literature was limited. The media and the magazine *Aviation Week and Space Technology* had chronicled in detail recent system-paralyzing congestion, gridlock, all-time-high delays, and the FAA's failed attempts to develop new technologies that would alleviate these problems, which left controllers working with obsolete 1960s technology. Books and articles were available about the air traffic controllers' strike of 1981 and Ronald Reagan's infamous firing of about fourteen thousand of them. The media coverage at the time had been extensive, before the strike and for the year after it. Few scholarly books had yet been written, however. Outstanding among them were Arthur Shostak and David Skocik's 1986 *The Air Controllers' Controversy* and Katherine Newman's 1988 *Falling from Grace,* which had a superb chapter about how those fired controllers survived the 1980s economic downturn.[30] However, I found no American social science scholars who had examined the work of US air traffic controllers in the workplace.[31] Worse, I had never been in an air traffic control facility.

In October 1998, I began the project with a low-key approach. Hoping to learn enough to write a research proposal to submit to the FAA, I signed up for a one-hour tour offered weekly at Boston Center, the large, high-altitude regional radar facility in Nashua, New Hampshire. There, 260 controllers worked traffic at thirty radar positions, three shifts a day, around the clock. I had a lot of questions: What was the physical layout and architectural design? Would it be possible to sit with them to see and hear what they were doing? What kind of access could I request that wouldn't disrupt the work? The tour was scheduled for eight o'clock on a Monday morning. Nashua was about an hour's drive from Boston. I arrived to the large, windowless concrete building located off the main highway on an isolated road. The parking lot was full of cars, but no one was in sight. I was the only person who had showed up for the tour.

Pete, the controller who met me at the door to lead the tour, was surprised. "What are you doing here? We normally get Boy Scout troops or senior citizens' groups." I began to explain what I wanted to do and why. He was immediately interested in my topic and we stood there talking for a while. Out of ignorance but to my good fortune, I had scheduled my visit for Columbus Day, a day off from teaching for me and also a slow traffic day for the controllers. As Pete explained, it was a holiday but not one for which people usually traveled by air, so traffic was less than usual. Consequently, Pete had more time for me. Our talk that had begun at the door turned into a two-hour conversation. He led me from the entry way around the corner to the center's large cafeteria, where he prepped me on what center controllers do and what I was about to see. I was fascinated. I was equally impressed by Pete's obvious enthusiasm for the job; he had worked air traffic at the center for over ten years.

Two hours later, he walked me to the brightly lit Traffic Management Unit (TMU), located at the entrance to the control room. TMU is the connecting link between all the facilities in the New England Region and "Central Flow"—the Command Center in Herndon, Virginia, responsible for regulating traffic flows throughout the US system. Pete was one of the TMU staff, all of whom were center controllers with extensive experience working traffic before moving over to TMU. When I explained to the TMU supervisor why I was there and what I wanted to study, he chuckled and said, "A monkey could do this job." I was not convinced by the explanations they gave me of TMU traffic-regulating functions, such as restrictions, rerouting, metering, spacing, expected departure times, gridlock, and delays that followed. But I did get a feeling for the system as a whole by descriptions of the daily routines of the Traffic Management Unit as the connecting node, funneling information between the New England regional facilities and the Command Center, then translating information from the Command Center into spacing programs that reorganized the traffic flows in the New England facilities, adjusting them in relation to events in the other parts of the system.

TMU controllers were talking to other controllers, not pilots, but it was dead reckoning nonetheless, and I saw that technology was necessary to every part of it. The TMU supervisor demonstrated the computerized Traffic Situation Display, which showed national traffic flows and the number of airplanes in the sky at a given moment. Over four thousand airplanes were airborne by midmorning on this Columbus Day. Each flight was a dot: Omaha had a scattered few; Chicago a dense mass. Clicking on one dot enlarged it to a tiny arrowhead representing a single aircraft, another click revealing its flight details. TMU controllers could watch hot spots, traffic congestion, ebbs and flows by time of day, and so adjust the center's traffic in relation to the national traffic flow. This was state-of-the-art technology in 1998. Where was the obsolete 1960s technology I had read about in the press?

When Pete had to go back to work, he led me the few steps from TMU to the huge, high-ceilinged, dark control room. As my eyes adjusted to the dark, I saw that infamous 1960s technology. Controllers were working traffic on sections of the sky that were divided onto thirty radar scopes clustered in five different parts of the room. Known as "areas of specialization," each area had its own airspace (their assigned specialization), an area supervisor, and six or more controllers working traffic at adjacent radar workstations, each displaying different airspace "sectors." The supervisor explained some of the basics while I stood beside him at his desk and watched for about an hour. I immediately sensed the intimacy of the space where controllers work. Working with the same people day after day, doing the same task in a small area, they know one another well. The supervisor asked if I wanted to sit with them. Feeling very much the stranger and intruder that I was, I sat with two controllers, Dan and Anita, who, seated side by side, were working traffic through the airspace sector represented on their radar scope.

Dan had the radar controller position (watching the radar, entering flight data on the computer, talking to pilots to control the airplanes in their sector), and Anita, the radar associate position (eyes on the same scope, she handled the landlines to other controllers and adjusted routes on paper strips, known as flight progress strips, each identifying a single aircraft, its route, type of equipment, altitude, speed, and destination). I sat between but a little behind them, listening on a headset plugged into their radio frequency while they talked to pilots, to each other, to controllers in the area, and to controllers in other locations. Afraid I might distract them, I was trying to be quiet and unobtrusive. But they began pummeling *me* with questions: "What are you doing here? No one ever comes to see us." "No one knows what we do or where we are. When we tell people we are air traffic controllers in Nashua, they say, 'Oh, I didn't know there was an airport there.' Everyone thinks all air traffic controllers work at airports."

As I explained my project, they immediately began volunteering information. They were masters of the interrupted conversation. In between flight instructions to pilots, entering codes and commands on the computer keyboard, and talking to each other about aircraft, with their eyes always on the scope they explained to me the "blips" (to me) on the radar screen (to them, airplanes are "targets" represented by "data blocks" with call signs and other information representing individual airplanes), what they were doing and why, and the technology and techniques they were using: "This is a slew ball . . . we have a checklist for the Position Relief Briefing . . . these diagonal lines on the screen . . . when we hand off an airplane." I noticed that Dan and Anita had a mental and manual rhythm going between them: each would automatically pick up the other's tasks when one was temporarily occupied with something (like changing a route, using the phone, marking a strip, talking to me).

In addition to the work basics and introduction to their specialized language, I had my first glimpse of informal meanings, norms, relationships, and standing in the group. They explained the "rules of separation" for high altitudes (the required five-mile horizontal and thousand-foot vertical spacing between airplanes), then demonstrated how they could use the computer to throw a ring around a target they wanted to watch, and then around two targets to monitor the spacing between specific airplanes. The ring represented the dimensions of that required high-altitude five-mile, thousand-foot spacing limit to keep planes safely separated. Then Dan quickly switched, hitting a key to show how they could use the computer to make the ring larger. "Here's a six-mile separation," he said. "We call this a 'sissy ring.'" Starting to laugh, he added, "It's also known as "Hinchcliff's hoop." He and Anita were both laughing. "Who is Hinchcliff?" I asked. The controller sitting at the next scope to the right said, "I am." It was playful, all three were laughing, but "sissy ring" was nonetheless a zinger, goading Hinchcliff about the quality of his work and deficient masculinity.

Such conversation was possible only when air traffic was slow. At moments when traffic suddenly picked up, I was forgotten. "A busy controller is a quiet controller," a supervisor told me. It was my introduction to the rhythm of their work—busy periods punctuated by downtimes. That rhythm was driven by airline schedules, consumer demand, and traffic patterns. I was witnessing my first system effect: how events in the institutional environment—here, airline competition and scheduling—affected the system and the work of controllers. Soon I witnessed another. While I was sitting with a different controller, the 1960s technology did its thing. A large rectangular block with a printed message suddenly appeared on his radar scope. Simultaneously, a controller on the opposite side of the area sarcastically said, "Gee whiz, my scope went down. What a surprise." They all continued working airplanes on their scopes, even though the screen had gone blank. In a few minutes, the screen returned to normal. No panic, no shouting, just the one comment. The controller I was observing told me that radar failures of this sort were routine. He said, "Not a problem. We can work the airplanes on the scope from memory for about fifteen minutes, and we have the strips [flight progress strips]. If it goes longer, we can stack them [the airplanes] and close the airspace, but that can get hairy." I had only a vague sense of what that even entailed, but it definitely sounded hairy to me.

Pete came by at two o'clock to tell me that his shift had ended. I asked to stay, and the area supervisor checked with higher-ups who gave the OK. Late in the afternoon I went to the near-empty cafeteria (the evening traffic rush had started, so everyone was working airplanes) to grab a quick lunch and get back before someone realized I was still there and threw me out. Only a few controllers were taking their breaks in the large-windowed lunchroom, too bright after being in the dark. I sat at one of the long metal tables alone, starting a conversation with a controller who, also alone, was reading the *Boston Globe* at a table in front of me. He turned around to face me as we talked about the Red Sox (a fan and *Globe* reader, my conversation opener), and what I was doing there. After a while he started laughing and said, "Did you hear that?" "Hear what?" And he repeated in its entirety a comic sequence that had played out on the large TV positioned high in the distant corner in the front of the cafeteria. During the fifteen minutes we had been talking, his back had been to it. Absorbed by what he was saying, I had missed it entirely. "How did you do that?" I asked. "We all can do that. My wife gets so mad at me because I never look at her when she is talking. But I don't have to. I can hear and repeat everything she says." What was this? They could all do it? Everything I had seen and heard indicated controllers had an amazing array of skills. Their technologies were essential, but I had seen a technical failure and how they worked through it, unperturbed, as if it were routine. Did they also possess unique hearing and memory abilities?

My visit did not end until ten o'clock that night. The area supervisor arranged to make it possible for me to come back again the next day. The sparse material available did not prepare me for what I saw and learned on those two days. Neither was I prepared for what was to follow: getting permission to do the study took another fifteen months. Getting access to a research setting is often described in textbooks as a glorious moment when the gates open and you are in. However, for this project, it was a circuitous, lengthy affair of negotiation and renegotiation, during which time I learned a lot about the FAA bureaucracy: hierarchical layers above the air traffic control facilities, the boundaries between facilities, and, unexpectedly, politics and union-management relations within and between each part of the structure.

System Effects on the Project

My time at Nashua was during the Clinton administration, a favorable climate for the National Air Traffic Controllers Association (NATCA), the union that succeeded the Professional Air Traffic Controllers Organization (PATCO) the original union so infamously decertified by the government after the 1981 controllers' strike. The program "Quality through Partnership" joined NATCA and management in decision making on many issues. Thus, my project proposal had to be approved by both union and management officials at each of my four chosen facilities *and* by the hierarchy above. In late November 1998, my proposal and I were directed to an official at the New England Region's headquarters. He said that I needed permission from both the FAA air traffic manager and assistant air traffic manager and their equivalents in the union, who were the NATCA president and vice president at all four facilities. Months later, with permissions I had obtained in meetings with the union and management leaders at each place, I returned to my original contact, the official at the New England Region headquarters.

Consistent with the partnership agreement, we were joined by his equivalent, the region's top NATCA official. They both were supportive of the project, even giving me information about NATCA and the region during our meeting, but the proposal still had two more layers to go for approval: FAA headquarters in Washington and the Civil Aeronautical Medical Institute, the research arm of the FAA in Oklahoma City. Alas, by the time these approvals were secured, the air traffic manager at the Center had been transferred, a new one installed, and new NATCA facility reps had been elected at Bedford. I had to renegotiate permission at those two places with the new officials.

This experience of visiting and revisiting the four facilities helped me frame the research design and methods to better capture the connections in the layers of the system, the cross-case comparison of the four facilities, and the variations in work and technology at each facility. Ultimately, I received permission, I was told, because of both my research topic and my research methods: qualitative, using both ethnographic observations and interviews with the people doing the hands-on work, to find out what makes air traffic control so safe. Both the NATCA and management officials were enthusiastic.

As the regional official put it, "When something goes wrong, everybody is on us, wanting to know what happened. On an ordinary day, when everything is going right, nobody comes around." I was also told that it helped that the proposal was short and straightforward ("no academic jargon").

Finally, I was not asking for a lengthy visit. When I wrote the proposal, I was not thinking about a book. I envisioned a scholarly article or two that would be relevant for specialists interested in organizations, technology, risk, and safety, and for people who were working in other kinds of organizations engaged in risky work. Also, air traffic controllers were used to surveys (they generally ignored them), but they were not used to ethnographers hanging around. Because I suspected that they might be concerned about a stranger loosed in their midst, possibly messing up the operation, I had requested an estimated research time at each place in weeks, or long enough for me to do enough ethnographic observations and interviews for an article or two around my initial questions. To compensate for the short duration I was proposing, I asked permission to work ten-hour days, seven days a week so I could see two different shifts of crews or teams working a day. I hoped that once I was there and controllers understood what I was doing, I might be granted more time if I needed it.

I had selected the four facilities on the basis of their differences. I expected variation but didn't know what I would find. So I proposed spending time observing at each facility, beginning the interviews later, after the observations had given me a feeling for each place, its people, and their work. That way I could ask questions in common but also could tailor others to local facility differences. Although ethnography was to be the guts of the project, interviews also would be important, because the work and the technology were complex and opaque to me. Also, sustained conversations between controllers about traffic while working were rare: "a busy controller is a quiet controller," after all. Air traffic controllers are known for their silent coordination with other controllers in the room and in other locations, due to shared knowledge and the many maneuvers that can be accomplished silently by computer entries alone.

Also, they do not have a plane in their airspace very long. At the high-altitude Center, for example, where controllers have an airplane in their airspace for longer than controllers in the other kinds of facilities, an airplane may be on the radar scope for only fifteen minutes at most, and typically controllers are talking to many pilots during that time. At Boston Tower, in contrast, they handle two planes a minute. Also, I would miss some things because of the technical language and the speed with which they did things, and my presence would suppress other things. Consequently, most of the cognitive processes, human-technology interaction issues, and organizational influences that constituted their work practice would be revealed only in interviews and informal conversations.[32]

Happily, and thanks once again to system effects, my fieldwork extended much longer than I originally proposed. Controllers were free for interviews only at times when traffic was low and staffing was high. This combination didn't happen very often. When it did, controllers could use their breaks for an interview and were able to stay longer than the usual twenty- or thirty-minute break. So while waiting for controllers to be free to break from work for interviews, my time for observations extended months beyond what I had requested. The wait for available interviewees was a gift. It allowed more discoveries based on my observations and more spontaneous conversations with controllers, driving the interview questions in unplanned directions. It was also fortuitous that I began at the center, because it was there, watching controllers sending and receiving traffic from the other facilities in the New England Region and between neighboring regions, and from observations and interviews with TMU staff, that I had a beginning sense of the dynamics of the system. My day-to-day challenge would be how to accomplish an ethnography of a large-scale socio-technical system.

From its uncertain slow start in October 1998, final project approval was given in January 2000. Beginning that March, during spring break, I did concentrated fieldwork in each facility, one at a time, then worked four-day weekends during the rest of that semester. During a sabbatical year from June 2000 to June 2001, my fieldwork was full time: seven days a week, from roughly eight in the morning to eight at night, so I could spend time with two shifts of controllers a day. I sat with controllers at work, listening to their conversations with pilots and other controllers on my headset, I spent time with them on breaks and during meals, and I had continuous opportunity for informal conversations. Uninformed as I was when I began the fieldwork, they had to teach me about their work. Some of it was planned. On my first day at Boston Center, they organized a combined meeting of about fifteen NATCA and management people, including supervisors and operations managers, so people would understand what I was doing there and have an opportunity to ask me questions. They had all read the proposal. I gave a summary, and after a round of questions, instead of the grilling I expected, it turned into a brainstorming session between them about how to initiate me into air traffic control and different ways that I might proceed in the control room to help me learn as much as possible.

The final decision was to give me a day of "training." First came a memorable morning of classroom instruction with a controller, an experienced trainer who taught me the fundamentals of how the air traffic control system works. This was followed by an intense afternoon session on a simulator, where two controllers gave me "simple" problems to work getting targets (data blocks on the computer) from point A to point B, with one controller playing the role of pilots. I mastered little of this and was totally overwhelmed, but I picked up some of their language, learned that the job was about separating airplanes and helping airlines cross airspace boundaries, and understood why simulation was an important stage of their training. When starting at each separate facility, supervisors introduced me to the technologies (both advantages and foibles) and gave me explanations of the controller work positions and the airspace, then each day I spent time plugged into the radio frequency to hear and observe controllers working each position. I never got it all—there are no geographic indicators distinguishing cities or bodies of water on radar maps or visible markers of flight paths out the window of a tower—but after some time, and with controllers' help, I could see the variation in airspace traffic patterns, the tasks at the various positions that controllers worked, and how the work changed as the position changed. Moreover, I had a close-up of the interaction between them, their supervisors, and the multiple devices in the room, and I heard their conversations with controllers and pilots in distant locations.

Much of what they taught me was informal, however. Because I was at each place for an extended period of time (several months, at the large facilities, and one month at Bedford Tower), and present ten hours a day, including weekends, I became a fixture. People would think of something about the operation and volunteer it. Or something was said to me in a hallway conversation or in the lunch line, or overheard while they were working, raising a question in my mind and a new subject to investigate, like my conversation with the controller in the cafeteria that first afternoon that showed his impressive hearing and memory skills. Some examples that were major turning points in my understanding: "Bradley [TRACON] has a funny personality." "This sector is a rat's nest." "See that guy sitting over there, looking like he's asleep? He's a natural. Born to it." "There's a lot of talk about competence around here." "Has anyone shown you our pad management system?" "The stress of this job is not the airplanes; it's the people you work with." "He's a nice guy [a supervisor], but don't let him touch anything." These comments and others, with their discovered meaning, led the research in new directions, elaborating my original framing with additional major themes and concepts that have come to shape this book.

At each facility, I had the freedom to roam, and repeatedly I received permission to expand the study in directions that came from insights gained after I was there. I interviewed controllers who were assigned to specialties, like the Office of Air Facilities, the Quality Assurance Unit, and the Critical Incidents Team. At the center, I sat in on the annual review of operational errors, in which NATCA and Quality Assurance Unit staff analyzed controller violations of the rules of separation to determine what caused them. I spent time with technical specialists in a large room at the center where the Host Computer System (HCS), for multiprocessing all radar and flight data, radar visuals, and radio recording system, is stored.[33] At the TRACON, I went along on chow runs and played cards in the break room a few times.

I observed the meteorologists in the center's National Weather Service Unit, and interviewed the head of Air Facilities, the unit responsible for technical upgrades, repairing technical breakdowns caused by lightning strikes, auto crashes into crucial cables, and breakdowns of computer equipment. I was there during low-traffic times of quiet conversations, gossip sessions, or study, as well as the high-traffic times and horrific weather situations when people would shut down with concentration or shout in frustration or both. I witnessed just about every traffic experience an air traffic controller can have, except an accident. Moreover, I was in the field long enough to see many system changes in procedures, architecture, and technologies. Without exception, every system change had visible effects on the work of air traffic controllers—they required new learning, new techniques, new routines, and practice—and when those changes were implemented in live traffic, they created stress.

From the four facilities, in addition to my books of field notes based on observations and conversations, I interviewed 133 controllers who volunteered to talk to me. I also interviewed supervisors, Traffic Management Unit personnel, facility air traffic managers, local and regional NATCA officials, and specialists in airspace design, cartography, training, radar, computers, meteorology, and quality assurance, totaling 174 interviews. A total of 158 controllers completed a two-page survey on their personal history, background skills, work history, and how they came to the job; with all personnel, that led to 191 surveys.[34] In addition, I tape-recorded telephone interviews with 22 PATCO controllers who had been fired by President Reagan in 1981; I contacted them by advertising on a PATCO website. These supplemented my interviews with the PATCO controllers in the four facilities, who had been allowed to reapply and retrain as controllers during the Clinton administration (see chapter 2).Never did I imagine that my interviews would nearly double in number as historical events, conditions, and actions by social and institutional actors impacted the system, changing it and the work of controllers. The result would be two ethnographic revisits, timed to record the effects of two major transformations of the system

The Terrorist Attacks of September 11, 2001

In June 2001, I reluctantly left the field to begin data analysis. I was in the early stages of transcribing and coding interviews when, on September 11, terrorists hijacked airplanes and used them as weapons against the United States. Two of the hijacked planes—American Flight 11 and United Flight 175—departed Boston Logan and were worked by three of the four facilities in my study: Boston Tower, Boston TRACON, and Boston Center. All four facilities were involved in the US systemwide effort to get the airborne planes on the ground in the hours immediately following the attacks, and then in gradually getting them up and flying again in the year after. Having spent the previous year and a half with those controllers, in the days following the attacks I could picture them on duty, maintaining their around-the-clock shifts, on position, idle, no airplanes in the sky—except for F-15 fighter jets controlled by the military air traffic controllers. I understood the enormity of these events to them and what they must be feeling and doing.

My permission to do the project had included an agreement that I could return later, to ask questions, and return again when the results were in draft form. After September 11, facilities were closed to all visitors. Beginning in 2002—again, one at a time, I received security clearance to revisit the four facilities to answer questions that I had from my fieldwork at each place, but also I was allowed to inquire about the effects of September 11. People would volunteer for interviews as before. Because of the emotional impact of the events, I would first explain why I was there, ask if they felt like talking about September 11, and if not, ask only questions related to the material I already had. Many volunteered; again I was there ten hours a day.

My revisits to Boston Center, the tower, and the TRACON were months apart during 2002. They occurred at different times for each facility, determined by their security needs, ongoing internal system changes, seasonal traffic flows, and breaks in my teaching schedule.[35] In June 2003, I received FAA permission to both observe and interview at the Command Center in Herndon, Virginia, for the first time. I asked controllers and administrators about their normal operation of systemwide coordination across all US facilities. Then, with that as a comparison, I asked them about the Command Center response and coordination of the system during the September 11 attacks and the year after, and how they and their work were impacted during this traumatic period. Not until March 2004 was I permitted to return to Bedford Tower. It had been delayed because controllers there were transitioning to their new tower, situated about twenty feet from the old one. Construction had been stalled for years for budget reasons. Now completed, the controllers were finally working in it.[36]

September 11 changed everything for them. It also changed everything for me, because I knew how they accomplished the impossible on that day and in the tumultuous year to follow. I understood how the system worked—its built-in ability to expand or contract as necessary, its resilience, reliability, redundancy—and how controllers worked—the years of training and experience, their interpretive skills and improvisation in unusual situations, the ability to negotiate boundaries and coordinate with one another, and to concentrate and work through a crisis. Moreover, I knew that at both the system level and the individual level, the development of these capabilities occurred incrementally over time in response to changing circumstances as the system evolved to meet new challenges.[37]

The final piece of the research frame fell into place: history mattered. It was not just *that* it mattered, but *how* it mattered. The duration of the project's life gave me unexpected insights into the connection between historical conditions—political, economic, technological, cultural—and the actions of institutional, organizational, and individual actors in the system, its technologies, the workplace, and the production of knowledge in it. Early on, I had noticed that, individually and collectively, controllers' awareness of history, time, and temporality were unusual, embedded in their occupation's normative structure and embodied as a function of their training in a system where efficiency—speed, delays, on-time arrivals, setting priorities—was a central concern.[38] Further, they spend a career in this occupation. Throughout my time in the field, controllers from different cohorts had told me how things had stayed the same or changed over time, which enabled me to see the visible effects of history on the system, the airspace, and the work of controllers in the present: the end of the Cold War, political and economic change, airline deregulation, aircraft equipment, pre-radar technologies, and the infamous 1981 air traffic controllers' strike were among those leaving a permanent mark.[39]

However, it was not just their stories of history. The past had left its mark in the workplace. Like controllers, daily I wore a headset from the 1960s, full of static and needing periodic repair. When I first visited the center, I sat watching controllers work in the original control room; when months later I returned to begin the research, I was watching in their new location, as they adjusted to a new control room equipped with new radar monitors, lighting systems, and changed acoustics, the old control room standing empty beside it. In all four of my facilities, a few fired PATCO controllers were back, being retrained by the controllers hired to replace them—this, courtesy of a window of opportunity provided by the Clinton administration in 1993. Most important, the time span between my first visits to each facility for permissions to do the study, my entry and departure from the field, and my post–September 11 return had unexpectedly allowed me to record dead reckoning before and after this historic terrorist attack. I knew the controllers' emotional work and their coordinated responses as they transformed the system by innovating, improvising, and adjusting work routines; creating new organizational repertoires and technological systems; and enacting major and lasting change in the system, at each facility, and in the technologies of coordination and control.

I felt compelled to set aside my original idea of several scholarly articles and a written report for my four facilities, instead writing a book that would more fully incorporate the effects of historical events and actors on the air traffic control system so that readers—including controllers—could see the causal mechanics of its evolution as a socio-technical system, the effects on the organization, its technologies and tasks over time, and the ongoing repercussions for controllers and the workplace. Readers, too, could understand how the system worked and how controllers brought the planes down that day. Since then, time has passed. Like the stalled construction of the new tower at Bedford, my construction of this account was also stalled. In 2003 NASA had its second shuttle tragedy: the loss of the space shuttle *Columbia*. Because of my *Challenger* research, I became involved in the official investigation, first publicly testifying and then researching and writing for the *Columbia* Accident Investigation Board's official report.[40] This work kept me occupied until 2005, at which time I changed universities and cities—yet another year away from my analysis and writing. I returned to serious engagement with the analysis and began writing chapters in 2006. A lot of life intervened. It is also the case that taking a situated action approach to examine a historical cross-case comparative ethnography of four organizations in a complex socio-technical system takes a long time.

As I wrote, I kept track of ongoing developments from a distance, finishing all chapters except the last one. In the interim, however, the FAA implemented a massive systemwide modernization project to upgrade its deteriorating 1960s system, automating the system and consolidating radar facilities. Simultaneously—and crucially—political events and actions external to the air traffic control system had created a serious shortage of controllers nationwide. Because these changed circumstances were essential to my research questions about dead reckoning, I returned to the field in the fall of 2017 to look again, asking, What makes this system so safe—or is it? And, What do controllers do that technology can't replace?

On Time and Discovery: Historical Ethnography and Socio-technical System History

This book is a historical ethnography. We can think of all ethnography as historical because it intervenes in the ongoing process of a place, revealing social life at a particular moment in time. The ethnographer intersects history, observes a slice of it for a while, then leaves the setting. Ethnography has an advantage over other methods because it captures time—it exposes both structures and processes as they unfold. It is a motion picture compared to the snapshot version that some other research methods offer. Yet it, too, is time-limited. After we leave, the people and the places go forward in time and change, unobserved by us. We can, of course, go back later for an ethnographic revisit—but at some point, it must be over or ethnographies would never be written.[41] We also miss the prehistory before our entry, although sometimes that is available to us through interviews, surveys, archival records, photos, the work of historians, and the like. Reconstructing the past is easier when the subject matter is a formal or complex organization because all of these sources may be available through letters, memos, minutes, official documents, photos, and other archival records, although sometimes not available to *us*, or at least only partially available. Thus, often we are in the dark about the links between the past and the present and what happens to the patterns we discovered after we depart.

Usually an ethnography's location in history speaks for itself, without specific articulation by us. History is captured by the language, characteristics of place, its politics, social organization, sciences and technologies, and what we know of the historical moment from what historians and experience tell us—war, peace, economic bounty or depression, race and gender relations, politics, national culture. So we do not always need to reconstruct history as part of the research because the patterns are clear and history is not necessary to the ethnographic explanation. The value lies not only in the patterns revealed at that specific moment in time but also in their resonance for the present: what has been uncovered years before has enduring value because it is repeated in other research settings.[42] Sometimes, however, the ethnography needs to be *intentionally* historical because the result of historical actions taken external to the system become an essential part of the explanation of the present, thereby allowing for comparison between time 1 and time 2 in the same setting. This situation can occur when, for example, a previous ethnographer has published an analysis years before, thus allowing a comparison with the dynamics of the same setting in the present,[43] or, alternatively, when archival records created by the participants show how a precedent begun in the past became established, repeating over time and affecting the present in ways that matter to fieldwork.[44]

My approach is different. Sometimes being intentionally historical calls for an ethnographer to locate an ethnography *between* the past and the present, as I do here. This allows us to track the ongoing process of change as it unfolds over time: we can study social change and the sequences of events that produce social transformation.[45] Technological and organizational innovations, structures and cultures observed in the present, do not develop in a vacuum. From interaction, they grow from preexisting innovations, structures, and cultures, enacted by individuals who often remain invisible to the historical record. In certain forms of social life—nation-states, bureaucracies, neighborhoods, families, local cultures—certain patterns carry over, perhaps changing slightly in form but persisting inexorably into the present and into the future. But patterns coexist with variation and historical contingency: unexpected events can produce unanticipated consequences that effect dramatic change, altering an unfolding trajectory, elaborating structures and cultures in novel ways, reproducing some earlier social form, or eliminating it altogether.[46] More typically, change is incremental, having a low profile, until something happens that makes us realize that we are in a different place. We must wonder, with Abbott, "If change is the normal state of things, how does anything ever stay the same?"[47] Thus, it is sometimes important to examine both patterns and variations across time in order to understand the mix of old and new in a social setting in the present.

Historical ethnography helps to explain why things happen in the present the way that they do.[48] The fieldwork for this research took place at the turn of the twenty-first century. That historical moment marks it as peculiar, in the same way that my selection of facilities on the East Coast, the terrorist attacks of September 11, the technologies available for both aircraft and air traffic control, and the economic and political conditions at the time mark it peculiar. However, locating this research within its historical trajectory reveals what is stable and persistent about air traffic control over the preceding years as well as the change and contrast between then and today. Moreover, we can explore the reverberating interactions of system effects across time because the kinds of social pressures that are beyond individual actors and cognition can be identified by empirical traces in institutional, organizational, and individual actions. The past manifests in the present in what people say and do in the present.

The framing and method of this book is atypical for ethnography: an occupation at work is situated between its socio-technical system history and the present.[49] We can see how sequences of events unfold, showing the causal links between actions taken in the past and the moment of research intervention. It shows system emergence and the ongoing making and remaking of that system's structures, processes, technologies, architectural arrangements, and socially organized work practices, allowing us to consider both institutional change and persistence across time. In addition, taking a situated action approach reveals these changing system effects on the workplace in everyday tasks, technologies, material objects, cultural understandings, and rules and procedures. Further, it reveals how system changes alter controllers' embodied sense making, problem solving, material practices, interactions with one another, their interpretive work, and the meanings that their work experiences have for them. Equally important, we see their responses. Far from a top-down model, this history displays the agency of individuals, both external and internal to the system.

This ethnography is also atypical because of the kind of system-embedded knowledge work controllers do: coordination of high-speed aircraft, "fast in and fast out" of an airspace; this is silent coordination that is culturally specific, embodied, time limited, computer based, and, post-9/11, also automated. Consequently, much of what they do is invisible and inaudible to an observer. Only rarely does their work lend itself to a typical ethnography of sustained interactions over time among people who become familiar to the reader, people whose verbal and physical behavior allow for a continuing, extended narrative. For controllers, such sustained exchanges are rare. I was able to capture only a few: a loss of contact with pilots due to a failure of all radio frequencies and backups at Bedford Tower on a slow Sunday morning; the drama of the runway change between Boston Tower and Boston TRACON, over in minutes, but which I could witness repeatedly from both the Boston Tower side and the TRACON side to capture the necessary coordination between them. However difficult to observe, sustained interactions are retrievable in interviews. Controllers described in detail interactions from mistakes, accidents, training, and other events so intensely experienced that they lodged permanently in their memory. The duration of a two-minute incident may seem fleeting to us, but controllers experience an intense incident as a long duration. History lives in the culture of the present: controllers even reexperienced emotions when they were recalling certain incidents in interviews years after. So it was—and still is—for controllers about September 11 and about their transformation of the system during the year after.

Keep in mind that air traffic control is an exotic occupation for the rare combination of characteristics that distinguish it from other occupations. With air travel, lives are always at risk, often more than three hundred in a single airplane. The responsibility for the safety of the flying public rests with a government agency. Air traffic control differs from many other complex systems that do risky work because the airplane is a developed technology, not an experimental one, and the work occurs within an international, standardized, socio-technical system in which coordination between the parts is essential and must occur in a standardized way to be effective. Consequently, it differs from others in the extent of formal rules and procedures governing work practice, intensity of socialization of new members, and interdependence of the parts of the system. Although rule bound like the military, in air traffic control, coordinating high-volume civilian traffic movement between the parts of the system is an around-the-clock activity, with each controller making thousands of decisions daily to direct the airplanes under his or her control in an ultra-short time frame. They work in small groups doing a coordinated task with the same people day after day.

Controllers' technology, seemingly simple and obvious to outsiders who observe it in the workplace, becomes exotic when one can grasp the scope and unfathomable variety of technological infrastructures that support the visible devices and work of dead reckoning that is ongoing in the work setting.[50] Beyond the single facility, the infrastructure coordinating tasks, organizations, and airplanes extend to the scale of the nationwide socio-technical system:

many thousands of controllers doing essentially the same task, in different locations, with the central need to coordinate with one another to move airplanes from controller to controller across great distances in a timely way. The technology that is visible in a controller's workspace is connected to multiple technical infrastructures that show what the human eye can't see about the movement of aircraft in the sky. On the ground, airport runways have their own technological infrastructure, supporting signaling systems of lights and beacons for pilots and radar surveillance systems to alert controllers to wayward airplanes on the ground. Other technical infrastructures support signaling systems for pilots and controllers across the country that are automated directional indicators and finders, and communication systems with satellite support and GPS. Yet with all this science and technology built into the system, none of it is perfect. Humans are fallible and so are complex organizations and technologies.

Another striking peculiarity is the requirement that everyone who works in the system, with the exception of the very top echelon of officials at FAA headquarters in Washington, DC, who are political appointees, must have been trained, certified, and worked as an air traffic controller. The rule exists so that everyone will have in common not only the rules but also experiential knowledge of air traffic, its ebbs and flows, and the characteristics and challenges of working in a highly politicized system typified by frequent change. Moreover, working air traffic is such a specialized task that those who survive the training tend to make a career of the job. Like a conductor who can't conduct without an orchestra, air traffic controllers need air traffic. The job pays well but also offers to the unsuspecting job applicant the benefits of deep bonds and an occupational identity that grows out of the sacrifice and emotional commitment made during the training and shared experiences over time. So, cohort after cohort, controllers stay. The history of the system is recorded in their memory of daily experiences and is a part of the common cultural script that they acquire by working there.

For research purposes, the occupation's exotic quality is an advantage because its extreme nature throws into broad relief aspects of institutions, organizations, cultures, technologies, and work practices that are less visible in other kinds of occupations. The time I spent in each facility led to major themes, unimagined in the beginning, that appear in this book. The starting questions I had about dead reckoning at the outset—system effects, how controllers identify early warning signs and correct them so small mistakes do not turn into accidents and disasters, the dynamics of human-technology interaction, and what controllers do that technology can't replace—remain central. However, during the fieldwork they became more elaborate in their scope, linked together with new insights I realized only by being there, leading to additional major themes and concepts that shape this book. These themes resulted as a consequence of the evolution of the research design over time. My progressive discoveries and revisits to the field changed the initial project from a time-limited empirical work into a historical-comparative, cross-case, multisited ethnography. This research is comparative in three ways: I compare the four facilities at the same time (the initial 2000–2001 period of the fieldwork); sequentially, across three periods of fieldwork (2000–2001, post-9/11, and two facilities in fall 2017); and I situate the fieldwork between the system history and the system present, allowing for multiple points of comparison over decades. Across time, this reveals how changing the system—its organization, technology, and culture—inevitably alters controllers' knowledge work and material practices.[51]

We know a lot about how institutions persist but less about how they change and the effects on tasks, technologies, decision making, action, and reaction in the workplace.[52] This book reveals how and why this socio-technical system has persisted and changed over time, making visible the local improvisation by controllers who must adjust their work, decisions, and actions in response to these multiple intersecting system effects. Building on research in the history of technology and historical sociology, the book shows the role of structure, technology, and heterogeneous assemblages of social actors in system emergence and its development and remaking across time and social space.[53] We can see how actions taken in response to changing external conditions resulted in the redundancy, reliability, and resilience for which high-reliability organizations are known, demonstrating how these capacities enabled survival even as the system was buffeted by politics, budget shortages, and major shocks. Thus the historical cross-case comparison captures the changing nature of work, showing the unanticipated consequences—both positive and negative—of replacing aging technologies and physical structures with organizational and technological innovations. In particular, the book reveals the little-known before and after of the introduction of automation in the workplace and workers' temporal process of physical and cognitive adjustment that follows, which is made more complex by the necessary redesign of the organization, its architecture, and the social arrangements of work.

Further, I was able to trace the formation of an occupational *habitus*—a common way of being and acting—of this professional group.[54] As a result, the book expands what we know about culture, cognition, and the embodied material and technical practices of work.[55] DiMaggio's work on culture and cognition, or the idea that cultures manifest in people's heads, and Hutchins's work on distributed cognition, or the idea that cognition is distributed across people and material objects interacting in a confined physical space, take on new meaning within this layered situated action approach.[56] In the workplace, we can see controllers' cognition and action being shaped by the melding of cultural beliefs that conform to the institutional mandates of the system, by others produced in interactions at the facility level in response to local variation in airspace and task, and also by individual interaction as controllers collectively create culture as they seek solutions to the problems they face in common.[57] Cultural sociology and distributed cognition merge in a sociological ethnocognition.[58]

A Cultural System of Knowledge: Ethnocognition and Boundary Work

Situated within these larger structural patterns, the ethnographic chapters all narrow in on dead reckoning in the workplace and the interaction between controllers and their multiple technologies of coordination and control in use. The human-technology interaction is not a neutral encounter but an interactive relationship: from devices, rules and procedures, tacit knowledge, and material objects, controllers construct risk and safety. Controllers must negotiate the intersection of standardization, cognition, and information that appears in standardized form but varies in behavior.[59] On the ground, in towers and radar control rooms, the crux of dead reckoning is this. Although airplane technology is standardized, the variation in airspace, airplane design capabilities, pilot technique and competence, airline policy, traffic volume, fleet mix, geography, weather, and the liabilities of technological innovation combine to give the sky a high degree of interpretive flexibility.[60] Interpretive work is what controllers do that technology simply can't replace.

The craft of dead reckoning consists of two distinct but interacting threads of interpretive work: ethnocognition and boundary work. Both concepts are developed in every chapter of the book and are elaborated beyond previous understandings. The concept of ethnocognition originated in anthropology. I have done the research and written the book guided by the anthropologist Clifford Geertz's influential notion that ethnography should reflect "the native view": ethnographers should approach a culture in order to capture the meaning that interactions, situations, events, and material objects have for the people who inhabit a place.[61] Geertz pays particular attention to what he calls "local knowledge" as a field of perception beyond local experience that is a *cultural system of knowledge*, peculiar to a particular time and place.[62] Here, I extend his concept of a cultural system of knowledge by developing ethnocognition as a sociological concept.

Controllers are *in* and *of* both system and workplace. The work they do is a process of working things out from moment to moment with coworkers, pilots, and devices, moving material objects across time and social space. Their cognitive, physical, and material practices are situated, embodied, local, distributed, and practical. At the same time, these practices are shaped by historical actions and actors, as well as by external political, social, technological, and economic factors, and by cultural beliefs that are passed on through training methods that produce durable embodied transformations of their thinking, noticing, hearing, vision, and emotions. Consequently, ethnocognition is not only distributed beyond the room across boundaries of time and space; it is also *layered*. Defined in sociological terms, ethnocognition is the situated enactment of expert bodily and sense-making techniques and understandings that work together in an active, interpretive thinking and doing in relation to others in the room; to the multiple devices, material objects, and socially organized arrangements that surround them; to the local situation; and to parts of the larger system in which they work.

Controllers are *in* and *of* a system of many boundaries. Ethnocognition is enacted in accomplishing the multiple varieties of boundary work—social, material, cultural, and symbolic—that controllers do.[63] Boundaries and boundary work demonstrate that human-technology interaction is not a neutral encounter. Fundamentally, controllers' task is to move airplanes across the boundaries of the sky safely. Doing so requires integrating interpretive work, improvising, and following standardized rules to keep those airplanes from colliding. Boundaries and boundary work are crucial technologies of coordination and control for controllers.

Note, however, that in response to changing traffic needs and political interests, authorities external to and above in the hierarchy also do boundary work, creating and changing system boundaries, both in the sky and on the ground, causing controllers to adjust both ethnocognition and boundary work practices. Further, the boundaries of the system on the ground also act on controllers. Controllers do boundary work in response to status differences within and between facilities, symbolically realigning the social boundaries of the system in which they work to distinguish themselves from others, correcting the inequalities in the system.[64] When enacted to restructure the status system symbolically or socially, boundary work becomes power work. Boundary manipulation is central to ethnocognition.

The Architecture of This Book

The organization of this book's chapters is designed to show the links between the past, the period of my fieldwork, and the present for air traffic control. History, time, and temporality are themes that carry through all the chapters. Collectively, they show the incremental development and transformation of system, its technologies, and dead reckoning across time, social, and physical space. Part I, "Beginnings," opens with two kinds of beginnings. This chapter, "Dead Reckoning," has introduced the research, its key questions, theoretical and methodological framing, central concepts and important fundamentals about controllers, their work, and the air traffic control system in order to prepare readers for what follows.

Chapter 2, "History as Cause: System Emergence, System Effects" is a formation story tracing the process of emergence, institutionalization, persistence, and change across eras, showing how the system developed the characteristics it has today.[65] It begins at the turn of the twentieth century, when air flight was in its infancy. It connects external conditions (the surge in the development of auto and railway transportation) with the agency of individuals (inventors, aeronauts, and institutional actors) who initiated the nascent air transportation infrastructure and subsequently the air traffic control system. History has a causal relation to some present only through the problem solving actions of social actors, acting at different times and locations, moving development forward, constraining it, or transforming it into something else entirely.[66] As the airplane's increasingly sophisticated abilities captured the imagination of the nation, assemblages of professionals and heterogeneous organizations became system builders, shaping and reshaping the nascent socio-technical system's boundaries on the ground and in the sky.[67] System development was uneven, shaped by historical contingency, the liabilities of technological innovation, and the unanticipated consequences of planned change.

Once controllers came into existence, every change had an impact on their work. Boundary work became a key strategy for safety; as boundaries in sky and ground were drawn and redrawn, controllers had to think and solve problems at the local level that were institutionally embedded. As airplanes flew higher, controllers' technologies of coordination and control multiplied; standardized changes had to be adjusted to fit local situations. Problem solving at the local level had a surprisingly large impact as informal solutions became a foundation for formal strategies that incrementally developed the system's resilience, redundancy, and reliability. Problem solving by political actors external to the system engaged in boundary work as power work, generating cycles of decline and repair for the system. Further, power work within the system due to conflicts between management and labor contested system directions and drove change, beginning in the 1970s. The chapter concludes at the turn of the twenty-first century, revealing the system as history had shaped it and the system effects on the workplace, controllers, and dead reckoning at the time I entered the field.

Part II, "Producing Controllers," shows the process of becoming a controller, tracing the development of ethnocognition and boundary work. Chapter 3, "From Skill Acquisition to Expertise," follows them through each stage of the intensive training. Beginners struggle to adjust cognitively and physically to the material objects, devices, and ways of doing and thinking required for their temporally driven work. Then, as they progress to apprenticeship and working live traffic, we see how trainers fine-tune interpretive practices and transform cognition into ethnocognition, shaped by system structure, goals, technology, history, and culture.[68] The trainer instills competing system goals as cultural scripts: in this case, the safe, orderly, and expeditious delivery of air traffic. The practical accomplishment is a highly developed cultural system of knowledge. In becoming experts, mind, body, technologies, and institutionalized cultural beliefs merge, such that tasks can be done automatically, freeing controllers to look for and deal with anomalies.

Chapter 4, "Embodiment: The Social Shaping of Controllers," shows how system effects produce a greater transformation. Through repetition and daily practice, "developmentals" are transformed in physical and mental abilities: the skills and habits of mind necessary to the job become embodied to the extent that they carry over into everyday life. Moreover, they are transformed in more fundamental ways, as persons. They not only achieve the expertise necessary to the profession; they also emerge from training with a common occupational *habitus*: group culture and personal history that shape body, mind, and action. Central to this *habitus* are their embodied sense of time, timing, prioritizing, and planning, all essential to coordinating and predicting the position of airplanes across time and social space in a system driven by being "on time." They become "controlling": of pilots, of airplanes, and of their own physical and emotional responses to the work. Controllers are reshaped into the kind of people who can meet conflicting system goals, and meet them predictably.

Part III, "Boundary Work: Airspace, Place, and Dead Reckoning," follows controllers from their training into the workplace. The four ethnographies show the inner workings of human-technology interaction and the deployment of ethnocognition and boundary work as controllers interact with one another and with their devices, material objects, and architectural arrangements that shape their work. This chapter shows how ethnocognition combines with distributed cognition, thus shaping coordination in the room and across the boundaries of the system to controllers in other locations.[69] The space in which they work is a "center of coordination," in which controllers problem solve moment to moment as they move aircraft across time, social, and physical space.[70] Chapter 5 juxtaposes ethnographies of Boston Center and Bedford Tower; chapter 6 compares Boston Tower with Boston TRACON. These ethnographies reveal a standardized system that is riddled with variation. The system effect here is the relation between airspace and place.[71] As a facility's assigned airspace and air traffic vary in volume and complexity, so does the place and its architecture, technologies, rhythm of work, and consequently, culture. Within this so-called standardized system, controllers must adjust dead reckoning to ways of doing and being that vary by place.

The boundaries of the system are sites of conflict and are a major challenge and source of stress, both at the system level and in the workplace.[72] Controllers do two kinds of boundary work: moving airplanes across the airspace boundaries in the sky and negotiating airplanes' movement with controllers in other locations who "own" the neighboring section of airspace. The result of ownership is turf wars in the sky and on the ground. For each facility, examples of boundary work in daily routines and emergencies show how the work varies from place to place. Moreover, the examples show how controllers safely negotiate boundaries by combining standardization, interpretive work, and improvisation, thus supplying the resilience, reliability, and redundancy essential to the system.

Part IV, "Emotional Labor, Emotion Work" shows system effects of another sort: the embodied emotional and physical effects of the job on controllers.[73] In chapter 7, "Mistake and Error: Emotional Labor," controllers distinguish the two and talk about their experiences of near misses and tragedies, recalling every detail of what happened. The strong emotional and physical impact of these experiences is an integral part of their narrative. Technology is a necessary aid, or they couldn't do the job at all, but in use it often falls short or fails as a result of controllers' reliance on representations of airplanes, not the objects themselves, the inability to design a technical system that fits the physical boundaries of air traffic facilities on the ground, and technological lag due to budget shortages. Human fallibility of pilot and controller also matter, but the liabilities of devices and technological representations is another challenge and source of stress in their work.

Then, in chapter 8, "Risk and Stress: Emotion Work," their accounts of risk and stress contradict their emotion-laden narratives in chapter 7. The chapter reconciles this contradiction, exposing the role of culture in emotion work. In response to the pressures of the job, controllers' individually and collectively create strategies that become cultural, easing their emotional labor.[74] These strategies normalize emotions, distancing them from the experience of risk and stress. Similarly, place—variation in architecture, devices, and technologies of representation, routines, rules and procedures—gives them the opportunity to redefine their experiences, and they *do* do this, mediating the experience of risk and stress so they can do the job.

Part V, "That Little Frisson of Terror," consists of three chapters that reveal the interactions of people, structures, and the multiple technologies of coordination and control that enabled the air traffic control system to bring planes in safely on September 11 and the system's transformation and persistence in the wake of the September 11 attacks. The interaction of these four facilities with one another and with other parts of the system reveals a microcosm of the processes and experiences of controllers across the system that day. Chapters 9 and 10 are based on the accounts of controllers at Boston Tower, Boston TRACON, Boston Center, Bedford Tower, and the Command Center in Virginia about the events of September 11 and after. In a rare close-up of controllers in a continuous interaction, the ethnocognition, boundary work, and emotional labor that were the subject of all preceding chapters are opened to view as managers, supervisors, and controllers in each facility were making decisions in conditions of unparalleled uncertainty.

These two chapters expose the interactions that animate the flexibility and porousness of system boundaries. They demonstrate the agency behind the resilience, reliability, and redundancy of the system, as controllers combine standardization, interpretive work, and improvisation. Chapter 9, "September 11," is a moment-to-moment description of events on that day and the first few days after, showing controllers at each of the four facilities as they experienced and responded to the crisis, clearing their airspace of all aircraft in an unrehearsed effort completed without incident in two hours and fifteen minutes. Chapter 10, "The War on Terror: Policing the Sky," describes another unprecedented challenge, that of getting the planes back up again. It reveals the accelerated remaking of the system and also controllers' job in the weeks and first year following the attacks as they took on a new responsibility: policing the sky. Collectively, controllers in local facilities everywhere refashioned the air traffic control system to meet a new threat.

Then, building from the crisis experience, chapter 11, "Symbolic Boundaries: Distinction, Occupational Community, and Moral Work," reveals how this system, riddled with variation, holds together as a system of interdependent parts in normal times and in crisis. It focuses on the structured status differences and inequalities that variation in airspace and place build into the system and how, nonetheless, controllers strategically construct symbolic boundaries that bridge status differences, forming a collective occupational community that binds them together, affirming their moral work and resisting the status and identity conferred on them by the social boundaries of the system.[75] Integral to this process, they elevate their status across formal system boundaries and within facilities, on the basis of the distinctive qualities of the work at each place.[76] An unintended consequence, controllers' construction of symbolic boundaries enables both collective action and institutional persistence.

Part VI, "System Effects, Boundary Work and Risk," traces changes that transformed the system from 2002 through 2017, triggering a third crisis. Historical actions, begun in the 1990s, created two trajectories of independent events that intersected, increasing system risk. The first was an FAA modernization effort, NextGen, which included automation, relocating and consolidating regional TRACONs, and streamlining the organizational system for efficiency. Beginning also in the 1990s, political battles reduced the FAA's budget, resulting in hiring freezes that combined with a wave of controller retirements to produce a severe systemwide staffing shortage. The system effects of the coincidence of these two trajectories undermined the very aspects of the system responsible for the successes in operations on September 11 and in the year after.

Chapter 12, "The Age of Automation: 2002–Present," revisits Boston Tower and the relocated, consolidated Boston TRACON. An unplanned comparison—of chapter 6, which presents the ethnographies of Boston Tower and Boston TRACON in 2000–2001, and chapter 12 in 2017—reveals the many differences before and during this crisis, in particular the effects of automation on dead reckoning and the unpredicted effects of modernization on the social arrangements of work. In common, controllers at both facilities struggled with cognitive, physical, and material adjustments to automation and architectural changes. Moreover, the staffing crisis, the NextGen organization changes, and two generations of controllers resulted in inequalities and mistakes. The workplace was fraught with conflict over airspace, architectural, cultural, and generational boundaries. Improvising tools of repair, NATCA officials, facility managers, supervisors, and controllers worked to adjust technological and organizational innovations to the local setting and at the same time preserve the cultural understandings, skills, and expertise that had been lost during modernization.

Chapter 13, "Continuities, Change, and Persistence," reflects on characteristics the system acquired over the life course to answer the original questions this book poses: What makes this system so safe, or is it? And what do air traffic controllers do that technology can't replace? Among the continuities, one pattern was system effects originating in the external environment—political, technological, economic, cultural—enacted by powerful actors with clashing agendas that gave the system an inherent vulnerability rooted in resource scarcity and uncertainty. Second, recurring technological and organizational innovation had unpredicted effects at the system level and in the workplace. As a result, change itself was a continuing pattern. Interrupted by two shocks, the life course otherwise was marked by eventfulness, contingency, and unanticipated consequences that led to periods of decline and instability when the system became more risky.[77]

So how did this public agency persist over time, maintaining its original form, rather than failing or being transformed into a corporate enterprise? The answer lies in controllers' ethnocognition, boundary work, and expertise that power the dynamic resilience, reliability, and redundancy of the system and enables them to become collective, coordinated change makers, improvising tools of repair following periods of decline. In the conclusion, "Dead Reckoning: Coordinating Action and Anticipating Futures in Complex Organizational Systems," I extract from this case to the dead reckoning of all complex organizational systems, reflecting on the inherent unpredictability emanating from the institutional field, the complexity of systems, and the liabilities of modernization, the changing nature of work, and the implications of this research for risk and what Evgeny Morozov has called "the folly of technological solutionism."[78]

2

History as Cause

System Emergence, System Effects

How does the past affect the present? This chapter takes up this inquiry, tracing the emergence, development, and operation of the air traffic control system from its beginnings until I entered the field in 2000. Across time, this history exposes the processes of formation and transformation of the system and simultaneously exposes dead reckoning

from the Wright brothers, to aeronauts, to pilots, to glass cockpits, to free flight;
from landmarks, to signal flags, to shrimp boats, to paper strips, to radar, to GPS;
from air balloons, to gliders, to props, to jets, to wide-bodies, to regional jets, to drones;
from landing fields, to lighting systems, to looped routes, to airports, to global system.

Across eras, we follow the development, persistence, and change of the air traffic control system.[1] Why does this matter? Because locating ethnography within its history allows us to trace both continuities and change across time so that we can understand the mix of old and new in a social setting in the present. In these circumstances, we can think of history as cause. History is not just a scene setter, sufficient to locate the research in time. Neither is history itself—the passing of time—a social actor. History has a causal effect on the present only through the agency of multiple heterogeneous social actors and actions originating in different institutional and organizational locations and temporalities that intersect with a developing system and through its life course in unanticipated ways.

Reconstructing the past shows how and why the air traffic control system became an error-reducing system, developing characteristics and strategic organizational repertoires in response to changing political, economic, technological, and cultural conditions, including shocks that threatened the survival of the system itself.[2] Second, the historical sweep of events captures the actions and reactions of institutional, organizational, and individual actors. We can see how, when, and why air traffic controllers came into existence and, subsequently, how the system's responses to its ever-changing institutional environment shaped and reshaped controllers' tasks, interpretive work, and material practices of dead reckoning as well as their responses to the changing circumstances.

Causal explanations of historical events, institutions, and outcomes are best understood by storylike explanations that capture the sequential unfolding of events in and over time, revealing the interaction of structures and social actions that drive change.[3] Crucially, when an action happens—its order in a sequence of occurrences—is more important than the fact that it happens. And how things happen is the explanation for why things happen.[4] These historical sequences have no predictable outcome or pattern. They are marked by contingency and particular combinations of causes coming together in different times and places.[5] Some historical narratives are path dependent, meaning they are made up of sequences of occurrences at one point in time that seem to lead inevitably toward a particular outcome in the future. Others dwindle over time, reverse course, or disappear altogether. Time and timing make each case unique. Despite this diversity, historical narratives tend to have two characteristics in common.

First, the sequential historical movement of events, conditions, and actions of social actors through time and social space helps us understand the relationship between the past and some determined present.[6] Second, the explanation of each case has identifiable patterns and variations that stand out across temporal settings. To sensitize readers to both the patterns and the variations in the story that follows—and that still characterized the system at the time of my revisit in 2017—here is a brief overview.

A Formation Story

A formation story is a narrative history of the life course of a social entity. It is inherently causal, explaining how a social entity comes into being and then becomes stable enough to have causal effects on the environment and the individuals and organizations in it.[7] Rather than a transformation of a social kind already in existence, it traces the formation of a novel social form and how it came to have the shape, vulnerabilities, and capacities it has.[8] Thus, it captures both system emergence (being subject to external forces) and system effects (becomes a force in itself). Tracing events as they unfolded, I found no pattern leading inexorably to some predictable outcome. Instead, the life course of the system was messy, typified by historical contingency, unanticipated consequences, and the unexpected convergence of multiple patterns of causal links.[9]

The origin, development, and operation of what was to become the National Airspace System was marked by continuing change, even experiencing several major shocks originating both outside and inside the system: the strike and firing of controllers in 1981 and the catastrophic events of September 11, the latter resulting in sudden disruptions and extensive major changes to the system. But, surprisingly, even changes from these most extreme shocks were absorbed by the existing structure, rather than eliminating or destroying parts of it, or changing its basic direction. Why and how this system persisted and changed over time is an important question that this chapter answers, helping us understand the air traffic system and its effects on controllers and their work at the time of the study and today.

Central to formation stories are the actions of social actors and people's own understandings of the social world.[10] From the earliest moments on, this history reveals both the creative and constraining actions that problem solving individual and organizational actors provided, slowing or speeding social transformation.[11] Assemblages of interacting social actors—material objects, artifacts, professionals, engineers, managers, scientists, and multiple heterogeneous organizations—became "system builders."[12] The life course of the system was alternately plagued and helped by changing political administrations external to the system[13] Moreover, the responses to these institutional factors by problem solving individual and collective actors within the air traffic control system were significant, affecting the system and the changing nature of work. Consequently, the formation story follows the development of controllers' work and workplace from the creation of the first air traffic controller through the life course of the system. It reveals the development of the social, technological, and architectural arrangements of work as well as the indeterminacy, interactions, experiences, decision making, and emotions that affect collective understandings, actions, and reactions of the controllers who work there.[14]

Although I necessarily condense history, events are sufficiently varied and detailed to surface incidents that show how small events can have seemingly large consequences.[15] Moreover, the details often correct well-known aspects of collective memory. So, for example, Charles Lindbergh's record-breaking 1927 flight from New York to Paris became the iconic representation of the successful achievements of American air flight during the 1920s, but the less-remembered 1929 Women's Air Derby reveals the hazards, hardships, and uneven development more typical of the fledgling system at the time. In the 1980s, the Professional Air Traffic Controllers Organization (PATCO) strike and the firing of the striking controllers in 1981 is a well-known part of recorded history, but not well known is how the FAA's attempts to fix the system afterward backfired, instead reproducing the conditions that caused the strike and consequently producing a new union, the National Air Traffic Controllers Association (NATCA).

This formation story is marked by four eras that show both patterns and variations in system emergence and development over time:

The Age of Innovators: The Diffusion of Ideas, Networks, and Infrastructure Formation, 1880–1930
The Age of Organizations: Controllers, Technologies, and Boundary Work, Ground and Sky, 1920–1950
The Jet Age: Congestion, Technological Lag, and PATCO, 1950–1980
The Age of Conflict, Decline, and Repair: The Strike, NATCA, and Technological Glitches, 1980–2000

The names identify the dominant actors—individuals, organizations, technologies, or their combined effects—of an era; the subtitles indicate the era's distinctive themes, so we can see the variation both within and across them. Then these four eras are followed by a transitional period that shows the system effects on the dynamics of the air traffic control system, the facilities, the workplace, and controllers when I entered the field in 2000.[16] Tracing contingencies and unanticipated consequences, "Dead Reckoning at the Turn of the

Century: 2000–2001" shows the effects of the past on the present as it was in 2000. Thus, the ethnography chapters that follow are located in system history, revealing the mix of the old and new in the workplace.

Although the invention and development of the airplane was a crucial factor across eras, other events—war, airline deregulation—were driving the technological innovations that changed the capability of the airplane. Indeed, the evolution of the air traffic control system had no major turning point that changed its direction; instead, the system had an incremental development over the life course.[17] The four eras were "event-full," following Abbott's identification of the life course of organizations: sequences of many events—"trajectories"—with events that were not equally weighted but were of greater or lesser import or impact on the system structure, processes, and controllers.[18] The history of air traffic control is marked by many turning points. Some were internal to an era, such that the turning point marking the transition to a new era seemed to be the culmination of a series of events—a trajectory itself, not a singular event.[19] Development was uneven and often halting, with unsteady progress, and turning points were not sudden, unexpected events but typically revealed themselves to be a slow process of varying duration—the exception being September 11 and the year following.

Even as this formation story reveals the variation within and between eras, it also illuminates the larger patterns of institutional change, persistence, and the changing nature of work across all eras. The following sections discuss patterns that overlapped and moved concurrently over time. Linear only in retrospect, system development proceeded in fits and starts, the repeating patterns standing out amid external contingencies and unanticipated consequences.

Precedent and Innovation: Boundaries and Boundary Work

Organizational and technological innovation characterized the system across eras, not only during the Age of Innovators. Often, a successful innovation in one era became a precedent that carried over into the next in an elaborated form. Many innovations began as an informal solutions to a local problem; when they were successful, they became formalized at the local level and often institutionalized throughout the system. Those innovations that survived became more sophisticated and their use expanded to cover new situations. This sequence of informal to precedent to formal to institutionalization not only was essential to system emergence but also was key to persistence and the ability to survive change.[20] The airplane, the aviation infrastructure, and the beginnings of the air traffic control system were all innovations that demonstrate the importance of precedent for innovations across eras. Even the system beginnings occurred in smaller forms of organizations—networks of aeronauts, small groups, professional associations, and early entrepreneurial business organizations—that were foundational to organizational infrastructure formation.

Initially, innovators were individuals responding to their local situations. The Wright brothers built upon and altered the work of earlier aeronauts, both in the design of their airplane and their testing equipment. The first air traffic controller was a local airport operator's innovation in response to the airplane's capacity to fly high enough to need some guidance from the ground. The first spacing patterns, in which planes are put into sequence to keep them separated as they approached urban airports, were independently initiated by other airport operators concerned about planes colliding. Unknown early innovators initiated a specialized language, some adapted from use in other modes of transportation (from highway to airway); other language was invented to match novel system characteristics. In combination, these actions were the beginnings of an elaborate system of boundaries in the sky and organizational boundaries on the ground that would make boundary work central to controllers' dead reckoning.[21]

Subsequently, however, the primary sites of knowledge production shifted to organizations. As air traffic increased, the FAA innovated in response to systemwide problems, many of its efforts successfully building on earlier precedents. But as the system structure grew complex, planned changes often had unanticipated negative consequences. Solving one problem tended to produce another.[22] Across eras, the creation of standards and their implementation led to controller resistance[23] The introduction of a new technology was complicated by design problems that surfaced during implementation and use. As technologies grew more complex, so did system effects. The advent of radar, then computers, then automation called for a technical infrastructure to mesh with the existing organizational infrastructure, creating tension between the need for standards and the need to customize according to local needs.[24] Too often the immediate effect was what Stinchcombe called "the liabilities of technological innovation": design problems created technological lag and added unpredicted costs into the system, initially complicating controllers' ability to do their work rather than making it more efficient and safe.[25]

The Sky as a Socio-technical System

The early local airport operators' innovation of spacing between airplanes at towers shows how small changes can have system effects with large consequences. The sky as nature was transformed into airspace, a virtual space constructed of artificial lines representing boundaries in the sky. It began simply. As air transportation increased, first the government, then the fledgling air traffic control system, responded to catastrophic air collisions by creating boundaries in the sky that classified airplane equipment into categories with similar capabilities to keep them separated.[26] Struggling to create order out of disorder as the airplane developed more sophisticated capabilities and air traffic volume increased, government actors made boundaries more refined, sorting airplanes by altitude and direction.

A precedent was set such that creating and moving boundaries in the sky became institutionalized as a key technology of coordination and control to improve traffic flows. The corollary development was that structures on the ground incrementally were constructed to hold the people and their devices that further enabled communication: first small structures, or "stations," then towers, then centers. As boundaries in the sky became more complex, specialized, and refined, so did the division of labor between and within the systems' organizational boundaries on the ground.

The standardization and divisions of the sky made it a site of contestation. Across eras was a continuing dynamic about where the locus of control of the sky should be: in the sky or on the ground. Gras and his coauthors brilliantly pose this as an ongoing power struggle between the "Icarus model," where the control is in the device in the sky, and the "mechanical bird model," where control is in devices on the ground.[27] The early eras show the shift of the locus of control from the pilot, flying in the sky and in nature (Icarus), to the air traffic controller and devices on the ground (mechanical bird). Then, over time, a reversal: later eras show the incremental transition of the locus of control back to the devices in the sky. The progression went thus: the invention of cockpit devices for pilots; then automation, which divided control between pilot, airplane, and controller, and the third was the late twentieth-century initiation of "free flight," aided by the Global Positioning System (now ubiquitously called GPS), which allowed the airplane to fly by itself, with pilots and controllers monitoring and intervening only occasionally. Always contested, both shifts were driven by trajectories of technological innovations, those driven in turn by political and economic conditions and the US standing internationally.

In response to changing external conditions, refining and elaborating the boundaries in sky and on the ground became key technologies of coordination and control.[28] Boundaries in the sky were refined and elaborated on the basis of classification categories that separated aircraft by equipment capability and direction. Classification categories were inviolable; boundaries, however, could be created or removed, permeable or shut, expanded or contracted, such that the system could be tightly coupled or loosely coupled, as necessary.[29] These and other developments were essential precedents to a system of interrelated parts. In combination, these technologies of coordination and control constituted a set of strategic organizational repertoires that incrementally lent the system the resilience, reliability, and redundancy that became a durable survival strategy, contributing to institutional persistence in the face of changing circumstances. A crucial development was the emergence of a supportive organizational field, including education institutions for aeronauts that would supply the system with future engineers, scientists, pilots, and research.[30]

The Changing Nature of Work

From the moment in the Age of Organizations that the job of air traffic controller was created, this formation story tracks the system effects on controllers and their work across eras. As aircraft equipment became capable of flying higher and higher, we witness the progression of changing architectural arrangements, technological innovations, and material practices as controllers moved from the airfield in all weather into towers, then into centers and TRACONs, where devices of representation proliferated.[31]

Dead reckoning incrementally changed. No longer working one-on-one with pilots, controllers' work consisted of human-technology interaction that required new cognitive skills, technical expertise, and material practices to cope with the changing boundaries in the sky and on the ground.[32] Distributed cognition expanded beyond the immediate environment.[33] No longer working alone in towers but in crews, the job entailed not only coordinated action between controller, devices, material objects, and other controllers in the room, but also coordination across greater distances, between controllers and pilots and with controllers in other locations. The training system was invented and reinvented to match the increasing complexity of airspace and ground structure: dead reckoning incorporated both ethnocognition and boundary work. Ethnocognition encompassed the constellation of experiential, cultural, interactional, and device-interpretive systems of sense making peculiar to the air traffic facility and professional niche they occupied. In addition, boundaries and boundary work in the sky and on the ground acquired social and symbolic meaning: divisions of airspace had become both territory owned by and conferring status upon controllers in the facilities that worked it.[34] Facilities worked to retain, increase, or keep their airspace territory—and hence, status and salary—from shrinking. Accompanying this transformation was a status shift across eras from airport worker to occupation to profession.

These patterns are visible within and across eras, affecting controllers, their knowledge production process, their work practices, and the interactional socio-technical dynamics within and between air traffic control facilities at the turn of the century.

The Age of Innovators: The Diffusion of Ideas, Networks, and Infrastructure Formation, 1880–1920

In the closing decades of the nineteenth century and far from Kitty Hawk, North Carolina, where the Wright brothers would make aviation history in 1903, aviation technology had advanced in Europe to the point at which lighter-than-air aircraft—airships, balloons, and gliders—were already capable of lifting people into the air. The earliest aeronauts traveled only short distances, relying on the pure form of dead reckoning—the Icarus model in which control rested with the pilot who interacted with the device in the sky, relying on human eye, ear, sense of smell, and attention to the wind, position of the sun and stars, shape of the clouds, and lay of the land.

Even at this early stage in modern flight, a pioneering European airman and balloonist recognized the need for special maps as an aid to dead reckoning for aerial navigators. Motivated by the unpredictability of flight and recurring injury and death, Hermann Moedebeck was the instigator of what would become one of the most important technologies of coordination and control for air navigation: aeronautical maps. Described as "an inspired and dynamic Prussian artillery officer and balloonist" and "zealous aeronaut" in the history of aviation cartography, Moedebeck actively encouraged the development of aeronautical maps to make flight more predictable.[35] Because of the low heights achievable in flight, his early advocacy for aviation cartography was solely for maps of the ground; the idea of a sky that could be mapped had yet to be conceived. In addition, as an artillery officer, he subsequently traveled around Europe, speaking about his balloon experiences, founding air navigation clubs in several of the cities where he was based and aiding their establishment in others. These local clubs were the first organizations dedicated to aeronautics.[36]

American had lagged behind other countries in contributing to the development of modern flight. In 1783—a watershed year for developments in France, the country most advanced in aviation—Americans were busy establishing their new country. A century later, America was in the midst of technological transformation.[37] Transportation had captured the national imagination. The first transcontinental railroad was completed in 1869, making the notion of a transportation system a reality in this country. The first design for an American automobile with a gasoline internal combustion engine was made in 1877. The historical moment was right: the national imaginary at the turn of the century increasingly embraced individual initiative, small businesses and a speed mode of transportation driven by energy, more like the train and car.[38] However, the technical knowledge for transporting people by air was still undeveloped. In Europe, the glider was viewed as the key to manned flight. Many aeronauts were engaged in soaring experiments, but few Americans were working on it.

In 1878, two children growing up in the Midwest, eleven-year-old Orville and seven-year-old Wilbur Wright, became aware of the potential for when their father, a pastor, brought home a small helicopter modeled on an invention of a French aeronautical pioneer.[39] When activated by rubber bands, it could lift itself into the air. When the much-loved toy broke, the brothers successfully built one of their own, but they failed when trying to construct a larger model that would fly. Growing up in a home environment where children were encouraged to pursue intellectual interests and to investigate whatever aroused curiosity, the brothers were inventors and voracious readers even in childhood.[40] Later, in their twenties—and neither having received a high school diploma—they started a printing business in Dayton, Ohio. In 1882, the Wrights opened a bicycle shop, manufacturing their own brand. Both their mechanical ability and their business acumen would bear on the iconic story of the invention of the airplane.

They remained fascinated with flight, reading newspapers and articles describing experiments with gliders and balloons by the earliest aeronauts in Europe and the United States. They learned from precedent. An important source was *Scientific American*, which published monthly the most recent patents and international developments in science and technology. Pioneer aeronauts were losing their lives in failed attempts to conquer the problems of flight. The tragic crashes made them heroes, provided a public record of the strengths and vulnerabilities of their designs, and led to changes in models and experiments. When the Wrights read of the death of Otto Lilienthal, a famous German engineer and inventor widely known for his soaring experiments who crashed in a glider of his own design in 1896, their attention returned to the problem of flight first encountered with their model helicopter.[41] In 1899, Wilbur wrote to the Smithsonian Institute in Washington, DC, about the problem of mechanical and human flight, receiving many materials back. Two were most important to their future work.

The first was Lilienthal's 1889 *Birdflight as the Basis of Aviation*, the result of twenty years of research and gliding experiments during which Lilienthal developed the principles of the curved wing and lift.[42] Running down the slope of a high hill, Lilienthal was pulled into the air by curved, bird-shaped wings above him, mounted on his shoulders, moving his body for balance as he dangled down below. He was the first to accomplish repeated successful gliding flights. Most significant for aeronautical inventors, his book contained careful systematic data and calculations on thousands of attempts. The second was Octave Chanute's *Progress in Flying Machines* (1894), a compendium of all international research on fixed wing, heavier-than-air research then available.[43] An engineer, inventor, and aviation authority, Chanute was a Paris-born American living in Chicago who corresponded with many aviation pioneers. Chanute became the central node in a network of aeronautical exchange. He translated parts of Lilienthal's book, becoming the social and scientific link between Lilienthal and early aeronauts in the United States and Europe, including the Wrights, who had written him for information after reading his book.

From publications—newsprint, articles, and books—innovators learned from others' successes and failures. Past invention—and aeronautical tragedies and triumphs—became the basis for further invention, both organizational and technical, a principle that would hold through the future decades of aviation. Lilienthal died when he lost control in a heavy wind. Although other pioneers believed the resolution to flight was to equip gliders with powerful motors that would keep aircraft aloft and enable heavier-than-air flight, Lilienthal's death convinced the Wright brothers that the central problem was controlling flight. Only if they could resolve it would any progress on motor-powered flight be possible. They started by building a biplane kite (two wings, one above the other) to experiment with wing position and lift. While retaining Lilienthal's interest in lift, they took a new principle from bird flight. Imaginations triggered by analogy and experience, they observed that just as bicyclists leaned their bodies to achieve balance and a change of direction, birds in flight were able to change the angle of the ends of their wings to make their bodies "bank" or "lean" into a turn. They mounted the two-wing kite structure on one of their bikes and began experimenting on the Dayton roads.

The Wrights' insight was in direct contradiction to the aeronautical innovators they read about, who, like Lilienthal, focused on using their body movements to keep from nose-diving. In multiple experiments with their "gliding machines" between 1899 and 1903, the Wrights repositioned the operator inside, flat on the body of the glider in order to lower wind resistance. Each design corrected problems of the last. Ultimately the Wrights created what they called the "three-axis model of control."[44] The glider experiments were at Kitty Hawk, North Carolina, where the brothers had relocated in 1900 for privacy and milder weather and winds. Both their 1900 and 1901 gliders had problems with lift and control. Returning to Dayton to work out the vexing problem of lift, they built a wind tunnel. Although their construction and use of the wind tunnel was seen as revolutionary at the time, they again were innovating from a precedent.

The Wrights had studied the work of Francis Wenham, who, in England in 1871, had first invented, designed, and operated an enclosed wind tunnel.[45] The Wrights copied his tunnel's structure, altering its size and aerodynamics to fit their problem. Based on detailed data from repeated tests of many wing shapes, made from scraps of metal and from the bicycle shop, they abandoned the Lilienthal data as incorrect and used their own wind-tunnel data.[46] From the results, in 1902 the Wrights designed a third glider that would ensure their place in history. It demonstrated a vastly improved ability to control flight. In March 1903, they applied for a patent on the 1902 glider with its three-axis flight-control design. Experts on the history of aviation declare that the patent on the design of the 1902 glider was equal to or even more important than the first powered flight that would come in December 1903 at Kitty Hawk, because "the 1902 glider essentially represented the invention of the airplane."[47]

The Wrights began to work on a powered model. Their work was again inspired by analogies drawn from personal experience. They concluded that a propeller had to be like a wing, the shape of the one they designed, but rotating in a vertical plane at the front.[48] Their engineering innovation becomes clear only in the contrast to their contemporaries, who were trying to achieve lift with horizontal propellers placed atop a machine.[49] Moving again to Kitty Hawk, they lived in tents and worked on into the winter. On December 17, 1903—a bitter cold and gusty day—Orville and Wilbur took turns with the new flying machine, named the Wright Flyer (fig. 1). They had four successful flights. The first was a twelve-second flight into a twenty-mile-per-hour headwind that reached an altitude of about 10 feet and a distance of 120 feet. The fourth was the best of day, a fifty-nine-second flight reaching an altitude of 10 feet and a distance of 852 feet. Although the fourth had the greatest distance and endurance, the first one was most significant to the Wrights. As Orville wrote, "it was the first time in the history of the world that a machine carrying a man and driven by a motor had lifted itself from the ground in free flight."[50]

Fame, however, did not immediately follow. They had few witnesses. The press was not present or impressed by the news that the Wrights circulated by telegram. Even the press back in Dayton had a lackluster response. It was as if nothing had happened.[51] However, the diffusion of their ideas through networks of aeronauts was already in progress.

Networks of Aeronauts: How Theory Travels

The Wrights' biographical trajectory reveals the relation between individual innovators, the diffusion of ideas, careers, networks, and infrastructure formation.[52] Like many other inventors who became prominent, the Wrights had worked independently and in isolation, free from organizational constraints.[53] Autodidacts, they engaged always with each other, but knowledge, material practices, and expertise can be acquired only by immersion in the society of those who already possess it.[54] Initially, they had learned about the work of others by reading. But they had also been in correspondence with the aeronautical inventor Octave Chanute, who took the lead in organizing small aeronautics conferences so that information might be shared—events that were foundational to the future development of aeronautical engineering as a profession. When they were near the end of experiments with the 1901 glider, Chanute had invited the Wrights to give a speech at a meeting of the Western Society of Engineers, of which he was president, on September 18, 1901.[55] Written and delivered by Wilbur to some fifty members, "Some Aeronautical Experiments" was a detailed account of the glider experiments conducted up to that point.[56] His presentation set forth both the steps and the missteps in the process and the engineering calculations, data, and systematic testing that went into each iteration of the design. Most significant, he made public the principles of the three-axis model of control that they would soon implement for the 1902 glider. Published in the conference proceedings, Wilbur's paper or reports of it were reproduced in professional engineering journals and aeronautical magazines. Consequently, the Wright brothers' theory of flight and engineering design traveled, becoming the basis for development by other aeronauts.

The Wrights moved their work back to a cow pasture close to Dayton for economic reasons, and in 1904 they built their second powered model, the Wright Flyer II, for which they had to actively work to keep the press away. Their 1902 patent application had not yet been granted, and so they talked openly and publicly to aeronauts about work already accomplished but would keep works in progress private until they had the result they wanted and the patent was approved.[57] Unfunded innovators working for the love of it, they needed money to continue their work independently. Chanute had offered them financial backing so they could drop the bicycle business and spend more time on the flying machine, but they had declined it, preferring to continue as they had, supporting themselves with the manufacture and repair of their bicycles.[58] In 1904, they achieved greater altitudes, making the first complete circle by a heavier-than-air powered machine. Still, there were control problems, the failures resulting in serious damage. By September 1905, for the Wright Flyer III they had altered the design to accommodate an operator sitting upright in a seat, all the better for working the controls and for vision. At the end of 1905, after 105 trial flights, the Wrights had achieved sustained flight for twenty-four miles, aloft for about thirty-eight minutes. The accomplishment was witnessed by several local people, one of whom took photos for the Wrights, but the Wrights did not make the photos public and kept reporters distant.

During 1906 and 1907, they did not fly. Now confident from tests of Flyer III, they were building six or seven machines in anticipation of selling them in order to fund further research. The machinery was so delicately related, part to part, that each model was an experiment that had to be tested and reworked, and they had to learn to fly anew. Still protecting their privacy, they nonetheless welcomed aeronauts—Chanute and others—who had read of their work, initiated correspondence, and traveled to Dayton and Kitty Hawk to see the Flyer III. Word of their work was spreading, but the brothers were not yet known outside of aeronautics specialists. During this period, they wrote to the US War Department to sell their flying machine.[59] They met with no success, brushed off by form letters, due to careless reading of their proposal and bureaucratic ignorance of their airplane's proven capability of flight.[60]

Trajectories of innovations begun independently by innovators in other locations led to the earliest steps toward organizational forms that would be the basis of an aviation infrastructure. In Europe, interest in aerial navigation was on the rise as an outcome of networks of aeronauts that Moedebeck's work generated. He had actively worked to attract interest, writing a handbook for aerial navigators and founding two magazines for aeronauts that attracted engineers, inventors, aerial navigators, and cartographers.[61] From interaction came networks and formal organizations. By 1905, aeronautical clubs had formed in Belgium, Germany, Switzerland, France, Spain, and Great Britain. The idea traveled to America: the Aero Club of America formed last, in New York. Also in 1905, these clubs banded together to form the Fédération Aéronautique Internationale. It was the first international organization created to encourage aeronautical science and sport. Continuing his safety advocacy, in 1906 Moedebeck published "Aeronautical Maps: A Necessity for Air Travel."[62] In its 1907 meeting in Brussels, the federation established the International Commission for Aeronautical Charts, headed by Moedebeck, which created a set of international symbols and produced the first international air map.[63] Fully international, the federation's existence and regulatory responsibilities lent international legitimacy to all aviation-related activities.

Going Public

Initially, although needing resources for their work, the Wrights were obsessed with perfecting their design, so they protected their privacy. They were not entrepreneurs in the Latourian sense: self-interested agents actively recruiting a network of interested others to support their invention.[64] Instead, as information about their successful invention spread informally from Chanute, aeronauts and interested others sought them out. Contacted in 1906 by the Aero Club of America, Orville and Wilbur sent a complete record of flights, winds, distances, and details of the motors since Kitty Hawk. In April 1906, *Scientific American* published a positive review of their work, reporting the materials they sent the Aero Club of America.[65] Representatives of both the British and French governments came to Dayton. The Wrights' patent application was approved in May 1906, freeing them to publicly demonstrate their design without fear of losing it and the routines that went with it.[66] Now they fully engaged as entrepreneurial actors. As subsequent negotiations with France and Britain fell behind, the US War Department woke up, advertising for bids for an airplane that could meet specifications matching those of the Flyer. In 1908, the Wrights' bid was accepted, and the price of a flying machine set at $25,000.

The Wrights received contracts from the US Army Signal Corps and a private French company in December 1907 and January 1908, both contracts requiring public demonstration of flight. In the summer of 1908, Orville traveled to Washington, DC, and Wilbur sailed to Europe for public demonstrations. Their respective Flyers were disassembled and transported, the one towed on an army wagon overland and the other shipped by sea, both needing extensive repair and reassembly on arrival.[67] On August 8 at Le Mans racetrack, Wilbur amazed observers with his technical skill and the controlled flight of the brothers' machine. Over several days he made complete circles and figure eights that impressed even the most skilled pioneer aeronauts. Thousands came to see him daily. His best performance was seven circles of the track in eight minutes on August 13, which was front-page news across Europe that week. After several short flights that began September 3, on September 9 Orville wowed spectators at Fort Myer, Virginia, where he made the first hour-long flight, breaking the record Wilbur set in France. The Wright brothers were world famous.[68]

Tragically, within days these achievements were followed by what history would record as the first deadly airplane crash. On September 17, 1908, with Orville flying another demonstration flight at Fort Myer, this time with US Army Lieutenant Thomas Selfridge in the newly designed passenger seat, a propeller shattered.[69] The airplane flew out of control. Selfridge was killed. Orville was pulled from the machine badly injured. The tragedy did not deter the course of history, however: the future of manned, powered, heavier-than-air machines was already in motion. Evidence that the world had become caught up in the potential of the flying machine had already been chronicled throughout the year in *Scientific American* issues. Before the crash, on February 29, two articles—one by Wilbur Wright, "Flying as a Sport—Its Possibilities," and a second by the editors, "Shall America Take the Lead in Aeronautics?"—had emphasized implications of controlled flight for recreation, commerce, and international scientific leadership.[70] In August, "The Wright Aeroplane Tests," reporting data from Wilbur's flights in France, was followed by a similar treatment of Orville's "The First Flight at Fort Myer" in the September 12 issue.[71]

Then, immediately following the September 17 accident, the September 26 issue of *Scientific American* contained two articles that strongly supported—indeed, furthered—the new enterprise. The first article in the issue, "Lessons of the Wright Aeroplane Disaster," took the position that although the disaster was tragic, involving death and serious injury, the enterprise must continue:

"The accident should not be allowed to discredit the art of aeroplane navigation. If it emphasizes the risks, there is nothing in the mishap to shake our faith in the principles on which the Wright brothers built their machine and achieved such brilliant success. The defect is one of structural detail. The breaking off of the blades of a propeller of an airship is comparable to the bursting of a tire on an automobile. . . . The accident should not be taken to indicate that the principles and design of the whole machine are at fault."

The second article, "The Construction of the Wright Aeroplane," included photos and diagrams showing the structure and dynamics of the design: how the wing surfaces warped, how the rudders operated, and how readily it all folded up for travel. The article noted that "the great simplicity of the entire machine is the most striking point about it and the one which most strongly evidences a real stroke of genius."[72] The article also included dramatic photos of the broken Wright Flyer III, with some people trying to extract Orville from the wreckage and others gathered around Selfridge.

In an exemplary act of "rhetorical closure," the influential *Scientific American* affirmed that the problem of flight had been resolved; alternative designs were not necessary.[73] Whereas the tragedy might have been a major setback to system development, the Wrights' contracts, successful public demonstrations, and the endorsement by *Scientific American* reinforced the legitimacy of aviation internationally and America's place as a competitor. Further, the possibility of aviation had become a cultural reality, a coveted normative aspiration in Europe and America.[74] The Wrights continued to be sought after and celebrated, and they received many invitations for demonstrations.[75]

The public achievement acclaimed at the time was the technology and the scientific innovation that went into it: design of the three-axis model—wings, engine, propeller—and the demonstration of the skill to operate it, which could be taught to others. It was not the series of single accomplishments, however, but the combination of them, linked together, that produced a complete technological system capable of flight.[76] Moreover, and crucial to how their theory traveled and obtained legitimacy, the Wrights' work conformed to the accepted standards of science.[77] The success of the Wrights was attributed to their "superb engineering skills."[78] Self-taught, they adopted the scientific methods basic to engineering: knowledge of the literature, design and construction of their own testing instruments, preliminary calculation repeated before implementation of design, analysis directing modification, and test. Innovation and learning from mistake called for constant adjustments and corrections in the next design. The Wrights had created a careful record of data, design and, most important, principles of flight that could ground future innovations.

The diffusion of ideas alone would not have had the same rapid effect without the demonstrated engineering skills. The Wrights' ideas were in scientific form and content, which not only gave them legitimacy as experts but also readily transformed into useful material practices for others. From Wilbur's first public paper presentation, they openly discussed ongoing efforts with other aeronauts, even inviting them to their work site to see how they were doing this, thus making replication possible.[79] Their success was not the simple iconic story of the heroic genius inventors, however. They were "heterogeneous engineers": economic, political, and social successes as well.[80] The immediate scientific, technological, social, cultural, political, and economic ramifications were enormous, as were the long-term implications for aviation and an air traffic control system.

Organizations, Technologies, Institutions, and Field

The aviation infrastructure, initially consisting of networks of aeronauts, aerospace clubs, and some interested engineering associations—began to expand and change shape as a result of the rapid development of formal organizations that would support aviation and the development of professional expertise. Scientific interest drew additional engineers and aeronauts to specialize in aeronautics, forming a basis for a new professional specialization. By 1910—just seven years after Kitty Hawk—both Massachusetts Institute of Technology and University of Michigan had ongoing design projects, gliders and wind tunnels, as well as informal programs by guest lecturers and not-for-credit classes in aeronautics. The first regular courses leading to a professional degree in aeronautical engineering were offered at both places in 1914 and 1915, respectively.[81] The Wrights' public demonstrations legitimated the flying machine as a multiple-use vehicle, with sport, commercial, and military uses. Their contracts with the War Department and private industry were proof that the flying machine was marketable. These small-town inventors, having previously turned entrepreneurs, then turned "professional entrepreneurs": using their business acumen and experiences dealing with contract offers, the Wrights formed a company in Dayton to continue their work and set up a flying school that would be the first of many throughout the country.

Moreover, they expanded the company to produce and sell their product, with branches in New York, France, and Germany. The scientific competition already written into aeronautics within and between countries multiplied, driven by market competition. In the United States, by 1911, the number of companies producing powered air machines was nearly a dozen. The market was small, as these companies mainly sold to wealthy individuals or the military. However, the flying machine also attracted the adventurous, who participated in demonstrations and races sponsored by aeronautics clubs in their pursuit of record-breaking, longer and longer flights. These events attracted large crowds, and as flying machines few over, the sounds of an engine in the sky had farmers stopping in their fields, people running out to their porches, and store clerks stepping outdoors to look up for their first glimpse of the flying machine.

Reinforcing the nation's aspirations, aviators became the new cultural heroes, turning heads on the street and being commemorated in literature. Caught up in the enthusiasm for flight, Sinclair Lewis wrote *The Trail of the Hawk* (1915), a quasi-autobiographical novel depicting the adventures of a kid who, like him, grew up awkward and shy in the Midwest.[82] Fascinated with flight rather than with writing, Lewis's hero became one of those aviator heroes. The book's depiction of flight and dead reckoning are persuasively realistic, suggesting that Lewis surely must have known fliers or been in a flying machine himself during those early years.[83] Crashes, injury, and death in aviation were normal and expected. Wilbur had written, "The sport will not be without some element of danger, but with a good machine this danger need not be excessive. It will be safer than automobile racing, and not much more dangerous than football."[84] In sporting competitions, the risks of flying remained patently clear, however. The Wrights trained a team of pilots to demonstrate their airplanes in organized flying demonstrations and sporting competitions with others. As many pilots died in fatal crashes, the Wrights canceled their team's participation.

The diffusion of ideas through networks often bore fruit in unexpected ways, with contingency playing a role. Interested others, aware but outside the main action, acting spontaneously and independently in the course of their routine activities, innovated in response to local situations, inadvertently having a large effect on the nascent aviation system. Among early speculation about the practical uses of the airplane was the potential for faster delivery of mail. Air transportation of mail by balloon originated in Europe and the United States in the eighteenth century, continuing sporadically (despite mail-laden balloons lost to wayward winds) in the nineteenth century.[85] In the United States, Fred Wiseman, a California native who worked for an auto dealership and raced cars on the side, was one of the many who visited the Wright brothers in Dayton to see their airplane. He returned home to have one built after their design. After a year's experience in flying competitions, Wiseman planned a trip from Petaluma to Santa Rosa, California. The flight became famous because he carried three letters, one of which was written by the Petaluma postmaster and addressed to the Santa Rosa postmaster. Hand delivering the letters on February 17, 1911, Wiseman's flight became the first officially sanctioned airmail delivery—official because in their written exchange, the two postmasters declared their letter a pioneer of US airmail.[86] Before the end of 1911, the US Army Air Service had taken up airmail.

At this moment in history, institutional actors began to give impetus and shape to the infrastructure foundations of aviation. Individual innovators did not disappear; rather, institutional actors also became prominent. The government aided the development of both the aircraft industry and aircraft technology. World War I had a tremendous impact on this development, creating a demand for aeronautical engineers and pilots. Racing to match the international effort and establish US power, the War Department estimated need for twenty thousand planes. Congress supplied the money, transforming the size of the American aircraft industry. By the war's end, a strong link had formed between the military and the aircraft industry that would continue.[87] Government policy after the war again boosted the aviation industry when in 1918 the US Post Office Department took over airmail service from the US Army Air Service. The first year the Post Office's Air Mail Service employed about forty pilots. Service was unpredictable that first year, with ninety forced landings, some due to weather, the rest to faulty equipment. Rail service was more predictable. Postal authorities decided that airmail would be more useful for serving longer routes.[88] The following year, a connection was forged between Chicago, Cleveland, and New York. Small networks of looped routes were created in other cities.

Aviation's unique social, organizational, and technical history was the basis for further development of an air transportation system. The informal looped route arrangements became formalized structures that were the rudiments of a nascent air traffic control system. Precedent mattered: ideas and structures were being diffused by analogy from the past. The first sign of a specialized language appeared. The concept of the airway was born: the equivalent of the highway for automobiles. The airways the Post Office's Air Mail Service created were based on landmarks, such as rivers, towns, and roads that retraced informal routes established by independent pilots, just as early roads had retraced well-worn carriage and cattle paths. The Air Mail Service joined the looped networks to follow the route of the Union Pacific railroad to San Francisco, relying on the nearby landmarks that pilots knew. War-hero aeronauts, sport navigators, and vagabond barnstormers who were scratching out a living in crop dusting, bootlegging, five-dollar rides, and aerial mapping found steady employment delivering airmail. The airmail route consisted of fifteen landing fields, each about two hundred miles apart, with pilots flying back and forth between particular pairs of fields.[89] The term *landing field,* or *airfield,* was literal: pilots were landing in open grassy areas. Moedebeck's aerial maps had not yet come to the United States. Some pilots used railroad maps, road maps, or pages torn from atlases. The challenge was holding on to them in open cockpits and flying at the same time.[90]

Technical innovation on the ground furthered development of aviation. The success of relaying mailbags across the country was handicapped because planes could not fly at night. Two technologies of coordination and control were specifically designed so that pilots could fly safely when the ground was not visible.[91] The first design began with problem solving by local agents. Two army lieutenants stationed near Dayton, Ohio, acting independently, created lighting for the airway loop between Dayton and Columbus.[92] The lighting comprised rotating beacons, flashing markers, and floodlights at airfields along the way so that pilots approaching from a distance could find their way to land. The head of the airmail service at the time had the idea to extend lighted airways across the country but didn't have the resources to do it. However, he did make changes that professionalized piloting and increased safety by requiring five hundred hours of flight, tests, medical exams, as well as instituting aircraft inspections for the US mail service, thereby bringing the first regulation into the fledgling system.

⁂

By 1920, key elements were in place that would be foundational to the future development of the modern air transportation system. This early era was marked by turning points in the emergence of an aviation infrastructure from which an air traffic control system would later develop. It shows the role of key innovators, their actions, their networks, and career moves that laid the groundwork: Moedebeck, Lilienfeld, the Wright brothers, Chanute, and Wenham. The diffusion of ideas through these relational connections brought about the genesis of a new kind of socio-technical expertise that would be foundational to the aviation profession. But contingency also mattered. Local innovators like Wiseman, the two Dayton army lieutenants, and the many others unnamed in this condensed history had an effect. This era also reveals the power of the written word—correspondence, newspapers, magazines, books, journals, conference papers—to shape the diffusion of ideas and the course of innovation. Untoward events—breakthrough flights and tragic crashes—spurred further development and improvements. The scientific and technical development of human flight was the driver of agency that would lead to other organizational forms: networks of aeronauts, including navigators, inventors, engineers, and cartographers, that were the essence of new aviation-related professions.

Dead reckoning was the domain of pilots. As technology increased the ability to control flight and airplane capacity for height and distance, these elements lead to the development of the earliest technologies of coordination and control to enhance dead reckoning: airways, airmail routes, and ground maps that established standardized ways of getting from place to place. Created from pilots' informal flight patterns, the routes became the basis of formal organizations on the ground. These were the initial structures and technologies from which an air traffic control system would grow. Building from the infrastructure that the US Army's airmail service began, the US Postal Department established a system of airfields, complete with beacons to guide the way. Stations at landing fields, with postal employees checking mail delivery between points, were the earliest official effort to track flight progress by people on the ground.

Legitimacy of aviation and a supporting aviation infrastructure proceeded incrementally, beginning with the Wrights' public demonstrations, contracts, and endorsement by *Scientific American*. Originating from multiple independent starts in different locations, additional new organizational forms included aeronautical clubs, private manufacturing companies, organized competitive aviation sport, and the first international association for the development of aeronautical science and sport. Speeding up the transformation, powerful institutional actors expanded and formalized infrastructure foundations. Universities initiated aeronautical and engineering programs. Government support, the quest for military power, and war pushed forward not only the production of airplanes, the formation of new companies, and subcontractors manufacturing special parts but also the production of pilot schools, pilots, aeronautical engineers, and mechanics.

At the end of the war, the Post Office Air Mail Service had instituted the first regulation. The nascent system of organizations on the ground was, in many ways, analogical to the invention of the airplane that spawned it: the design structure was built from previous structures, each version a springboard for further innovation and elaboration. Not yet a fully connected system of interdependent parts, by 1920 the aviation infrastructure was firmly ensconced in a supportive organizational field that would reproduce it. After the war, fliers and flying increased because airplanes were readily available and inexpensive, pilot schools flourished, and unlicensed pilots took to the sky.

The Age of Organization: Controllers, Technologies, and Boundaries, Ground and Sky, 1920–1950

During the second era, the Age of Organization, from 1920 to 1950, effects of the independent inventor, individual agency, and networks were obscured by organizational and institutional actors. The legitimacy and technological development of the airplane resulted from the intertwined interests and resources of airlines, military, and government, leading to interdependence of the three. As one grew and became more complex, so did they all, solidifying the organizational field. At the same time as the nascent aviation system flourished because of the activities of these powerful actors, it became dependent upon them, reactive rather than proactive. Indeed, boundary work—a key element of the resilience that would later characterize the air traffic control system—began during aviation infrastructure formation as a response to increasing air transportation and accidents: the sky became "airspace," an object with boundaries that would be organized, reorganized, and regulated; also, the first physical structures on the ground for flight would be expanded or contracted as needed. As aircraft equipment became more diverse and sophisticated, these initial boundary divisions would become increasingly specialized classification systems that sorted airplanes in the sky, which in turn would determine the division of labor between and within facilities on the ground.[93]

The external factors that had shaped the aviation infrastructure continued to affect it. Tragic accidents, sensational flights, war—and political administrations and national and international competition—drove technological innovations that advanced the capabilities of the airplane. Development continued to be marked by contingency and unanticipated consequences. System emergence was typified by problem solving by agents—both individuals and organizations—leading to solutions that, when effective, set precedent for future situations.[94] Fatalities were a concern, especially near airports, where planes were coming together. Independent local agents introduced informal safety practices in the sky and more elaborate organization structures on the ground. Some initiatives came from the private sector, as airport owners and airline representatives responded to local conditions. Originating independently from multiple locations at different times, across the era we can trace the coincidence of trajectories of organizational and technological innovations as they elaborated the nascent organizational infrastructure and its organization field.

The imagined analogy of the contest for locus of control of the sky between the Icarus model (devices in the sky) and the mechanical bird model (devices on the ground) materialized in the nascent system.[95] The coincidence of new technological and organizational innovations incrementally transferred the responsibility for control of the sky away from pilots flying in nature to air traffic controllers on the ground. Controllers controlled the skies by virtue of simple material technologies that made possible communication across social space. Initially used only near local landing fields, these communication technologies next connected the looped routes between airfields, a first step toward a true air traffic control system of interconnected parts. Further, as the amount of air traffic and its altitude capabilities increased, controllers relied upon newer technologies that were material representations of moving aircraft otherwise invisible.[96] These and other devices were heterogeneous, and they multiplied throughout the era. The shift in control from pilot in the sky to ground was institutionalized by the Civil Air Regulations, which created the concept of controlled airspace and mandated that pilots do what controllers ask.

By the end of World War II, distributed cognition between individuals, material objects, technologies, and places had become a defining characteristic of work in air traffic control.[97] Controllers' boundary work, organizing sky and ground, had initiated resilience within and between its interdependent parts. The concepts of traffic and congestion had come into being in relation to the sky. The blurred lines of responsibility for air flight and safety between the private sector and government early in the Age of Organization were clarified over time by the creation of formal regulatory organizations, and the initiation of procedures, rules, and law to introduce reliability and redundancy into the system. The increase in aircraft's speed and altitude capabilities had incrementally combined with additional ongoing technological innovations and the informal and formal actions of individual, organizational, and political actors to solidify and institutionalize air traffic control as a complex sociotechnical system of interdependent parts that participate in an international system

⁘ ⁘

World War I brought the realization that airplanes were sufficiently developed in Europe to be vehicles of international commerce. The start of regular transportation of passengers and freight between London and Paris made this a reality. However, US airplane technology was not ready. Pilots in the United States were flying two hundred to five hundred feet above the ground, navigating by following roads and railways, their dead reckoning supplemented, if at all, only by magnetic compasses. Entrepreneurs had developed businesses focused on flying people and goods between cities, but schedules were interrupted by weather, breakdowns, and crashes. Fatalities were routine.

Ambitious pilots saw the future in passenger travel, found investors, and started airlines, but those newly created airlines struggled. Now with more resources, the Post Office Air Mail Service began lighting the airways across country, setting up a coordination system on the ground that would be fundamental to the progress of aviation and a future air traffic control system. Known as the "Highway of Light," beginning in 1923, flashing acetylene lamps were positioned every three miles on the ground to guide pilots along the routes.[98] Steel towers with beacons visible for a hundred miles were built at airfields where pilots would land to pass on mailbags to the next flier on the route. Next, the Post Office installed the first airmail stations along the route so that progress of the mail could be tracked between airfields. These structures on the ground, both organizational and technical, were the first occupied by a person—a postal employee—who monitored progress of flights in the air on the basis of face-to-face conversations with arriving and departing pilots.

By 1924 the one bright spot for airmail service was that the routes developed and run by the Post Office were beating railroad mail delivery by two to three days.[99] Europe had nothing to compare. Fearing a Post Office monopoly of mail service and loss of revenue, railroad executives lobbied Congress and won a major victory that began to separate air commerce from government. The result was the Kelly Contract Air Mail Act of 1925, which required the Post Office to release its mail routes to private air carriers.[100] From a weak start with small companies, the bidding for air routes caught the imagination of the wealthy. Henry Ford, William Rockefeller, Cornelius Vanderbilt Whitney, and a Seattle plane builder named William Boeing competed to buy airmail routes. Boeing took an early lead.

By the mid-1920s, more people were flying greater distances, which led to the earliest attempts to organize airplanes in the sky. Pilots' dead reckoning was aided by the cross-country airmail beacons, which by then consisted of revolving, motor-driven lights atop sixty-foot towers. Beside each tower was a shed containing a generator and an operator, its roof painted with the beacon number and an alphabetical abbreviation for the route and nearest landing field. Near each structure, a huge concrete arrow was laid into the ground pointing the route direction (fig. 3).[101] Known as "inland lighthouses," by 1926, the light beacons were under the control of the Bureau of Lighthouses. President Calvin Coolidge signed the 1926 Air Commerce Act into law, giving the Department of Commerce regulatory responsibilities over the "inland lighthouses." Also, the department authorized the new Aeronautics Branch to establish safe altitudes of flight and rules for the prevention of collisions. The rules of the road, so to speak, from maritime navigation were applied to visual navigation—"see and be seen"; keep to the right; do not to take off until there is no risk of collision with landing aircraft; wait until preceding flights clear the field. To reduce hazards, cartographic maps of the ground became imperative. The Army Air Service initiated photomaps of air routes that were systematically marked with indicators on the ground.[102]

The independent, individual inventor was displaced by organizations: units of all branches of the military, private manufacturing companies, and university research institutions were rapidly developing new airplane technologies.[103] Driven by the resulting increases in air transportation, the aircraft industry grew. By 1926, there were eight airlines in the United States.[104] Between them, the total number of available seats for passengers was only two hundred. Surprisingly, about 5,800 passengers took flights, surely a statement that the public's romance with aviation remained strong, for passengers likely did not have a pleasant experience.[105] Airplanes were flying at a speed of 100 miles per hour and landing at 50 miles per hour on grassy fields. Passengers suffered through cramped quarters, uncomfortable seats, bumpy rides and landings, unbearable cabin temperatures, and the penetrating noise of propellers. Many passengers opened windows to get some air, but many opened them to throw up, and some planes had to be hosed out after landing.

By 1927, government had become both a builder of the industry and its regulator. At the research facility at the Army's Langley Field in Virginia, government engineers made tremendous technological advances. One result was the "Langley cowling": a removable cover for air-cooled engines that allowed a new high speed of 177 mph.[106] The cowling subsequently became a standard feature in both military and civilian aircraft, increasing their capacity for both altitude and speed. These new capabilities opened up the possibility of international flight from the United States. The dramatic flight of Charles Lindbergh in 1927 across the Atlantic propelled the transformation of commercial aviation, much as World War I had transformed military aviation.[107] Like the Wright brothers, Lindbergh seemingly came out of nowhere. He had done some stunt flying, was an army pilot, then flew airmail between Chicago and Lambert Field in St. Louis. In France, a competition was set up with $25,000 to be awarded to the person or persons who could fly nonstop from New York to Paris. Several who preceded Lindbergh, flying with a navigator or a crew, either were injured or died in the attempt. Lindbergh, who would fly alone in *The Spirit of St. Louis,* chose a single-engine Wright Whirlwind because he couldn't afford the highly esteemed trimotor.

In his memoir, Lindbergh described how he found his way. He relied on a mix of nature, devices, and documents: his dead reckoning holistically combined the terrain, position of the sun, shape and color of the clouds, water movement, indications of wind and weather, two magnetic compasses, and terrain and coastal charts on which he laid out a route in advance.[108] He kept a log. Overland, he flew low, between two hundred and six hundred feet, at approximately 90 miles per hour. To stay above the clouds over the ocean, he flew between 7,500 and 10,000 feet. The struggle was to stay awake. About twenty-six hours into the journey, he descended when fog cleared, spotting land he identified as Ireland. He was two and a half hours ahead of schedule and fewer than three miles off course. Lindbergh landed at the Le Bourget Aerodrome, in Paris, after 33 hours, 30 minutes, 29.8 seconds, and 3,610 miles. He received a hero's welcome and national and international acclaim.

As with Kitty Hawk and the Wright brothers' public demonstrations there in 1908, Lindbergh's daring accomplishment proved the successful development of the technology, this time, for international transport: the Whirlwind's air-cooled engine had kept Lindbergh's plane aloft for thirty-three hours, affirming the legitimacy of the enterprise. Moreover, the flight had a system effect: the arrival of "air traffic." After Lindbergh's flight, aircraft production, which had been about one thousand in 1927, rose to over six thousand by

1929. The number of paying passengers shot up from just under six thousand to over four hundred thousand by 1930.[109] Another effect was that competitive air sports also flourished. Lindbergh's flight success had many pilots, excited by his achievement, engaging in competitions for time and distance. As aircraft designs changed, pilots shattered endurance and speed records. For aviation, endurance and speed were crucially important for legitimizing the airplane as a reliable, high-performance device, not a novelty.[110]

Spurred by European developments and the economic promise of aviation, the Commerce Department's new Aeronautics Branch had initiated a plan for the federal airway system. It completed the installation of lighting for airways that the Post Office had begun. By the time of Lindbergh's transatlantic flight, the United States had 4,121 miles of lighted airways. The transcontinental airway system was made possible by the simultaneous development of other technological innovations, including beacons to guide pilots, and teletype and radios for ground communication. These in turn made necessary the erection of new physical structures—stations—to house these new technologies and the technical specialists necessary to operate them. By 1929 the skeleton of an air traffic control system was physically visible in the aviation structures on the ground. The transcontinental airway system had grown to include "92 intermediate landing fields, 101 electric beacons and 4,117 acetylene beacons. Also included were 17 radio stations. Personnel involved in the transfer included 45 radio operators, 14 maintenance mechanics, and 84 caretakers."[111]

Risk

Not all developments in aviation were as smooth as the relative ease of the Lindbergh flight and the seeming steady progress of the aviation system in retrospect. Development in some areas was uneven. Not all landing fields were linked to the transcontinental airway. Many pilots were flying off route. Some airplanes had radios and some pilots bought the charts then being sold, but many flew without these aids. Getting from point A to point B was unpredictable and risky. Illustrating the hazards of flight was the Women's Air Derby, the first official women-only air race that took place during the 1929 National Air Races and Aeronautical Exposition.[112] For over a decade, American women had been barnstorming, flying in air circuses, making their own mechanical repairs, and setting altitude and speed records. Racing, however, was reserved for men. To qualify for the national races, women had to meet the same criteria as men: one hundred hours of solo flight, including twenty-five hours of cross-country flight. However, they were required to fly planes with horsepower "appropriate to a woman"; one plane was excluded by the judges as being "too fast for a woman to handle" (despite her own experience racing it). On August 18, 1929, twenty women—Amelia Earhart among them—took off from Santa Monica, California, for Cleveland, Ohio. The race was in legs, with overnight stops. It was, for every one of the twenty, harrowing.

On the first leg, one pilot who was sitting low in the pilot's seat, despite being in an open cockpit, breathed in carbon monoxide from the engine, barely managing to land before passing out. She recovered, the next day repairing the problem by using a pipe to channel a flow of fresh air toward her, and continued the race. Several made emergency landings. One detected a fire in her luggage while in flight, landed in the desert, tore out the wooden flooring, and put out the fire with sand. Two were forced to land while blinded by a dust storm. A few lost their way and landed in the Mexican desert. Everything possible happened, including sabotage, because of the many who opposed women's racing. Fifteen of the pilots landed at Cleveland Municipal Airport nine days later. Amelia Earhart finished third. Louise Thaden, who was the one who had suffered carbon monoxide poisoning on the first leg, finished in first place. The pilot Marvel Crosson succumbed to carbon monoxide poisoning, crashing in the desert of Arizona's Gila Valley, apparently having blacked out.

The difficulties these women pilots encountered reveal one aspect of uneven development in the aviation system. Safety and avoiding collisions was clearly a concern, especially at airports. Government involvement in airports had been forbidden by the 1926 Air Commerce Act in order to encourage private enterprise. Following script, local governments and private companies that wanted to attract airlines to their cities took over airport building, expansion, and operation. Wanting to reduce injury and death and encourage commerce, some of these airport operators innovated, providing an early form of air traffic control based on visual signaling. The first official air traffic controller—an iconic figure in the history of aviation, still heralded by air traffic controllers—was Archie League (fig. 2). Hired in 1929 at Lambert Municipal Field in St. Louis (now St. Louis–Lambert International Airport), League was a former licensed pilot and a licensed engineer and aircraft mechanic.[113] Holding the position of "flagman," League directed traffic from the ground with flags. He positioned himself near the runway threshold to monitor wind direction and runway conditions. His wheelbarrow full of equipment included an umbrella for protection against the summer sun, a beach chair, and lunch. His technologies were a notebook, a red flag to signal pilots to "hold," a checkered one to signal "go," and the wheelbarrow, for changing his own position on the airfield as the wind changed.

The contrast between the uneven development of the aviation system on the ground and the rapid development of airplane capabilities was stark. Early on in the Depression, the budget for the Aeronautics Branch was cut. However, the US military was eager to catch up with the Europeans' war air capability, and so energized the airline industry with lucrative contracts and demands for more sophisticated equipment. Interdependence between the aviation industry, government, and the military grew. The strong competition within the industry for these military contracts sped up technical innovation.[114] In the few years between 1929 and 1933, a number of technical developments converged that would lead to a new generation of aircraft that were precursors of the modern airliner. Seeking to use military innovations to acquire a larger, faster commercial plane that would beat the competition, United Aircraft and Transport contracted with Boeing for a ten-seat passenger plane. Boeing produced the 247, a remarkable advance over the trimotor, which was the dominant model at the time for commercial air travel. Falling behind the others, United got back in the game with ten twelve-passenger Douglas DC-2s, faster than the 247. And so it went. Between 1930 and 1931, incorporation and mergers in the airlines industry led to domination by Pan American, the (later renamed) United Airlines, Eastern Air Transport, American Airways, and Transcontinental and Western Airways (aka TWA).

The airline industry's main problem was competition with rail travel for passengers. Because aircraft could fly only when the weather was clear, air transportation could not maintain the fixed schedule that railroads could. With the exception of those few airports that had followed the lead of St. Louis and instituted the use of flagmen like Archie League, all flight was guided by the pilot's dead reckoning. The rules that existed were visual flight rules (VFR)—"see and be seen"—which stated the rules for pilots proceeding from one point to another by means of contact with the ground and visual contact with other aircraft in the airspace.[115] When pilots couldn't see, they couldn't fly. However, new aircraft speed and altitude capabilities made clear that VFR would not work for all types of aircraft. Technological and organizational innovations, some long in the works, some created in response to changing conditions, ensured the continuing survival of the airline industry.

From Icarus to Mechanical Bird: Devices, Distributed Cognition, Ethnocognition, and Boundary Work

Originating from different social locations, multiple independent trajectories of innovations would intersect to transform the control of the sky from the pilot flying in nature to the controller and devices on the ground.[116] The transformation was incremental, proceeding in sequences of small events initiated independently from different locations, that intersected with the development of the system at different moments to culminate in a system-wide change.[117] No major turning point distinguished the transformation. It began, inconsequentially, it seemed, with the St. Louis airport owner's act that brought the job of air traffic controller into existence, assigning Archie League to the task of guiding pilots to land and depart. At the same time, several innovations were ongoing, following trajectories that would set off actions and reactions to elaborate the organization structure and its field. One crucial innovation was the adoption of radio for use in air transportation. By 1931 the Aeronautics Branch had placed intermediate landing fields from thirty to fifty miles apart with a network of radio beacons to connect principal cities.[118] Once two-way radio communicate on was installed in airplanes, airport owners constructed the first air traffic control towers at the busiest airports. Controllers—including Archie League and the many like him—moved from the airfield into the tower. By 1933, in poor weather with low visibility, airline radio operators and tower controllers were able to guide pilots along a route using two-way radio communication (fig. 5). Communication was indirect: airline operators conveyed messages between pilots and controllers. The radio signals were not always clear, and some pilots were not equipped with radios, so tower controllers sent messages to the pilot by flashes from the Aldis lamp (fig. 4), or "light gun."[119]

Another trajectory of innovation was the development of a cockpit device that provided "instrument flight" capability, so airplanes could fly with no visual earth references as guidance.[120] Cockpit devices had been in the works as early as 1914. With the installation of cockpit devices and the two-way radio, pilots had the ability to follow air traffic controller directions to fly during poor visibility. Consequently, instrument flight rules (IFR) were developed to govern flight under conditions in which flight by outside visual reference was not safe. The further advance of instrument flying was stimulated by the advance of weather forecasting, weather reporting, and the development of greater passenger comfort in commercial aircraft.[121]

The fourth trajectory of innovation during the 1930s was the creation of boundaries in the sky to separate airplanes. The development of instrument flight capability had solved one problem but created a new one: how to avoid collisions between pilots who could not see each other. Both government and airlines responded to ensure the separation of airplanes. The Aeronautics Branch created boundaries dividing the sky—formerly open to all human flight and limited only by airplane capabilities—into two parts: one for those aircraft flying at low altitudes (up to six thousand feet), using VFR, and one for those flying at higher altitudes (up to ten thousand feet) and/or under conditions of poor visibility, using IFR. Pilots were required to master the instrument flight rules and be licensed in order to fly.

However, the most dangerous airspace was at airports because pilots using either set of flight rules combined in low altitude during takeoff and landing, raising concerns about collisions. Alarmed airline radio operators, in separate locations, acting independently, problem solved by initiating informal practices using radio communication to coordinate actions across social space. If two airplanes were in possible conflict, radio operators of the two involved airlines would coordinate flights with each other and with the tower air traffic controller using a local airport intercom. Making notes with pencil and paper, they organized planes in relation to one another, maintaining space boundaries between them so they would arrive and land one at a time. This was the beginning of flight sequencing.

In a move that formalized this informal arrangement, in 1934 American Airlines initiated a "flight-following system" at Chicago Municipal Airport (today's Midway International Airport). American Airlines operators sequenced aircraft for arrival when they approached within one hundred miles of the airfield.[122] Initiated by former pilots Earl Ward and Glen Gilbert, the success of the airline's coordinated sequencing lead to formal agreements among major airlines to coordinate traffic around Newark, Chicago, and Cleveland.[123] These agreements included the first traffic separation rules, designed to ensure that all airline pilots adhered to the same rules about allowed distance between planes. The accompanying response was organizational innovation: the creation of new facilities on the ground to house the people and devices that worked the fast-changing airspace. By 1936, the airlines had built airway traffic control centers to facilitate this coordination, corresponding to the coordinating function of the tower. The new responsibility for ground personnel was sufficiently complex to require specialist training.[124] Standardized training was the official beginning of air traffic control as a profession. Moreover, this was a major step toward air traffic control as a true system because the new communication technologies, when effectively linked, would make possible fully coordinated actions across all facilities.

By the mid-1930s, the result of the coincidence of these several trajectories of organizational and technical innovations was that pilots flying by IFR became fully dependent for dead reckoning upon controllers on the ground who worked in organizations that functioned as "centers of coordination."[125] Dead reckoning—the prediction of the positions of objects in space and time without benefit of direct observation or direct evidence—had been incrementally transformed. Condensed as the history is here, we don't see the scientific discoveries, negotiations, networks, politics, and economic interests that drove it, but assuredly they were there.[126] As dead reckoning changed, there was a rush to develop material objects that could represent what controllers and pilots could not directly observe. Aviation cartography, by then an industry, had to change. Boundaries dividing lower airspace from upper airspace meant that existing topographical maps used by pilots for visual navigation had to be supplemented by maps of the boundaries in the sky used by controllers and pilots alike so that flight paths could be coordinated.[127]

Distributed cognition evolved as a problem-solving response for air traffic control. Additional devices gave controllers a means to "see," or visualize and measure, striving for more accurate spacing by simulating the positions of the airplanes in the sky. Controllers used metal geometric compasses, blackboards, small boat-shaped metal objects called "shrimp boats," and chalk. It was the combination of controllers and these devices that was the basis of the new dead reckoning. Information was still relayed indirectly: pilot to airline dispatcher to controller to airline dispatcher to pilot. But the process became more systematic and complex. Controllers, pilots, airline operators, and their material objects interacted, moving traffic via distributed cognition: local knowledge and memory were generated in ongoing exchange between individuals, objects, and tools in the immediate environment of material workplace as well as across space to others.[128] Controllers' dead reckoning had come to consist of both ethnocognition and boundary work.

Distributed cognition worked like this: Pilots filed flight plans with their airline operators or dispatchers, who checked the plan and passed messages on to controllers, who had no radio contact with pilots. Wearing mandatory white shirts and ties by the mid-1930s, controllers wrote flight data representing each plane, its identification number, and route on large blackboards, sequencing them to better predict possible conflicts (fig. 6). To visualize the

Early Devices for Dead Reckoning

Figure 3: Beacon, steel tower, shed, and concrete directional arrow, beginning Transcontinental Air Mail Route, 1925
Dreamsmith Photos, www.arrowsacrossamerica.com

Figure 4: Archie League with signaling light, Lambert Municipal Field, St. Louis, Missouri, 1933
FAA History Office/National Archives (237-G-183-8)

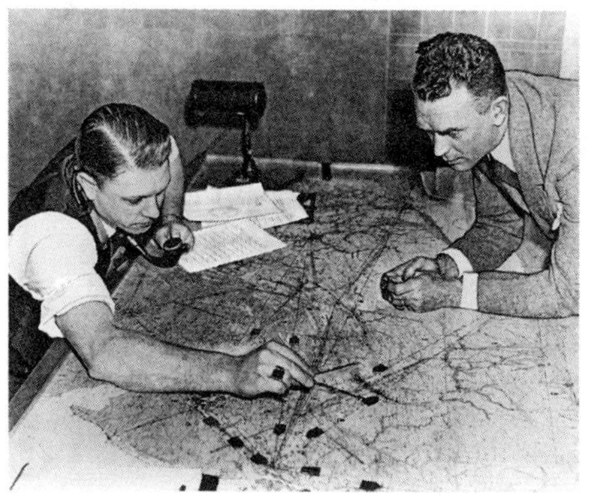

Figure 7: Sequencing en route traffic with compasses and moving "shrimp boats" on table maps, Newark Airway Traffic Control Station, 1936; left, Earl Ward organized the air traffic center; right, R. C. Eccles
FAA History Office/National Archives (237-G-65-2)

Figure 8: Women controllers, who replaced men during the war, sequencing en route traffic with flight progress strips, replacing blackboards, early 1940s
FAA History Office/National Archives (237-G-65-34)

Figure 5: Air traffic controller in radio-equipped tower connecting to airline dispatchers, Newark, New Jersey, 1936
FAA History Office/National Archives (237-G-183-15)

Figure 6: Controllers using blackboards, maps, and phones to airline dispatchers to sequence en route traffic between airports, Newark Airway Traffic Control Station, 1936
Smithsonian National Air and Space Museum (NASM 95-2882)

Figure 9: Radar arriving and controller following blips on upright screen, Washington Air Route Traffic Control Center, 1948
FAA History Office/National Archives (237-G-65-48)

Figure 10: Controllers use flight progress strips to sequence en route traffic and move shrimp boats on flat radar, Washington Air Route Traffic Control Center, 1955
FAA History Office/National Archives (237-G-65-42)

position of planes in the sky in relation to one another, controllers translated the flight data on the blackboard onto material objects that they could manipulate. A small, flat metal object—the shrimp boat—represented each airplane and was moved along the airplane's route on a table map. The identification number and flight plan of an airplane was written on a note card that was clipped to the back of the shrimp boat, the device analogous to the plane in the sky. As a plane moved in the sky, the route of the plane was duplicated by physically moving the shrimp boat on the surface of the map. Controllers used a metal compass to estimate distance and angles, which led them to make any corrections to routing (fig. 7). To avoid possible aircraft conflicts, controllers relayed information about aircraft direction, route, and moves to the airline dispatcher, who passed it on to the pilot. Every fifteen minutes this process repeated, and aircraft positions were updated. It was progress, but like all dead reckoning, it was imperfect. Not all pilots filed flight plans; shrimp boats' positions were representations, not the real thing; movement directives went indirectly from dispatchers to pilots, and instructions from controllers were considered advisory, not mandatory, so not all pilots followed them.

The final fundamental changes to dead reckoning in the 1930s were legal mandates institutionalizing control of the airspace by controllers and the organizational system on the ground. During the Roosevelt presidency, increased government involvement resulted from dramatic failures of flying, the Depression, and New Deal work programs.[129] Government plans were in the works to "establish a uniform and centralized system of airway traffic control . . . to direct and coordinate the progress of all flights . . . to insure the maximum safety . . . to prevent collisions and to direct traffic so as to insure arrivals at airports in an orderly manner."[130] In 1936, the government took control of the Newark, Chicago, and Cleveland airway traffic control centers from the airlines and by 1937 had established five more.

The airlines welcomed government intervention that aided the industry and reduced their costs. At the end of 1936, US civilian airlines had transported over a million passengers in scheduled flights in a single year.[131] In response, the government requested greater regulatory powers. The Civil Aeronautics Act of 1938 created a new agency, the Civil Aeronautics Authority (CAA), a division of the Department of Commerce. The CAA immediately established standardized practices, coordination, and consolidation among the various parts of the system. However, the rules and procedures already in existence were discretionary. For safety, the CAA sought to convert these into legal mandates by creating the Civil Air Regulations that codified these rules.

Perhaps most important, pilots were mandated by the Civil Air Regulations to obey controllers' directives. The very idea of controlled airspace—that is airspace regulated by controllers—was controversial. Many pilots were used to going where they wished. But more control followed. Airspace boundaries were refined, beginning a sophisticated classification system that sorted aircraft in the sky by speed, altitude, and destination.[132] Using as precedent the Aeronautics Branch's creation of separate airspace for flights using VFR and IFR,[133] the new Civil Air Regulations required that airspace be divided into routes and intersections, with aircraft movement restricted by traffic rules and controller directives. For high-altitude air traffic, the first "rules of separation" were codified. Even and odd cruising altitudes also were designated. The rule was northeast, odd; southwest, even. For planes heading northeast, the assigned altitude was thousands odd plus five hundred feet (e.g., 21,500); if heading southwest, their assigned altitude was in thousands even plus five hundred feet (e.g., 20,500).[134] The result of this was that airplanes approaching head-on would always be separated by one thousand feet of altitude and five miles of distance. Air traffic controllers were to preserve these rules in controlled airspace, but pilots were mandated to follow them even when not under air traffic control jurisdiction. The shift from pilot dead reckoning to controller dead reckoning was complete. Further, although development of this aviation system had been uneven, distributed cognition and standardization finally had the air traffic control system functioning as a fully connected system of interdependent parts.

The Legacies of World War II: Resilience, Reliability, Redundancy, and Radar

The system was complete, but it was dynamic, and consequently dead reckoning continued to change in response to events, conditions, and powerful institutional actors that affected the organization of the system, its technologies, and the work of controllers. Further driving these changes, after Germany invaded Poland on September 1, 1939, beginning World War II in May 1940 President Roosevelt called for the production of fifty thousand airplanes a year, anticipating the US entry into the war.[135] Resilience of boundaries on the ground and in the sky was key to the persistence of the system in the face of its continuing to be buffeted by external factors. The number of towers and centers were increased "in the interest of National Defense," and the CAA took over the construction, maintenance, and operation of towers.[136]

In 1941, in preparation for the influx of military traffic, the airspace was divided in yet a third way: airspace boundaries were changed as chunks of one airspace were taken from one air traffic control facility and given to another. Setting precedent for future strategy, the CAA redistributed the airspace between centers and towers to reduce centers' workloads in some locations, thereby affecting the work of controllers in both places.[137] The shift in jurisdiction delegated authority to the towers and enabled coordination of authority between the two. The government takeover of tower operations made possible the consolidation of the system and also improved communication and coordination between its parts. To secure this goal of improvements, between 1941 and 1942 the CAA established seven training centers for controllers. When controllers were called to serve in the military during World War II, for the first time women became air traffic controllers, going through the same training and receiving the same salary as men. At their peak, women represented well over 40 percent of the controller workforce, a condition never since repeated.[138]

To handle the increase in military traffic, technologies for controllers' dead reckoning had to be altered. At centers, the shrimp boats remained, but the traffic surge rendered the blackboard method of tracking and sequencing aircraft no longer viable.[139] The blackboards were replaced by what would become an enduring technology: movable flight progress strips. A small, rectangular printed paper strip represented each aircraft, identifying its flight number, starting point, route, and destination. Controllers marked the strip with changes of altitude and direction along the way, to match the plane's movement through the sky. These strips were placed in plastic "strip holders" and placed with other strips in "strip bays," slanted boards for stacking the strips. Controllers sequenced strips—that is, airplanes—in relation to one another by time and location (fig. 8).

To manage strip sequencing, the architecture of the workplace in centers had to change. To accommodate the sequencing, six to eight air traffic controllers sat in a row at a counter in front of the slanted strip bays, each bay holding strips and twenty bays to a row. Controllers physically organized, sequenced, and reorganized the strips on the basis of telephone communication. Some controllers took a shift as runners who relayed information between the controllers sequencing strips "on the boards" and the controllers working the shrimp boats, in an attempt to keep the visual representation analogical to the organization of the flight progress strips. Controllers left extreme spacing between airplanes to compensate for the unknown difference between the true position of an aircraft and the controllers' material representation of it.[140]

The war's additional technological legacies included radar and direct pilot-controller communication in both towers and centers.[141] First developed in Britain, radar was used extensively during British military operations in World War II. After the war, radar was brought into regular use in US centers as well (fig. 9). Initially, it was used only to provide redundancy to the system of the shrimp boats. Because the ability to represent planes by alphanumeric indicators on radar had not yet been developed, flight progress strips remained absolutely essential to the operation. Foreshadowing technological advance for decades to come, the combination was an odd patch-up of old and new. Tabletop maps were eliminated; instead, flat, circular radar scopes were set into tabletops. Controllers sat or stood around the tables, placing shrimp boats on top of the radar scope (fig. 10). To track the flight, they moved the shrimp boat markers through the route atop the appropriate blip on the scope, relying on flight progress strips for identification. This required little change in distributed cognition. To better see the radar images, controllers worked under small tents that provided them with the necessary darkness but also kept their cigarette smoke from escaping.[142]

Later, international standardization reinforced system reliability. Wartime cooperation between countries gave renewed international impetus for common rules, procedures, and practices. Under the auspices of the United Nations, the International Civil Aviation Organization (ICAO) became a reality on August 4, 1947. The ICAO advanced international technologies of coordination and control to increase the reliability of air travel both within and between countries. Long-distance flights across multiple airspace and national boundaries risked miscommunication between pilots and controllers whose national languages, ways of pronunciation, and meanings differed. English was declared the international language. The ICAO generated three-letter airport and airline codes, alphanumeric aircraft-type codes, standardized phrases for exchanging information between pilots and controllers, and a phonetic alphabet of codewords to represent letters and numbers, assigning common pronunciations in order to increase accuracy in radio and telephone exchange.

By written agreement, these were incorporated into member countries' flight regulations. Standardization of the US system had already proved effective. Fatalities were no longer normative. Aircraft equipment had improved; the air traffic system had improved. The CAA focus on refining air traffic procedures, rules and regulations, expansion of facilities, training of controllers, regulation of pilots, and aircraft inspections had increased system reliability.

At war's end, the air traffic system consisted of 113 towers and 24 centers, operated by 1,800 personnel. The fatality rate had improved dramatically, averaging one death per hundred million passenger-miles.

The war's final legacy was congestion. Postwar air travel went from 6.7 million passengers in 1945 to over 12 million in 1946.[143] The large city airports were handling millions of people a year.[144] For safety, tower controllers were spacing traffic so that only one airplane landed every ten minutes. In response to this new problem, the CAA had been testing and perfecting an instrument landing system that allowed pilots to control their own landings, guided by instruments in the airplane and technologies on the ground: a glide slope beam, an approach lighting system, and a series of regularly placed radio markers that indicated the path and slope that would align the aircraft with the runway at the proper angle, direction, and down. This new technology allowed for "precision landings"; moreover, fifteen planes could land in an hour. At the end of the 1940s, sixty airports had instrument landing systems. These systems did not solve the congestion problem, however. The diagnosis was that, despite tremendous advances, the air traffic control system was falling behind. It was still saddled with uneven development, and technology for dead reckoning was either antiquated or lacking altogether. The system was in transition, a mix of the old and new. Heppenheimer itemized the system's weaknesses: "It still relied on radioed course and position reports, with controllers pushing shrimp boats on a table top. And while control towers could communicate directly with the aircraft, the air traffic centers still were hampered by the limitations of the 1930s vintage radios that remained in use. Controllers still had no direct contact with pilots because there were not enough radio channels. Radio relay operators guarded the few that were available and acted as middle men, receiving messages that pilots and controllers would be sending to each other and passing them on."[145]

Despite rapid development, its technological and organizational achievements, and the increase in safety, air transportation had advanced beyond the air traffic control system's ability to manage it.

The Jet Age: Congestion, Technological Lag, and PATCO, 1950–1980

The air traffic control system was a system under stress. The very factors that had brought it into existence were testing it. By this point a fully institutionalized system, the Jet Age, from 1950 to 1980, was marked by the system's growth and elaboration in response to changing institutional conditions. Every external factor that impacted the system had effects—system effects—on controllers and their work, making it more complex and challenging. Rapid advances in airplane technology increased the volume and complexity of air traffic: dramatic air collisions drew attention to flaws in the system, making safety an urgent national priority. The arrival of subsonic and supersonic jets led to continuous expansion and reorganization of boundaries of the sky and the boundaries of the air traffic facilities on the ground. The system's resilience was the key to its persistence. Whereas in previous eras recorded history revealed the effects of external conditions, actors, and actions on the system in broad scope, for the Jet Age and the eras to come, the archival record allows us to narrow in on the effects on controllers collectively and individually in dynamics of the workplace.

The compatible goals and easy interdependent relationship between the military, the airlines, and the air traffic control system were transformed into conflict, as the military and the airline industry became "users" with competing demands for airspace. With the arrival of postwar congestion, however, the resilience gained by changing boundaries was no longer enough. To increase safety, the government expanded the powers of the regulatory apparatus by replacing the CAA with the Federal Aviation Agency (later Federal Aviation Administration). Using its greater authority, the FAA produced a flurry of standardized rules and procedures to control the growing diversity of aircraft capabilities and the complexity of the airspace.

The Jet Age was an era of far-reaching political, economic, and cultural change. The system became a center of political struggle that originated both outside and within it. Vietnam vets, African Americans, and women entered the controller workforce. Airline deregulation, a response to inflation, brought a deluge of traffic, making controllers' work more arduous. Congestion made delays a nationwide problem. The rapidly increasing technical sophistication of airplane equipment, leading the FAA to its first efforts to incorporate automation into the system. Automation refers to a device or system that either partially or fully accomplishes a function that previously would have been executed by a human operator. The purpose of automation was to increase the speed and accuracy of dead reckoning and, at the same time, increase the amount of traffic that each controller could safely handle.

However, controllers' technologies remained mired in the past. Relieving congestion was stalled by "the liabilities of technological innovation":[146] the skills and knowledge to design an automated solution to fit a unique organizational architecture and airspace did not exist. The time needed to produce a workable automation prototype was uncertain; the task proceeded by learning from mistake and making incremental adjustments. The effort was similar to what MacKenzie called "inventing accuracy": even as dead reckoning became more automated, accuracy was an ever-elusive goal.[147] Promised new aids to dead reckoning also were slowed by contractor problems and competing national political and budgetary priorities. During budget shortages, the FAA tried two new survival strategies: quota systems regulating flight departures from airports and controller hiring freezes. Unsurprisingly, the combination was self-defeating; the traffic improvement from quotas was canceled out by limited controller personnel. The unanticipated consequences of changing boundaries and budgets sparked system change from the inside. This time the workforce took up problem solving to improve the system, which had been domain of government agencies and the top FAA hierarchy. The burdens of working traffic in an increasingly complex, crowded, and changing sky with inadequate technology fell upon controllers, who initiated a movement for unionization, setting in motion the preconditions of the infamous 1981 union strike of the Professional Air Traffic Controllers' Organization, or PATCO.

⁘ ⁘

The Jet Age made clear the effects of changing political administrations on the system. Both the Roosevelt and Truman administrations strongly supported the development of the airline industry, military air strength, and the air traffic control system. In 1952, the last year of the Truman administration, there were no fatalities in the year for the first time in history. However, when Eisenhower became president in 1953, military spending immediately went up and domestic budgets—including federal aid to airports and CAA appropriations—were slashed. The CAA had a revitalization plan that included long-range radar and better navigational aids, but the appropriations cut left the plan on the table. No fatalities again in 1954 gave the appearance that the system was working well enough. However, the warning signs were there. When the weather got bad, large numbers of pilots opted to fly by IFR. The airways and airports jammed up. In September 1954 in the northeastern United States, the weather closed in. Airplanes that were up couldn't get down. Known as Black Wednesday, over forty-five thousand passengers were delayed as much as a full day.[148] In June 1956, it happened again. The New York airways were hardest hit. These two incidents focused national attention on congestion in urban areas.

At the end of that same month, a collision in the open skies over the

Grand Canyon redefined the problem. In uncongested airspace outside the boundaries that air traffic controllers watched, a Trans World Airlines Super Constellation and a United Airlines DC-7 collided while flying by VFR—"see and be seen." One hundred twenty-eight people died. With both planes flying by VFR, the DC-7 pilot had tried to get out of the way of the newer, faster TWA, but the DC-7 didn't have the maneuverability to do so in time. The accident investigation found that the system had no method for separating slow-moving from fast-moving traffic. The problem was not crowded skies; it was the air traffic control system. The fatality-free years of 1952 and 1954 had obscured the long-term record: sixty-five midair collisions between 1950 and 1955.[149] The long-range radar that would have shown a controller the two planes' positions was unavailable, the proposal having been scrapped by the budget cuts. Within days Congress approved appropriations. By 1957 the CAA had ordered long-range radar, doubled the former number of navigational aids, and forty additional airports got control towers.

The issue of uncontrolled versus controlled airspace was more difficult to resolve. Many aeronautical experts advocated for an extension of "positive control": expanding the amount of airspace under the control of controllers. Controllers could watch more sky if the CAA required IFR over many parts of the airspace no matter the weather conditions. Changing airspace boundaries was by then a go-to solution. So the CAA expanded airspace boundaries into the upper altitudes—above twenty-four thousand feet—where positive control would prevail regardless of the presence or absence of airways, good or bad weather, throughout the US airspace. The change expanded all centers' high-altitude airspace responsibilities. However, immediately there was a classification problem. Separating the slow-moving from the fast-moving traffic got stuck on what to do about military jet traffic, which could fly through all airspace, crossing both VFR and IFR boundaries in the interests of national defense.[150] Although the CAA had the legal authority to establish controlled airspace, the military vigorously resisted any form of civilian control.

Another tragedy broke the deadlock. In 1958, a US Air Force jet fighter collided with a slower United Airlines DC-7 over Las Vegas in uncontrolled airspace on a clear day. It was the Grand Canyon incident all over again. The CAA again expanded airspace boundaries: the area of positive control was extended through thirty-five thousand feet, citing the "extreme closure rates of high performance aircraft." Creating airspace boundaries to separate aircraft with different—and conflicting—capabilities was not limited to differences in military jets versus civilian aircraft. As these boundary changes went into effect, the Boeing 707, the first US-made turbojet airliner, entered scheduled airline service for Pan American. With technology that made it possible to fly long stretches without having to stop for fuel, the 707's 500-miles-per-hour, 181-passenger capacity grabbed the lead in high-speed aircraft away from the Douglas DC-8, its chief jet competitor. Opening up international markets and lowering airfares, in 1958 Pan Am made the first commercial nonstop flight from New York to Paris.[151] Other airlines jumped into the competition for faster transcontinental and transoceanic flights. The category "long-distance carrier" was born and speed records for passenger travel often broken. The Jet Age had even brought economy class, beginning the golden age of travel.

The system effects of the jet airliner on the work of air traffic facilities and controllers were major. Airport owners—local governments and private companies—began to lay new runways that were long enough to accommodate high-speed jet landings. The runways changed the flight patterns of arriving and departing aircraft, also changing controllers' dead reckoning. Moreover, adding the jetliners to the airport traffic mix of small, less predictable general aviation and older, large but slower air carriers, increased safety risks. The CAA moved to increase system standardization and reliability. Responding to the increasing complexity of controllers' jobs, the CAA closed the older regional training centers, opening a single central training facility in Oklahoma City. The government also established a more comprehensive and influential rule-making body. In 1958, Congress approved the Federal Aviation Act, which established the Federal Aviation Agency (later the FAA). The CAA—part of the Commerce Department—was officially eliminated. The FAA was given new rule-making powers, and for the first time, air transportation would be the domain of a free-standing independent federal agency.

The Cold War handicapped system improvements, however. High-profile projects to capture international leadership in space, war, and air transportation deflected attention—and the FAA budget—from air traffic control system weaknesses. The end of the Eisenhower administration and the beginning of the Kennedy administration was forever marked by the Soviet launch of Sputnik, the first man-made earth satellite. In 1958, Eisenhower signed the act creating the National Aeronautics and Space Administration (NASA). Kennedy followed with an appeal to take control of the space race by putting the first man on the moon. NASA's Apollo program began.[152] Military aircraft further jammed the airways. In the 1960s, about twelve thousand air traffic controllers were working the nation's airspace. Scheduled airline flights carried about fifty-eight million passengers annually.

Air traffic controllers were struggling to handle the increasingly complex boundaries of the sky.[153] As traffic mushroomed, some facilities were still using radar equipment from World War II naval vessels. Overworked and understaffed, radar controllers were still saddled with the laborious system of radar, shrimp boat, and flight progress strips, with tabletop scopes, another legacy of World War II.

The expectation had been that radar would increase the speed and accuracy of controllers and increase the number of aircraft they could handle. The proposed Automated Radar Terminal System would help radar controllers handle more traffic by relieving them of the tediously slow work of pushing shrimp boats along on horizontal scopes and also increasing accuracy. Further, by automating routine tasks, the system would free up controllers to concentrate on separating airplanes. The shrimp boats (by now plastic with grease-pencil markings) would be useful only until radar was joined up with computers capable of displaying on the radar scope the alphanumeric codes identifying each aircraft by flight number, altitude, speed, and location—essentially the same information conveyed on a flight progress strip.

These "identification tags" would appear on the radar next to the hard-to-see blips. FAA experiments with computers that could display alphanumerics on a radar screen began in the 1960s. However, information had to be sent from the airplane. To that end, research was ongoing for a "transponder" that would be part of airplane cockpit equipment. The idea was that the transponder would send out a code that would be received by radar instrumentation on the ground, decoded by a computer, and displayed on the radar scope as an identification tag so that a controller could see it beside the airplane's blip and track the airplane's position, altitude, and call sign. Contracts had gone to IBM for the software and Burroughs for a transponder.

In 1965, the agency began an eighteen-month field test with two prototypes: one at Atlanta Tower; the other, at Indianapolis Center. These were successful. However, in what had become an unfortunate precedent, budget cuts left the agency without the money to go forward with building these systems. Worse, at this crucial time, the FAA issued a controller hiring freeze because of shortages related to the Vietnam War. These conditions hit hard upon the already-insufficient number of air traffic controllers in New York and other high-traffic metropolitan areas, which had controllers regularly working overtime, six days a week, without scheduled work breaks and with overtime pay restricted by civil service prohibitions. Unsurprisingly, given inadequate equipment and staffing, the period between 1962 and 1965 saw a spike in fatal crashes and the number of near collisions around airports.[154]

The Union: The Unanticipated Consequences of Boundaries and Budgets

Work conditions and their outcomes triggered activism among controllers, leading to the spontaneous formation of small groups that would eventually be the foundation of a union. In a January 1962 action lost in US history books to the Bay of Pigs invasion and Cuban Missile Crisis, President Kennedy signed an executive order that guaranteed federal employees the rights to join organizations "having as a primary purpose the improvement of working conditions among Federal employees" and to engage in collective bargaining. Ironic in retrospect, the then FAA administrator argued unsuccessfully that air traffic controllers should be excluded because they served a national defense function.[155] In scattered air traffic control facilities across the United States, small local unions formed. Small organizations within the parent FAA, these unions took local job actions in response to congestion and work overload, but they did not interact with one another. In January 1968, moved by the spirit of protest and growing nationwide skepticism toward the government, a group of dissatisfied air traffic controllers in the New York area met to begin a nationwide organization, the Professional Air Traffic Controllers Organization. The name was to establish that they were, first and foremost, professionals organized to have a national voice.

It makes sense that PATCO was started by controllers in the New York area. the geographically small airspace served a dense population that contained Newark, LaGuardia, and Kennedy airports—also Teterboro, with its airspace clogged with general aviation traffic. Here, more than other areas, controllers in the facilities were struggling with congestion. Then, ironically, an FAA change to the New York system boundaries sparked the opportunity for a collective plan. In response to changing traffic conditions, in 1968 the FAA initiated boundary changes to create better, safer conditions with fewer delays. The FAA consolidated the three major New York air traffic control facilities to form a new one: the New York Common Instrument Flight Rules Room at Kennedy Airport. This room, known throughout the system as the Common I, brought together radar controllers working intermediate-altitude airspaces at the three main airports into one facility. Later, the Common I would become the New York TRACON, known for working the most heavily congested airspace in the United States.

The FAA's logic behind the creation of the Common I was sound: it was the solution to a troubling boundary problem. Previously, the intermediate-altitude airspace worked by each facility had airspace boundaries separated by buffer zones from the others in New York. Because controllers' technology to communicate across buffer zones was so poor, and because the boundaries were inviolable, controllers had no flexibility in dealing with changes in traffic flows. With the three facilities combined in the Common I, controllers working different areas of airspace were in easy communication, giving greater flexibility to the New York–area operations. Inadvertently sparking the beginnings of PATCO, the consolidation also brought together into the one facility a critical mass of controllers, leading to awareness among them that they shared many grievances. Within six months, PATCO had a national membership of over five thousand.

In July 1968, PATCO's first constitutional convention was held in Chicago, prompting the new union's first appearance on the national stage about working conditions.[156] The union began flexing its political muscles. "Operation Air Safety" was a job action to maintain FAA rules of separation, which PATCO argued were repeatedly being violated by FAA supervisors in order to accommodate the high levels of traffic. During "Operation Air Safety," PATCO controllers nationally would go by the exacting separation procedures in the air traffic controllers' manual. Because going by the rules would have airplanes flying farther apart, it was, in effect, a work slowdown. Implemented during the heavy-traffic summer tourist season in the New York area, it kept almost two thousand aircraft from taking off or landing, some for as long as three hours. The work slowdown had traffic repercussions from coast to coast and in Europe.

The FAA called a meeting with PATCO representatives and union lawyers to stop the job action. PATCO had a list of demands to reform the air traffic system. Each of the demands required legislation. Within four months, the FAA had met all PATCO demands. In a related move to reduce congestion, the FAA innovated. The agency established hourly quotas on all IFR operations at five of the nation's busiest airports; such flights were required to make advance reservations, which became known as "the slot system." Thus, airplanes would depart with an expected arrival time that the airlines could meet. Planes would still be delayed, but on the ground, not in the air. The number of aircraft in the sky that each controller was juggling would be more manageable. Jubilant, controllers everywhere were emboldened by the victory.

Recognizing the system's problems, in November 1968 the president-elect Richard Nixon vowed to strengthen the air traffic control workforce, improve work conditions, and provide new equipment. To that end, the Nixon administration created the Airport and Airway Development Fund, taxing the system's users in order to raise the funding.[157] It got quick results. The project was reactivated, and in 1969, the New York Common I was the first facility to receive the Automated Radar Terminal System (ARTS) following successful trials at Atlanta and Indianapolis. A major organization change followed. To reduce congestion, in 1970 the FAA established the prototype Central Flow Control Facility in DC to coordinate traffic flow nationwide. Linked by teletype and telephone to all twenty-one Air Route Traffic Control Centers, Central Flow detected potential trouble spots and suggested traffic restrictions, spacing, or rerouting to centers. Within a few months, it had proved its worth in reducing delays. Soon after, the FAA replaced it with the large Command Center in Herndon, Virginia, to handle the national flow control. The new name reflected both the military heritage of the site and its authority to shape patterns in every region's airspace. For controllers, Central Flow was an affront. In the interest of improved system flow, they had to cede their autonomy over traffic patterns in their facility airspace to people who had never worked it.[158]

The Liabilities of Technological Innovation: Dead Reckoning

After the Command Center in Herndon came the first official suggestion that automated technology might be able to function in place of some controllers. A 1970 Department of Transportation report darkly predicted that the crisis of congestion would continue. It recommended that striving for an even higher level of automation with future ARTS versions would enhance dead reckoning by adding spacing, sequencing, and conflict prediction programs to "enable the system to handle two or three times the 1969 traffic with the same controller workforce."[159] The FAA was receiving a barrage of criticism for failing to prevent accidents, which were occurring more frequently.[160] During the mid- to late 1970s, the development of many automated technologies was in progress in order to streamline the lagging system.[161] Although each was important in its own right, the major hope for rescuing the system from congestion rested on ARTS.

The arrival of ARTS was hailed as an advance as important as the postwar introduction of radar. Indeed, in the long run, it was, but extended periods of adjustment and tinkering were necessary. As centers slowly began receiving and installing ARTS, the change also required major boundary alterations to airspace and ground organization: both became more complex. The sky was sliced up to create an intermediate altitude, between tower airspace and center airspace, which led to the creation of new intermediate altitude facilities—Terminal Radar Approach Control, or TRACONs—and special training for controllers assigned to them. TRACONs would handle transitions: the arriving and departing traffic flows between the high-altitude centers and towers. The basic architecture of the facilities settled into three: towers, TRACONs, and centers. The alphanumeric tags on the radar blips would help controllers better track aircraft in the complex altitude space between that of towers and centers that was packed with crossing arrivals and departures.

By the time a new FAA technological innovation was being tested in a facility, it had already been through a long process of investigation, analysis, and massive amounts of data and trails involving aircraft. These developments proceed under stringent FAA or internationally formulated ICAO controls. Nonetheless, fitting them into the existing system was an entirely new and uncertain process. The agency was not prepared for the liabilities of technological innovation that make the development and implementation of prototypes unpredictable. The practical ramifications of automating air traffic control were not well thought out. The technical problem was unprecedented: how to program the software for dead reckoning to mesh with the boundaries of the organization, the sky, and existing technology of the air traffic control system on the ground? Planes were being handed off between three types of facilities. The blips and their identification tags had to be able to move from radar scope to radar scope as an aircraft traveled its route across the sky. The technology had to "recognize" the airspace boundaries between sectors of airspace that would be represented on a different radar screen operated by a different controller in another location.

Moreover, controllers' cognitive processing had to change. They were used to working with material representations of aircraft that they could move and manipulate, making mathematic estimates upon which to base the changing locations of airplanes, keeping ample spacing to make up for error. Controllers referred to ARTS as "three-dimensional radar" because of the alphanumeric display of aircraft altitude, direction, speed, and location. On the radar scope, however, controllers saw a one-dimensional object with its tag, or its data block, moving on a route on a flat screen. It was controllers who had to read the alphanumerics and cognitively convert the moving image into three dimensions, predicting the position of the aircraft in space while preserving separation above, below, behind, beside, and in front of it. Instead of visualizing an aircraft from a shrimp boat that they moved along a route on a flat tabletop radar, or from a printed route on a flight progress strip, with the new automated system ARTS, controllers explained to me that they visualized it either as a moving three-dimensional block or an aircraft passing through tunnels in the sky.

By 1975, all Air Route Traffic Control Centers plus the sixty-three TRACONs were using ARTS. After the ten years from drawing board to full implementation, controllers finally had upright scopes and a technology to make the shrimp boats obsolete. But the liabilities of technological innovation descended with force upon the workplace. Despite extensive testing before installation, glitches inevitably happened. Writing about the future of the system in 1973, Glen Gilbert identified ten flaws that could affect the efficiency of radar, among them false targets, false emergency alarms, false data readouts, and false identification responses from aircraft.[162] As he predicted, all of these went on to occur, requiring incremental adjustments of the technology in the facilities after installation.

The Short-Term Conflict Alert System was incorporated into the radar in all centers by 1976. For radar controllers, the conflict alert system provided a warning on the radar scope when two airplanes were close to violating the rules of separation—the required spacing between airplanes. The computer threw a flashing circle around the two airplanes to warn the controller in order that he or she take avoidance action. However, its accuracy was unpredictable. The conflict alert system was still giving false alarms until the year 2000, when radar backup systems were perfected. Worse, the much-heralded computers failed regularly, leaving controllers at their scopes with no alphanumerics attached to aircraft blips—or worse, no blips—so controllers had to fall back on full manual control using flight progress strips. Over the years, ARTS became more reliable, but radar still would unexpectedly fail.

Political Demands on the System, External and Internal

In the late 1960s and early 1970s, the FAA and the air traffic control system were subject to political pressure from some unexpected sources. The ideological shifts that materialized in social movements and political protests gained force, ultimately effecting permanent change on the system. Hijackings seemed to become common in 1968, when hijackers diverted twelve US airliners and six general aviation aircraft to Cuba.[163] In January 1969 alone, eight airliners were hijacked to Cuba. Federal sky marshals were assigned to ride on airplanes, metal detectors and other surveillance procedures were installed in airports, and passenger baggage was inspected. In addition, US citizens, who first became incensed when the arrival of jets introduced sonic booms, began protesting FAA efforts to build new airports and modernize existing ones in order to accommodate jet airliners.[164]

Residents of surrounding communities objected to the jet noise and kicked-up dirt that increased airport capacity would bring. In response, the FAA changed tower and TRACON airspace, developing new "noise abatement routes" in and out of urban airports to protect neighboring communities for as long as possible from the roar of arriving and departing flights. The new routes helped, but it was an unresolvable situation because planes had to go up and come down. Protesting communities were joined by the environmental movement, forming an effective voting bloc against new facilities. The FAA Office of Noise Abatement was replaced by the Office of Environmental Quality. As the environmental movement grew, the strength of public opposition was so great that in 1975, after the opening of Dallas/Fort Worth Regional Airport in 1975, no new airports opened until Denver International in 1995. Congestion went unrelieved.

Political pressure on the FAA and the air traffic system also originated in-house, as PATCO continued its organized resistance to work conditions. Working overtime had become mandatory at many facilities. Responding to renewed PATCO grievances, in 1970 the Department of Transportation's Air Traffic Controller Career Committee recommended reducing overtime and the number of consecutive hours on position to two, as well as setting an optional early retirement at age fifty after twenty years of service.[165] However, some proposals were highly controversial, such as deploying experienced controllers to high-density facilities and developing incentives so they would go there, which put the needs of the system above those of the controller. Disputes arose and were settled in informal agreements with the FAA, which the FAA then negated because they were informal. In March 1970 about three thousand PATCO controllers, feeling betrayed, staged a sick-out to demonstrate their objections to the FAA's transfer of three controllers who had not wanted to change facilities.

In a harbinger of things to come, the Department of Transportation viewed the sick-out as an illegal strike against the government.[166] Subpoenas were served to participating controllers. Under court order, PATCO agreed to stop the sick-out. The FAA suspended nearly one thousand controllers and fired fifty-two. Legal fees bankrupted the union. In 1971, the Department of Labor temporarily took away PATCO's status as a labor organization, pending a statement promising no illegal job actions before it could apply for recognition. Six months later, in June, PATCO was authorized to again apply, finally certified as the sole bargaining unit for air traffic controllers in October 1972. In 1973, PATCO successfully negotiated with the FAA for its first labor contract. Conflict between the FAA and PATCO continued, with a second contract agreement reached in 1975. After Nixon's resignation, the Ford administration proposed a federal employee pay increase so low as to be, in effect, a pay cut. All federal employees, including air traffic controllers, were being hit by inflation.

PATCO decided to circumvent the civil service prohibitions against federal employees bargaining over pay by demanding that the government reclassify controllers to higher grades on the civil service salary scale. Upgrading controllers' grade would automatically give them a pay increase. With support from a new FAA director, the Civil Service Commission began a study of air traffic controllers' job classifications.[167] Then in 1976, President Ford took a strong anti-union stance to stave off the advances made by conservative California governor Ronald Reagan in the campaign for the Republican presidential nomination. The Civil Service Commission report rejected the PATCO proposals for reclassification, in fact downgrading the status of some controllers. In a strategic, attention-getting job action, PATCO staged a nationwide rolling work slowdown that slowed the entire system.

PATCO itself was experiencing internal conflict. The historical social and cultural shifts in the country between 1968 and 1978 had the PATCO leadership struggling with a divided air traffic controller workforce. Traditionally controllers were male, white, and working class, and they came to the job from the military. They brought with them aspirations for a solid middle-class lifestyle. The civil rights movement, feminism, and the Vietnam War altered air traffic control facilities in the same way it altered the rest of the country.[168] A small number of African Americans, women, and Vietnam vets who grew up in the 1960s entered air traffic control facilities with a very different worldview from that of those who were already there. Although also working class, they were against hierarchy, challenging the military-derived traditions of the air traffic control system and the people who built it. They were dissent-minded, against government, against authority, against white shirts and business shoes, and against military haircuts. Facility managers who ruled with a military mind-set faced mini-rebellions daily from long-haired younger controllers who believed things could be done in a different way. Few in number, shunned, and not welcomed, African American and women controllers joined PATCO, but they also formed their own professional associations.[169]

PATCO members protested to change work conditions. They were not targeting the pervasive racial and gender inequalities that permeated air traffic control facilities. Black controllers began to organize themselves, forming the Coalition of Black Controllers. The FAA mobilized, creating its Civil Rights Office to lay out guidelines for implementation. Progress was slow. By 1973, there were eight hundred African American air traffic controllers. At PATCO's inception in 1968, twenty-seven women controllers were scattered through the system. By 1978, they numbered 1,500. These women, spurred by feminism, organized and formed the association Professional Women Controllers, which was recognized by the FAA in 1980. They also joined PATCO, but racism and sexual harassment there remained pervasive. Across this diverse mix, in the late 1970s a militant PATCO coalition formed around certain issues. Old-school controllers with military experience united with the new generation of younger controllers who grew up during the Vietnam War in opposition to government power and inequalities in controllers' pay grade, outnumbering the votes of more conservative members.

In 1977, Jimmy Carter was inaugurated as president. Two Carter administration actions had major impact on the air traffic control system, moving PATCO toward a strike action. Responding to the display of power by PATCO's rolling work slowdown, Congress approved the PATCO wage reclassification. In a stinging contradiction, Carter's appointed FAA administrator was dedicated to keeping PATCO in line and refused to put the upgrades through.[170] Soon to follow was Carter's signing of the Airline Deregulation Act of 1978. The act was part of the president's campaign to halt inflation. Air transportation had been heavily regulated, but the Civil Aeronautics Board's control over fares, routes, and mergers had ended. The act allowed immediate fare reductions of up to 70 percent, and new airlines could automatically enter into routes formerly "owned" by other airlines. By the end of the year, 248 new airline routes had been awarded to applicants.

Deregulation had major unanticipated consequences for the airline industry and the work of controllers. First, deregulation initiated a flurry of aircraft design innovations. Fledgling airlines filled the sky with new, smaller short-haul aircraft: regional jets. Not as dramatic as the jet revolution but as important, the arrival of the regional jet was made possible by the invention of efficient jet engines for smaller aircraft.[171] A small number of regional jets had entered the short-haul feeder-line market in the 1950s and 1960s, but with deregulation, both new carriers and the "legacy" airlines were using them to develop untapped markets. Smaller and faster than other aircraft in the sky, they were flying on air routes unfamiliar to controllers.

Second, the standard route structures of existing air carriers also changed nationwide. Following the "hub" precedent set by Federal Express in 1971, for greater efficiency the major airlines reconfigured their route systems from point-to-point air transportation to a hub-and-spokes system. Rather than flying direct from city to city, air carriers began operating in and out of hub airports that were home to large numbers of flights. Passengers reached their destination by changing planes at the hub to travel on to a spoke. Pilots and controllers alike had to relearn the system. Controllers working at hub airports were the hardest hit. They experienced huge increases in flights departing and arriving at the same peak morning and evening travel times, creating major ground congestion because more aircraft were changing gate positions simultaneously.

Another PATCO contract renewal agreement was reached in 1978, by which time the union was seventeen thousand members strong.[172] The PATCO membership was sharply divided about the contract provisions and angered by the Carter administration's failure to support the union, which the election of a Democratic president had led them to expect. More disappointment was to come. Carter was still struggling to control inflation. He proposed a salary increase for federal employees that trailed the inflation rate, and in each year of his administration he continued this practice. Federal unions wanted the legal right to bargain over wages. The Carter administration acted against unions on this issue in the Civil Service Reform Act of 1978. PATCO was seeing no movement by the Carter administration's FAA on the key issues of wages—salaries were suffering from inflation—and hours. Chronically overworked and understaffed, controllers were also struggling to adjust to the technology brought about by effects of deregulation. Internally divided about strategy and in a weakened position with the FAA, controllers were concerned about successfully negotiating the next contract, due in 1981. In late 1979, the same coalition of old generation and new generation PATCO members began to construct a strike plan.

The Age of Conflict, Decline, and Repair: The Strike, NATCA, and Technological Glitches, 1980–2000

Each age in the life course of this system has drawn attention to the changing dominant actors over time—individuals, organizations, or technologies—that emerged as the drivers of change, transformation, and persistence. In contrast, this era was typified by the reverberating interactions of all three, moving system development forward, constraining it, setting it back, or aiming to transform it into something else entirely.[173] As a result, the Age of Conflict, Decline, and Repair, from 1980 to 2000, was marked by periods of decline that weakened the capability of the system, followed by periods of efforts to repair and rebuild the system to deal with risky conditions in the present and prepare for challenges of the future. Problem solving emanated from sequential presidential administrations that had contrasting political ideologies about how government should work. Their appointed FAA directors then enacted policies based on different ideas of what *repair* meant and therefore what constituted the tools of repair. The cyclical dynamic of decline and repair was propelled by the response of problem-solving actors in the system, as power struggles between labor and management produced conflicting efforts to repair and rebuild the system throughout the period. Organizational and technological innovations themselves were important actors, as early efforts at automation and redrawing boundaries in the sky and on the ground had system effects, changing relationships between the parts of the system as well as the nature of controllers' work within it.

Many have written about how the historic 1981 air traffic controller strike, President Reagan's firing of all striking controllers, and the demise of PATCO altered the course of American labor history forever.[174] Less well known are the effects upon the air traffic control system and the work of air traffic controllers that would carry into the twenty-first century. System effects produced the strike, and, in turn, the strike had system effects, ushering in another era marked by unintended consequences. Although the FAA immediately began to enact an extensive plan to hire and train controllers and correct the conditions that led to the strike, the effects of the strike rendered the system unable to respond effectively to changing conditions. Instead, it reproduced the very conditions that created the strike. Air traffic control remained a system under stress, affected by politics, externally and internally.

The external factors that contributed to the emergence and development of the air traffic control system, and then began to overwhelm it during the Jet Age, continued. The workforce was depleted. Changing political administrations—Reagan, Bush, then Clinton—faced domestic and international problems that cut into FAA budgets, reproducing hiring freezes as a solution. In response to technological advance of aircraft equipment and the continuing ramifications of deregulation, new controllers were pressed into service with inadequate training. In the facilities, labor management relations deteriorated with a return to the military management strategies that existed before the strike. Overworked, understaffed, and underpaid, the workforce responded as before: they built a new union, the National Air Traffic Controllers Association (NATCA).

To solve the worsening problems of understaffing, congestion, and delays, the FAA turned again to its favored technology of the past: changing boundaries in the air and on the ground, and also to the newest technology—the hope of the future—automation. However, the FAA was applying these strategies to problems of larger scope, far beyond their original beginnings in local settings. The agency aimed for large-scale automation efforts that could be patched onto the existing system. The liabilities of technological innovation produced technical glitches that not only were dangerous to pilots and passengers but also made work more difficult for controllers. The FAA became the subject of public ridicule for its poor decision making, waste, and big-budget technical failures. Beginning in 1989, the attention of the George H. W. Bush administration went to the bombing of Pan American Airways Flight 103 over Lockerbie, Scotland, on December 21.[175] Action focused on international security from terrorist attacks and emergency measures for Civil Aviation Security. However, the administration proved predictably conservative, perpetuating the Reagan lifetime ban of controllers from the FAA, and endorsing privatization.

Early in the Clinton administration, NATCA saw Clinton's moves toward privatization as a threat to safety and their very jobs. However, as the administration began taking steps to correct past failures and improve the system, an empowered NATCA had victories. Moreover, at the end of the 1990s, the FAA's planned airspace boundary changes and tentative beginnings of automation grew into a modernization plan. The FAA began building its structures and systems for innovation, and to design and test new technologies. Nonetheless, technological glitches were still the norm, making work for controllers harder, as did the vast reorganization of the airspace, which affected ethno-cognition and boundary work. Crucially, airspace ownership had social and symbolic status meaning for them. The era ended as it began. As airspace was redesigned, it triggered territorial conflict between facilities, this time pitting controllers in one region against another.

⁂

It is not clear in retrospect why PATCO—a traditionally Democratic union—would endorse a conservative Republican presidential candidate. But it made sense to them at the time. They were disillusioned with Carter. Reagan was labor, himself a former president of the Screen Actors Guild who led actors in a strike against film studios. As governor of California, he had walked a careful line, not laying down a clear pattern of actions either for or against labor unions. Indeed, he had avoided a punitive policy and demonstrated restraint with strikers.[176] He was actively courting unions in his campaign, and they were endorsing him. The clincher was a letter he wrote PATCO on October 20, 1980, promising support in exchange for endorsement: "In an area so clearly associated with public safety, the Carter administration has failed to act responsibly. You can rest assured that if I am elected president, I will take whatever steps are necessary to provide our air traffic controllers with the most modern equipment available and to adjust staff levels and workdays so that they are commensurate with achieving a maximum degree of public safety. . . . I pledge to you that our government will work very closely with you to bring about a spirit of cooperation between the President and the air traffic controllers."[177]

Believing that Reagan was in their corner, PATCO's demands for the new FAA contract were far beyond anything ever asked for by federal union employees and what the law allowed federal workers to negotiate.[178] They were even beyond what some PATCO members believed they deserved in the high-inflation period of the 1980s. Of the union's ninety-six demands, the top three were for facilities fully staffed with controllers, the replacement of outdated equipment, and the right to negotiate rights in collective bargaining equal to or greater than those of postal employees, including the right to strike under some situations. But some of the other demands stymied a settlement and publicly branded the union as greedy: a $10,000 salary increase for every controller, semiannual cost-of-living raises one and a half times the inflation rate, a four-day workweek of thirty-two hours (controllers elsewhere in the world were working between twenty-nine and thirty-eight hours a week, while US controllers in some places were running six- to seven-day workweeks plus overtime). They also demanded a *secure* retirement plan: because of insufficient staffing, controllers' retirement was not guaranteed.[179]

As negotiations stalled, PATCO threatened to strike on June 22, 1981, at which point Reagan offered the union what was possibly the most generous set of concessions made to a federal public employee union in government history.[180] FAA personnel were stunned by the capitulation. PATCO members were divided. The offer was much lower than what they asked for. Combined with the insults they had already suffered from the FAA and the Carter administration, the offer was one more. Confident of Reagan's support in spite of his unwillingness to make further concessions, the union set a second strike date for August 3, 1981. On that day, nearly 85 percent of PATCO's membership—about 12,300 controllers—walked out.[181] Four hours into the strike, President Reagan took a strong stand against the illegal strike, ordering the strikers back to work in forty-eight hours or be fired.[182] Some controllers went back to work but 11,345 were fired.[183]

PATCO membership believed that they would win because a strike would shut down the system. "They can't fire us all," they said. They were wrong. The FAA had prepared. Immediately the agency invoked an interim air traffic control operations plan, "Flow Control 50," finished in the hours before the strike.[184] The plan allowed the FAA to limit the number of aircraft in the national airspace. The plan was expected to handle 83 percent of the normal volume with 50 percent of the air traffic controller workforce.[185] It would achieve this by resurrecting strategies used in the past to relieve congestion: the quota system and boundary work. For example:

- Drawing on precedent, the innovative 1968 New York quota system that assigned departure and arrival time slots to reduce congestion at towers was reactivated at major towers nationwide, reducing the number of planes in the air at any time. Takeoff and landing slots were scheduled at major airports by national priority (economic, political, defense, medical) as related to the flight's activity. General aviation was severely restricted. Some small towers were closed.
- The largest airlines agreed in advance to cut peak-hour flights by about 50 percent.
- Because the greatest number of controllers would be lost at the centers, the FAA moved large portions of airspace from centers to TRACONs. With

> personnel ranks decimated, high-altitude centers would compensate by increasing horizontal spacing between aircraft from the usual ten to thirty miles.

(Inconceivable in 1981, twenty years later this plan would also serve as precedent for the gradual, high-surveillance return of the more than four thousand airplanes that controllers had grounded on September 11, 2001, to the emptied sky.)

The first week was dire: about 4,200 controllers showed up for work. Eventually, about ten thousand people were assembled, exceeding agency expectations. In-house were the nonstrikers who chose to stay on the job, the striking controllers who returned before the forty-eight-hour deadline, and supervisors, who previously worked traffic as qualified controllers. About eight hundred military controllers were moved to air traffic control facilities, and the FAA hired about one thousand from the people who lined up on the street for the job.[186] With no time for formal FAA training, this skeleton crew was a ragtag operation. Supervisors were rusty. For military controllers, the switch to civilian airspace, its traffic, and aircraft equipment was very difficult. The new hires off the street were able to act as support help but couldn't work traffic. Pilots, temporarily laid off because of limited flights, were hired as runners or for administrative tasks. Sixty-hour workweeks continued for several months. Rules were relaxed. Food and drink in towers and radar control rooms, usually forbidden, were brought in out of necessity. Within a month of the firings, over one hundred thousand applications flooded the agency.[187] The FAA Academy in Oklahoma City brought in extra staff to train cohorts of new controllers. Their place in history marked them: they would ever after be known in air traffic control facilities as "the replacement hires."

It was the end for PATCO. In October 1981, the union was decertified—stripped of its legal standing—by the government. It was also facing bankruptcy due to fines, lawsuits, and legal fees related to the strike. The government filed criminal complaints in federal courts in eleven cities against twenty-two controllers. The FAA boasted that "law and order" had been restored.[188] The impact on individuals was devastating.[189] Fired controllers were forbidden to return to the FAA and to work for the federal government in any capacity in their lifetime. They were not eligible for unemployment benefits, food stamps, or welfare. With recession in full throttle, public opinion against them, and no other marketable skill, they were unemployable. Many survived in the short run, hoping things would change. Some scrambled to get back in.

Refusing to accept the outcome, the rest met, planned, discussed. They could not grasp the finality of it, or the full consequences to the union and to them personally. As many said twenty years later, it was not only that they lost their jobs—they lost their identity.[190]

Pouring salt into the strikers' wounds, the FAA began to fix the very aspects of the system that the strikers complained were broken. Two weeks after the strike, Reagan's transportation secretary created the Jones Committee to study labor-management relations in the FAA.[191] To deal with staffing shortages, salaries and benefits were increased in hopes of attracting new hires. The capstone of the plan was the proposed twenty-year National Airspace System Plan, a $12 billion effort to update and replace aging technologies. It included the Advanced Automation System, which would replace the IBM 9020 computers from the 1960s; new software, consoles, and displays for controllers; and—at long last—a proposed secondary radar system to back up the primary system in case of failure. The plan looked like a cure-all, but its implementation was fraught with obstacles.

How the System Reproduced Itself

The system effects of the strike and the firing of controllers were vast and enduring. It would take years for the FAA to recover. The structural constraints affecting the system remained the same. The replacement hires, many inadequately trained in the rush to get people working, were facing the same system limitations the PATCO controllers had. By the end of December, reports got to Congress that instructors at the FAA training academy were falsifying test scores in order to get more "developmentals," or controller hopefuls, into the facilities.[192] Once there, the not-yet-competent developmentals were training the new developmentals who followed them. The inexperience was producing mistakes, and even with traffic reduced, the new controllers couldn't handle the workload. The FAA couldn't remove the quota system and increase traffic levels because the newly qualified controllers were qualified for only average traffic. Many were not sufficiently competent to handle traffic in thunderstorms, for example.

Out of necessity, the agency cut back by closing about eighty small towers, moving those controllers to the understaffed larger facilities.[193] By 1984, the FAA had reopened most small towers, but in an experiment driven by staffing scarcity, it transferred operations of nine of them to private contract companies. These "contract towers" set a precedent that controllers saw as dangerous. Indeed, "privatization," initiated as a temporary, emergency fallback measure, became a repeated threat, later proposed as a formal option to improve system efficiency. In 1985, with highly experienced PATCO controllers still struggling to find ways to make a living, the air traffic control system was operating with about two-thirds the number of qualified controllers that existed in 1981 but was transporting 50 percent more passengers. Overtime pay was mandatory.[194] In 1984 the agency reported a 65 percent increase in midair near misses than in 1981.[195] Still, the airport quota systems remained in place.

The situation grew worse. After the first year of common, cooperative struggle between controllers and management in the facilities, as more replacement hires came in, the old authoritarianism returned, and with a vengeance. In a 1982 report, the Jones Committee on labor-management relations declared that autocratic management practices were the problem, predicting recurring problems with employee relations.[196] And so it came to pass. Major discontent was stirred among the controller workforce when FAA efforts to reform the system by forming cooperative local committees comprised of managers and controllers were crushed by facility and regional managers, whose automatic response to controller recommendations was to reject them.

Discontent was fanned into action in 1983 by an FAA plan for "structured staffing" at centers, which would limit the number of certified full-performance-level controllers, so that new hires could not move up until a vacancy occurred. Moreover, new controllers with some college education would be given priority in promotion over others, even those with prior experience, such as military controllers.[197] Displaced in priority, angry senior controllers saw this as an FAA attempt to bring younger, non-union-leaning controllers into the system. In 1984, these rampant inequalities combined with management practices, mandatory overtime, inadequate technology, and work overloads to spark the formation of a new union, the National Air Traffic Controller Association (NATCA). Begun as a local at Washington Center, which controlled the congested New York–DC corridor, it gradually spread to other facilities.[198] The NATCA founders vowed to create a different kind of union from its striking predecessor by emphasizing professionalism.[199]

The FAA also was trying to reinvent itself. During the waning years of the Reagan administration, the FAA began to implement its 1982 poststrike National Airspace System Plan to deal with congestion and delays. The FAA awarded a contract to IBM to develop the proposed Advanced Automation System, which would modify (not replace) the existing computers from the 1960s, providing greater speed, reliability, and storage capacity.[200] The second part of the National Airspace System Plan was a major reorganization of the sky. Changing the airspace boundaries to match changing traffic needs had become normal. But this proposed change was different. Called a "realignment," it reorganized airspace at such proportion that, given the interdependence of the parts of the system, it altered the airspace of controllers in several contiguous regions. A phased change, realignment would take years. And the technique, formerly used to solve smaller problems, was being used to control congestion across huge areas.

Begun in 1986 as the East Coast Plan, initially the airspace change was limited to adding new departure routes to ease congestion and noise in the New York–DC area.[201] However, the plan's designers quickly realized that the problems would remain the same unless the parts of the system that fed into and away from the New York–DC area also were modified. So in 1987, the East Coast Plan became the Expanded East Coast Plan: a massive, phased realignment of airspace from Maine to Florida and west to Detroit and Chicago. The New York metropolitan area airspace also would be redesigned. The final phase, to be initiated in 1988, would realign airspace to ease traffic flow between New York, New England, Philadelphia, Baltimore, and DC. The work of controllers in all relevant regions would be affected, so the realignment was also massively expensive. Controllers had to be trained to work their new portions of airspace, which meant traveling to the facility in possession of it before the switch happened. The transition to the realigned airspace would be gradual but not likely smooth. NATCA protested strongly against the FAA's failure to include in the airspace redesign effort any air traffic controllers who were actively working traffic and therefore most familiar with the airspace.

Then, a NATCA victory. On June 19, 1987, the Federal Labor Relations Authority certified the new union as the exclusive representative of all controllers.[202] NATCA activists and organizers had worked steadily and on their own time to accomplish this achievement. They acknowledged the important contributions of both former PATCO controllers who had passed on lessons learned and helped plan strategies, and the Marine Engineers Beneficial Association, the resource-rich union that funded their organizing efforts. But they also credited the FAA. The FAA, they said, was "a self-propagating system for managerial incompetence."[203] All FAA managers were drawn from the controller ranks. Many also had a military background. Because as controllers they were trained to control, and to control authoritatively, managers did not know how to manage people. When in a supervisory position, it was the rare manager who could actually manage team members—also trained to control—without alienating controllers.

The Reagan administration was stunned at the arrival of a new union so soon after PATCO had been destroyed. PATCO controllers, many still actively fighting to get their jobs back, were stung by the victory and its negative implications for their return to work. The external factors that had combined over time to precipitate the strike conditions persisted in new incarnations. Still far below optimum staffing levels, controllers were struggling to keep up as traffic increased. The negative consequences of airline deregulation were reverberating throughout the workplace. In a 1988 assessment, Alfred E. Kahn, instrumental in developing the deregulation policy, included among its deleterious system effects "the explosion of entry, massive restructurings of routes, price wars, labor-management conflict, bankruptcies and consolidations . . . and the increased congestion and delays that have plagued air travelers."[204]

Another unexpected effect of deregulation on controllers was that it led to new generations of aircraft that were increasing airspace complexity. The wide-body jets, capable of carrying large numbers of passengers, were filling the skies: the 747 was joined by the 767, 777, and the Airbus. A newcomer in the wake of deregulation, the regional jet had become a mainstay of the hub-and-spokes system and spread into longer routes. The regional jet, though, created problems for air traffic controllers because it was too fast to fly in airspace with other aircraft of its size and too small and slow to share airspace with the long-distance jet carriers. The sky had to be changed to accommodate them. Regional jets were given their own layer of airspace and new routes in the already-complex classification system in the sky. Controllers at understaffed centers struggled to keep up with the changes. In May 1989, the first year of the Bush administration, NATCA and the FAA signed their first labor agreement. Recognizing the budget problems in the system, the Bush administration passed legislation allocating special funds for the FAA and establishing funding for development of technologies to pursue the Reagan-era modernization plans.[205]

Tools of Repair: Liabilities of Automation and Boundary Work

Instead of correcting the systemwide equipment and staffing problems, the FAA again turned to technological and organizational innovations. This time the technological innovations were automated devices for the airplane cockpit. This move reversed the tradition begun in 1929 with Archie League and the advance of radio communication that shifted dead reckoning from pilots to air traffic controllers on the ground. Now the agency reduced the control of controllers, instead locating it in the pilot's cockpit devices.[206] The Traffic Alert and Collision Avoidance System (TCAS, pronounced "tee-kass") was a cockpit warning system designed to keep planes from colliding. The system's in-cockpit radar scans the nearby skies for aircraft and, when two planes come close, gives the pilot an automated voice warning that indicates an immediate avoidance action ("Climb!"). By 1991, TCAS was mandated in all airliners with more than thirty passenger seats. In addition, an FAA task-force report urged that a global navigation satellite system using an in-cockpit GPS was "the greatest opportunity to enhance aviation efficiency and safety since the introduction of radio communication and navigation."[207] The GPS satellites promised to give the pilot a more precise identification of the aircraft position and greater security in landing in low-visibility weather.

However, the promises of automation were undermined by the realities of implementation. Like the post–World War II problem of putting alphanumerics onto radar-screen blips and moving them from scope to scope, new technologies were slow to develop because of unexpected glitches that could be worked out only incrementally, typically after installation. TCAS was nothing but trouble for controllers, who had had little input into its development. TCAS often gave false alerts or alerts that contradicted controller instructions, causing pilots to rapidly ascend or descend into the paths of other aircraft, leading to near misses. At the same time, the automation program was moving haltingly forward. Design and testing problems had the much-touted IBM Advanced Automation System behind schedule. Software deficiencies and other problems added fourteen months to the already-delayed timetable. Making matters worse, in the last year of the Bush administration, the FAA compensated for out-of-control cost overruns on new technology by imposing a 1992 controller hiring freeze that would go on for more than a decade.

Meanwhile, the FAA coped with the crowded sky by enacting its by-then-well-honed strategy of reorganizing boundaries on the ground and in the sky to relieve congestion in the New York area. However, in yet another action that at the time defied credulity, the FAA added to the problem by transferring six hundred thousand square miles of oceanic airspace from Miami and Boston Centers to the understaffed, overcrowded skies of New York Center. In addition, boundary work on the ground went in a new direction: consolidation. After years of expanding, in 1991 the FAA consolidated five California

TRACONs into one. Plans were in the works nationwide for consolidating TRACONs and centers in the same region into one facility. Then another temporary solution was resurrected from the past. In a promise of efficiency made during the last year of his presidency, and little noted in the press, George H. W. Bush signed an order that facilitated the privatization of airports and other public assets built with federal assistance.[208] Privatization threatened air traffic controllers due to the implications of corporate control for prioritizing efficiency over safety by reducing the number of controllers. With privatization a looming threat, cockpit automation giving pilots greater autonomy, and the 1992 hiring freeze, controllers were staring into the face of a grim future.

The Clinton administration began in January 1993. For NATCA, that year was a confusing mix of FAA policy reversals. With the controller workforce at 14,071, and 10,600 of those NATCA members, a new collective-bargaining agreement went into effect between the FAA and NATCA on August 1. Then two weeks later, in a surprise move, the Clinton administration announced that air traffic controllers fired for participation in the PATCO strike could apply for reemployment in the FAA. It was a "window of opportunity" with a defined start and end date. For PATCO controllers, initial excitement was soon tempered by a harsh reality. Twelve years had passed. Many had successfully built other lives.[209] Others were unable to geographically relocate, were bitter, or were not physically up to the job's challenges.[210] For some, the window had opened far too late; several, never having recovered socially, economically, or psychologically, died by their own hand. Still, thousands had spent that time waiting for this moment. Some had refused to give up the work and had continued working air traffic in the military or foreign countries. About five thousand controllers applied.

But the FAA bureaucracy thwarted their efforts to return. The process of application, review, and hiring was confusing, arduous, lengthy, and plagued by lack of communication. Some PATCO controllers never received initial notification of the opportunity. For many who wanted to come back and even went to great lengths to expedite the process, the window of opportunity closed. Worse, the opportunity came immediately following the FAA's 1992 hiring freeze to cover cost overruns for automation. The window of opportunity was narrow indeed. The FAA said that once the freeze lifted it would hire fewer than two hundred controllers per year. The first PATCO group would not begin training until January 1995, with twenty-six in the first cohort and fourteen others rehired during that year.[211] Eventually, about eight hundred reentered the system and began training.

In another unexpected policy reversal in 1993, the Clinton administration took steps to privatize air traffic control by increasing the number of small, low-traffic towers to be contracted out, thus resorting to the practice of closing of small towers and diversifying into "contract towers," which had begun as a temporary measure during the poststrike controller shortage. Once again, an informal arrangement set a precedent for the future. The Clinton administration's blatant contradictions in policy objectives had NATCA protesting that the FAA had no intention of bringing controller numbers back to prestrike levels (seventeen thousand) and was jeopardizing safety. This concern was reinforced by the slow but steady increase in the use of contract towers and the beginning of outsourcing other services, including the expansion of contractor roles in designing and maintaining air traffic technologies.[212]

The FAA had not been able to circumvent the liabilities of technological innovation and budget-induced resource scarcity. The year 1993 ended as it began, with the FAA dropping another major plan for automation. Having already invested huge amounts in its development, the agency junked IBM's multibillion-dollar Advanced Automation System. The media had a field day with the FAA failure and the costs to the public. One part of the plan would continue, however. All existing controller radar workstations at centers would be redesigned to receive a Display System Replacement (DSR)—a color monitor for radar display being run by newer computers that would replace scopes from the 1960s, with instant backups in case of radar failure. But there was no new software. Fifty years after their development, flight progress strips and strip bays were still a mainstay of operations.

Finally, in desperate hope to save time and money on the DSR, the FAA yielded to the efforts of NATCA and brought in four controllers as technical consultants for the upgrade.[213] They found that the design flaws were so obvious as to be laughable. The new DSR monitor for centers had to accommodate flight progress strips, but the new strip-bay design would partially cover the radar scope. The extended development time would delay the first DSR installation for an unknown period. The FAA's technological glitches stood in stark contrast to ongoing innovations in aircraft technology: in mid-1994, the Boeing 777 became the first US jetliner to incorporate a "fly-by-wire" automated technology, an advance over pilot's former cockpit stick control. In response to these industry innovations, the Clinton administration began to move forward the modernization of the system, pushing FAA reauthorization acts through Congress that supplied funds for airport improvements, technological advances, and, in 1997, $8.3 billion for FAA programs, including the hiring of hundreds of controllers, maintenance technicians, safety inspectors, and security personnel. There were steady improvements in GPS, necessary to the future of free flight, which shifts the control of the airplane to satellite rather than radar, reducing the reliance on air traffic controllers. Another shift in dead reckoning from ground to sky, pilots flying by IFR at high altitudes could deploy automated cockpit GPS devices to chart their own routes rather than adhere to standard routes assigned by controllers. The promise was reduced costs to airlines via more direct routes; controllers, though, would also pay a price, as they still had to monitor to prevent accidents but under conditions of greater unpredictability.

Also in progress for radar at TRACON was what was known as STARS, or Standard Terminal Automation Replacement System, which, like DSR, included a full-color monitor for use in a dimly lit room. In an advance far beyond the imagination of that early champion of cartography Hermann Moedebeck, STARS included two hundred digital maps. Again, controllers were not involved in the design until late in the development. Again, they found fundamental flaws: pop-up menus on the radar scopes obscured critical aircraft data blocks and other flight information. The first STARS would not be installed and tested in a TRACON until 1999. Even when the money was there, technological advance occurred by learning from mistake.

NATCA Empowered

NATCA had been quietly building and growing strong. In 1992 the union had returned to the wage-reclassification issues that PATCO had pursued. NATCA formed its Reclassification Committee, which included expertise from PATCO ranks.[214] To increase salaries, NATCA first had to design an acceptable salary scale that could replace the flawed General Schedule salary scale.[215] The FAA had adapted the General Schedule to controllers by matching it to a five-level ranking system that matched facility traffic volume. The pay system was riddled with inequalities. NATCA wanted a scale based upon *airspace complexity*, so controllers would be paid in accordance with the level of difficulty of working traffic at a specific facility.[216] Complexity included not only traffic volume but also geography, fleet mix, weather, traffic patterns, type of equipment, number of other facilities in the airspace, and other factors. After visiting facilities to gather data, a NATCA team came up with a formula and a twelve-level ranking system for determining the salaries of controllers working in facilities at the same level of difficulty.

Crucially, the right to negotiate wages was even more complex, but NATCA had several things going for it. By 1995, membership was up. The union was sufficiently financially secure to begin stashing away $30,000 a month toward a fund to build a union headquarters. Moreover, NATCA had created one of the stronger lobby efforts in Washington. It was a Democratic administration, and, as one of the smaller federal-sector labor unions, it supported an industry that contributed $3.5 trillion to the world's economy.[217] Then, in a stroke of political luck for NATCA, in 1997 Clinton nominated—and the Senate approved—Jane Garvey as head of the FAA, for an unprecedented five-year term. The former head of the Massachusetts Transportation Authority, located at Boston's Logan International Airport, Garvey had a reputation for cooperative leadership and negotiation.

In 1998, NATCA won a historic contract and affiliation with the AFL-CIO, solidifying its strength. That same year, the union won the ultimate triumph: the legal right to bargain over pay for controllers and other FAA employees. This right was unique in the federal government.[218] Winning the fight begun by PATCO, NATCA's wage-reclassification plan was approved. Subsequently, the FAA and NATCA negotiated an unprecedented pay agreement that would result in $200 million in salary increases for controllers over a three-year period, amounting to nearly a 30 percent raise for each controller. The individual amounts would be determined using the new NATCA-developed formula for airspace complexity, according to the new twelve-level facility-ranking system. Garvey promised to support the agreement if she got what she wanted from controllers in exchange.[219] What she asked for was their increased participation in improving the system—exactly what they wanted. First, she required controllers to have active input in new technology development and problem resolution, including the upcoming Y2K computer crisis, system modernization, and realignment of airspace. Second, to help pay for their salary increases, she asked that controllers take over more supervisory responsibilities, which would allow the FAA to reduce the number of supervisors, saving millions of dollars annually. Finally, both parties agreed that in the following three years, the total number of air traffic controllers would be increased to fifteen thousand, with a fixed percentage growth in the two years following that. Enthusiastically, the NATCA membership ratified the agreement.

In the last few weeks of his administration, Clinton signed Executive Order No. 13180, "Air Traffic Performance-Based Organization," which reversed his first term action supporting privatization.[220] In the document, he declared air traffic services "an inherently governmental function," ordering the establishment of the government-run Air Traffic Organization. An effort to modernize, the proposed new businesslike organization emphasized performance, equivalent to what many proponents of privatization hoped to accomplish by moving air traffic operations under the control of the private sector. The outsourced small "contract towers" remained.

Dead Reckoning at the Turn of the Century: History, Boundaries, and Turf Wars in the Sky, 2000–2001

By the year 2000, the air traffic control system was an aging, large-scale socio-technical system. The overarching entity was known as the National Airspace System. The airspace itself was a system of many boundaries, with a structure had settled at nine geographic regions. Within each region was one center, sometimes two, and a number of towers and TRACONs. Each facility was a socio-technical system, and within each facility, each controller position—where a controller interacts with pilots and other controllers close at hand and in remote locations—constituted the smallest of these nested and interacting socio-technical systems. Regulating the national traffic flow was the Air Traffic Control System Command Center at Herndon, Virginia, a major structure with technologies of coordination and control that adjusted flows of all regions. To this end, the Command Center had an internal arrangement of regional divisions that was analogical to the regional structure of the system it was designed to regulate.

The National Airspace System was an essential component of national economic well-being, international competition, and military power. It was also a monopoly interacting in an dynamic organization field dependent on the government, the military, and the airline industry. The reach of overland routes to countries immediately adjacent to US boundaries and to oceanic routes linked the United States to the global air traffic system. A major player, the National Airspace System was handling one-third of the world's traffic. That said, in the United States it remained reactive, vulnerable to the effects of the unfolding actions of the powerful institutional actors that provided its funding and shaped its trajectory.

At the turn of the twenty-first century, the effects of problem-solving actors both outside and inside the air traffic control facilities were visible in the mix of the past and the present in those same facilities. By 1997, only eight hundred fired PATCO controllers had been accepted back into the system. Although many of the eight hundred had failed the training before the year 2000, the survivors were in the facilities. Some had officially certified as controllers and now were working traffic beside the replacement hires; still others were in training. The original military influence over air traffic control remained in language and everyday routines: the watch desk, stand-up briefings, the Command Center, use of phrases like "roger" and "on the boards." Even the airspace was a mix of old and new: some airways followed rivers or railways, as they did in the era when dead reckoning was the responsibility of low-flying pilots; others were recent, reworked after the 1970s deregulation to follow the major airlines' change to the hub-and-spokes system, or later, realigned in accordance with 1988 East Coast Plan.

This mix of old and new in the facilities was pervasive, and sometimes deeply ironic. The most recent automation effort—in 2000, the Display System Replacement—was gradually being introduced into centers and worked by controllers who were still relying on technologies older even than many of them: flight progress strips, computers, and headsets from the 1960s.At small towers, typically lacking in computers, controllers were using pencil and paper to sequence airplanes, as did those airline operators in 1933 who first initiated the sequencing of tower arrivals. At the major towers, the latest radar designed to monitor runway incursions was being installed on airfields, even as pilots still positioned their aircraft on a "compass rose" painted on ramps near where they park airplanes, slowly rotating the plane to adjust their compass headings before takeoff. The compass rose, a device for dead reckoning that originated in medieval times to determine direction in relation to wind and the poles, still appeared on every airfield surface and nautical chart. At the same time that the FAA turned strongly to automation to upgrade the aging air traffic control system, technological glitches were more the rule than the exception. And even as FAA plans and design efforts were ongoing for a future dead reckoning when pilots would engage in free flight, guided by cockpit GPS, the busy TRACON in Oakland, California, was still using the flat tabletop radar from the 1940s.

System Dynamics of Boundaries and Boundary Work

As the year 2000 began, the air traffic control system's many boundaries—those in the sky and those on the ground—were both the site of political conflict and the cause of it. Once boundaries were created, boundary work became a systemwide characteristic: individuals, groups, and organizations had to negotiate their differences across boundaries. Internally, the boundaries between the systems' two parallel structures—labor and management—were sites of ongoing conflict, negotiation of differences, compromise, and even co-operation. In each facility throughout the system, the morning began with a meeting of the air traffic manager, assistant air traffic manager, and NATCA's president and vice president to make a plan for the day. They reviewed data on facility performance the day before and discussed the new day's traffic and weather, any operational updates from Washington, FAA mandates, union grievances, and other labor-management issues. The ongoing conflicts and power sharing were jarring to management. In the third year of their three-year raise, controllers were wielding their new influence. Air traffic facility managers and supervisors, smarting at the newfound power of the union voice, complained, "Garvey gave away the system."

Airspace boundaries, too, became sites of political pressure and contention. The historical problems of congestion, delay, technological lag, and inadequate controller numbers that first plagued the system after World War II had become virulent systemic problems; they were out of control. The number of daily air traffic operations—the number of airplanes that were worked through the airspace—began a steep, unprecedented climb in 1997. In an attempt to cope, the FAA changed airspace boundaries at the very highest altitudes over the North Atlantic. The change lowered the designated vertical spacing between airplanes from two thousand feet to one thousand feet, which required extra training for pilots and cockpit equipment changes. In 1999 and 2000, US air traffic operations reached an all-time high of 150 million operations annually.[221]

Paradoxically, the airline industry was in trouble. An economic downturn was affecting business travel. Airline industry representatives and the public were putting pressure on the FAA to resolve the problems of congestion and delays. Although the 1988 Expanded East Coast Plan to ease congestion in the New York region brought some relief, by 1997 congestion in the New York metroplex—Newark, Teterboro, LaGuardia, and Kennedy airports—was greater than ever. Delays had become intolerable for many, often holding up departures from the West Coast. At each major facility, air traffic managers were spending huge amounts of time negotiating across system boundaries with the airlines and with the general public. The airlines were aggressively demanding action. Environmental action groups and angry residents in airport neighborhoods affected by the increase in noise and dirt were calling for change from beleaguered air traffic managers at towers small and large. As air traffic increased, delays got more frequent and longer, creating pressure on controllers in understaffed facilities to deliver more traffic with greater speed.

Scrambling, the FAA returned to its tried-and-true boundary maneuvers: reorganize the airspace and the structures on the ground. However, the scale this time was unprecedented: a massive realignment of airspace boundaries known as the New York/New Jersey/Philadelphia Airspace Redesign Project. The redesign would have extensive system effects on controllers' work, calling for new learning and new material practices for those in the involved regions, with ramifications for traffic flows extending well beyond. Planning began in 1998, with an expected start date in 2008. This mammoth undertaking challenged both the system's ability for complex boundary reorganization and the strength of the new FAA-NATCA partnership. Moreover, the massive realignment set off a systemwide political battle for airspace ownership, pitting controllers in one region against those in others.

The airspace redesign would happen one facility at a time. This extensive realignment called for a level of expertise and systems understanding that only experienced controllers from all affected regions could put together. Design teams comprising NATCA and management representatives from all the involved regional centers would redraw the airspace boundaries of the system to change traffic flows into the New York area.[222] Redesign started with New York's TRACON because it controlled traffic ascending and descending all the metroplex airports. TRACON controllers figured out how the flows of traffic into their airspace needed to be adjusted to relieve congestion, for example, by shifting approaches from the northwest to the west. As New York TRACON redesigned traffic flows, adjacent facility flows would be affected. New York Center would be the first. Ultimately, the airspace redesign would alter the traffic flows of Philadelphia TRACON and Boston and Washington Centers, and in the Midwest, the Cleveland, Chicago, and Indianapolis Centers. Controllers at each regional center would need to redesign the flows between and within each internal airspace sector that its controllers worked. This sectorization, as it had become known, was a procedural term that had been formalized during implementation of the 1988 Expanded East Coast Plan.

Changes of the boundaries on the ground necessarily followed. New York TRACON and New York Center would relocate to a new facility. Joined together, six hundred to seven hundred controllers would work there. The expected benefits were enormous. The airspace would be reconfigured into one airspace, not two; thus, the more accurate radar of the TRACON would govern the entire airspace and increase safety. Also, delays would drop because the center's five-mile spacing could be reduced to the three-mile spacing required for the TRACON's smaller airspace.[223] The proposed new facility—officially named the New York Integrated Control Complex—immediately became known as the Crystal Palace. The nickname was typical controller humor, aimed at the unprecedented size of the new combined facility and its expanding airspace, but also for New York controllers' reputation for big egos—they worked a small, complex, high-traffic airspace so tight and congested that controllers in adjacent regional centers were forced to adjust their traffic to conform to New York traffic needs. Now, the airspace and the work of every adjacent regional center would have to be reorganized to meet New York's needs.

For controllers, airspace boundaries had social, economic, and symbolic significance that led to conflict between the regions.[224] Begun with a spirit of cooperation, the design teams' negotiations quickly turned into turf wars in the sky. The NATCA wage-reclassification victory also had unanticipated consequences.[225] The new formula for determining a controller's salary was based on complexity of the airspace that a facility worked. In some cases, changed traffic flows would result in one center losing airspace to another center. No facility wanted to give up airspace because the loss would reduce that complexity level and, accordingly, salaries at that facility would go down. But as negotiations began, the outcome was uncertain. All involved centers were large, so the loss of a airspace sector or two might not have much effect on salaries. Even so, it was a matter of status. The two New York facilities had the highest complexity ratings of any facility in the country, and were proud of it, so the Crystal Palace would maintain their status. For the adjacent regions, however, every boundary was contested—primarily because changes in each region's traffic flows changed controllers' work in consequential ways. Also, though, the "runway people"—as controllers refer to TRACONs—did not understand national flows as the "en route people," the centers, did. What's more, the possibility of reduced salary and status, and New York as "the enemy" in general contributed to the uncertain outcome as well.

The system effects on controllers and their work were hitting hard. The reorganization of major traffic flows had been settled at the national level. The detailed local airspace sector reorganizations were beginning in the regional facilities. As controllers ground out agreement after agreement, uncertainty still clouded the outcome. Once redesigned, how were they going to "turn on" the new flows?[226] The timing was not likely to be optimal. Controllers, used to working in perennially understaffed conditions, were awaiting the onslaught of summer traffic. In addition, they were facing a wave of retirements as replacement hires, by then in their forties, approached retirement age. Although the Garvey agreement with NATCA promised that the number of controllers would be increased to fifteen thousand with additional increases after that, it takes years to train a controller, so skillful replacements would be a long time in coming.

George W. Bush was campaigning for president, and NATCA worried that the privatization of small towers begun by George H. W. Bush would expand, leaving the crowded sky in the hands of an ever-shrinking FAA labor force. With another Bush in office, controllers anticipated reductions in influence, salary, and benefits. Although NATCA was successful at getting Clinton to sign an executive order in December 2000, stating that air traffic control was "inherently governmental," which meant that it was not subject to privatization or contracting out, the union was only too aware that if elected, Bush could cancel that executive order.[227] And although new automated technologies were being worked up to help with the routine problems of daily traffic, the direction of automation troubled the NATCA membership. The global navigational satellite system and GPS cockpit devices signaled progress toward an automated future of dead reckoning that was threatening to operate without them.

Air traffic was at historic highs. In the facilities, the pressure to reduce delays and errors was constant. The FAA routinely gathers and distributes statistics on traffic volume, collisions, and delays. These statistics take material form at the organization level, conveying symbolic and social meaning. Here again, facility status in the system was on the line. The system mandate for safety is reinforced in statistics on operational errors: violations of the rules of separation that specify the amount of space between aircraft moving in the sky and on the ground. The greater the number of errors, the lower the status of the facility in the system and in the facilities, the status of the controllers who make them. The system mandate for expeditious handling of traffic—speed, schedule, and efficiency—was reinforced by FAA statistics reporting the number of operations and delays. The total daily number of operations (i.e., "the count," or number of airplanes moving through the airspace) for each facility is reported to FAA headquarters. The greater the number of operations, the greater the status of a facility and the controllers who work there. The status of every facility, its air traffic manager, and controllers was measured and evidenced in these several publicized measurements of performance, tying the standing of individual controllers in each to the performance of their coworkers.

The past had left its mark on the present. Funding and staffing shortages were continuing problems. Despite ongoing efforts to update its aging structures and technologies to meeting changing conditions, the liabilities of technological and organizational innovation had not only added to internal divisions and conflict, but had left the system behind the times. This was the state of the air traffic control system and the general effects upon air traffic controllers and their work at the turn of the century. Based on ethnography and interviews, the rest of the chapters locate controllers within the context of history in progress and the socio-technical system they inhabited. The chapters reveal the connection between the system, controllers' technologies of coordination and control, and their cognitive, physical, emotional, material, and cultural practices during the year 2000, thus provide the context for understanding their collective and individual responses during the events of the September 11 terrorist attacks and the year after.

PART II

Producing Controllers

Who are the people who take on the job of controller, and how do they develop the skills to engage in dead reckoning? The cultural imagery of air traffic controllers depicts them as having a uniform personality—competitive, aggressive, risk taking, macho individuals with big egos. It also depicts them with a uniform lifestyle—hard drinking and hard driving. In confirmation of this, as the opening credits roll in the 1999 film *Pushing Tin,* the only woman controller working traffic is wearing a muscle shirt and doing bicep curls with a ten-pound weight as she talks to pilots while on position at her radar scope. The two male leads are highly competitive, both on the job and off, and for fun one enjoys lying down on a runway and getting tossed about by the wake turbulence of a departing jet. Although obviously exaggerated, these characteristics nonetheless make some sense for people who do risky, high-stress work in a fast-paced, highly competitive global air traffic system.

The 2006 film *United 93* challenges this popular imagery, showing controllers under stress but professionally and systematically doing their jobs during the unprecedented crisis of September 11. A mystique understandably surrounds air traffic control as an occupation. With the exception of pilots, most people don't see or meet controllers at work, as we do teachers, pharmacists, or the police. Even the public tours of FAA air traffic facilities, ended by September 11, preserved the mystique because the job itself is a complex combination of rapid mental, verbal, and physical maneuvers that are not immediately visible or comprehensible to an observer. Controllers enjoy the intrigue and cultural myth that surround their occupation, even as they acknowledge and distance themselves from the distortions.

As exposed as everyone else to the cultural imagery, I did not know what to expect when I began this study in the facilities in March 2000. As I spent time with controllers in the workplace, again and again I was surprised by the variety of people who hold the job. It was a controller at the center who recommended I read Patrick O'Brian's great series of historical novels about sea captain Jack Aubrey, the stories woven around battles at sea and filled with historically accurate information about marine navigation—dead reckoning, through and through. A voracious reader, he had read all the volumes on his work breaks. Some controllers were health conscious, working out, bringing Lean Cuisine lunches or vegetarian meals prepared at home. Some were quiet and shy, leading quiet lives. Some were athletic, like runners, or regularly engaged in seasonal sports. Others, not so physically active or fit, enjoyed NASCAR, the lottery, the Patriots and Red Sox, and going drinking with friends, fitting the cultural imagery to a greater or lesser extent. However, many of the replacement hires—those hired after the 1981 PATCO strike—said that in their early years they regularly went drinking with coworkers after work. But once they were married with kids and in their midforties, they went home. Many controllers live far from their assigned facilities, so headed out of work for an hour or more commute.

Observations about people's interests and characteristics in the workplace are not enough, however, and likely to be misleading. Nonetheless, my survey of these four facilities confirmed a diversity of controllers.[1] At the time of the 1981 strike, controllers had similar common backgrounds: the typical controller was a white, working-class man with a military background for whom the job was a springboard into the middle class.[2] Over the years it has remained primarily an occupation for white men, but when I entered the field, controllers had more education and represented more social classes. Several controllers said to me proudly that they come from all walks of life, and they do.

Although a high school education then was the requirement for the job, during my study, 83 percent of controllers had more than twelve years of schooling. Of this group, 41 percent had received degrees from four-year colleges. Moreover, 8 percent undertook professional or postgraduate education after their degrees. The controllers at these four facilities, in fact, were educated beyond the national average for air traffic controllers—nationally, 78 percent had more than a high school education, 26 percent had degrees from four-year colleges, and only 2 percent had postgraduate education.[3] However, they fit the national pattern for controllers on race and gender.[4] Both the national distribution and that of the New England Region for controllers were 86 percent men and 14 percent women. In my surveys, however, 82 percent (154) of respondents were men and 18 percent (34) women.[5]

These controllers came to air traffic control in hiring waves that created distinctive age and gender patterns. These patterns are system effects, the result of three political events: the 1981 PATCO strike, Equal Employment Opportunity legislation for women, and the 1993 Clinton administration decision to open a three-month rehiring window of opportunity for fired PATCO controllers, who had been banned from federal service for life.[6] Those who came to the profession during this time had different social histories. They came by different routes and for different reasons. Some people grew up dreaming of becoming controllers. They lived near airports and loved biking over to watch the planes through chicken-wire fences. The lucky ones got invited up to visit the controllers in small towers. Others fell in love with aviation because they went flying in a small plane with a relative or friend. A few had flying instruction. The largest percentage of controllers (26 percent) learned about it because of family or friends who already worked in aviation, either as pilots, mechanics or technicians, or airport personnel. For many, the national news focus on the PATCO strike first introduced them to air traffic control as an occupation. Some became aware while in the military because they knew a military controller or were themselves assigned to the job. One person went to take the civil service exam for a Post Office job, got in the wrong room, took the air traffic controllers' exam by mistake—and passed.

Many controllers described some triggering mechanism that made them start thinking seriously about it as a job possibility. For example, some of the replacement hires went straight from being a military controller to a civilian controller. Still others were drawn to the profession because the work seemed exciting and mentally challenging. Finally, many (32 percent) said the strike opened up jobs for them at a time when they were looking for a job and being a controller offered good financial compensation and security. A few found it attractive for personal reasons that were tied to other occupations. For example, one said that the job entailed "no heavy lifting"; another that it kept him "from being a gunner on a helicopter." A former Jesuit priest explained his decision by saying, "I was used to working in a rule-bound hierarchical authority system."

For both men and women, personal contacts first provided the introduction to the profession, then the triggering mechanism prompted them to actively seek out an occupation in air traffic control. Network ties are known to be associated with obtaining all types of jobs.[7] Because the air traffic controller community has a history of being dominated by white men, informal recruitment to the job through personal relationships is more likely to occur between white men than it is between white men and men of other races, or between white men and women. Through network ties, the racial and gendered occupational structure of air traffic control reproduces itself. The effect of military participation on women's limited presence in the occupation is clear: although several women were pilots, no women, in contrast to thirty-six men, identified the military as the trigger that led them to consider air traffic control as a job possibility.

I had thought that surely controllers would have some prior common talents or skills that would indicate a cluster of capabilities necessary for the job. Were they better in particular areas, like math or physics, and did being good at these things push certain people toward this job? My survey showed this was not the case. Many said they were good students, but just as many said they weren't: "We're a bunch of underachievers," one controller told me. Asked about their best and worst high school subjects, many said math was their best subject, followed by English, but almost as many controllers said math and English were their worst subjects. I had thought that doing well in science might be a good indicator of skills necessary to becoming a controller, but only 9 percent said that science was their best subject—and 11 percent said it was their worst. Many said they had excellent memory capacity in high school, but just as many said their memories were poor. Fully 64 percent could not identify any hobbies, activities, or interests in their background that suggested to them that they had skills necessary for the occupation. Those who could identified activities that gave them useful background knowledge: in descending order, pilot training, sports, manual tasks that required physical coordination, or aviation interests.

I had thought that perhaps current leisure activities would show some common interests, hobbies, or skills. In their time off, controllers tend to be busy people. Those with children don't have leisure time. Air traffic controller couples with children opt to work different shifts; two were handing off their children in the parking lot at shift change. Many did volunteer work: Little League coaching, charity drives, voluntary associations, community organizing. Others were self-employed in addition to their controller job. Some were working on advanced degrees after hours (e.g., several master's students and a law student during my research). Two were teaching air traffic control in community college programs. Across the four facilities, forty-two were licensed pilots: thirty-seven men and five women. At each facility, the core group of controllers elected as their facility's NATCA officials devoted extensive time during the workday and on days off to union responsibilities. At the center, a number of controllers—NATCA members—were involved at the regional or national levels: among them, crisis counseling with controllers traumatized by an accident; redesigning the national airspace, and modernizing air traffic control technology.

The bottom line is that my research turned up nothing that controllers would have had in common before coming to the occupation. Yet in conversations and interviews, they volunteered a set of core traits that all good air traffic controllers have, including "multitasking, the ability to do many things at once—listening, talking, watching";[8] "type A personality, someone who thrives on stimulation and lots of activities, outspoken, aggressive, not afraid to take charge and go out there and do it and make a decision and stick with it"; "excellent recall, short term memory is extremely high, long term, variable . . . we can recall it all"; "great vision and hearing"; and "common sense." The characteristics they named were all individual attributes: psychological, physical, and cognitive capabilities. They said that individuals bring something to the training that the training can enhance, but not everyone can be trained. Years of education could, in fact, be a handicap. Several times I heard the story (no doubt told pointedly) of the PhD holder from Massachusetts Institute of Technology who didn't make it through the training. "You can't be overanalytical," they said. Watching them work, I saw the individual attributes they said they had in common—the ability to do many things at once, excellent memory, great vision and hearing, decisiveness, and other traits they identified as "type A." While all these traits were essential, controllers emphasized that the characteristic most essential to controlling airplanes was common sense. This shared belief came up repeatedly, not only in the interviews and surveys but also in spontaneous remarks by controllers in hallways or while they were working air traffic. What do they mean by common sense? Common sense, one explained, was "the ability to avoid messes." A mess happens because of an anomalous incident, bad weather, or a technical problem. Common sense helped "avoid a mess in the first place, and get out of a mess whatever way you can."

Common sense, in everyday parlance, is defined as "something that is evident in the natural light of reason and thus common to all; . . . good judgment or prudence in estimating or managing affairs, [especially] as free from emotional bias or intellectual subtlety or as not dependent upon special or technical knowledge."[9] In contrast, for Clifford Geertz, common sense is a loosely connected body of belief and judgment, not just what anybody properly put together cannot help but think.[10] He argues that common sense is a cultural system, a frame of mind, that both differs from place to place and takes a characteristic form. Controllers indicated that "much of traffic management is rote"; it "becomes second nature." It becomes second nature because their cognitive activity is embedded in a cultural system of knowledge from which decisions can be made and coordination effected automatically, without conscious calculation, so that physical and cognitive attention and action can be directed to matters that need to be dealt with in the immediate situation. Collectively, controllers' cultural system of knowledge is a set of embodied repertoires—cognitive, physical, emotional, and material practices—that are learned and drawn upon to craft action from moment to moment in response to changing conditions. In constructing the act, structure, culture, and agency combine.

Common sense, then, is a frame of mind and a capacity to act that is shaped by the system in which controllers work. Although they perceive it as a cognitive trait that each individual brings to the job, common sense is a product of their occupational training that is continuously reinforced in their daily work practice. The core of dead reckoning, common sense enables them to make sense of what they see, to identify anomalies and warning signs early and correct them, to make decisions and coordinate actions without thinking too much on it, and to meet the conflicting mandates of the air traffic system. How does this happen? The explanation lies in the connection between culture and cognition and the way cultures outside the individual at the collective level affect the interpretations, meanings, and actions of individuals and groups. Cultures are multiple and varied, are often inconsistent, and are brought into play when triggered by cues in the environment, then used strategically.[11] *In Craft and Consciousness*, Bensman and Lilienfeld explain how the techniques and methodologies of many occupations and professions affect the ways of thinking and behaving of those who pursue them.[12] The craft-based procedures, assumptions, and practices in handling specialized materials and symbols create habits of mind that give each occupation its distinctive character. For Bourdieu, skill acquisition includes the development of an occupational *habitus*: individual predispositions and habits of mind are tied to their location in larger structures—in this case, an occupation confined to the air traffic control system.[13] In the two ethnographic chapters that comprise part II and those that follow, we can see how these predispositions materialize in the workplace, affecting the interpretation, meanings, and actions of controllers at work.

The patterns found in much of social life derive from the taken-for-granted quality of cultural belief systems—the framework of rules, roles, and authority relations that govern organizations, occupations, and professions. They exist as common understandings, deeply embedded, seldom explicitly articulated. Organizations reinforce these cultural scripts, refining and elaborating them to meet the local situation. They become embedded in organization structure, culture, technologies, language, and symbols that are passed on to new members in the form of rules and procedures as well as informal practices. These cultural scripts are then perpetuated by the daily activity of doing the work itself. A well-known fact of organizational life is that organizations spend time and money socializing and training new members in proportion to their need to achieve coordinated, predictable, and cost-effective outcomes. When there is extensive risk of costly harm, the investment tends to be higher. The FAA investment in training air traffic controllers to engage in dead reckoning is very high, indeed. However, it is not formal training alone. Being there triggers layered cultural understandings: the system of enacted rules, language, physical and social positioning, formal and informal relations, technologies, and the task specific to a time and place.[14] The result is a layered cultural system of knowledge that is enacted in everyday interaction at work and passed on to new generations. Remarkably—and in contrast to other ethnographies that show how people with similar backgrounds develop sets of dispositions in common[15]—air traffic controllers are exceptional because they come from all walks of life and different social histories, with little in common at the outset.

The crucial institutionalized mandates of the air traffic control system are three: the safe, orderly, and expeditious delivery of air traffic. Safety is the primary goal—the first-duty priority is separating airplanes and giving safety alerts to avoid collision—as explicitly stated in the first two sections of "the 7110.65," or "the Book":

> **ATC Service**
>
> The primary purpose of the ATC system is to prevent a collision between aircraft operating in the system and to organize and expedite the flow of traffic. In addition to its primary function, the ATC System has the capability to provide (with certain limitations) additional services. The ability to provide

additional services is limited by many factors, such as the volume of traffic, frequency congestion, quality of radar, controller workload, higher priority duties, and the pure physical inability to scan and detect those situations that fall in this category.

a **Duty Priority**
Give first priority to separating aircraft and issuing safety alerts as required in this order. Good judgment shall be used in prioritizing all other provisions of this order based on the requirements of the situation at hand. Note—*Because there are many variables involved, it is virtually impossible to develop a standard list of duty priorities that would apply uniformly to every conceivable situation. Each set of circumstances must be evaluated on its own merit, and when more than one action is required, controllers shall exercise their best judgment based on the facts and circumstances known to them. That action which is most critical from a safety standpoint is performed first.*[16]

For controllers, to expedite means meeting airline needs for speed, schedule, and cost efficiency. And orderly delivery of air traffic is essential to both safety and expediency. Research shows that in many organizations that have similar priorities, safety, schedule, and efficiency are conflicting goals. Typically, however, the tasks related to safety and those related to production goals are segregated, managed by different people located in separate units in the organization's structure. Air traffic control is different. The first line of responsibility for safety, schedule, and efficiency is combined into one job: the air traffic controller.

What is important to recognize here is that these institutional mandates not only are reproduced as cultural scripts in the organization and in technologies used to train controllers but also are reinforced in controllers' daily work practices by the techniques and methodologies of the occupation, as Bensman and Lilienfeld have found.[17] Accomplishing these system goals requires that people of different personality, lifestyle, socioeconomic class, education, skill, age cohort, gender, race, and ethnicity are brought together and molded to work and respond in a predictable, coordinated way in a standardized system in which safety, time, and cost-effectiveness are of the essence. Necessarily, the focus must be on the interaction of controllers with one another, with pilots, with technologies, and with material objects during their training and in routine activities during the rest of their career.

The chapters in this part examine the production of controllers. The two chapters are the first of several that demonstrate system effects: how the cognition, actions, work practices, and experiences of air traffic controllers are shaped by the institutions, organizations, technologies, concepts, and methods that make up their professional lives. In the United States, the training of controllers happens the same way throughout the air traffic control system: the principles taught are the same, the stages of training are the same, the goals are the same. The strategies and content have varied somewhat across the years, but at the time I entered the field, the principles, stages, and goals of the training had recurred across all facilities for several generations of controllers. The result was that controllers from different facilities could coordinate activities with one another, a capability essential to the functioning of the system. Consequently, the four facilities in this study, chosen because they represent the variety of facilities and kinds of work controllers do, give us insight into the production of controllers throughout the system.

Drawn from field observations and interviews during 2000–2001, these two chapters show the incremental process by which people with no skills in common become knowledge workers able to act in concert. Revealing the acquisition of skills necessary to dead reckoning and how those skills are elaborated into expertise, these chapters follow controllers chronologically through the three phases of their training. First, at the FAA's training academy, hopefuls must master the basic techniques of the task, the fundamentals of the international system's rules, standard procedures, and—less obvious to them—the cultural scripts of the system. Second, they are assigned to a facility where they learn the rules, procedures, and local knowledge specific to that facility. Then moving from the classroom to the apprentice phase, the learner begins working live traffic with a trainer who passes on experiential and intuitive knowledge, situation discrimination, and sense making. Finally, with practice and experience, learners gradually develop tacit knowledge, with the result that they can accomplish tasks quickly, without pausing to calculate and think.

This does not mean controllers are nonthinking. Tacit knowledge refers to knowledge that cannot be written down, either because of its characteristics or because it is embodied. It is a habit of mind: as one said, "I can't explain it; I just do it." With the acquisition of tacit knowledge, routine tasks become automatic, freeing controllers to think about the higher-order matters of dead reckoning, such as planning, anticipating, recognizing anomalies, and responding to surprises that call for improvisation.[18] When learners progress to the stage of tacit knowledge, they have reached the final stage of skill acquisition: expertise.[19] Expertise in this case is the ability to assess and respond to situations in the moment, based on past experiences, without articulated purpose or intentional decision making.[20]

Expertise must be explored by focusing on task and devices within the social, cultural, material, spatial, organizational, and conceptual arrangements that make it possible.[21] The layered acquisition of skill begins at the institutional level, becomes increasing specialized at the organizational level, then is refined during apprenticeship. At each stage, expertise develops in interaction with the individuals, groups, and devices that constitute the training process, either formally or informally, who through language or symbolic actions pass on what they know. Tacit knowledge can be acquired only by immersion into the society of those who already possess it.[22] The tacit knowledge that air traffic controllers acquire is what Collins calls "collective tacit knowledge,"[23] acquired from interactions with others in the organizational spaces in which they are trained and do their job.

Controllers master the technologies and methodologies of dead reckoning, until having gained the experience of being in multiple situations, the use of these skills and the myriad technologies of coordination and control that enable them—maps, runways, fixes, keyboards, radar, checklists, flight progress strips, computers, simulators—are literally incorporated, taken into the body, where they become a bundle of taken-for-granted practices that can be done automatically. Central to transformation from beginner to expert controller is this embodiment: the incorporation of skill, both motor and intellectual, into the body. In learning and doing the task, mind and body adjust to the social conditions of work.[24] In addition to the physical and cognitive skills that controllers learn, the system's cultural scripts reinforcing the safe, orderly, and expeditious delivery of traffic become embodied habits of mind that controllers enact intuitively.

In combination, these two chapters reveal the components of controllers' culture system of knowledge and how they are acquired. Consequently, the chapters introduce what controllers do that technology can't replace. Collins observes, "With collective tacit knowledge, it is the brain and body's unique capacities that allow it to acquire tacit knowledge from the collectivity in the way that no machine can imitate."[25] Possessed in common, thus giving new meaning to *common sense*, collective tacit knowledge enables controllers to coordinate activities across social space, often silently. The air traffic control system works because controllers are able to combine standardization, taken-for-granted understandings, and the higher-order interpretive work and improvisation that are basic to expertise. Intuitively, they draw from their repertoire to match their action to a situation. However, the system effects an even greater transformation. Those who survive the training are fundamentally changed in physical and cognitive abilities, personality, and cultural understandings, possessing in common the characteristics necessary to dead reckoning. Their work transforms them. The depth of the transformation makes the system resilient and safe, if not fail-safe, because controllers can be counted on to fill the gaps in the system.

Occupational training is a technology, a people-shaping technology. Intensely and effectively deployed in the production of air traffic controllers, the result is the development of an occupational group that is working at the upper ranges of human cognitive capacity and who can engage in dead reckoning "without thinking"—skills essential to a system with ever-increasing traffic volume and complexity and never enough controllers. Their interpretive work enables them to go beyond the limits of standardization, organization, and technologies as they predict the positions of aircraft in the sky and the responses of other controllers on the ground, effecting coordination across distances and differences, simultaneously meeting the contradictory goals of the system. In the end, the two chapters demonstrate that the FAA technology for training controllers is one of the most important—if not *the* most important—of the technologies of coordination and control in the air traffic control system.

3

From Skill Acquisition to Expertise

Historically, all applicants had to pass the civil service exam, then the official process began at the FAA training academy in Oklahoma City, or "Oke City," as controllers call it. Before the 1992 hiring freeze, applicants came to Oke City either "off the street" or from the military. Only a few came from co-op degree programs at community colleges, like Daniel Webster College in New Hampshire or Embry-Riddle in Florida, where students received a liberal arts degree with majors in aeronautics-related subjects and served as interns in an air traffic control facility. The academy training required residency in Oke City and could take anywhere from two to six months, depending upon program variation the year a class entered. In training they learned the international, standardized rules and procedures of the system and the basic mechanics of applying them in laboratory situations. Across cohorts, the academy training always was rigorous, stressful, and tough. Those who came before and immediately following the PATCO strike had a more militaristic training experience than those who came later, but military influence never entirely disappeared. The academy program was pass-fail: half the people washed out. This was so even for those who were in training immediately after the strike, when the FAA was desperately in need of controllers to replace those fired. If and when applicants mastered the skills taught at the academy, they were assigned to an air traffic facility for the second phase of their training.

When the 1992 hiring freeze was finally lifted in 2002, the academy opened again to people off the streets and military controllers. However, because the length of time necessary to train controllers was high, and the immediate need for controllers was, too, the first people hired to facility assignments came directly from college co-op programs. These programs included facility training time, and students attended the academy for only a few days of tests, or a shorter stay. In 2002, as my field research was concluding, the co-op hires were only beginning to arrive and train at the four facilities I was studying. So almost all controllers working during the period of my fieldwork had spent several months training in Oklahoma City, under the academy's unforgiving pass-fail program. Those doing facility training in the four facilities during my study came from a mix of backgrounds: the few graduates from the academy, returned PATCO controllers, fully qualified controllers who elected to transfer from other regions to change locations or types of facility (e.g., from TRACON to tower) and controllers who had failed or dropped out of training at other facilities and were starting anew at a smaller place. Regardless of background, including previous experience as a controller, they all viewed the training as one of the most stressful experiences of their careers.

The Academy: The Screen, the Game, and Survival of the Fittest

The FAA academy taught air traffic management by the book—specifically, the 7110.65. Teaching the system's institutionalized formal rules and procedures was the stated goal of academy training. However, controllers complained that their academy training had little or no relation to the everyday work of air traffic control:

> It wasn't about teaching you to do air traffic control, it was about weeding you out to see if you're competent to develop to do air traffic control. (Tower, M)

> So it was just exercises. It had nothing to do with aviation. I mean, ostensibly it did, but being a pilot, I knew it didn't. (Center, F)

> I don't think the academy had much to do with what we do, except that you had to not completely fold on the job and seize up. (TRACON, F)

Yet when they began describing the academy training process and their responses to it, their accounts disclosed that it *did* prepare them for the everyday work of air traffic control. In addition to teaching standardized formal rules and procedures, academy instruction also began developing both the physical and the cognitive traits necessary to the job. Moreover, it instilled the cultural scripts of the air traffic system: safe, orderly, and expeditious delivery of air traffic. Academy training built upon the native abilities that applicants brought to the training and developed those they didn't have in order to shape them into people who could handle the conflicting mandates of the system. The extent to which this training developed new traits or reinforced or exaggerated traits they brought with them to the situation, we can't know. But looking at what they say about their experiences shows the repetition of principles essential to their survival in the academy training and, not coincidentally, in the system in which they wanted to work.

Describing their academy experiences, the different cohorts of controllers all said that it was the most stressful time of their lives. And, consistent with one of the traits they gave as a characteristic of a good controller—"if it is a significant event, we can recall it all"—the details of events were fresh for them years later. They call the academy experience "the screen." The stress began immediately, as the entering class met together for the first time in the auditorium:

> The first thing they say to you is look at the person to the left of you, look at the person to the right of you, and by the end of this class you won't see those people here. So they gave you a possibility of about a 33 percent success rate. Some classes didn't even make that much. So you've got this oh my God, I'm the one that is going to fail, or I'm one of the two-thirds that are going to fail. That's an initial pressure right there. And it's constantly pounded into your head you've got to pass, you've got to pass. And instead of just instructing and finding out if people can do the work or not, it's the added pressure of the training. I don't know if that's what they were trying to do, to see whether or not you would get crushed under pressure, but it's not the same kind of pressure you get working the job. The threat of livelihood is not the kind of pressure you have down here [at the center]. (Center, M)

> It wasn't training. It was a screen. "Let's see if we can scare, intimidate, frustrate these people enough. And the ones that are leftover, maybe they'll make good air traffic controllers." (Bedford, F)

> I'd have to say the best comparison I'd give would be between kindergarten and a concentration camp, the way they treated people down there. (TRACON, M)

> I hated it. It was very hard. I don't think it was particularly tough on women, I don't. The female part of it didn't seem to have any difference, being a female versus a male didn't seem to matter. I think it was old-style FAA training . . . the same as marine boot camp. I wasn't a marine but . . . we're going to keep hitting you until you break, and if you remain standing after ninety days of that, you win. That's where it was. It was horrible. (Center, F)

The need to drop out of their regular life to attend added to the stress. They made sacrifices. The sacrifices were both social and emotional—leaving known places and relationships for strangeness and an uncertain future:

> Most of us gave up everything we ever had to start. It's almost like joining the military. I mean, you cut all your ties, you quit your job, you go out to school and start a new career, if you don't pass that course, then you're done. Some of us had safety nets that we could go back to what we were doing before, but a lot of [us] didn't. It was a very fast-paced environment at the academy and you had to keep up or you got left behind. And when you got left behind, it's not like the military where they keep giving you chances, you get left behind and you're, you're gone. (Bedford, M)

> My whole life hinged on making it through. Because if I failed, what I had been wanting to do since I was sixteen years old was gone. There was no turning back. There was no second chance. It was all over. So it was do or die . . . It was like . . . what I wanted to do all rested on this four months. So I was going down there for four months and I could care less what the sacrifice was. I was ready to sell my soul to get through this school. All the stress of four months of just constantly, total 100 percent dedication to one thing. (TRACON, M)

The goal was to "pass," but passing was not easy. They were tested frequently. They were allowed to drop some low scores, but everyone's scores were so low that instructors would tell students that a 60 was a "keeper." The final score—the cumulative average—had to be a 70. Neither performance on the civil service exam, military experience, a college co-op program, or walking in off the street predicted success. Even former military air traffic controllers did not all make it through.

The academy training was divided in two segments: academic learning and laboratory application. For those who had not been military controllers, it was their first exposure to the concepts and equipment that would be fundamental to their tasks. The academic segment called for mastery of the formal international standards and procedures. It required rote memorization. Standardization and rule mastery are the first steps in skill acquisition, and foundational to the development of expertise. Primary source material was the 7110.65, the Book—the three-and-a-half-inch-thick, information-packed manual of air traffic procedures. It included the complete "phraseology": the formal standardized international words and phrases required in exchanges with pilots and with one another in official communication. Phraseology covered weather, pronunciation of numbers (3, "tree"; 9, "niner") and phonetic alphabet letters (*a* = AL-fa, *b* = BRAH-vo), instructions for clearance, takeoff, altitude change (climb and maintain flight level one eight zero), vectoring (left [compass] heading one seven zero). It gave specifics for landing instructions; route assignment, fuel dumping and other emergency procedures, airspace maps, airfield markings, rules of separation, types of aircraft equipment, technology—in short, everything.

They had to master seemingly limitless details in a limited amount of time. Here is one person's first introduction to airspace and airspace boundaries:

> It was very intense, the learning, the pace that they expected you to learn at was incredible. I remember feeling so defeated the second day I was there. They handed us this (airspace) map. I don't know if you've ever seen it, the "aero-center map," it's called. The map with all of this crap you have to memorize on it. And you get a test seven days later. It's like a road map, with all the different symbols and stuff. And they give you basically a blank piece of paper and you have to draw it from memory. And I was terrified when I opened it up. I thought this is impossible. There is no way I can learn this in seven days. These people must be geniuses if they're doing this, you know? I called my wife, who was staying with her folks, . . . saying, "I don't think I can get through this." And she said, "Well, just try." So I sat there that night and spent about studying that map and was amazed how quickly you learn something if you have to. You really memorize it. (TRACON, M)

Most found the rote memorization very tough going. It called for mastering new language, concepts, symbols, and altering perception. For example, the "aero-map," designed for training purposes, contained no markings found on road maps—no place-names, no north and south. It required translation: cognitively reimagining the sky as sections of airspace boundaries full of artificial lines indicating airways, and material directional technologies and system facilities on the ground marked with unfamiliar names and icons. Memorizing the information in the 7110.65 was straightforward, they said—"number of miles separation is number of miles separation"—the challenge was the amount they had to master in a short time: "We had to learn two years' of material in two months." They felt harassed and degraded by some of the instructors, who taught with a boot-camp mentality. Even those with admirable instructors felt harassed by the demands of the material and schedule. Officially, they were learning the rules and procedures essential to a safe system and developing memory skills. Unofficially, but equally as effectively, they were being instructed in the cultural system of knowledge necessary to the job: the ability to complete a heavy workload quickly; persistence under pressure; and the importance of speed, schedule, efficiency, rules and rule following, precision, and concentration.

The Game

Equally stressful was the second classroom segment: running air traffic control problems in a laboratory environment. "The labs" were the practical application of the formal rules and, with the exception of those who were previously military controllers, their first experience with spatial thinking. They had to guide imaginary aircraft through an imaginary airspace in a simulation with no visual clues. To keep planes from hitting each other, they had to preserve formal spacing boundaries between them. To accomplish this task, they had to adjust mind and body to the technology, the first of many challenges of this sort. They "ran problems manually," using paper flight progress strips, as it had been done in the 1950s and 1960s before radar came to air traffic control. Running problems manually in the labs was known as "the game." The game was the foundational layer of ethnocognition and boundary work.

Each paper flight progress strip represented one aircraft.[1] One student would take the role of pilot, and another (the one being graded) was the "controller,"

DAL 1835	14Ø4	BOS	+BOSOX BDL BDL255 VALRE		EDCT	
T/B738/E	183Ø		V157 HAARP LGA+		1847	
677	16Ø		BOS BOSOX V419 BL*** LGA			

Figure 11: Flight progress strip
Illustration adapted by Noah Arjomand

making decisions about aircraft spacing, speed, heading, and distance based on information printed on flight progress strips. The strip represented the pilot's plan. The controller had to go through the process of departing aircraft—giving them clearance to depart, putting them on a route—and then separating the departed aircraft from other aircraft, on the basis of the information on the strips and no other visual aids. The lab exercises called for adjustment of cognition, physical skills, and timing. To be successful, upon picking up a strip the student controller had to mentally translate the printed information, visualizing the airplane in three-dimensional space with other airplanes in order to keep them from hitting each other. The technology was minimal: mics, headsets, flight progress strips, and marking pen. They also had to take into account what the pilot was saying. The controller would amend the original plan, marking the flight progress strips with any route, altitude, or speed changes given to the pilot, so that each strip corrected the route, so each strip was analogical to the airplane's progress. Like the early navigational practice of dead reckoning, they were learning to predict the position of objects in space by deduction, without benefit of direct observation or direct evidence. The controller had to be able to cobble together what he or she had learned with the accumulating flight progress strips and construct a three-dimensional picture of all the planes and their relation to one another while moving in space. This was only the first step. In the next stage of training, these three-dimensional visualization skills would grow to encompass elements of social context in the room.

The game called for controllers to master what they experienced as multitasking. Cognitive skills had to be combined with physically manipulating devices and material objects. The manual simulation problems call for listening and talking to pilots, marking on strips, visualizing in three dimensions, instantly recalling information on the flight progress strips, and managing the trajectory of several aircraft at the same time. Each problem presented a different air traffic situation and was timed and graded. Students were also graded on applying the standardized international rules and procedures learned in the classroom to common air traffic situations. The scenarios not only introduced situational diversity but also forced them to put together pieces of knowledge, thus developing basic response routines upon which they could incrementally build skills in handling greater situational diversity.

Moreover, the game simulated the institutionalized cultural scripts and socio-technical conditions of work. To pass, they had to master the fundamentals essential to meeting the three conflicting mandates of the air traffic control system: the safe, orderly, and expeditious delivery of air traffic.

Many commented on speed, time, and efficiency. One said:

> It is just a big game. They tell you how to play it, and they say, "These are the rules, here you go. We'll set the game up this way, and when we say go, you play. You get marked off if you do this, this, and this, and if you don't do this, this, and this you will do well"' It was all about getting in a given amount of time the hang of the nonradar game, so to speak, that they were teaching us . . . they teach you how to play the game and you learn it in X amount of time or you're not going to be a controller. (Tower, M)

Trainers were strict, and every controller action was scored. On persistence under pressure, precision, and, again and always, the import of speed, time, and efficiency:

> If you're working a problem and you make a mistake, you have to keep going. You can't just sit back and go "I can't do this." You have to keep working the next thirty, forty minutes on the problem knowing you messed up. You've got somebody standing behind you with a clipboard, writing notes. And you see them out of the corner of your eye writing stuff down, which means they're writing down mistakes. So they were looking at everything from how you wrote on the strips. If your [number] two didn't look like the proper two, they could write you up for not having a good two. Or if you didn't write the number in the right box, the number outside the box, they write you up for that. But it was that type of stress; that you knew you had to work on the clock. The clock was going to be running. Airplanes, even though they're on paper, they're still moving. And as you're working, there is somebody else pretending to be the airplanes, calling you up, wanting clearances out of airports and you have to work, you can't stop. (TRACON, F)

On rule following for safe, orderly traffic delivery, a controller shared:

> You weren't actually learning air traffic control, you were learning how to follow the rules. And that was the hardest thing because it didn't always make sense. There were people who had actually been controllers, in the military or whatever, knew how to work in the real environment. But in the lab environment, they got marked down. If you said something that wasn't exactly in the book, even if you said "good afternoon" or "good evening," some of the instructors were marking people off for that, just for being nice to somebody.

Just minor errors. And for a lot of them, that difference took just a few points off a test, flunking them out of the academy. (Center, M)

The instructors conveyed the lessons with tough language and reinforced them by deducting points for mistakes. All students experienced the training as harassment. Diversity of background was inconsequential to this. To survive, people bonded together in the face of the enemy: the FAA. They learned from each other in their rooms at night and at meals in continuing conversations about the assignments, the lab problems, tests, the instructors, and shared successes and failures. Interaction inside and outside the classroom spurred the development of collective tacit knowledge.

> Most people were helping each other get through because there was no competition. If one hundred of us went and one hundred of us graduated, one hundred of us got jobs. It wasn't the type of thing that there are only ten jobs that we're all competing for. We were competing against the standard, not each other. (Tower, M)

True, they were not competing against each other for a job, as he said, but they were *competing*—against the standard, against the odds, and against time. Could they master the amount of information they were given in the time allotted? Knowledge was cumulative—every day the lessons got harder, they said—and the stress also was cumulative. The emotional costs were high. Some said it took a physical toll: hair loss, nausea, sinus attacks. Several who never had headaches had headaches every day. They described uncertainty and a mounting fear of failure:

> Ninety percent of our grades were revealed to us only in the last ten days. And the level of concern grows and grows. You think you're doing pretty good but you don't really know. So that was really hard on people. Sometimes there was a lot of anguish. People thought they were not doing good and gave up even though you might not be doing bad. I think the FAA did that on purpose. (TRACON, M)

> Confidence was a big issue. Oke City was designed for you to never be able to catch up. I was always feeling like I can't do this. This is going to be so hard. I'm never going to pass. I've been here for a couple of weeks and this should be easy now. I should have this mastered, but it was never easy. (Center, M)

> I would call my husband on the phone in tears because I couldn't understand why I couldn't get it, or why it was taking so long to understand it. But I think the same was true with everybody. Even the men would call their wives in tears because . . . they pushed you as far as the stress. They had time tests, and we had to do things in a certain amount of time. I had never failed at anything in my life. And I really thought I was going to fail the academy. I was always honor-roll student, very good grades. Every job that I had I always excelled at it. But I was not going to make it through the academy. And I really thought that the embarrassment of all of that was going to kill me. . . . We had like a 50 percent washout rate. And I was like, this is ridiculous. I might not even make it through this. (TRACON, F)

Controllers still remember the score that they had to get on the final test in order to make it through. And they didn't know until the last day whether or not they did. Emotions ran high. The uncertainty and stress at the beginning were sustained through the very last minutes of the experience, when they returned to the auditorium where they had assembled on that first day. This controller recalled:

> When we were done with the labs, we got a score. And that score was so much of your final grade. Then you did a final exam, and they would tell you what you needed to score in that final exam in order to pass the course. I needed a 30 on the final exam. So, I mean, I was basically in. Other people needed a very high score. You took the exam Friday, you didn't know that weekend whether you were going to make it or not. And then they called everybody into the auditorium. And then they started calling names, starting with the highest score in the class. When they called your name, you would go to the front of the class, go up to the stage, get your diploma and walk out the side door into the hallway. When they stopped calling names, if you were still sitting there, you didn't make it. And you'd be outside in the hallway waiting for your friends, who you cared about, who you knew were borderline whether they were going to make it. And when they closed that door, that was it. I mean, there were people that I haven't seen, that were in my class, that we just never saw again. (TRACON, M)

Those who graduated from the academy had absorbed some of the fundamentals of ethnocognition. They had mastered sufficient knowledge of the formal institutionalized rules and procedures of the system and practiced applying them. They had acquired a stock of knowledge in use, a beginning set of air traffic situations and responses that could be compared. They had practiced talking and listening in new language with someone pretending to be distant in time and space. They had begun the process of adjusting mind and body to the simple technologies and material objects. Using them had developed their cognitive ability in short-term memory, three-dimensional thinking, and visualizing the sky as airspace. These they recognized during the process and after, telling me so in the interviews. They also knew—all too well—that the academy was "screening" for people who could work under pressure, concentrating and controlling their emotions, which the controllers understood to be part of the job.

Unacknowledged was their acquisition of the cultural knowledge essential to the air traffic system: beliefs, ways of thinking, talking, deciding, and behaving that were unarticulated and not formally taught but still acquired in the training process. Those applicants made it through who learned to accomplish a heavy workload in a short time, to follow rules, to recall detailed information from memory, to persist under pressure, to multitask, to have self-discipline, and to attend to speed, accuracy, precision, time, and schedule. Those who studied in groups to help each other through began the essential work practice of coordination and cooperation with people "from all walks of life."

During the academy training, these principles prepared future controllers to act in a predictable manner in a system that placed the responsibility for the conflicting goals of safe, orderly, and expeditious delivery of traffic primarily on their decision making and actions. These formal institutional rules, cultural scripts, and material practices were the foundational layer of their common sense: the "ability to avoid a mess in the first place and get out of a mess whatever way you can." All who passed possessed in common the basics of the cultural system of knowledge uniquely shared by members of this occupation.

There were additional system effects on those who graduated from the academy training. First, the accomplishment developed self-confidence. This sense of accomplishment was pervasive, but especially significant to those who came to the academy as self-defined underachievers in school. Second, they made an emotional commitment to the next phase of training and the job.[2] They made personal sacrifices to get to the academy, and once there, the fear of failure and sustained stress were emotional costs they experienced daily. Having so invested, they continued. Third, being trained in cohorts in a game that was analogical to the cultural and socio-technical conditions of work was a first step in developing the collective tacit knowledge that would enable coordination across system boundaries. Fourth, the shared experience laid a foundation for a common occupational identity. One explained:

> You see academy bonds. You see people in this building who are friends and at first glance they have nothing in common, why would these people be friends? And then you find out they were at the academy together and that's a lifetime bond. Even with people at other facilities. It's a hardship we all went through, even if we didn't do it together, we all went through the same hardship to get to where we are. (Center, M)

There was one exception to this emotional roller-coaster at the academy. During the 1992–2002 hiring freeze, the FAA academy was closed. In 1993, when the Clinton administration created a window of opportunity for the fired PATCO controllers to return, the academy temporarily reopened to train only them. Having waited and hoped for twelve years to get back in, the desire of some PATCO controllers to retrieve their lost occupation was so great that, despite harboring bad memories of their former training, they could not wait to get there. But this time it was different. It was not an emotional roller coaster in the same sense as other cohorts, but it was a deeply emotional experience nonetheless. On the way to Oklahoma City, they ran into former PATCO controllers at airports across the country, in waiting areas for a plane, or waiting at a curb at the airport for transportation.

Whether they knew one another before or not, they recognized one another by age, look, and destination. They hugged and swapped stories. At the academy, they were welcomed and respected. Their skills were rusty, for sure, and they were not as fast as they had been. Far from it—yet they were not subjected to the same training harassment as the replacement hires and the controllers who would come later. It was a reunion of people joined together first by the occupation, then by their mutual firing, and again by the desire to return. Although they struggled to regain skills back, all admitted that the academy training was made easier by the fact that system was standardized in the same basic ways. They already understood the principles being taught. Moreover, they joked, the technology had not changed. The FAA was still using the same 1960s technology that they had struggled with before the strike.

For the returned PATCO controllers, the completion of the academy training brought an extraordinary sense of accomplishment and strengthened bonds. Most had longed to return to their original facilities, or at least to facilities of the same level of difficulty, but in a location nearer to where they were living more than a decade after the firing. But like all other graduates, they would have to start low, at small facilities, and work themselves up through the system. Few in numbers, they would be dispersed to different facilities, losing the support of their cohort but bearing its identity: they were now "the old guys." They would be trained by the people who had replaced them, and the next stage would not be as cozy and welcoming as the first. Once in their assigned facility, they got the same treatment as the other "developmentals" (the new academy graduates), except they were older and tended to be slower, which affected how they moved through the facility training stages and the response of their coworkers.

The Facility: The Apprentice and the Trainer

At their assigned facilities, the academy lessons were refined to suit the local situation: "This is how we do it here." Geertz wrote that "the shapes of knowledge are always ineluctably local, indivisible from their instruments and their encasements."[3] Facility training builds on the institutionalized rules, procedures, and cultural scripts that developmentals learned at the academy, adding sets of organizational procedures and practices that are particular to that facility and its distinctive air traffic patterns. As one controller said, "You have to learn the facility—what are the patterns—before you can start paying attention to other things, like talking to people in the room, being able to cooperate with them, share tasks, do the job." When they enter a facility, they still are learning the basics, but in time they begin to absorb the detail and situation discrimination necessary for the transformation of skill acquisition to expertise. They acquire local knowledge in stages of increasing refinement: formal knowledge learned in the facility classroom, knowledge acquired from observations of working controllers, experiential knowledge taught by trainers in an apprenticeship, knowledge gained in informal conversations with other controllers, and the solidification of knowledge gained by their own practical experience that is cumulatively integrated into self.

Like Oklahoma City, facility training begins with two classroom segments. They begin with academic learning of formal rules and procedures, then progress to practical application in lab exercises. As before, the academic learning called for rote memorization. First came the local airspace map. Then the style, language, symbols, and icons were familiar—their memory skills were already developed, so only the distinctive markings and patterns of a facility's airspace were new. The markings, though, took on meaning: soon the new people would be working real boundaries. The local formal rules and procedures appeared in memoranda of agreement that defined the parameters of exchange of airplanes with the other facilities whose airspace bounded theirs. Next, the developmental progresses to observing controllers working traffic in the "tower cab" at a tower, in the "control room" or "on the floor" at a center, or in the "radar room" at a TRACON. It is there for the first time that they are exposed to the workings of a team (or crew, at some places) assigned to the different positions of the same airspace. This is the team and airspace to which they will be permanently assigned.

They can observe and listen to the operation of the group as a whole. The observation period itself is a kind of dead reckoning, as they begin anticipating the movement of their own bodies through the positions they will be assuming and the actions associated with those positions. They sit beside and observe one position or controller at a time, often sequencing and marking the strips to assist. By assisting first on less complicated positions, developmentals learn the traffic patterns of airspace associated with each position and have a chance to see how others do it, anticipating the apprentice phase of their training. These observations are always a language-mediated exchange between the two: the controller working the position explains between transmissions with pilots the peculiarities of the airspace, the traffic, the moves he or she is making, and the dealings with other facilities and also answers questions.

Equally important are the social conditions of work. Both observations and language are essential to the development of expertise. They learn from talking with the others while observing, on break, and during meals or carpooling. They begin to absorb how to be a controller on the job: how others do the task, use the technology, coordinate activity despite differences in technique and skill level. They learn norms, individual controller histories, standing in the group, and how each fits in with the others, as a personality and as a member of the team or crew, and also how collectively members react to one another, the supervisor, and the facility's air traffic manager to traffic situations, mistakes, and errors.

Then, it's back to the classroom for lab exercises. For radar facilities—TRACONs and centers—simulator problems are essential. The simulation segment is different from the simulation at the Oklahoma City "screen." There, developmentals were working air traffic using flight progress strips with no representational visual aids, known as the "manual method." Back in the classroom, though, the developmental can "see" the traffic projected on a computerized dynamic simulator (DYSIM) as it appears on a radar scope: a data block represents each airplane, or target. This advance calls for still another alteration in perception: translation of a two-dimensional object on the screen spatially distributed with other objects in a three-dimensional sky that matches the information and markings on the strip. As at the academy, a controller role-plays the pilot; the developmental takes the role of controller. Each problem is timed and graded. The DYSIM exercises begin with simple problems—three or four targets. Over time, the simulation increases traffic volume and complexity until finally developmentals successfully can work more traffic than they would be expected to handle in real life on the job. In the process, they increase their stock of knowledge of possible situations and begin to be able to discriminate between them. Only then can they begin training on the radar positions with live traffic.

Surprisingly, neither in interview nor in casual conversation did anyone complain to me about the facility classroom segment or the labs. Their silence was in stark contrast to the strong feelings expressed about the stress of the academy. As it turned out, the definition and experience of stress lay in the comparative experience. The stress of the classroom segments was nothing in comparison to what had gone before and what would come next. An apprentice is a person who is learning a trade, art or calling by practical experience under skilled workers for a prescribed period of time and rate of pay that usually will increase when the apprenticeship is concluded. An apprentice system makes sense in certain occupations in which the practical application takes a dexterity of mind and/or touch that can be acquired only by doing and developing the necessary skill incrementally. It makes especially good sense when, in addition, the occupation entails risky work, when the costs of mistake are so high that they must be avoided at all cost, thus making close supervision during the learning period essential—think health-care professions, police, firefighters. In air traffic control, this stage is called on-the-job training.

The trainer is the single most important person in the production of controllers and the transformation from skill to expertise. The foundation has been laid. In the training stage, the lessons are crucial to ethnocognition and boundary work. The trainer, who does not have an easy job, is an experienced controller on the team to which the developmental has been assigned. The trainer instills the fine points of the craft, enhancing the apprentice's interpretive work by helping him or her to develop the distinctive physical, cognitive, and material skills that all controllers must have. It is a subtle refinement of knowledge, a passing on of manual and mental techniques and finesse, meanings and noticings, that are crucial to prediction in dead reckoning and the ability to quickly innovate to save a situation. Further, in the hands-on imparting of all they know, trainers reinforce and fine-tune the institutionalized formal rules, cultural knowledge, and local knowledge necessary to meet the conflicting requirements of the air traffic control system: the safe, orderly, and expeditious delivery of air traffic.

The apprenticeship begins with training on one of the easier positions in the airspace during a low traffic period. The trainer, with headset plugged into the position's radio frequency, sits or stands close behind or beside the developmental. With the trainer watching and instructing over the shoulder while wearing a headset to hear both sides of a developmental's transmissions to pilots and to other controllers on the ground lines, the developmental begins working live traffic. Visually and aurally, working live traffic is the supreme exemplar of what controllers experience as multitasking, even though experts on the brain say the human brain can't multitask: it works sequentially. The trainer talks steadily while the developmental watches the sky or radar, talking to pilots, coordinating with other controllers, working the keyboard, and marking flight progress strips. The trainer takes notes during the training session—mainly on what the developmental does wrong—and during breaks, the trainer gives more feedback in the break room. Then they go back and train some more.

The developmental must be certified as competent on one work position in the designated airspace before moving on to the next. Gradually, skills build up. The training begins with the simplest task, increasing the complexity each time the developmental demonstrates that he or she has mastered a task and the position. Certification of competency for each position happens after a "check-ride": a supervisor does an over-the-shoulder evaluation of performance during a busy traffic period. When checked out on all positions in the designated airspace, the developmental becomes a certified professional controller. Becoming certified takes different amounts of time, depending on the size and complexity of the facility airspace, the number of controller positions necessary to work the airspace, and how quickly the controller masters them. Becoming certified can take anywhere from several years at a center to a few months at a TRACON or tower.[4]

No matter how well prepared by classroom instruction and observation, controllers described their apprentice period as a completely new challenge and extremely stressful—they had to develop skill at dead reckoning with moving aircraft and lives at risk. The layers of knowledge are cumulative, and months of work precede this stage of the skill acquisition process. One described it:

> It's a difficult mental process because there's so much information. You have to build a base. It's like a pyramid. You have to have a large base of information before you can even sit down at the radar scope. When you sit down, there's even more things to learn that you have on top of this base . . . it's a constant memorization testing process. However, when you sit there at the radar scope that's when the stress really comes because that's when you're proving whether or not you can do it. (Center, F)

The trainer starts them off with the application of the basic book work and skills: taxiing, departures, landings, how to get airplanes in a certain location, turn them to certain headings, slow them down, speed them up, change altitudes. The initial challenge for a developmental is to put all the pieces together:

> Say you've got ten watts of brain power to apply towards something, right? So usually running the key pack might take a quarter of a watt. And knowing the frequencies for Manchester East, that would take an eighth of a watt. What's the handoff again? What, Manchester is Delta 4? Maybe you use 20 percent of your attention on trying to remember the frequencies or the procedure. Then you're thinking, what altitude am I supposed to be here? What do I own here? What's he going to do with this guy? What's his call sign again? So then you've only got maybe half of your ability to separate airplanes and come up with a plan. But the more you get that under your belt and make it routine, so you're not even thinking about it, then the more attention you can put towards making your plan and the better you're going to do. (TRACON, M)

To refer to this process as simply socialization obscures the importance of interaction, observation, and language in the development of expertise. Cognitive processes are usually invisible, available only in interviews, which are retrospective. However, during training sessions, the trainer's running commentary and the apprentice's response make visible the cognitive and material processes and practices that are being passed on. To give a sense of what a session is like, next, drawing on my observations, I describe the instructions of a trainer at Boston TRACON to a developmental in the first week of her training on a new sector of airspace.

This developmental had certified on all positions except this one. So it is her final position, but also it is the TRACON's "final" position: the landing aircraft are in the sector of airspace belonging to the "final approach" controller, who receives planes by taking handoffs from other controllers in the room, makes the moves necessary to descend and sequence them, then turns planes coming from several directions into an evenly spaced line, "like Rockettes," headed toward the runway, continuing the descent, slowing, slowing, then finally handing each off to the tower one at a time for landing. This is the TRACON position that everyone covets: to master it, joining ethnocognition and boundary work, craft and timing are everything.

To follow the session I describe below, the context matters. When training on a new position, the trainer dictates every action. The developmental is trying to do exactly as told. Training for any new position always begins when traffic is "slow," meaning light. This trainer gave the following instructions in a five-minute time span. When imagining "slow" and "five minutes." remember that the play-by-play from the trainer is a guide to what is happening, but it is only about a half or a third of the action, because you can't see and hear what the developmental is seeing, doing, and saying. The action and interaction happen too fast to write it all down and I was not allowed to tape record. After each instruction, the developmental will relay it to the pilot. As the trainer instructs, this developmental is "multitasking": listening to the trainer, talking to pilots, typing on the keypad, scanning, and coordinating with other controllers on the ground line. Here are the Boston TRACON trainer's running instructions to the developmental:

> Take the handoff on USAir. Turn right on that Eagle. Very good, very good. Take the Delta and vector him [give a compass heading for a turn]. Lufthansa bye-bye [handoff to the tower]. USAir vector and decrease speed. See what you're about to do to yourself on the Learjet? OK, now—tight turn if you can. OK—USAir, cross NAV at 3,000, clear ILS, maintain 17 knots. Now, would it help us to put the commuters together? Now, Eagle to tower [handoff to tower]. Watch the speeds. The front USAir slowed up. Now play your altitudes—he's at 3, he's at 4, he's at 4, he's at 5. We need them at 3. Get back to Tango Juliet to verify speed. He's still at 21, you said 17. Now—she's ruining the hole. You need to tighten up his turn and slow her down a little.

The intensity of a training session is palpable. It comes from the concentration and focus required of both people. With the addition of the trainer and live traffic, the work becomes more complex. As the developmental becomes more proficient, the trainer instructs less but becomes more reactive and critical. Adding to the developmental's stress, learning takes place in public, as it did at the academy, but in the facility, pilots and experienced controllers are the audience. They react, too. Handling the audience response is another part of the job and is crucial to success. Learning to talk to pilots on the radio frequency was a big deal. Many experienced "mic fright" the first several times they plugged their headsets into the frequency and spoke to a pilot. They lacked confidence: Would they make the best decision? Would the pilot do what they asked? Their voices gave away their uncertainty and insecurity. The words often are halting at first, the pitch elevated from the stress.

Trainers teach impression management. Developmentals are taught to control their emotions. The trainer instructs the developmental that when a controller's voice sounds uncertain, pilots recognize it and, apparently needing reassurance, respond by asking more questions. "Like feeding piranhas," they all get on the line to talk to the controller. "Talk fast, and keep control of the frequency": developmentals must sound confident and in control or the pilots in their airspace jam the frequency and the developmental loses control of the traffic. They are taught that maintaining a consistent, professional tone helps pilots to stay calm in emergencies. They are instructed to conform to the standard phraseology and disguise any emotions they might feel, controlling the tone, rhythm, and speed of their voice. Be decisive, assertive, in control, and if they can't yet, they still have to sound like it. They develop a "frequency voice." With practice, the "type A" characteristics of decisiveness and control become incorporated into the body. As one said, "A way of talking becomes a way of being."

Crucial to the process is feedback, positive and negative. In addition to the trainer and pilots, the reaction of the in-house audience has its effect. All the controllers assigned to work the same airspace hear the developmental's transmissions and see what is happening in the air or on the scope. Mistakes become the subject of jokes that travel throughout the facility and may follow a controller through the rest of his or her career. It is a ritual hazing ceremony endured by generations of controllers. Hazing is a public attribution of failure that leads to deep feelings of embarrassment, caring, responsibility, remorse, and ownership of mistakes, a quality that the expert must develop.[5] All controllers talk about "getting hammered" by airplanes when they have a busy session with a lot of traffic; when discussing their training, they talk about getting hammered by their coworkers. The hammering is gender blind:

> If you didn't know the answer when they asked you a question, they would tear you apart. Make fun of you. Call you a dumbass or something like that. You suck. Get out of here. (Center, F)

> While you're going through it, it's the worst. I would say probably 80 percent of the people that go through it get very, very aggravated, and that 20 percent that don't get aggravated are those, for the most part, those are the artists. (TRACON, M)

But the developmental's most important audience is the trainer. It is the trainer who determines how quickly—or whether or not—the developmental makes it as a controller. The trainer's daily evaluation reports are forwarded to the supervisor. Thus, the trainer's power over the developmental's future is another source of stress. As one developmental said, "I'm sitting there trying to do all this stuff, wondering, *Am I doing things exactly the way that he or she does them? And is he or she happy?*" Usually not. The attention primarily is on mistakes. Indeed, if there is a single guiding principle for producing controllers, it is their learning from mistake. There is no training for trainers, and some are naturally better than others. Sometimes there are personality conflicts between an apprentice and a trainer, which can be devastating to the career of a developmental. Also, trainers have different training styles, usually having adopted the style of the trainer who taught them. Training is always harsh. But those controllers who were replacement hires—as many were during my fieldwork—were trained in an especially austere and harsh environment:

> Right after the strike there were a lot of hard feelings here. The guys that stayed didn't really like the new people and it was intimidating because if you didn't know your stuff or if they didn't like you, they could just get rid of you. I mean there was no union, nothing, so if they said that the moon is made of green cheese, then it's made of green cheese. (Center, M)

> Back then, there were very few trainers who wanted you to succeed. They had spent all of their life trying to get into this position, Boston, and make the money that's here. You walk in with anywhere from four to six years of military experience, and you're walking into their pay grade. That's something they might have worked for ten, fifteen, sixteen years to get to, and you are going to have that in two years. So there was a little bit of resentment. (TRACON, M)

> The training back then was atrocious. It seemed to be a game to see if they could make you break or wash you out. I'm in class for like three days and the team I was going to be on had a team meeting so they tell me to go in and meet my crew. I said hi, shook their hands, and all that. They laughed and said, "The last eight people that have come on this crew have washed out. What makes you think you're so special?" (Tower, M)

But the tradition has never totally died out. Even during the late 1980s and mid-1990s, before the hiring freeze and after, controllers said that half of those who passed the Oklahoma City training failed in facility training. During my fieldwork, some controllers still trained in that harsh poststrike style, some adopted parts of it to carry on the tradition, and others rejected it entirely. For the developmental, it was luck of the draw. One shared:

> Some instructors like to yell at you and scream at you, ask you what are you doing, and treat you like an idiot. Others are pretty mellow. . . . You could have some instructor that will sit behind you, whacking you on the head with the clipboard while you are training. Others will sit back there and you know, quiet, look at you afterwards and go, "Did you learn something from that?" So it depends. (TRACON, F)

> I've had wonderful trainers. When I first started training, I'm like, "This just isn't right, I'm not getting it." He said, "Trust me, come to the roof, watch the airplanes depart." So we went to the roof, looking at them. I watched the airport to see where they were headed. And I got a better picture of what's going on here. (TRACON, M)

> [On training at Boston Tower:] I couldn't do anything right, no matter what I did. A perfect example was one day training, running the aircraft to a certain runway a certain way. She wrote up everything that I did wrong that day. The next day, the same hour, the same airplane, the same call sign, I ran it exactly the way she wanted me to run it the day before. And she wrote me up for doing it wrong. When I questioned her, pulled out the training form from the day before, and said, "This is how I ran it yesterday, which wasn't right. This is how I ran it today, which was the way you told me to do it yesterday, and you're saying it's not right. How should I do it?" And she looked me in the eye and said, "You will never get it right." (Bedford, M)

Training is stressful for the trainer as well. Perhaps the single most important skill all controllers must learn is to concentrate and persist in even the most difficult situations. "When that fight or flight response kicks in, we learn to work through it." The trainer teaches this skill by making developmentals dig themselves out of a difficult traffic situation, taking over control only at the very last minute. Learning from mistake is a key training principle. Trained to control air traffic, the trainer is in a position of formal responsibility for the developmental's air traffic but unable to actively control it. Trainers' own certification is at risk: a mistake "is on my ticket," they say. Possessing the expert's fine-tuned ability to discriminate among situations, the trainer can see a problem unfolding long before the developmental can. But trainers struggle to restrain themselves from intervening, letting the developmentals get into trouble and figure out how to work out of it. They do intervene, taking over the frequency. When to do so is the problem. Trainers described their experience as follows:

> You come to know, as their trainer, what to expect—say, snow. This is all new to her, the conditions are poor, She's probably going to be a little tentative, her working speed is going to be slow. You have to be able to let this person get to the verge of out of control but you have to be able to get back in there and control the situation, correct it in time because instilling a little fear in somebody all of a sudden they realize, oh boy, I won't be doing that again. You learn by your mistakes. Unfortunately, we don't make mistakes in air traffic control, or we're not supposed to make mistakes, but you have to let someone get to the point of making a very, very critical mistake and then learn by that. (Tower, M)

> It's your ticket that they're working on. It's hard to think of them rather than just thinking of your survival and that's what it is a lot of times. It's you surviving what they do because if they do something wrong you're responsible for getting them out of it and there's different tolerances as to how far a trainer can let a developmental or developmental go. You know, you can only go to the edge of where you can recover. (TRACON, M)

Trainers sometimes hit the limit of their tolerance:

> You're chomping at the bit behind them, training them to do this thing that you think is number one priority, and they're doing what you think is number three or four priority and it's driving you crazy. You want to smack them. Your

> mind can take over. You've got to sit back there and try to be cool, or you're not being a good trainer. (Center, M)

> A couple of weeks ago I guess, one of the developmentals upstairs got their headset ripped off and thrown across the room. At night when it gets busy, the tension gets higher. Things go flying. And that's happened to me, too. When I was in training here I was getting towards the end of a checkout and it was very busy throughout the whole session. One of my trainers reached over and took all my strips and threw them all on the floor and told me to pick them up and that was kind of degrading, you know. Here I am ready for checkout and somebody's telling me to clean up somebody else's mess. (Tower, M)

The public degradation of the developmental by trainer and coworkers has a high emotional cost for them, reinforcing memory and lessons learned. Sometimes, a pilot will call in to complain. The public humiliation deepens the developmental's emotional commitment and ownership of mistakes. Publicly, developmentals get slammed for their mistakes, but after a training session, out of public view, a coworker may pass on a helpful suggestion about how something might have been handled differently or better: "I saw you do this; this is another way to do it." In some cases, they might give supplementary instruction, as in this example:

> I was having a hard time with this particular sector trying to train and get proficient and the supervisor one Saturday morning, we had a lot of people, sent me to the DYSIM lab. Which he didn't have to do. He sent me up there with two or three controllers from the floor and they just sat there with me, all day long, going over things and scenarios. It wasn't real you know, it wasn't talking to airplanes. But they were throwing all these real-life scenarios at me and that really helped a lot. (Center, F)

The scenarios add to developmentals' stock of knowledge of possible situations, allowing them to compare them and develop a repertoire of responses that match situations.

The Subtleties of the Craft: Dead Reckoning

Situation discrimination is built from many experiences, beginning in the early apprenticeship. It comes in many interactional forms: the crack on the head with the clipboard, the trip to the roof to watch departing airplanes, a group of controllers rehearsing scenarios on the DYSIM, the training session for the first experience of working in snow, coworkers' bashing of a performance. These informal and formal conversations continue to build collective tacit knowledge: knowledge that comes from the group. At some point in the apprenticeship, the trainer begins to sharpen and refine the apprentice's ability to discriminate among situations. Even as the developmental is striving to master the basics of maneuvering airplanes, the trainer begins teaching the subtleties of the craft. The trainer was trained by somebody else, who was trained by somebody else. And all of that experiential knowledge just gets passed along.

This phase of the apprenticeship advances the transformation from skill acquisition to expertise by instilling the ability to notice and give meaning to material objects and social conditions, a sophistication of sensibilities essential to the predicting, anticipating, enacting, and reacting to social and technical situations early to preserve safety. In the process, the trainer further alters the developmentals' physical and cognitive capabilities, reinforcing and deepening the transformation begun at the academy. Moreover, the trainer passes on the system goals of safe, orderly, and expeditious delivery of air traffic by teaching the techniques necessary to achieve them. The characteristics controllers cite as typical of "good air traffic controllers"—"multitasking, the ability to do many things at once—listening, talking, watching"; "type A personality, someone who thrives on stimulation and lots of activities, outspoken, aggressive, not afraid to take charge and go out there and do it and make a decision and stick with it"; "excellent recall, short-term memory is extremely high, long term, variable. If it's a significant event, we can recall it all"; "great vision and hearing"; and "common sense"—all are sharpened by the trainer.

Controllers must pass medical exams, including vision and hearing tests, before their academy training and annually after becoming certified. However, trainers refine developmentals' natural hearing abilities. Trainers teach them to recognize differences in pilot speech patterns and tone of voice and respond accordingly:

> I try to teach people to understand what they're listening to—how someone answers, how long it takes them to answer—those nontangibles we were talking about. Some people have a knack for it, some you try to point it out so that they develop a knack. I don't think you can just teach it. It's not like you can

> tell someone "They answered this way, so that means this." But you can point it out and say, "Listen to how they said that and now watch," they're not going to be able to accomplish what you just asked them. (Bedford, M)

This Bedford trainer conveys the difficulty of teaching tacit knowledge: he cannot give specific instructions for the "nontangibles," but he can point out what to notice. Beyond what the pilot is saying in the official phraseology, the pilot's rhythm and tone of voice have meaning. Developmentals learn to distinguish the voice of the professional pilot when in control—cool, crisp, rapid response—from the pauses and slightly elevated pitch of when something is going wrong; the hesitations and tone of voice of the "weekend warrior," or the less-experienced, irregularly flying pilot of a small plane. In recognizing foreign pilots by their accents, developmentals learn that some master enough English to use routine phraseology but do not have the language skills for nonroutine situations. So developmentals are instructed to speak more slowly, to repeat, to pay extra attention.

Dead reckoning includes predicting the behavior of coworkers on the same shift.[6] This, too, calls for ethnocognition and boundary work. A skill essential to coordination and safety, trainers teach situation awareness, or room awareness: listening to the voices of all the controllers in the room. One controller's troubling air traffic situation affects the traffic of the others. Attending to what coworkers are saying and doing helps controllers predict and prepare for what is coming their way. It is crucial for coordinating with others:

> Your trainer might say, "Did you hear that? Did you hear what so and so said?" or "Did you hear what the supervisor said?" Or "Did you hear what was coming over the speaker?" if another facility was calling you. If you didn't, it kind of . . . teaches you a lesson that, OK, I have to listen to the speaker a little bit more. It's another thing you just kind of get proficient at, listening to what's going on around the room. (TRACON, F)

> Sometimes people that don't need that [room awareness] for their jobs don't get to fine tune it. . . . When you first start this job, you're constantly told, "You've heard the people in training, pay attention to everything in the room." . . . If you sit in a factory and just put this piece on top of there, this piece on top of there, you might not be aware that the guy down there is five pieces behind or something. (Center, M)

Room awareness also depends on more indirect cues and intuitive, experiential knowledge. Because airplanes are handed off from one controller to another, "learning the people in the room" is essential to a smoothly coordinated system. This means understanding both coworkers techniques and how they react in certain traffic situations:

> It helps with the job when you know people really, really well, like we do. You know when you work next to somebody what they want, what you can do to help them. You know their quirks and idiosyncrasies. You know what they're going to do. You can adjust the way you work traffic, depending on the person you're working with. You know who's going to be able to help you out a lot, or not. (Center, M)

Indeed, controllers are known for their skill at silent coordination. They have to learn the people in the room well enough to grasp changes in their behavior that reflect their ability to manage their traffic. It is another form of situation discrimination in which observations are mediated by language—or sometimes the absence of it. Coworkers' voices may become rapid, the pitch elevated, louder, or they may become uncharacteristically quiet, indicating that they have become exceptionally busy, signs of stress for them as well as for pilots. Alternatively, quietness may mean a coworker is tired, so they need to "feed airplanes accordingly."

In addition, trainers alter and enhance developmentals' visual acuity. Trainers teach the "scan": look everywhere, keep eyes constantly moving, so you miss nothing. The scan works to correct the natural tendency of developmentals to develop tunnel vision. As a potentially problematic situation is unfolding, developmentals tend to concentrate on it, to the exclusion of other things going on. When that happens, they can miss something happening in another area of the sky or ground. It is crucial to identifying early signs of things going wrong and fixing them. This, from a radar controller:

> You're taught from the beginning, keep scanning, you've got to keep checking out every plane and go back to every plane. As soon as you take that handoff, you know where he's going. He goes off that flight path, you know he's off course. You know he's assigned that altitude, all of a sudden it looks different. It doesn't look right. You know his altitude is wrong. "Where's he going? Who gave him direct?" you'll ask . . . And you're always looking at your boundaries.

> You're looking for people coming in [to your airspace] that shouldn't be. It's something out of the ordinary you're looking for. (Center, F)

Part of the subtlety of the craft is learning to predict an airplane's response to a controller's instruction. Developmentals are taught to always have in mind a plan A, a plan B, and a plan C in case things do not go as expected. Pilots' voices are one indicator, but trainers also teach how to figure in the effects of wind and weather on aircraft movement and what different kinds of equipment can do: which equipment is slow to climb in the summer, which is slow to turn. Airline policy differences and destinations also matter, a point we should all remember when booking our next flight:

> There are airlines that will put out specific guidelines as to how the pilots will fly the airplanes. For instance, Northwest, United, and American are a little bit more conservative in what flight crews will do for you than, say, USAir. USAir gets a bonus for getting in early. Now it doesn't always pertain to the tower but for instance in the TRACON where if they have one overhead and it's a clear day, they'll say, "Can you make a visual approach from there?" and USAir says, "Yeah, I can do it," and they call it the "slam-dunk." They just dunk them in front of somebody that's quite a ways out. That's to fill the space up on the final so there's not wasted space. Where traditionally United or American or Northwest wouldn't do that because of discomfort to the passengers. You know, passengers might get a little bit nervous about an airplane making a steep dive or a turn, but USAir will do that.
>
> That's just an idea of what I mean, but you can also make reference to where the airplane's going [to predict how an aircraft will respond to an instruction]. If there's an American MD-80 going to Dallas, it's not going to take off very fast like a USAir going to Rochester because of the weight of the airplane, the amount of fuel, the people that are usually on for such a long flight. So those are all things that have to be taken into account. I mean it's not just an airplane that's going to do something; it's an airplane with probably a conservative flight crew that has a lot of fuel on board, or a lot of cargo to passengers so you have to take that all into account . . . and that's not something you come in here knowing. It's something you have to be taught. (Tower, M)

To develop the ability to predict an airplane's response, developmentals have to translate the standard information on the flight progress strip and/or the data block into a social interpretation of an aircraft's technology that is idiosyncratic, differing from other aircraft equipment, even that of the same model and manufacturer. With experience, this, too, becomes automatic because equipment becomes associated with the airline's schedule: controllers know, for example, without even looking at the strip, that the MD-80 departing at noon is going to Dallas and know from experience the equipment capability, amount of fuel and baggage and what that means for aircraft responsiveness.

The manual simulation problems at the academy "labs" laid a foundation for multitasking, but it is harder once on the job. Developmentals must simultaneously talk to and listen to pilots, enter commands on the keyboard and/or mark flight progress strips, coordinate actions with other controllers, and respond to the trainer, who is instructing, correcting, reprimanding, asking questions, and giving feedback in an ongoing, rapid-fire commentary as they work. Moreover, the additional tasks are more complex and numerous:

> You have to be able to talk and listen at the same time. Even while you're transmitting to an aircraft, somebody in the back of the room might be yelling something out. So you kind of have to be able to multitask, type on the keyboard while you're talking. It's just very difficult to learn to do. You'll key up to talk to the aircraft. You're talking and your hands automatically stop. Your trainer is always behind you: "Type and talk, type and talk." Because you have to, in order to keep up with the amount of traffic when it starts to get busy, you have to be able to do several things at once. It's another thing that you just kind of get proficient at, like listening to what's going on around the room. (TRACON, M)

Even the physical skills cannot be learned only by observation. Observation has to be mediated by language: "Type and talk, type and talk," or "Did you hear what he said?" The trainer instructs on the physical practices—scanning, typing, talking, listening—keeping at it until repetition replaces the old pattern of performing them separately with integration that is done automatically, "because you have to, in order to keep up with the amount of traffic," said one controller.

On Time and Timing: Safe, Orderly, and Expeditious

System needs are driving the training. The trainer is both master of the craft and rule enforcer, aiming for room awareness, adherence to the rules, strict attention to phraseology, and the conflicting institutional mandates of the system. Trainers teach the expeditious delivery of traffic by reinforcing the cultural scripts of speed, schedule, and efficiency. Trainers criticize moves that are a "waste of space" or "a waste of time." About the rules of separation, one said: "you can't run them any closer than this, but you need to run them as close as this or you're wasting space." At centers, fuel and time are saved if "you give pilots the most direct route whenever you can." Speed, schedule, and efficiency matter at all facilities, but expediting traffic is perhaps most important at towers, where traffic is most dense and departure and arrival times loom large in the minds of controllers. At Boston Tower, where controllers are handling two airplanes a minute, trainers teach that saving seconds matters and teach the techniques to do it. These trainers describe the challenges:

> I tell them it's much more efficient if you do it like this. If you do this you can get two airplanes out instead of one. They sit there and look at you and say, "Yeah, but it wasn't wrong; yeah, but I got one out." It would work that way, but that's not the point. The point is there's a different way to do it that's much more efficient. (Tower, F)

> I try to beat this into his head when I train him. You've got to come up with some phraseology. Have one way you're going to say it and say it the same way every time because you can say it four different ways. If you're going to say something like that four different ways you're going to stumble on it, so don't do it. It's the rules, get it out of your mouth on initial contact before you unkey the mic, then all that stuff's done. You've got things to do. There's no time to play around here, there's always something to do. (Tower, M)

Expedience calls for setting priorities:

> After you've learned the basics, you increase your working speed, then you learn your priorities. For instance, a guy that called me is ten miles out. He doesn't need to be cleared to land right now, he's not going anywhere, he's still going to fly to the runway. He wants to be talked to, but he's OK and your priority is to get this guy off the runway because you've got another guy on short final and you also have to load the runway because you need to clear a guy for takeoff and you have a helicopter calling you . . . and you have to do everything chronologically in the correct order of sequence. You talk to the wrong guy, you're not going to be able to do six things in this period of time, you're only going to be able to do three, and if you talk to a second wrong guy,

> you mightn't even be able to get one thing done. You hear me tell people up there, come on, come on, wrong transmission. Because of that wrong transmission, you didn't get another nine departures out cause you would've had about three or four more seconds, you would've been able to have another couple of thousand feet. (Tower, M)

This last trainer's example reveals that in order to set priorities, the developmental must *cognitively* be able take into account all activity in the airspace, predict the path of each aircraft, and have an embodied sense of the time each type of equipment will consume in its flight trajectory. The developmental has reached a sophisticated level of dead reckoning but is not yet an expert.

Controllers describe a moment when everything came together and they realized that they "got it." They say that the moment is captured as "when the light bulb goes on" or "when you get the big picture" or "getting the flick," all expressions used by controllers to describe having a mental model of the current and future positions of air traffic in a section of airspace. Controllers are able to visualize the paths of multiple aircraft in terms of positions, altitude, trajectory, and speed. One said:

> It's hard to explain. You understand all of a sudden how everything works. It's almost like being enlightened. But I just remember getting the picture. And it was like, from then on I couldn't fail. I knew I'd be able to do it. They call it the big picture. I don't know how to explain it, but it means you can work it now; you kind of understand how everything flows. It's almost like it opens up your mind and you can figure it all out now . . . Everything just clicks right and everything flows right and now you got it. You understand how you can manage all of it. And you never forget it. (Center, F)

The light bulb goes on when controllers have enough time and experience that the pieces they have struggled to remember fit together, having become so much a part of them that they do the work "without thinking"; it becomes "automatic" or "second nature." Formal institutional rules, local knowledge, and experiential knowledge combine, together constituting collective tacit knowledge. Controllers who have achieved expertise cannot describe how they have done so, but they can explain how they learned it:

> [For miles separation] I don't do mathematical calculations or measurement. If I need two and a half miles at this point, I just see it. When I came here, the

> hardest thing for me was getting used to that spacing, incorporating the radar, what's three miles, what's two miles, what's five miles. So a couple of things happened for me to make the transition. My instructor—I'm very up front, I said, "I don't know this, can you help me here?" And he sat down with me in front of the radar scope and said, OK, this is one mile, this is two miles, this is three miles, this is four miles, this is five miles. He would take me to the TRACON, we would do it up in the tower, or he would draw it for me during briefings. So then at home I would take time either to visualize it or just draw little airplanes and draw little lines, this is what four miles looks like. If it's a heavy behind a heavy, I need four miles and this is what it will look like. You've got to do a lot of homework. A lot of these things you've got to visualize in your head because that way it becomes second nature to you. So at this point a lot of things are just kind of second nature to me. I just look at them and I know this is not going to work, this is going to work. (Tower, M)

The controller's mind has adjusted to the equipment; the skill has become embodied as tacit knowledge. With repetition and experience, controllers achieve such deep absorption of details that the work is no longer the details: they grasp it holistically. Similarly, in his study of championship swimming teams, Chambliss wrote: "The acquisition of superior skill is really a confluence of dozens of small skills or activities, each one learned or stumbled upon, which have been carefully drilled into habit and then are fitted together in a synthesized whole. There is nothing extraordinary or superhuman in any one of those actions, only the fact they are done consistently and correctly."[7] Chambliss described how, once they have achieved this level of mastery, swimmers can concentrate on the extraordinary moves necessary during competition. Analogically, controllers' skills are acquired and practiced everyday so that they become an ingrained habit, an ordinary part of everyday life. Performing them is a mundane act, no matter how abnormal the situation. The result is a cultural system of knowledge consisting of habits of mind and scripts that minimize choice, making air traffic control "common sense," in controllers' view. When everything they've learned comes together, is common sense, they are freed up to concentrate on the essence of dead reckoning: anticipating events; coping with anomalies, spacing, technique, speed, and efficiency; and making a plan that innovates in the moment. Structure, culture, and agency combine.

At some point, the trainer says the apprentice is ready for the final checkride. How do trainers determine when a developmental is ready? Developmentals must have themselves, the air traffic, and the position under control;

be able to fit smoothly into the system, coordinating actions with coworkers, other facilities, and pilots; and be able to anticipate problems and correct them. A crucial indicator is whether or not they can use all the skills that constitute their common sense "to avoid a mess in the first place and get out of a mess whatever way you can." A trainer describes the moment and its significance:

> There was one point when he, as we call it, lost the picture, his plan. I know he did. I could see him groping. And then it all clicked and he got his picture right back. It may sound crazy, but this is almost what you look for in a controller. To see if he can lose it and stick with it enough to get it back. I've seen some, they lose it, the trainer takes over, and they just stand there, they can't get back in. Some days you are going to lose it. You just get overwhelmed with airplanes. You can't look at one of the other guys and say, do you want to take them, because they can't. So you dig yourself back out. (Bedford, M)

A second criterion, and equally important to the trainer, is are they sufficiently decisive, confident, assertive, and aggressive to handle the job? In other words, have they become the kind of people controllers identify as type A personalities? One said:

> You feel pressure on ground control when traffic's not leaving the airport on time, or, not on time, but you're just trying to get as many out there as you can and it's still backed up. You've got to get to the point, when I'm training somebody I always tell them you'll be ready for checkout when on ground control you feel like you could turn around to clearance delivery and say, "No more airplanes." Like normally, if you turn around and say that in training, people make fun of you 'cause it's like you can't handle it, blah, blah, blah. But when you're ready for checkout, you know that you're at the max for the airport, it doesn't make any sense to take any more airplanes, it's not a pride thing. It's like you turn around and say, "Look, no more airplanes," and people will kind of realize, hey, you've gotten over this thing of worrying about what everybody else thinks about you, you know what's going on. (Tower, M)

> I was quiet, you'd never believe it now, but I was quiet, I was shy. I had a very hard time telling someone no and/or what to do. In fact, they finally decided that they would check me out on my first radar position after I told the guy off who was training me. They weren't going to check me out. They were just going to keep me training until I "showed some balls," as they put it. And he was

> yelling at me one day, and I unplugged and I said, "OK, you do it, asshole!" and I walked out of the building. Next day they checked me out. He told me later they were waiting for me to get tough. I sort of realized later the yelling and screaming at you was so that you could handle it when it really got hectic. They couldn't tell otherwise whether you could or not. (Center, F)

The final check-ride is always on the most challenging position in the facility, and at a high-traffic time. The other controllers are paying attention to their airplanes but—courtesy of room awareness—also listening to the test. The trainer is plugged in behind, strangely quiet, watching. The supervisor is standing, too, evaluating performance with a checklist. For the developmental, this is a major life event. It is the culmination of a long, hard quest that will be resolved in two hours of drama. Whether new to air traffic control, a transfer from another facility, or a PATCO controller, everyone goes into the check-ride knowing full well that after all this time and work, failure is a possibility. The drama is greater if the developmental tried a check-ride on this position before and didn't demonstrate proficiency. When developmentals pass the final check-ride, both officially and unofficially they are elevated in status. They have "checked out," earning the title of certified professional controller (CPC). Their salary goes up and so does their ego:

> When I was training, I can recall sometimes sitting at my sector looking across the room at another sector that's busy. And I remember specifically one or two times thinking to myself, "Man, I really don't want to go over there. It would bother me." Shortly after my checkout date, I would walk into the area looking for that sector. Let me at 'em. I can handle anything. I'm as good as anybody in here. Let me at those airplanes. (Center, M)

They have survived a long training period, which they experience as a prolonged hazing ceremony. Many before them have failed. They feel, because of the rites of this particular passage, that they have joined the ranks of an elite group. The more difficult the position, or rite of passage, the greater the loyalty to the group the person has joined. It is not the FAA that earns controllers' loyalty; it is other controllers. The occupational bond that began at the academy is reinforced by the facility training hardships that they endure. The years of academy and facility training created an us-versus-them attitude in controllers toward the FAA administration that is a permanent dynamic in the organization in which controllers work.

4

Embodiment

The Social Shaping of Controllers

Becoming a certified professional controller doesn't end the learning period. Newly certified controllers must "season": learning to work on their own without a trainer, developing their own strategies for getting out of messes, accumulating more experiential learning. Moreover, the system effects that produce controllers continue to reinforce the lessons. The airplanes keep coming. It's constant repetition: the occupational skills, techniques, multitasking, memory, hearing and visual capabilities—all are reinforced daily. Further, training is never truly over for any controller. Controllers are required every month to update their proficiency with computer-based instruction, each time being refreshed on a different skill. Every change in the system—the introduction of new rules, alterations in technology, new types of aircraft equipment, changes in the airspace—requires renewed instruction and practice, and cognitive and physical adjustment on their part. Moreover, many controllers transfer to other facilities. They do not return to the academy, but they must go through facility training again to learn local knowledge: "this is how we do it here." Learning a facility is no easier the second time. Returning to apprentice status, with its ritual hazing and assignment as the coffee gofer, is hard. Yet they do move; some of them several times during their careers. Repetition of the learning and doing of the daily tasks maintains the physical, cognitive, and technical skills that are the foundation of their expertise.

Furthermore, the institutional mandates of the air traffic control system for safe, orderly, and expeditious delivery of air traffic are continuously reinforced. In each facility, controllers are reminded that the system mandates are important because they are embedded in local technologies and practices: facility routines, procedures, documents, and material objects continuously reinforce the cultural predispositions gained in their training. The system mandate for safety is reinforced in FAA statistics on operational errors: violations of the rules of separation that specify the amount of space between aircraft moving in the sky and on the ground. A detailed report of individual operational errors for the region is entered into a binder in each facility known as the "Read and Initial" file, required reading when signing in for work to ensure that every controller is aware of the who, how, and why of an incident. The facility response to individual operational errors is uniformly experienced as punishment by controllers, as we will see in chapter 5. The mandate for safety is conveyed through official accident reports. Also videos of mistakes are routinely used to upgrade controllers' skills in handling emergencies and learning from mistake, reminding controllers of the seriousness of the job, how easily things can go wrong, and the human consequences of error. Tape recordings of controller conversations with pilots and other controllers are reviewed for precision of phraseology and procedural conformity. Checklists are ubiquitous: a mainstay of technologies of coordination and control, controllers and pilots used checklists decades before they were implemented in hospitals and publicized in the media.[1]

The system mandate for expeditious handling of traffic—speed, schedule, and efficiency—is reinforced by FAA statistics reporting the number of operations and delays. Posted in prominent places in all facilities, controllers notice the number of operations an hour, day, or session. Seasonal differences in the count for their facility they know by heart. Those controllers who can handle the most traffic safely are accorded a prestige by their coworkers that those who handle less do not receive. Delay statistics publicly announce the failure to achieve the expedient delivery of air traffic. Centers escape direct scrutiny, however, because delay statistics are based on flight departure and arrival times at terminals. Controllers are well aware that the public holds them accountable for delays. They protest that they are wronged by this practice: delays are routine because airlines compete by scheduling many departures and arrivals at the same times. The problem is the system. As one controller told me, "Too many airplanes, not enough concrete."

In each facility, I witnessed—and was awed by—the intensity of the training and the repetition and reinforcement of the physical and cognitive skills as controllers trained and worked traffic daily after checkout. I began to wonder whether the package of attributes that constituted common sense had any effect on them outside the facility. I vividly recalled my conversation that first day at Boston Center with the controller in the cafeteria who, with his back to the TV while we were talking, repeated everything said on a program, then said, "We all can do that. My wife gets so mad at me," indicating not only that controllers shared this ability (which I later learned was situation awareness) but also that the skills necessary to dead reckoning went home with them. Even in those early weeks, the centrality of speed, schedule, and efficiency showed in their reaction to my research. After a month at the center, a surprised supervisor said, "Aren't you done yet? We are used to finishing something and moving on. We never take work home. This would drive us nuts." At Boston Tower, when I told the air traffic manager that I had been there three weeks so wanted to make an appointment with him to report in, he exclaimed, "My God, has it been that long?" For ethnographers, three or four weeks in a place is nothing. I was still getting my bearings. All jobs have tasks that must be performed on schedules and deadlines that have to be met. The differences between the tasks, time, schedule, and productivity in air traffic control and my academic training and experience indicated these meanings were culturally—and occupationally—specific.

As my fieldwork continued, I noticed that controllers at all facilities were so preoccupied with time and timely matters that their preoccupation was visible even when they were not on position, actively controlling traffic. The annual bidding for work schedules was a big emotional issue. In dead reckoning of a different sort, controllers planned ahead for the coming year, bidding for shift schedules and days off (their "weekend"). When traffic was low, they used the slow time to search the current schedule for times they want to swap a day or week with another controller. At breaks and shift changes, they must sign in and sign out, recording their times. They eagerly anticipate breaks by watching the break rotation list and cannot leave for break unless they have permission from the supervisor. Then they go on break for some allotted time—twenty or thirty minutes, sometimes more, depending on traffic volume and number of controllers on duty. Most take seriously their responsibility to take only the allotted time and get back so the next person can take a break, even carrying timers set to remind them. But not everyone is eager to conform. For some, the import of speed, schedule, and efficiency showed in their attempts to resist it.

Perhaps like workers in other regimented jobs where time matters, controllers steal time: finding small ways to reduce the time spent working traffic. Some are chronically, purposefully, and publicly late returning from break, infuriating their coworkers and supervisor because being late means someone else's break is delayed. Some organize their break times during the day so that when the last break falls near the end of the day they can go home early. They call this "getting timey." Calling in sick is another form of resistance. Stealing time is most common at the center, where traffic volume is uneven because of slow days and seasonal rhythms that do not require full staffing. At Bedford, where traffic has high peaks in good weather and virtually nothing when it is bad, the small number of controllers—fifteen, five to a crew—seems to make more salient the burden on others caused by stealing time. If it is busy, a controller even may return to the tower from break early. But when traffic is low, sick leave is high. When traffic unexpectedly surges, the supervisors hit the phones trying to find controllers who can come in. At Boston TRACON and Boston Tower, where traffic is always up and the demand for controllers doesn't let up, such resistance was less common.

In addition, controllers' collective preoccupation with time and timely matters also manifested in a shared sensitivity to aging. The retirement age for controllers is fifty-six. Most controllers I met during my fieldwork started at the center after the strike, so in 2000–2002, they were in their early forties. Yet aging and retirement were on their minds. Their definition of aging is occupationally unique: they define forty as the beginning of aging, because it is when they first begin to feel changes in their ability to work as many planes as fast as they did earlier. Having seen the effects on the controllers who proceeded and trained them—"the old guys"—the fortysomethings mimic their strategies. They begin to adjust their technique to match their ability, making decisions and moves earlier, not leaving things until the last minute, not "running them as close."

These activities occurred consistently across the four facilities. Because they generalized from working traffic to working nontraffic situations, it seemed possible that the skills and ways of being instilled in them to do the job were ingrained in them so thoroughly that they affected them outside the workplace. To pursue the possibility, I added two questions to my interviews: The first was "Do you think that any of the things you learned in training or do on the job as a controller carry over into your everyday life outside the facility?" Then, the more serious possibility: "Are there any ways that you think being a controller has affected you more fundamentally, changing you as a person?"

Carryover into Everyday Life

Reflecting about whether the skills and ways of being that are necessary to the job carried over into their everyday lives, all controllers talked about some work-related trait or traits that seemed to come automatically in their off-site lives. Some were minor. For example, they described how they sometimes slipped, using phraseology in conversations: "say again," "affirmative," "unable," always saying "zero" instead of "oh" when giving addresses or phone numbers. On a phone call, some said that they occasionally said "stand by" rather than "hold on, I'll be right back." When ordering pizza, they often end the call with their operating initials. These, they say, crop up "accidentally" and were controllable. "I never say 'niner' other places. People think you're a loser. A real geek." Air traffic controllers' brand of humor also stays behind at work:

> The sense of humor that people utilize, I'd say that stays pretty much in here, and to a limited extent comes out. Because a lot of people outside of the aisle wouldn't appreciate some of the jokes. I think this is a unique environment people work in. People need to have the ability to communicate amongst themselves. And unfortunately not all that is communicated in here really is appropriate to be communicated outside of here. (Center, M)

Our ability to present different aspects of our self to different audiences is illuminated in Erving Goffman's 1959's classic, *The Presentation of Self in Everyday Life*, in which Goffman draws an analogy between social life and the theater.[2] We actively construct our persona in social interaction with others. We have multiple selves, electing to show different aspects to our family than we do to work associates or former high school classmates.

However, the carryover controllers' emphasized as significant to them involved skills and ways of being that had become part of their nature, were habitual and hard to control: their physical and cognitive skills of hearing, multitasking, scanning, and memory skills—those core traits they said that all good air traffic controllers have in common. Moreover, the cultural understandings so integral to dead reckoning also go home with them, incorporated into the body as habits. In answering these questions, controllers drew analogies between tasks at work and tasks in everyday life. The following comments are not the isolated observations of the few controllers that are quoted but are representative of patterns in the responses of all controllers who answered the question. Some of their answers are collectively framed using the word *we*, indicating that they recognize the characteristics are not theirs alone but an occupational pattern. Collectively possessed and expressed outside the facility, these habits of mind attest to the durability of their socialization, embodiment, and cultural system of knowledge across social settings.

Controllers describe how they routinely deploy the physical and cognitive skills of dead reckoning in other activities:

> I think maybe the ability to do multiple things at one time, and I'm always scanning, going to the mall, restaurant, home, always investigating things. But at the same time I'm listening, or I'm writing, or playing a game on the computer. That's the biggest thing that I take home from the job. (Bedford, M)

> One thing I've noticed is short-term memory. You learn to keep things in your memory bank for a very short time, which I think hurts me in some respects. You just remember, OK, I gave this guy 3,000. I gave this guy 4,000, This guy is doing 190 knots. But you only have to remember that for a very short time because once he goes through your sector, you're done with him, you really don't have to remember him. So outside I need to pay attention more. I find myself writing a lot of notes. (TRACON, M)

Their everyday activities outside the facility were also marked by the occupation's cultural mandates for the safe, orderly, and expeditious delivery of air traffic. For example, driving is dead reckoning on the highway:

> Yes, especially driving down the road. You base it on separation. When you are working final, you look at the first aircraft and the third aircraft, and you build a hole and you put the second guy in there. I do that on the highway. I'm not looking at the guy right in front of me. I'm looking at the guy in front of him, because I know if he starts slowing this guy is going to be slowing right in front of me and I'll adjust to keep a normal flow. Going on a roundabout, scooting up so you can tuck in behind a guy, instead of leaving a big old space, wasting space. I mean, it's wild. (TRACON, F)

> As controllers we all tend to drive fast. I think all the people upstairs in the tower cab drive fast. The left-lane people sitting there and they won't pull over. All right, so I start looking for plan A and plan B. Or on two-lane highways, you see someone coming on the on ramp. You start to thinking, I did this a lot when I was training, learning how to do traffic calls. I'd take a twenty-minute ride on the freeway and start calling traffic. "Traffic one o'clock, two miles same direction, a Mercedes." Sequencing. Do you speed up to get ahead of them? Do you slow down to let them in? Do you move over a lane, so that they can move over, to keep your spacing? (Bedford, M)

The analogy between sequencing air traffic and road traffic draws on the application of similar techniques and the need to combine physical, cognitive, technical, and cultural knowledge to do it. Visualize material objects in space; separate but don't waste space; scan constantly to keep the picture; anticipate, prioritize, and predict; strive for speed, efficiency, time, and timing. The analogy is made powerful because mind and body join within the technology of the car, so the controller is inside, controlling self and car position in relation to the others, rather than outside the process.[3] Mind, body, and technology cohere as a result of practice.

Time and timing are consistent themes. At home, they prioritize and organize, developing a hierarchy of tasks:

> If something is not organized, it bothers me. And I think maybe that is something that comes from this job. This whole job is about organizing. You sit and organize airplanes all day long. (Center, F)

> I always think of things: If I do this, what are the consequences going to be or think of my moves on the outside ten steps ahead. Like if I'm working around the house, or helping somebody, I will do this, this, this, and kind of have a game plan. I put everything in a game plan, subconsciously, how I'm going to do it. Then I go ahead and tackle it. And if for some reason it doesn't work, then I'm—boom, I'm ready with plan B. (Center, M)

They describe striving for precision, accuracy, and avoiding loose ends:

> I'm extremely anal retentive, which in this job is good. In fact, they call me Al the Anal. When I was trained, I was trained by Steve "Mr. Perfect" Morris, Wolfgang "Little Hitler" Lietz, Bobby "Every Time You Screw Up You Owe

> Me a Drink" Donnelly, and P. J. "I Just Told You That Ten Minutes Ago" Clawson. And they were very hard on developmentals, for a very good reason. You can't coddle a developmental and say, "That's OK, you almost put that 737 and that L-1011 together, but that's OK." I think you have to be hard on the developmental so that you can weed out the weakest, you know—EV, eternal vigilance. You know, just hammer that into them. (Center, M)

> I don't like to leave things undone. I want to make sure things are complete.... I think my biggest criticism may be of other people, be it family members or not, I have somewhat of a low tolerance for incompetence and stupidity. I've always had a job that you had to be responsible for something, and a lot of it affected human life. So you at least couldn't go, "Oh, I'll take care of it later." You have to take care of it now. (TRACON, M)

Controllers' emphasis on efficiency doesn't always transition well to everyday life because other people's responses do not match the expectations controllers develop on the job. Time matters, and they are impatient with delays:

> Controllers have very little patience. You want something done immediately. Before we went to our new computer, years ago, you'd make a computer entry into the data block and it would take forever. And you'd make it again, and you'd make it again. And that was so frustrating. There were more people yelling and screaming in here because they weren't getting the instant gratification. But you've got to have that in this job. You've got to do that so it's done and you go onto something else. (Center, F)

> We are used to using very specific phraseology with pilots, brief and to the point. I tend to be short with people. You ask a simple yes-or-no question and you don't get the yes or no answer. "Come on, this is not a hard question," you're saying to yourself. It does get irritating. (Bedford, M)

Fundamental Change: Becoming a Type A Personality

When I asked about characteristics that carry over into their everyday lives, controllers described physical and cognitive changes—multitasking, hearing, scanning, memory, and ingrained habits that reflect a preoccupation with precision, order, accuracy, speed, schedule, and efficiency. When I asked whether their work and requirements of the job had brought about more fundamental changes that changed them "as a person," their answers were dramatically different. Without exception, they talked about personality changes that conformed to the characteristics that they identified as belonging to a type A personality. Both men and women said that they became (in declining order of frequency) more assertive and aggressive, confident, controlling, and developing big egos, thick skins, and decisiveness. Remarkably, of the sixty-five people whom I asked this question, fifty-three, or fully 82 percent, said that being a controller had changed them as a person in at least one fundamental way.[4] The general sentiment among controllers was that fundamental personality change was intrinsic to making it as a controller. Based on his experiences as a trainer, one expressed it thus:

> In the beginning it was like, I really know this job backwards and forwards. I can teach anybody. And then a few more years go by and a lot more [developmental] failures and I go, not everybody's for this job and not everybody could be trained and I think that's pretty much where it stands now because I think it's beyond the training. It's more about the personality change and that person that desires to do this has to find it within themselves to make a personality change as well as to accept the training. (Tower, M)

Men and women said they had fundamentally changed in nearly the same proportions (table 1). However, when it came to the extent of fundamental changes, more women than men described major change that resulted from the occupation. The responses of those who said that they had fundamentally changed seemed to divide naturally into four categories: major change, some change, little change, or change honed by air traffic control experience (i.e., if they already had the characteristic but air traffic control brought it out, realized, or increased it; table 2).

Of the sixteen people who experienced major change, five controllers, two men and three women, said the change had been total: they were no longer the same person they had been before. I quote four of them here, two men and two women, one from each of the four facilities:

> My personality to that point was a little bit laissez-faire. I didn't have a decisive personality in that I wanted to take control, I'm controlling my life, dictating my circumstances, I didn't have that. In other words, people around me created the circumstances of whatever happened to me . . . in my childhood I was more quiet, more accepting, introverted and not confident. I had to build my

Table 1. Responses of air traffic controllers to whether the job has fundamentally changed them "as a person"

	Men	Women	Total
Yes	34 (81%)	19 (83%)	53 (82%)
No	7 (17%)	3 (13%)	10 (15%)
Don't know	1 (2%)	1 (4%)	2 (3%)
Total	42 (100%)	23 (100%)	65 (100%)

Table 2. Extent of the fundamental change experienced by air traffic controllers

	Men	Women	Total
Major	4 (15%)	12 (63%)	16 (30%)
Some	23 (68%)	6 (32%)	29 (55%)
Little	2 (5%)	1 (5%)	3 (6%)
Honed by ATC	5 (12%)	0 (0%)	5 (9%)
Total	34 (100%)	19 (100%)	53 (100%)

confidence, I had to not necessarily accept, I had to question. I had to solve and I had to be very strong minded and very decisive and even persistent to develop all these areas, perseverance, persistence, confidence, decisiveness. You know, these changes came about . . . there was a day in training when I wasn't going to take any more. I decided to stand up for myself and from that day on I was no longer the same person. I started making all of the calls, making all the rules, dictating all of the circumstances that I could possibly dictate and that was it. I never changed, never turned back. And that was a total personality change. (Tower, M)

I am a much different person because of this job. I'm much more decisive, much more, I don't know if it's controlling, but, how would I describe this? I think I have a much stronger personality, I don't mean stronger in a better way, but stronger in an A type of personality way because of this job. [*Q: What is an A type personality?*] Like strong, strong willed. You know the person that is going to always speak up, the person in the restaurant who will send the dinner back, or the person not afraid to, you know, control. Say something happens, someone has a heart attack in the mall. The type of person that is like you do this, you do that. That's how I picture the type A personality. . . . A job is a huge

> part of a person's life, whether they admit it or not. And once you start telling yourself I've got to be this way at work five days a week, over a couple of years you kind of get the hang of it. . . . I'm this way because that's what I was taught at work. (Center, M)

> I am the person I am outside work because of who I am and what I do inside of work, definitely. (TRACON, F)

> One hundred percent. I'm not even close to being the same person I was before I came here. . . . I think everybody here changes. Everybody starts out a much nicer and gay person. And then in the end everything is so harsh in the training. And you have to dish it out as a like defense mechanism. When you get it, you dish it out. In the end, you get a person that's got a very thick wall around them. And they're not quite as giving and caring as they were when they first came here. You can see it. And people talk about it. There is a definite change in everybody to a certain degree. Some people more than others. And I think the women are affected the most. 'Cause they are the softest when they come in. And they become as hard as you know, as the men, for the most part. [*Q: Do you mind?*] I don't mind. [*Q: Is it a change for the better?*] Oh, no. (Bedford, F)

In contrast to those who talked about total personality change, most controllers identified certain altered characteristics. Although a greater proportion of women experienced fundamental change than men, men and women alike changed in the same directions. All of those who said they experienced fundamental change talked about developing one or more of the attributes that they identified as a type A personality.[5] Most people named one characteristic, but some mentioned two, and a few named a cluster of three or more traits. They indicated that the impact was transformative, even if they named a single trait. One trait could represent a major change, depending on the person's description of that trait's influence and significance in his or her life.

Here I provide some examples of traits controllers identified as accounting for fundamental change in persona. Traits are in order of declining frequency of mentions in controllers' responses. Differences in the number of mentions by gender are in parentheses. Sometimes more than one trait appears in a quote because the controller seemed to see the two as interrelated. As important as identifying the patterns in their experiences is preserving the idiosyncrasies and variation within a pattern. Consequently, I've selected more than one quote for a trait to show variation. The training and work can develop the same trait in people, but as individuals they remain very different: so, whereas confidence gained can give one person the courage to speak up in the grocery store, for another it turns into an ego problem. Instead of balancing quotes to show gender similarities and differences, sometimes the gender numbers alone make such a strong point that I thought showing both male and female was not as important as showing the variation within one gender:

Assertive or Aggressive (F = 11; M = 8)

> I was a pretty upbeat, good-natured person. I find myself more assertive, slightly aggressive in my interpersonal relationships now than I was four, five years ago. And again, I don't mean to be an asshole to people but, I mean, be assertive, be aggressive, things like that. To be successful at the bigger facilities, like the center, the TRACON, you almost have to have an almost dominant type A personality. I don't have one of those. But I'm starting to get that way. [*Q: What exactly is a type A personality?*] Very controlling, banging, it's my way or the highway, rah, rah, rah. You know, the kind of people you don't generally like normally. (Bedford, F)

> I am more of a bitch than I was before. I can be a lot more aggressive. When I was first training I was this little sweet, right-out-of-college, didn't know anything and all I ever heard from everybody was I wasn't talking loud enough on the frequency or being forceful enough with the other controllers when I needed something. "Be more of a bitch, be more of a bitch. You have to be a bitch." Well, now I'm more of a bitch [*Q: By that you mean you're . . . ?*] More aggressive, a little impatient, "this is what I want and I want it now" attitude. (Center, F)

Both men and women spoke of developing confidence after a history of shyness:

Confident (F = 4; M = 7)

> If anything, definitely the shyness factor. It has probably obliterated that part of me as far as one-on-one or public speaking. That's the biggest, definitely. Probably even the perfectionist thing helped with that. It helped me to raise standards of how I do things and my expectations of myself. Because when you come in, if you haven't noticed it already, people ex-

> pect a lot of you just because of the sensitivity of the job here, what's involved with a potential mistake. (TRACON, M)

> Before I had this job I was very shy. Like when I was younger, just to walk up to somebody and ask them a question in the grocery store, I never would have done that. And then working with people who are always teasing you and joking with you, that kind of opens you up because you can't just sit there and not get back at them when they're saying things back and forth. I don't want to say really outgoing because I don't party or anything, but once I had this job, you know, going up to somebody in a grocery store and asking them a question, or asking a salesperson at the store isn't as hard as it used to be. (Tower, F)

On controlling outcomes as an effect of the job on them, the man was one of several men who spoke about this trait but she was the only woman. Other women spoke in terms of aggressiveness or assertiveness:

Controlling Outcomes (F = 1; M = 6)

> Eight hours a day, five days a week, I'm telling people what to do. I have a thirteen-year-old son. Do I need to go any further? . . . I equate you to pilots. I ask you the first time, I tell you the second time, and then the third time you're going to do it one way or another. (Center, M)

> Yes. I used to be very soft spoken and just, you know, oh well, there is a problem. Hopefully things will get better. But I am not like that anymore. I'm just like, they will get better because I'm going to be a pain in the ass about it. I'm tired of this bullshit. In that kind of way yes, I guess I'm a little more forceful and don't just say, "Maybe it will change." Now I'm like, "It will change. I'm going to do something to make it change." (Center, F)

Whereas both men and women expressed an increase in confidence, only men discussed it as a fundamental change in ego:

Big Egos (F = 0; M = 6)

> Maybe arrogance. That's for sure. People have to have things their way. Definitely comes from the job. I think it's probably a requirement. I think

> once you get to the point that you can do it, then you have an ego. I'm as good as anybody here. I didn't always feel that way. (Tower, M)

> It can be a hindrance, too, what we do. We as air traffic controllers, we talked earlier about self-confidence, sometimes that can turn into an ego. It really can. And that's probably not a bad thing in some cases, but your interaction with people has to be cautious because you're expecting the same type of response out of everybody (as with pilots), and you don't get it. Doing this, the environment that you're in can really let you think you're a little bit better than you are, and you have to be careful about it. (TRACON, M)

Several commented that the harshness of the training—the hammering from the trainer and coworkers during the apprenticeship—developed their ability to tolerate the heat and stress of the job and criticism generally. Others described it as "becoming hard":

Thick Skin (F = 3; M = 2)

> I was very quiet when I got here and stuff and I turned around and became completely different once I checked out . . . you have to be able to take criticism, even this yelling that used to go on and somewhat still does, it's an intimidation. (Tower, M)

> I wasn't supersensitive, but coming from the mental health field I was sensitive. And I think I took things personally. I've grown not to. [*Q: What happened?*] You just get told what you are doing wrong when you are training . . . that's mostly all you hear. Your training sheet is everything you did wrong. You've got like ten things written up that you did wrong. So you don't get the positive in here . . . It's gotten nicer and nicer every year in the area but by now you have gone through all the years of the do this, turn this plane, in your training. By the time you are done, you have a thick skin. (Center, F)

The capacity to act quickly and decisively was for some the most significant change, but facility differences mattered in the way controllers developed the trait. So at the tower (and TRACON), it was driven by the relentless pace of arrivals and departures; at the center (and Bedford), it was more as a result of the situation or problem solving:

Decisive (F = 1; M = 4)

> Here you have to make decisions rapidly. And the minute you talk to an airplane, you've got to be able to make the decision right then. You have no time to think about it. You have to turn them, climb them, clam up, then get rid of them. You have no time to say, "OK, well, you know, where is he going to fit in." You're like, the airplane comes off the ground, "Fly heading, climb and maintain." Next airplane, "Fly heading, climb and maintain." Because they keep coming, keep coming, and they're not going to stop. (Tower, F)

> A lot of us controllers are obnoxious. We're used to telling people what to do, you know. We're used to making really quick judgments, they are actually judgments about situations and solving it and this is it . . . somebody told me once I was the most opinionated person they'd ever met. I took it as a compliment. (Center, M)

A Cultural System of Knowledge: Expertise, Embodiment, and Ethnocognition

Controllers come to their profession with different personas, lifestyles, socioeconomic backgrounds, and different skill sets, talents, and interests. Nonetheless, they recognize that they have traits in common. Although the traits they name are individual psychological, cognitive, and physical attributes, they are system effects: products of the air traffic control system in which they work. Regardless of the individual differences controllers brought to the occupation—and still have—they say that these traits are common to all controllers: great vision and hearing, the ability to do many things at once, excellent recall, a type A personality, and common sense.

Common sense, though, is a cultural system of knowledge comprising institutionalized rules and procedures, standardized throughout the system, that are acquired at the academy; local knowledge is gained in the facility, as formal knowledge from the classroom or experiential, intuitive knowledge taught by the trainer and absorbed from the social conditions of work. Also, throughout the training, controllers develop the skills to meet the system mandates of safe, orderly, and efficient delivery of air traffic. Daily, these cultural scripts of the system and the physical, mental, and technical capabilities necessary to the task are inculcated and reinforced. With repetition over time, these various and layered forms of knowledge become sufficiently mastered that decisions can be made and tasks can be executed automatically. Controllers develop tacit knowledge. Crucial to managing traffic volume and complexity, tacit knowledge frees controllers to notice the nonroutine and spontaneously craft refined strategies in the moment to take care of an anomalous incident.

Polanyi described tacit knowledge as our ability to perform skills without being able to articulate how we do them.[6] His well-known analogy is the skill of riding a bicycle. He pointed out that no amount of reading or study will enable a novice to ride immediately, and the skilled rider is usually unable to describe the dynamics involved. Merleau Ponty explained how skills and knowledge become incorporated into the body as habits, the perceptual habit becomes a motor habit, so tasks can be enacted without specific decisions being made.[7] Tracing the training of air traffic controllers has shown the foundations of tacit knowledge: how it is acquired, how it becomes embodied, and its composition. Contradicting the idea that the invisible can't be studied, this chapter demonstrates that although tacit knowledge cannot be explained or written down, it is codified in local practices and communication. Consequently, it can be observed in what people say and do.[8]

It is uncanny how closely the chronology of air traffic controllers' training follows the process of skill acquisition to expertise developed by Dreyfus and Dreyfus, who incorporated both embodiment and development of tacit knowledge into their model.[9] Notice, however, that Dreyfus and Dreyfus treat expertise as a characteristic acquired by individuals. In contrast, taking a situated action approach shows how much the layered social context of training and experience matters.

For Dreyfus and Dreyfus, the process begins with the novice stage, during which learning is rule based and action is slow because the struggle is to remember and master the rules. Next, with experience coping with real situations, the beginner becomes an advanced beginner, having sufficiently mastered the basics to be able to start noticing relevant details of the situation. The third stage is competence, at which stage experience has expanded such that the learner has the ability to more finely discriminate, but the number of situations the learner is able to recognize is so great that the situation feels out of control. To cope, the competent learner begins to prioritize and make a plan. They must develop their own additional rules—rules of thumb—sorting through the different situations they encounter. They regularly experience both success and failure, thus practicing the skill has emotional costs that lead them either to drop out or to become emotionally committed to the learning process and so they stay.

The fourth stage: proficiency. Learning from mistake is essential to progress. It reinforces some responses and eliminates others. The learner shifts from operating on sets of rules to practices that discriminate between situations and are associated with appropriate responses. They begin responding on the basis of intuition and habit, without having to reason out a response to every situation. They acquire more speed at the chosen task and so can accomplish more.[10] This is the beginning of confidence. But the proficient learner still has to decide what to do. That is, action is not yet fully intuitive and automatic.

The fifth stage is expertise. With experience, the learner is capable of situational discrimination, associating different tactical decisions with different clusters of situations. The learner has a repertoire of the immediate, unreflective, intuitive responses to each situation that is characteristic of expertise. The relation between perception and action, both motor and intellectual, becomes a matter of intuition or habit, so actors carry out tasks with great speed, freeing up the mind for other work-related things that might occur in the moment. In achieving expertise, both motor and intellectual skills have become embodied. Mind and body join to produce the action.

The analogies with controllers' achievement of expertise described in chapters 3 and 4 are clear. However, the process goes beyond the Dreyfuses' narrow focus on individual motor and cognitive skills. Expertise is the acquisition of the cultural system of knowledge essential to working the air traffic system: techniques, beliefs, and ways of thinking and doing, talking, deciding, seeing, and behaving that constitute ethnocognition and boundary work. Consider, first, the role of the immediate social context. The development of expertise necessarily includes the embodiment of the technologies: the controller must adjust to them, mentally and physically, so that use is informed habit; like driving and dead reckoning, mind, body, and devices merge so that tasks are done automatically. Similarly, controllers must adjust to the architecture, spatial arrangements, and material objects of the workplace. Moreover, a subtle aspect of the transformation from skill to expertise is the role of the embodied emotions associated with their tasks. Benner, in her study of nursing, found that emotions—taking risks, fear, the remorse of mistake, and the highs of accomplishment—are crucial to commitment to the learning process.[11] For air traffic controllers, these emotional experiences begin at the very first academy training experience, weeding out many. For those who remain, emotional highs and lows continue throughout the training process and the career, affecting not only commitment to the task and new learning, as Benner showed, but also the shared experiences bind them to their occupational group.

For air traffic controllers, the socialization process plays a major role in the production of professional expertise. Collins and Evans point out that expertise is necessarily interactional and language mediated.[12] For air traffic controllers, learning from books of rules and standard procedures is the foundation, but the trainer enhances the developmental's natural physical abilities of perception: the scan expands range of vision, overcoming the natural tendency toward tunnel vision; room awareness produces acute hearing that enables controllers to take into account multiple actors located in various positions in near-range physical space. In addition, the trainer alters *perception itself* by changing the meaning of events in the sky and on the scope that controllers must act upon by pointing out what to notice ("Northwest [Airlines] turns like pigs"; "Listen to how they said that and now watch, they're not going to be able to accomplish what you just asked them") and what they miss ("See what you're about to do to yourself with that Learjet?"). Although Dreyfus and Dreyfus acknowledge the learning of subtleties and development of increasing refined categories of subclasses of like situations, here we see the mechanism at work behind such refinement: the interactional, language-mediated, socialization-altered perception that enhances controllers' interpretive work.

What transpires between the apprentice and the trainer is crucial, but it is not all there is to it. Collins, for example, identified the importance of "collective tacit knowledge,"[13] or an unspoken set of cultural understandings about how the work is to be done. At the FAA academy, controllers are trained in cohorts by instructors, and language mediates every learning experience. Even out of the classroom, they are learning from one another. Once in the facility, in on-the-job training and apprenticeship, developmentals absorb the social conditions of work—among them, how other controllers work traffic, use the available technologies, pass on history, help and/or mock, and express capabilities and preferences in coordinating traffic, standing in the group, and reactions to pilots and traffic conditions—all this constitutes collective tacit knowledge that affects perception and thus not only how developmentals do the work but also the meaning of the work and its concepts to them.

Finally, consider again how system effects play out in controllers' interactions in the workplace. Although we correctly think of expertise as belonging to individuals, the process of developing expertise is explained by the larger social context as well as the immediate social context of work. Expertise includes goes beyond the body and the practices of the individual. Eyal observed that expertise cannot be fully understood by the practices alone but must incorporate the social, material, spatial, institutional, organizational, and conceptual arrangements that shape them.[14] Specifically, "Expertise is composed of a network of other actors, devices, instruments, concepts and institutional and spatial arrangements distributed in multiple loci yet assembled into a collective coherent agency."[15] In the workplace, we can see how all these conditions combine to produce expertise. In writing about the occupational *habitus*, Bourdieu pointed out that a person's position in an occupational structure results in predispositions and habits of mind that have an effect on choices and actions in everyday life.[16] Further, he pointed to the larger structures that affect the occupation—like the externally generated pressures for safe, orderly, and expeditious delivery of air traffic that originate in the air traffic control system's competitive, economic, technological, and political field. It is these effects on the system that drive the nature of organizations and of work, and thus the shape and content of expertise. As we have observed, these factors, too, can become embodied to the extent that they can change the physical, cognitive, and personality traits of controllers. Such shaping by the structures we inhabit does not preclude us from actively constructing and choosing in our daily lives. However, they do create patterned tendencies in people who occupy the same structural position.

It is well known that all organizations attempt to recruit and socialize their members to effect coordination and control so that employees know what they are supposed to do and do it. Some organizations devote more resources to the process than do others. Goffman describes a "total institution" as one in which members' lives are routinized and regimented in the company of others of like kind.[17] He uses as examples prisons, mental hospitals, boarding schools, monasteries, the military—institutions where individuals reside around the clock and the institution provides all their basic needs. Goffman concludes that the most important factor in forming the member is the institution, not individual characteristics, so the reaction and adjustment of the individuals to one institution is analogous to those members of the others.

In this chapter, we see how, to greater and lesser extent, controllers are transformed by their job. Air traffic control is not a total institution in Goffman's sense. Controllers have days off and go home at the end of their shift. However, the standardization of the air traffic control system leads to a regimented, extended training period and a standardized work practice that ensures similar results: the most important factor in forming the member is the institution, so the reaction and adjustment of individuals in one facility is analogous to those members of the others. This outcome ensures that controllers will behave predictably in a risky system that requires a standardized response and coordination. However, organizations vary and system effects on individuals will vary. Air traffic control is probably an extreme case of system effects because, like the military, it is a large complex bureaucracy, internationally standardized thus requiring coordination across long distances, with huge amounts of resources devoted to training and retraining. The system's need for standardization, predictability, and coordination is high: the secret to safety in air traffic control is controlling the controller.

The quasi-total institutional quality of air traffic control may be the extreme example, but the extreme case allows us to identify system effects otherwise invisible: the acquisition of ways of thinking and being that go beyond simple adjustment to the place and the task. Bensman and Lilienfeld have documented the relationship between craft and consciousness: how occupations, their techniques, and materials can affect individuals, including their cognitive and physical abilities.[18] We all have our own experience that verifies that occupational training and practice change us all, perhaps building on characteristics already there or developing characteristics where none existed (or were not recognized) before. However, air traffic controllers are changed in both cognitive and physical capabilities that carry over into everyday life; moreover, most describe fundamental personality changes in directions that match the needs of the job. Training is a transition, a period of becoming that, for those who persist, changes their essential nature.

I am not claiming that air traffic controllers are all alike, or that they always respond in such uniform ways thus functioning as standardized cogs in a standardized system. Although the air traffic control system produces common traits, the extent or presence of these qualities varies across individuals. Moreover, controllers nonetheless retain their uniqueness, their individuality, and their ability to innovate. At the same time that they conform to the mandates of the air traffic control system, two controllers will never respond to the same traffic situation in identical ways. Further, the system is standardized but incompletely so. Each facility has its own local peculiarities—characteristics that arise from geography, airspace, fleet mix, traffic volume, runway layout, winds, weather, and changing seasonal conditions that standardized information and practices alone can't cover. These irregularities demand local knowledge from maps and manuals as well as cultural understandings and experiential learning that can be passed on only by a trainer and other controllers in both formal and informal venues. Moreover, a globally standardized system cannot handle every eventuality in a standardized way. As one controller said:

> You need neat and orderly, but you also need flexibility, and you also need the ability to just pull something out of the air. And you have to because there are going to be times in your career when neat and orderly is going to go out the window and you're going to have to pull an idea out of your head. And it had better be there. (Center, F)

Adjacent to the cultural system of knowledge that all controllers acquire is experiential knowledge acquired only through the improvisation that they generate when confronted with the unexpected.[19] Part III of this book shows air traffic controllers at work, demonstrating that the preservation of the uniquely human, intuitive, and creative in the midst of the routine and systematic is equally crucial to the mandates of the system.

PART III

Boundary Work

Airspace, Place, and Dead Reckoning

When controllers talk about their work, the word *personality* comes up a lot. They say how important it is to "learn the people in the room"—the work styles and traffic preferences, skill strengths and weaknesses, strategies, emotional triggers, and physical tics of coworkers assigned to work traffic in the same airspace. Knowing the personality of each person in the room, they say, allows them to predict that person's response to a traffic situation and organize their own work accordingly. But controllers also refer to the personalities of controllers at other facilities. One day during a quiet traffic moment at Boston Center, the controller whose session I was observing started to explain to me something he had just done. "Bradley [TRACON] has a funny personality," he began. When I asked what he meant, he described how controllers at Bradley want traffic to come to them, how the responses to requests by Bradley's controllers differ from those of other TRACONs in the center's airspace, as well as idiosyncrasies that affect how he sends them airplanes and how they send them to him. Controllers also refer to the different personalities of crews and teams in their facility. Commercial airlines, too, have personalities. At Boston Tower, USAir would do the "slam dunk" and Northwest would not because of company rules, procedures, training, and rewards that shaped the work practices of flight crews in characteristically liberal or conservative ways. Whether referring to individuals, other air traffic control facilities, or commercial air carriers, by the word *personality,* controllers referred to how the work is organized: ways of moving air traffic that are characteristic of a person, crew, place, or organization. Knowing personality is a major factor in the prediction essential to dead reckoning.

The airspace a facility "owns" bestows it with the distinctive characteristics that form its personality. The sky is standardized, divided by invisible lines—routes and intersections, identified by names and numbers, like a road map. These divisions are man-made and artificial, but they have system effects that are material, geographic, cultural, social, and symbolic. Each of the twenty-one regions owns a piece of the nation's airspace, and each region's airspace is further divided among its facilities, and within them, divided into yet smaller pieces. At radar facilities—TRACONs and centers—these smaller pieces are known as sectors. Sectors can be finely layered, divided by altitudes, often with smaller internal "shelves," that are owned by the individual controller who is assigned to work that sector of airspace. At towers, the airspace goes down to the ground. It is divided into pieces by the movement of airplanes within it: in the sky, on the runways, alleyways, or pushing back from the gate.

Ownership is a central safety principle of the system. As an aircraft moves through the sky, it cannot cross the boundary between one chunk of airspace and another without the owner's permission. The artificial boundaries in the sky are real in their consequences: they materialize in real boundaries on the ground. As airplanes cross the boundaries in the sky, they also cross the physical boundaries of formal organizations in the air traffic control system—towers, TRACONs, centers. In each facility, the airspace it owns is reflected in the division of labor, architectural design, and technologies of the workplace (fig. 12). Ownership rests with the controller holding responsibility for each stage of an airplane's movement, the responsibility shifting from controller to controller as the airplane travels across boundaries. Airspace determines place, and place affects the social organization of work, technology, and the work practice of the controllers who own it.

The system effects of this simple principle—the link between airspace, place, and dead reckoning—on controllers and their work are all encompassing. Standardization exists for safety and to make coordination possible. However, the relationship between space and place produces variation across this highly standardized system. Each type of facility—tower, TRACON, or center—has similar interior architectural layout of controller workspace, division of labor, technology, and administrative arrangements as do others of the same category. For example, all en route centers have a large airspace requiring a large workplace where more than two hundred controllers come and go; its high-speed, high-altitude traffic requires radar, and controllers are totally

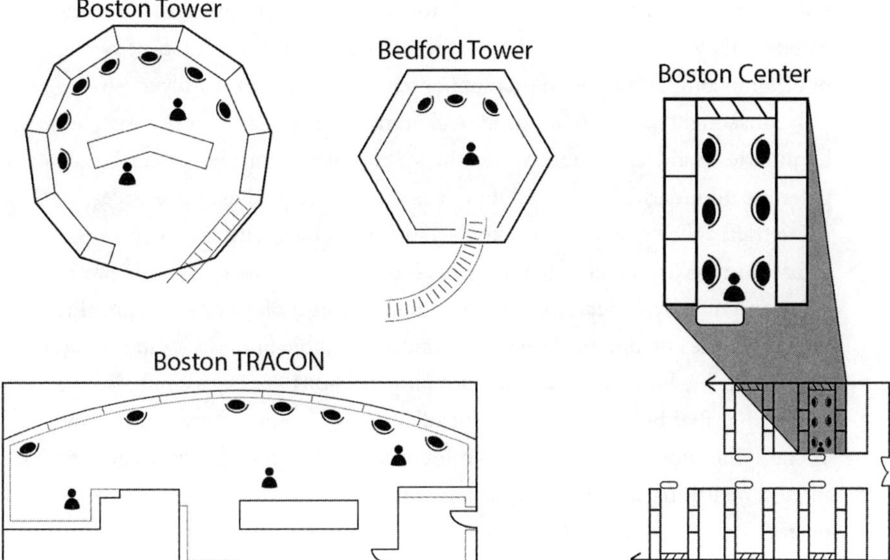

Figure 12: Facility architectural layouts, 2000
Illustration by Noah Arjomand

dependent on it. TRACONs have the middle altitudes between centers and towers, crossing traffic, ascending and descending between the other two, and those passing through its airspace, more controllers than a tower but many fewer than a center, and total radar dependence. Towers have less airspace, low-altitude traffic climbing and descending the airport, and thus fewer controllers, smaller workspace, and different technologies, including radar, but they are less radar dependent because they watch the sky and runways in real time. Nonetheless, in large towers, radar is essential in bad weather and high traffic as a redundant backup for what they cannot see.

But no two towers or TRACONs or centers are alike. That's because the peculiarities of the airspace a facility owns—the fleet mix, traffic volume, geography, weather, region, number of cities it serves—make working traffic at one facility different from all others of the same type. As airspace varies, so does the place, the task, and the social organization of work. Even within the same facility, the section of airspace one controller owns at the moment is different from the one owned by the controller in the next position, which requires different strategies of working traffic. Standardized language, formal rules, and local procedures dictate how controllers must hand off an airplane across each airspace boundary. But therein lies the challenge: how to move the traffic across sections of airspace that, for reasons of differences in air traffic patterns, traffic volume, altitude, weather, clashing controller airspace needs, or other incompatibilities, permanent or temporary, do not readily mesh.

Consequently, much of dead reckoning is boundary work. Airspace boundaries build vulnerability into the system. It is controllers who, by negotiating boundaries between different airspaces, bridge these differences. To move traffic, they engage in two kinds. First, in keeping with the early navigational form of dead reckoning, they must predict the positions of objects in space and time by deduction, without benefit of direct observation, managing the trajectories of aircraft across the artificial, invisible boundaries in the sky. These invisible boundaries include not only divisions of airspace but also the required spatial boundaries that controllers must preserve between moving aircraft. The latter, governed by the rules of separation, are the equivalent of a three-dimensional assured clear distance ahead on the road and necessary to prevent collisions. Second, for controllers to manage the trajectories of aircraft across the invisible boundaries in the sky, dead reckoning takes on the dimensions of a social physics: controllers must predict the actions and reactions of other controllers in facilities on the ground that use and own the airspace. Both these forms of boundary work involve spatial thinking—thinking in and across time and space—and are essential to coordination local and distant in an interdependent, standardized system that, when examined, proves rife with variation. The interplay of standardization, interpretive work, and improvisation becomes an everyday part of controllers' cognitive and material practices, as these chapters demonstrate.[1]

Keep in mind that what controllers think of as personality is itself a system effect. The personality of a place, for example, originates with its airspace, but the airspace design is a result of contested decisions of FAA headquarters, the airline industry, government, and NATCA. Boundaries are always contested.[2] Decisions about the airspace, including its boundaries, ownership, and use, and decisions about place, including resource allocation, architectural design, technologies, hiring policies, and the consolidation or expansion of parts of the system—all these are the outcome of political contests in the air traffic control system's institutional political, economic, and technological environment. In chapters 5 and 6, the impact of these kinds of external conditions on the facility structure, technologies, and processes are visible in the social organization of work and in controllers' social, cultural, and material practices.

The two chapters in this part consist of four ethnographies that follow controllers at work, showing how they engage in dead reckoning, deploying expertise, common sense, ethnocognition, and and boundary negotiation in the midst of system variation. The comparison exposes this variation, revealing the "personalities" of Boston Center, Bedford Tower, Boston TRACON, and Boston Tower. The four facilities operate on the standardized rules and procedures of the international system, yet the ethnographies show how different each place is, and therefore the job at each place. I present them in the order of my fieldwork as I moved from place to place between the physical boundaries of the system. Unintentionally, because of the timing of each facility's permission to start the research, rather than some extraordinary prescience about how insight productive this path would be, I went from radar to tower, radar to tower, and from dark to light, dark to light. This ordering made the differences and similarities between them vivid and compelling to me. To produce the same effect, readers also go from dark to light, dark to light.

Chapter 5 addresses Boston Center and Bedford Tower; chapter 6, Boston TRACON and Boston Tower. Although both chapters pair a radar facility with a tower, Boston Center and Bedford Tower are extreme opposites, so I present them separately in the chapter. In contrast, despite great variation, Boston TRACON and Boston Tower's contiguous airspace and colocation at the Logan International Airport give them many commonalities. Consequently, I begin chapter 6 by combining them, showing the effects of their common terminal location, and then separate them to show how they enact ethnocognition in dead reckoning. I then join them again to show the boundary work between them because that is the primary boundary that each must negotiate.

Following controllers at work shows how local knowledge varies by place. It reveals the multiple contributions of human cognition to air traffic control: what controllers do that technology can't replace. The craft of dead reckoning consists of predicting the positions of moving objects across the boundaries in the sky by deductive calculations based on technological representations of those objects, not the objects themselves.[3] As they work, we see controllers' meaning making, technologies, and formal and informal practices within the larger social context. The ethnographies show vividly how external events alter the air traffic control system and thus have an impact on controllers' dead reckoning. They reveal how the institutional and cultural rules of the system are enacted and reinforced in daily work practice, yet they also show improvisation to be central to that daily routine. In the workplace, we can see how controllers recognize and correct anomalies "to avoid messes"; accomplish silent coordination; bridge the boundaries of the system; and, when standardization and technology fail, "pull an idea out of the air." Collectively, the four ethnographies show how the relationship of airspace, place, and the task of dead reckoning produces distinctive cultures at each facility.

System Effects: Culture, Ethnocognition, and Distributed Cognition

In *The Interpretation of Cultures,* Geertz wrote that cultural analysis provides a detailed context that shows the meaning of events, context, and interactions to individuals in a social situation.[4] The distinctive cultures of each facility become palpable in these chapters. Culture is reflected in every interaction. Even in the most casual behaviors, such as what controllers do on their breaks, we get a sense of the experiences and meanings of work for controllers at each facility. These cultural differences are experienced by controllers throughout the system. As a controller who had worked traffic at Boston Tower then moved to Boston Center put it, "The work changes when you change facilities. It causes you to be a different person." Local practices shape not only collective tacit knowledge, ways of acting, material and interpretive practices but also collective ways of being. The notion that culture is "the solutions people arrive at as they resolve the problems that they face in common" rings true.[5] The solutions become cultural because the experiences of work are coordinated, repeated, collectively experienced and common to the group, creating a facility subculture that is a local variation of the wider culture of air traffic control acquired during controllers' training.

Why do these cultural differences matter? First, they create a boundary problem. To coordinate traffic effectively within a facility, new controllers must master not only technical and material skills but also the embodied cultural understandings about talk, signals, and how the work is done—within the facility as well as in coordination with other facilities, new to them.[6] As these chapters (and the rest of this book) show, when system-mandated changes alter airspace, architecture, or major technologies, in the short run they disrupt work: controllers must adjust their embodied cognitive, physical, and material practices. Thus, during a major system change, a place undergoes not only a work transition but also a cultural transition that impacts ethnocognition. Air traffic doesn't wait for controllers to make these adjustments. When technologies of coordination and control—rules and procedures, material objects, devices, and architectural arrangements—are new, initially tasks cannot be done automatically, without thinking. In the short run, the impact on controllers and their work is uncertainty and stress, no matter how much they rehearse in advance. Most compelling are the before-and-after examples from Boston Center and Boston TRACON showing how historically, system-level changes to improve the workspace, the technology, and the accuracy and efficiency of controllers work dramatically altered the work arrangements, creating stress in the short term and altering the culture within the same facility.[7]

Second, these chapters show how *ethnocognition*—controllers' embodied interpretive work and cultural understandings—*affects distributed cognition.* In his classic *Cognition in the Wild,* Hutchins's breakthrough was to extend cognition beyond the individual mind to the interaction of mind, body, and material objects in the immediate environment.[8] Memory, for example, is not simply in a person's head; it is distributed between people, technologies, and the organization of the task.[9] Most research on distributed cognition has examined interaction in small settings, such as teams of airline cockpit crews or navigational teams in one work location. Although air traffic controllers work in small spaces, these ethnographies reveal how distributed cognition works in a complex socio-technical system. They show that many factors outside the immediate environment affect distributed cognition.[10]

Experience—socialization and enculturation—is a factor.[11] Controllers bring symbolic meaning into their reading of the actions of technologies, coworkers, airplanes, and other material objects. Also, they actively modify the immediate environment to facilitate recall, using practices passed on from their trainers that also strategically send signals to others in the room about what they are doing. A flight progress strip is offset from others in a strip bay. A pencil becomes a memory device when a controller runs it down a series of strips in a strip bay as a reminder of the traffic to come, and the adjacent controller notices too. They make pencil marks on a flight progress strip in shared symbol language to record any changes in the plane's position in space for others see. When a strip passes from one controller to another, the strip acts as a physical reminder of who has control of the airplane to the controller and to the group.[12]

Moreover, material objects act. Buildings are not just scene-setting descriptive background. Architecture, design, and spatial arrangements actively shape the social organization of work and dead reckoning. Place itself—physical structures and their internal arrangements—channel interactions and communications in some directions and not others.[13] Design and layout are not neutral in effect. Place "works" in each facility to influence relationships and exchange. Notice in these chapters that across the four facilities, the spatial and architectural arrangements of the work place reinforce the routinization of the work and existing hierarchies, affecting dead reckoning. For example, the presence or absence of a Traffic Management Unit in a facility makes a difference, and when present, that unit's physical location in relation to controllers' workspace also is consequential: the farther away, the more conflict between controllers and the unit's staff. Similarly, distributed cognition is affected differently depending on whether supervisors are standing at consoles in the midst of the workspace, or seated at desks on the aisle, or working at desks in another part of the building. The simple addition of a boomerang-shaped podium to locate new technology and relocate the supervisors at Boston Tower altered distributed cognition by changing the movement of controllers, their patterns of scanning of sky and ground, and their interaction with devices.

Finally, system effects matter in the production of knowledge and action. Dominating these chapters are controllers' enactment of common sense—the cultural system of knowledge that combines FAA institutional rules, local knowledge gained at the facility, and tacit knowledge, acquired by experience. Hutchins located distributed cognition in a team, documents, operations manuals, device layout, training materials, and procedures in a defined physical space. By showing boundary work between airspace sectors and system parts, the ethnographies in these chapters demonstrate how controllers' distributed cognition spans social and geographic space, interacting with controllers and facilities in the same region, in other regions, and with pilots. In doing boundary work, controllers draw on collective tacit knowledge.[14] These include the embodied cultural understandings of the system goals for safe, orderly, and expeditious delivery of air traffic and standards rules and procedures at the institutional and organizational level.

In addition, the system effects that give the system its dynamics motivate cognitive and material practices. Here, we can see controllers' response to changing circumstances. Most research on distributed cognition explores its smooth operation. In contrast, this research shows how distributed cognition changes during a technical or human failure and emergency, as interactions speed up and parts of the system become more tightly coupled. System effects are, by definition, routine because they are regular occurrences, built into the system. Historically, controllers' work has been disrupted regularly by FAA changes made to keep up with traffic demands, technological change, technical failure of one sort or another, and by political pressure from actors internal and external to the system. The four facilities are in the same regional airspace, managing aircraft trajectories, exchanging aircraft, and necessarily coordinating activities internally and with one another; thus, they are a microcosm of the variation in the national system and the two kinds of boundary work necessary to dead reckoning in air traffic control. These ethnographies effectively show that despite the constraints imposed by system effects, or perhaps because of them, human cognition, coordination, and enterprise remain central to achieving all system goals.

5

Boston Center and Bedford Tower

Boston Center

Airspace, Place, and Work Practice

Surrounded by a high chain-link fence topped with barbed wire, Boston Center is an imposing square structure on several acres of land in a sparsely populated, isolated area just off the freeway near Nashua, New Hampshire. In the air traffic control system, it is officially known as Boston Air Route Traffic Control Center (Boston ARTCC), or ZBW; the *Z* is part of the identifier for all centers, the *B* for the Boston region, but the meaning of the final letter no one at Boston Center could recall. The center is responsible for all high-altitude traffic in the New England Region, which includes the New England states, part of New York, and northeastern Pennsylvania: a total of 165,000 square miles of airspace, plus the oceanic tracks extending more than two hundred miles out to sea. The center handles all traffic at fourteen thousand feet and above, so center traffic is primarily instrument flight rule: high-speed jets—commercial, private, and military.

Necessarily, the primary technologies in use are radar and radio. Annual traffic count is 1,800,000, with an average of 5,000 flights ("operations") a day. About 260 controllers monitor thirty radar positions in three shifts around the clock. The center moves traffic across more boundaries than do the other three facilities. They are responsible for the separation of overflights and the separation and sequencing of arrivals and departures for the Boston metropolitan area, New York metropolitan area, and others. At the same high altitudes, the regional airspace abuts three adjacent US centers and three Canadian centers. Below them, within the region's airspace, they exchange traffic with fifteen TRACONs, several military TRACONs, and many towers. Within the center, the regional airspace is divided into thirty-five smaller sectors that controllers work sending and receiving planes across boundaries.

A gray structure with few windows, the center has an austere look and feel. Entry is through a physical barrier monitored by a security officer in a small guard building. Only FAA identification badges allow people to pass through. Visitors are prescreened by the FAA, receive parking permits and temporary IDs, are stopped at the gate, and must be met at the door and escorted into the building. From the parking lot, the hum of an adjacent structure holding the facility's radar cooling equipment can be heard. A few controllers on break smoke outside or walk around the inside perimeter of the fence. Occasionally, a controller on skateboard whips around between the cooling equipment and the building, hidden from view of the road. In the back is a baseball diamond, but they are forbidden to use it because, according to facility lore, in years previous a controller playing on a center team was seriously injured there. The competing story, from the administration, is that controllers are civil servants, and because citizens pay their salary, when they drive by they expect them to be working, not playing baseball.

The center's locked doors abut the parking lot and can be opened only by employees. Just inside the main entrance is the cafeteria—which they all call "the caf"—a large windowed room filled with rows of long, gray Formica-top tables, with a few circular ones in the back. Controllers take breaks and meals there, playing cards, reading, or watching an oversized TV screen in the corner. By default, the cafeteria is the social center: controllers are required to stay within the confines of the barbed wire during their shift. In a typical eight-hour day, each controller has two hours on, then a half hour break, and lunch is taken during a break. The airspace governs the rhythm of their work: the number of controllers in the cafeteria rises and falls with the center's traffic. However, the cafeteria is seldom empty because of the many (nearly two hundred) administrators and technical specialists in the building, which gives the false illusion that the center is a place where controllers are always on break. Opposite the cafeteria entrance are two long bulletin boards covered with announcements that would be conveyed by word of mouth in a smaller facility: coming facility events, for-sale ads, pleas for pet charity drives, and searches for babysitters or "job swaps"—inquiries from controllers at other centers wanting to trade jobs with someone at Boston Center to expedite a slow-to-materialize transfer request. Sometimes snapshots of controller parties or new babies are tacked up, or cartoons related to air traffic control, the latter usually slamming the FAA.

The center's architecture reflects the size and complexity of its airspace and the number of people needed to deal with it. The building is a maze of floors, stairs, and corridors. Administrative offices of the air traffic manager, assistant air traffic manager, and five operations managers are in a short corridor of the first floor. The rest of the administrative offices, cartography department, quality assurance, and training classrooms—the map room, computer-based instruction room, and DYSIM lab—are on the second floor. When I was there in 2000–2001 and again in 2002, these training rooms were empty. Due to congressional funding problems, the FAA had not been hiring controllers for several years. In a first-floor hallway is the busy NATCA office. The union has a high-visibility and active presence here, with NATCA officers involved in national and regional union activity, and many controllers working with the FAA on the redesign of airspace, developments in technology, quality assurance, and other national and local projects.

Adjacent to the NATCA office is a small game area with video games and foosball, situated off a long hallway lined with men's lockers. Only once did I see anyone in the game room. The hallway ends at the men's break room. A long narrow space, it is jammed with recliners, usually filled with controllers watching TV, reading, or sleeping (not allowed). Of the 260 controllers at the center, only 20 are women. Their break room was in another location and of a different design. Rather than separate areas for lockers and breaks, both areas are located in the women's restroom. The sitting area seats about six and to the extent that it is decorated, the women did it themselves. Neither break room has adequate space or privacy. To find it, some controllers take their breaks in their cars, where they make phone calls, read, or eat, taking beepers for contact with their area supervisor or, in case they fall asleep, setting timers to wake them in time to return to work.

Another hallway ends at a stairway that leads down to a locked door at a lower level that opens into the guts of the center's technical system: a 1960s radar host computer that translated signals onto controllers' radar screens. Installed in all en route facilities around that time, the host computer was approaching the end of its expected life cycle in 1995. The age of the equipment is greater than the age of many of the controllers working in the control room above. Because replacement parts were no longer available, the host computer received many "patches" over the years to keep it functioning and increase its capability to match increases in air traffic. It reproduced a curved North America on a flat screen by reading radar data from eleven radar sensors spread throughout the northeastern United States. The center's host computer applied algorithms to calculate the position of each airborne object. Further processes placed a composite radar picture at each appropriate sector or controller workstation. Redundant technology is the goal: if one system fails, another is immediately available, so airplanes, time, and information are not at risk. The communications system at the center was fully redundant, thanks to a center controller who developed it, oversaw its installation, and supervised its maintenance. In contrast, the center's radar system did not have a redundant backup until 2000, the year I began my fieldwork.

Devices keep controllers under a microscope. Conveying the fullest sense of "technologies of coordination and control," these technologies surely make controllers one of the most—if not the most—accountable of professions. The host computer recorded and stored all controller actions for fifteen days: twelve cartridge tape drives archived all radar data and controllers' keyboard inputs. Another archiving system stored all communications: pilot and controller conversations and telephone conversations (on ground lines) to other controllers at the center and between facilities. A random hour of each controllers' conversations (controllers at all facilities are identifiable by their radio sign-off initials) with pilots and on ground lines were downloaded at intervals to review their use of phraseology. Supervisors went over these tapes with their crew members to achieve accuracy and to ensure standardized behavior that conformed to rules. In case of accident, controversial incident, or suspected operational error, the radar archives quickly reproduced the trajectory of all planes in the sector immediately preceding the incident and the radio archives retrieved all relevant controller conversations. Although these same technologies of coordination and control operate in all facilities, the center's size, complexity, number of air traffic operations, and number of controllers resulted in greater administrative reliance on all formal bureaucratic means of control than in the three smaller facilities.

The heart of air traffic control operations at the center is the control room. Here the relationship between airspace and place takes distinctive material form in the internal architectural arrangements and technologies that shape distributed cognition and work practice. The center's airspace is divided into five segments, each owned by one of five "areas of specialization," known locally as A, B, C, D, and E. Each area's airspace is further divided into seven smaller air sectors, which also divide the area's airspace geographically and by altitude (low, high, super-high). Each sector is represented by a workstation with a radar scope, radio, mic, phone, and computer keyboard. Sector numbers and names (usually some identifying characteristic of the airspace, often an air route intersection or a city, such as Sparta, Albany, Cape, Bosox) and maps are displayed at the top of each workstation. Each airspace sector has its own personality, calling for different controller strategies and skills. For example:

> Kingston—they're flying out of Newark. It's high altitude and high speed. They need two thousand feet (separation). It's hard to turn them if it gets crowded. (Center, M)

> Danbury is our least-sought-after sector. Flights are fast-moving, going out to sea. Traffic is slower in the winter, but it's still busy. Horrendous in the summer. No one wants to work it. It's got an extra strip bay. Lots of paper work [flight progress strips]. (Center, F)

> Do you see this? It's a rat's nest. Small airspace, nothing but ascending and descending traffic in crossing patterns. (Center, M)

Reflecting the size of the airspace and its boundaries, the control room's architectural and technological arrangements make visible the National Airspace System as systems within systems within systems. The workstation in a single sector with a radar controller working alone or with a radar associate is the smallest system. It is located in an area with other controllers, each working a sector and coordinating actions with each other, only one area among five in the center. As controllers move aircraft across boundaries in the sky, technology enables interaction within and between these small systems, crossing relevant boundaries as dictated by the airplane's routing, and thus distributed cognition extends beyond the immediate environment within the center to controllers working in other facilities.

Three crews consisting of fourteen controllers and a supervisor are permanently assigned to each area to staff the three shifts per day. The area supervisor is an "extra pair of eyes" and the connective link between his or her area, the other areas, the Traffic Management Unit (TMU), and the operations manager's "watch desk," which in turn connects all center air traffic operations with headquarters in DC and the Herndon Command Center. A controller's typical schedule over five days moves across days: on the first day, from 3 pm to 11 pm; on the second, from 2 pm to 10 pm; on the third, from 6 am to 2 pm; on the fourth, from 7 am to 3 pm; and on the fifth, from 11pm to 7 am. Then follows a controller's "weekend," whatever days of the week that may fall upon. They call the change from the shift starting at 2 pm to the one starting at 6 am "the quick turnaround" for the short sleep they get before having to turn around and go back to work at 6 am next day. Working the midnight shift ("midshift") is not easy either. Many myths surround air traffic controllers, but chronic fatigue is not one of them.

Airline schedules and weather determine the number of crew members on position at a given time and the rhythm of the work throughout a day. The center's rush periods are from 6 am to 9 am, from 11 am to 1 pm, and from 4 pm to 8 pm. Redundancy in personnel is essential: enough controllers have to be available to handle known patterns of heavy traffic and unexpectedly difficult situations, such as severe summertime thunderstorms. During the daily rush periods, each airspace sector is managed by two controllers, the radar controller and the radar associate. Interdependent and redundant, they sit close together at one workstation so each can see everything the other is doing. The two of them interacting with each other and their devices are among the smallest units in the system. Collective tacit knowledge enables either person to take over from the other in an instant. The radar controller talks to pilots, enters commands that track a flight on the computer, and manages the aircraft route through and out of the air sector. The radar associate works communications with other controllers on the ground lines and manages the flight progress strips, entering changes so that each strip is analogical to the movements of the plane in the sky, supplying redundancy if the radar fails.

Resiliency shows as the system expands and contracts with traffic conditions. As many as four controllers may be called to work one sector during a thunderstorm, two seated, two standing behind. However, staffing is reduced when traffic is low: between the weekday rushes, all day and evening on Saturdays, and Sunday mornings until the rush begins later, at about 11 am. The radar technology accommodates these fluctuations. When traffic is light, two or more airspace sectors can be consolidated on one scope, so one person can handle two sectors or more. Most airlines have their planes on the ground by 11 pm. During the midnight shift, from 11 pm to 7 am, all sector boundaries disappear: the center's traffic is consolidated on one scope and managed by two controllers and a supervisor. Everyone else goes home.

In contrast to my other three facilities, the salient feeling of this place was one of top-down imposed rules, regulations, confinement, and surveillance.

Controllers complained about the harsh environment, rules, and deprivations, with comments like "They treat us like children"; "We're not allowed to have food in the area and drinks only in spill-proof cups"; "We can't leave the facility to bring back food or coffee"; "No music in the areas." They resented the overreach and unwarranted surveillance by facility management. Although not allowed to leave the site during their shift, or perhaps because of it, controllers also feel more isolated by this rural location. Despite these differences from the other three facilities, in common with the others, Boston Center is a busy place and no two traffic days are alike. It offers a mix of conflict, stress, boredom, and fun, and is inhabited by people whose differences are bound together by an active NATCA, the survival of daily traffic challenges, and common workplace problems. Although not all would agree, many said, "Our best friends are here. We come to have fun with them." These shared experiences notwithstanding, at this place I also heard more complaining, alienation, and resentment toward the FAA and the facility's administration than at the other facilities. Also, what they said was matched by what they did. For example, NATCA officials in the facility filed many work grievances with facility management on behalf of controllers. Supervisors complained that in comparison to other places they had worked, absences due to sick leave were high.

These displays of discontent may partly have been due to the punitive, bureaucratically minded air traffic manager (ATM), the leader of the center at the time: the day that the announcement was made that the ATM was being transferred to another facility, someone played a recording of "Ding-Dong! The Witch Is Dead" on the speaker system, which carries sound throughout the building, including in the administrative wing. However, a focus on leadership style ignores the extensive evidence of system effects on the center that produce visible signs of alienation. Rule-mindedness is normal for all air traffic control facilities, but it is greater at centers than at other facilities for three reasons, all related to the relations between space and place. First, airspace size calls for a larger facility, and given the military roots of the system, a greater reliance on rules and enforcement of rules to coordinate the number of people and work activities. Second, the pressure on the FAA to reduce operational errors mainly falls on centers because they are the origin of data on numbers of errors per operation, which show up in national FAA statistics as measures of safety. Administrators respond to those pressures by greater reliance on rules and enforcement, and thus there is more attention and pressure brought to bear on individuals and punishment. Third, centers have more airspace boundaries than towers and TRACONs, thus increasing the daily challenges of crossing divisions of airspace, adding stress to the job. Every type of facility has its boundary problems, but here the intra- and interfacility conflicts about airspace boundaries seemed to bear especially heavily on controllers.

Place, Technology, and Ethnocognition

The timing of my fieldwork at Boston Center provided an unexpected opportunity to observe dead reckoning during the implementation of a major system change to upgrade the 1960s technology that also necessitated architectural redesign. When I first visited the center in 1998 for the guided tour in preparation for writing my study proposal, Pete, the controller who had met me at the center door for the tour, took me first to the brightly lighted Traffic Management Unit. TMU is staffed by experienced center controllers who coordinate the region's air traffic with the Herndon Command Center (known as "Central Flow"), which regulates national traffic flows. TMU also engages in dead reckoning. Reflecting both resilience and interdependence with the rest of the system, TMU adjusts regional traffic as necessary in relation to local and national situations, rerouting it during thunderstorms or other conditions, and spacing departures and arrivals to avoid congestion. Both the TMU supervisor and the operations manager work in TMU. The "ops manager" oversees all day-to-day operations in the control room from the watch desk and is responsible for operational resources, including personnel, and for serving as the connecting link to the center ATM, TMU, regional offices, and the Command Center.

In 1998, the Boston Center TMU was in an enclosed structure at the top of the control room that had easy access to it through open passages on either side (fig. 13). Stepping through, we entered the control room, where controllers worked in near darkness. The control room was a large, dark, two-story, windowless room bigger than three gymnasiums in which controllers' workspace occupied only a small part, with much unused space above them and extra open space on the floor, as if awaiting the arrival of more controllers and workstations. Air conditioning kept the room sweater-cool because the radar generates heat and must be kept cool, producing a sense of being underground in a huge tunnel or cave, despite the high ceiling and open space.

Visually, technology dominated the control room: the controllers seated in front of radar workstations seemed small by comparison, diminished by the height and mass of the workstations and the size of the room. I could not see

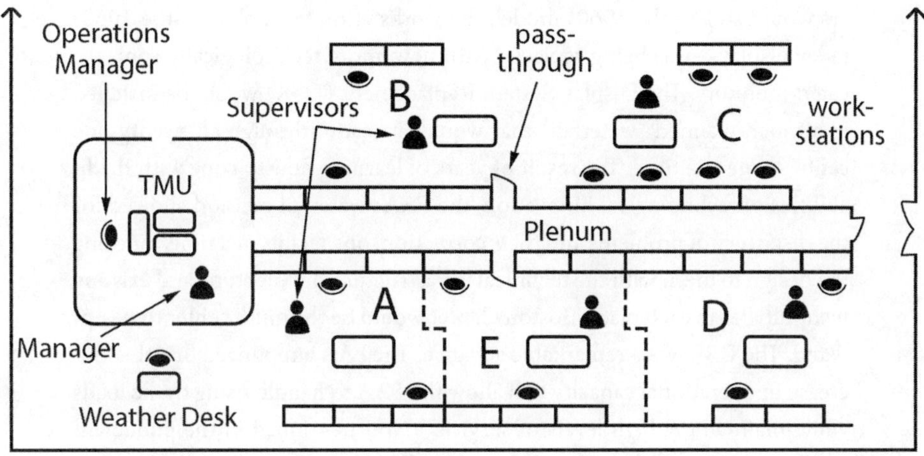

Figure 13: Boston Center architectural layout, control room, 1990
Illustration by Noah Arjomand

all the controller workstations because they were divided by a large rectangular structure higher than the workstations and extending from TMU to the back of the control room. Above the workstations, the structure's wall became a steep slope to a flat top almost the height of the ceiling. Called the plenum (a construction term, from Latin, meaning "full"), it contained the supporting technologies for the workstations and the air conditioning. It was designed to access the workstations on either side from behind for repair. Access doors for technicians and engineers were at the rear, a passageway through the middle, and from the front, through a door in the TMU space. Controller workstations on both sides of the plenum were arranged in areas of specialization: A, E, and D were on one side; B and C on the other.

On both sides, the areas were roughly end to end, with varying amounts of empty space separating them. In each area, controller workstations faced away from each other, with an aisle between each row. The inside rows abutted the plenum. Supervisors' desks were office desks of the era, usually in the aisle. The only illumination in the control room came from the green surface of the radar scopes, a small amount of recessed lighting at each workstation that illuminated keypads, buttons, and touch pads, and small gooseneck lamps at supervisors' desks. In each area, five to twelve controllers talked to one another and pilots in normal conversation tones; the radar hummed. But it was quiet. The room swallowed the sound.

However, change was already in the works. Although the host computers would still be the 1960s model, the workstation technology—the 1960s radar display—was being replaced with new, more technologically sophisticated monitors. The Display System Replacement (DSR) would be installed in a much-planned "waterfall" that would "cascade" through all twenty-one centers, one at a time. The result of years of learning how to cope with the liabilities of technological innovation, the FAA's waterfall method allowed for the discovery of problems and their correction, one facility at a time, tailoring the design to mesh with the technical infrastructure, architecture, and existing technologies in each place. Boston Center would be the ninth center to be updated. The DSR was a remarkable advance. The FAA announced that this "increase in operational capacity will allow the FAA to handle rising traffic loads while maintaining high levels of service."[1] Not mentioned in the public announcement, controllers would be able to handle more traffic faster because the keypads no longer stuck, they had more computerized information and options at their fingertips, and, of supreme importance, if the center's forty-year-old host computer failed, controllers were not faced with blank scopes, as before. The DSR had an instant backup connecting to another computerized radar system that instantly reproduced the planes on the scope. In preparation for the change, all twenty-one control rooms at centers and their workstations were being redesigned.

By the time I arrived in March 2000 to begin my research, Boston Center controllers had just moved into their new control room, adjacent to the old room and its equipment (fig. 14). The differences between old and new were vast. During the third week in the new location, the effects of the architectural redesign and the new technology on dead reckoning were visible. Controllers struggled to adjust to the changes. The center's airspace and number of sectors for each area was the same. However, the new control room was one story high, not two. The plenum and two wide aisles on either side were gone. Now, one center aisle ran the length of the new room, with the areas and other work units opening off of it for easy access. Maintenance and repair of workstations was through access alleys between each area. The walls separating the areas did not extend to the ceiling but were open a few feet from the top. The control room was long and narrow, not long and wide, as before.

To enter the control room was to enter the center aisle. To its immediate left was an open and spacious area housing the facility's new National Weather Service Unit, with an in-house meteorologist and state-of-the-art technology. Next on the left and adjacent to the weather unit, separated from it by the

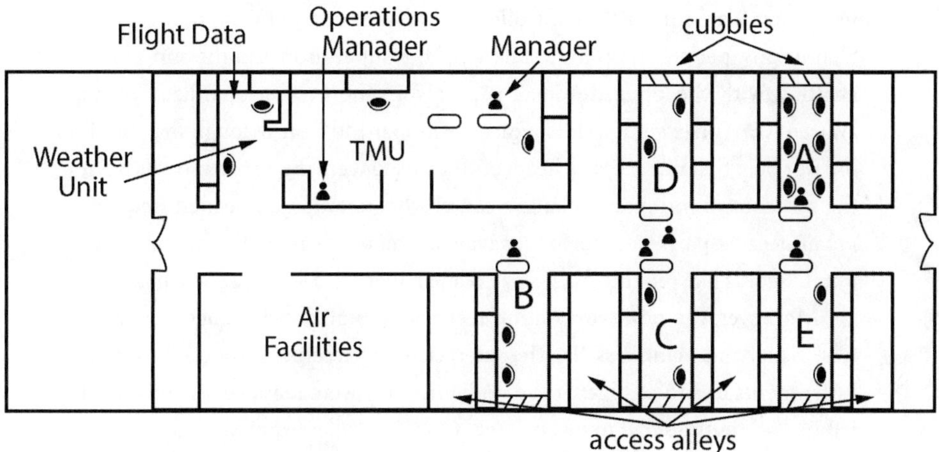

Figure 14: Boston Center architectural layout, control room, 2000
Illustration by Noah Arjomand

operations manager's watch desk and from the center aisle by a short-of-the-ceiling wall, was an enlarged Traffic Management Unit.

Across the aisle from TMU was the equally large Office of Air Facilities, responsible for maintenance and repair of all control room technologies and beyond. After Air Facilities, extending to the end of the control room's center aisle were the five areas. Rather than being end to end as in the old control room, each area was perpendicular to the aisle. Each area was twenty-seven feet deep and eleven feet wide, six controller workstations on each side wall. The airspace owned by each area and the relation between them in terms of patterns of traffic exchanged remained the same, but the areas were positioned differently in relation to one another. Supervisors' desks were inside the area, with supervisors seated on the aisle, facing in. The closed end was filled with four-by-six-inch drawers: controllers' "cubbies" for mail and headsets.

The effect of the change on ethnocognition and boundary work was dramatic. Acoustics were poor. The room captured sound—and there was more of it—rather than swallowing it. The sound of voices floated from one area over the top of the wall to the neighboring area. Also, now the areas were open to the main aisle. Controllers in the area were still on opposite sides, back to back, but closer to each other than in the old room. For the first time, controllers were exposed to conversations from the new heavily trafficked center aisle. Controllers' acute hearing skills picked up conversations of controllers in the area across the aisle, controllers going in or out on break, tour groups, technicians, people stopping to talk with the supervisor, and the supervisor chatting with the opposite area's supervisor. The noise jeopardized room awareness. Worse, some pilots' voices were hard to hear through the din. It did not help that many of the headsets they used were the 1960s models from the old control room and, like the center's host computer, needed frequent repair because parts were no longer available on the market. As one controller put it, "It was like moving from an airplane hangar to a shoe box. We hated it."

Moreover, the new workstation technology initially challenged controllers' visual and manual skills. The interaction between controllers and their instruments had to change. Speed and efficiency were reduced, as embodied habits and routines had to be adjusted. One controller expressed the consensus, drawn from many similar experiences in the past:

> Some new technologies save lives, but [change] sucks. So many operations here rely on the human component. When you introduce a new technology, it completely throws off the operation. (Center, M)

The changes forced controllers to convert tacit knowledge from memory back into active learning. Despite extensive training on the new equipment before the move, actions formerly done without thinking now had to be thought about. Nothing was exactly in its usual position. Many mechanical switches and knobs had been replaced with on-screen controls. Controllers were familiar with the location of the options keypads and information aids but not familiar enough to use them without thinking; instead, controllers had to use precious seconds to search, interrupting their scan. Previously, scopes were all green, now they were blue or black, square instead of round, and their placement angle in the workstation was different.

Further, the main lighting in the new control room was elevated from near darkness to dim, which was sufficiently light to recognize a person standing at the far end of the center aisle—a remarkable difference from the darkness of the old room. A refinement was that each area could choose its own light level. Adjust, work a week, vote, repeat. Experimenting to find a level agreeable to the majority of controllers in an area was another variation in the work environment. At some point during the transition, someone typed two quotations on a fragment of paper and tacked them on the bulletin board outside the caf. Typical controller humor, they captured a dominant aspect of place for those doing shift work in the control room's now varying shades of darkness:

> Light, God's eldest daughter, is a principal beauty in a building.
> —Thomas Fuller

> Space and Light and Order. Those are the things that men need just as much as they need bread or a place to sleep. —Le Corbusier

Center controllers had looked forward to the technological advances and to the new room. They complained that the old room was dirty from decades of use and smelled of smoke from the time when smoking was common, but more urgently, they wanted out because it was insulated with asbestos. The need was acute; planning for design and implementation was extensive. But the DSR installation was an example of dead reckoning gone wrong.

However, the FAA had done everything right. It was a mammoth, complex, and well-coordinated undertaking. The FAA had consulted with NATCA controllers about the changes. The FAA had predicted controllers' difficulty in adjusting to the DSR technology and thus had devoted time to training and practice on the equipment. Also, the FAA waterfall implementation allowed the administration to learn by doing, making adjustments to get the bugs out of the system as it went. Further, the controller workforce in each center gradually transitioned from the old control room to the new, keeping the ability to return to the old system if anything went wrong. As in the past, the implementation of organizational and technological innovation had unanticipated consequences. The FAA did not correctly predict the effect of *the combination of architectural, spatial, and technical changes* on controllers. In the short run, the changes interfered with the embodied skills essential to dead reckoning, threw off controllers' timing, and added stress to the job.

The new technologies—architectural and technical—eventually fulfilled their promise. Center controllers rose to the challenge: there were no catastrophes, no increase in operational errors—but, as is typical with changes in the air traffic control system at the operations level, controllers absorbed the cost. Controllers say they are "change resistant." That is because they thrive in a stable work environment where much that they do is routine and can be done "without thinking," freeing them up to predict, plan, and move aircraft in normal and unexpected circumstances. Any time the FAA makes a system change—major or minor—in attempt to keep up with changing traffic demands or new technology, the change has a short-run, negative effect on the work practices and stress level of controllers. In time, controllers' physical and cognitive abilities did adjust so that they again were capable of working at their previous skill level. Indeed, the next section shows them after the adjustment, deploying their expertise in the first kind of boundary work: interacting with pilots and technology to move traffic across the invisible boundaries in the sky and maintaining separation between airplanes.

Signals and Interpretive Work

At the start of a session, a controller stops at the supervisor's desk, signs in on the computer, and relieves the controller who is next on the break list. What happens then is a standard operating procedure at all facilities: transition in responsibility and ownership of an airspace sector. Cognitive redundancy and coordination is built into the process: as "the relief" gets the briefing, so do the seated area controllers, updating the collective tacit knowledge of the crew. Already wearing a headset, plugged in, and listening to the pilot-controller conversation on the sector frequency, the "relief" stands behind the seated controller, watching sector traffic while receiving an oral "position relief briefing." The seated controller, still working the airspace, reports sector conditions using a standardized checklist of items that convey traffic, weather, and other conditions that need attention: "CA271 wants direct"; "Watch this guy"; "Reports of turbulence at [flight level] 250." When the relief has "got the picture"—a cognitive map of what's going on—he says, "I've got it." They swap positions. The relief begins talking to pilots even as he or she is taking a seat. Now the other stands behind, monitoring, waiting for questions. Because both know well the sector airspace and traffic patterns, the entire swap is completed in a few minutes.

Initially, the relieving radar controller is a whirlwind of activity. While actively working traffic, he quickly adjusts the scope technology to personal preferences for "screen enhancement": the keyboard versus traction ball (slew ball) to move the cursor, directional lines, screen color, contrast and brightness, font size, amount of data displayed, location of information aids on the screen. At the same time, the radar controller is giving pilots instructions that "clean up the mess on the screen." Not that it is a true mess, in the sense that the previous radar controller has gotten into a jam (although sometimes this is the case). Typically, the new radar controller has a different plan and technique, so adjusts spacing, speeds, and altitudes. "Every artist has his own canvas," one said to me, as he sat down. Then, having imposed a new order on the virtual reality on the scope and thus in the sky, the controller settles into the traffic rhythm.

With my headset plugged into the sector frequency, I sat to the rear, between the radar controller and radar associate, or beside the radar controller when traffic was low so no radar associate was necessary. I had a clear view and could hear well. Initially, however, nothing made sense to me. On the radar scope I saw only blocks of data and meaningless symbols. The scope shows routes, intersections, ground technology, and data blocks but has no orienting city locations or state names. It was chaos: I saw no patterns. The headset was no help. The language was unfamiliar, professional pilots and controllers talk very fast, two pilots sometimes talked on the frequency at the same time, and occasionally static nearly obliterated pilots' calls. But after a few weeks, and aided by controller explanations during low-traffic periods, I began to recognize some basic patterns, and with more observations, teaching, and interviews, I could grasp some of what they are doing, seeing, and saying.

Controllers do not see individual blocks of data and individual flight paths, they see a gestalt. Their scan takes in the entire screen at a glance. It is not chaos to them: it is organized. Dead reckoning is about prediction and foresight. Common sense makes this possible by operating as a cultural system of knowledge: institutional rules (standardization and cultural understandings), local knowledge (this is how we do it here), and tacit knowledge (experiential knowledge) congeal into habits of mind particular to their task and system needs. Here is how their cultural system of knowledge—ethnocognition—works in practice, showing how standardization, interpretive work, and improvisation combine as controllers move traffic from moment to moment across boundaries in the sky.

Dead reckoning calls for them to visualize in three dimensions from a two-dimensional screen image. Each airplane is represented on the scope as a data block: a square of numbers and letters. Center controllers say they visualize an airplane as a block moving through space, having a destination from the moment it enters a sector. In surveys and interviews, most controllers said this was not a natural talent that they brought to the profession. Instead, the ability to convert a square of information on a flat radar screen into three dimensions was developed in training. Flight progress strips are central to the process of predicting the trajectories of airplanes in space. The strips have physical properties that are essential to controllers' cognitive and material practices. The waiting strips in the strip bay call attention to the sequencing of the planes in the controller's airspace sector. Touching or holding the strip sharpens the controller's focus on it and signals other controllers of a coming action. The controller simultaneously takes account of the information on the strips and scans the radar to follow all relational movement of planes on the scope in relation to the pilot's voices and actions, and at the same time, the controller's peripheral vision and situational awareness absorb the bodily movement, voices, and actions of supervisors and other controllers in the room.

Institutional rules provide the foundation for this embodied perceptual skill. Standards also act, affecting individual cognition and material practices. Controllers told me that their third-dimensional ability was aided by the rules of separation: one thousand feet above and below, or five miles in between each aircraft, gives height, depth, and width to the target. Surely another factor, not mentioned by them, is the formal system of position rotation. During a shift, when one controller relieves another, the reliever is sent to a sector of airspace that he or she hasn't worked yet that day. Because Boston Center sector airspaces are at different altitude levels (high, low, super-high), position rotation means not just moving to different positions in the area, but also into and out of different dimensions of airspace. In this way, controllers' physical movement between air sectors on the ground has a cognitive effect.

Also, standardized airways and intersections are guidelines by which controllers can predict traffic patterns, while standardized flight data tell them departure and destination points. Center controllers maintain separation by being able to predict the flight path of an aircraft in the sector and get it through the other traffic patterns that typify that sector. So standardized are the airspace routes that controllers develop routines to handle them, improvising when they need to:

> Keep in mind that sectors are set up, kind of, a lot of sectors are set up to separate aircraft on their own. Other sectors are actually designed to put planes together [crossing traffic, ascending and descending]. Depending on what scenario that you have, there is routine stuff that you see day, after day, after day, and you separate them using the routine way that you've done it. And every once in a while there is something else that comes into your equation, actually, most of the time there is something else that comes into your equation, so you just work around it. It's dynamic. (Center, F)

The effect of ethnocognition on distributed cognition is visible as controllers enact their trained capacity to identify anomalies and fix them, avoiding a "mess." Controllers' interpretive work gives meaning to the technological representations on the screen, what they hear on radio, and information from other devices. They are familiar with refinements in patterns ("Learjets don't normally fly that way"; "Northwest [Airlines] turns like pigs"), and experience educates them in equipment capability, which, together with air speed, allows them to predict and make the changes that keep planes moving to their destinations and avoid collision. For example:

> Here are two flights, a jet and a prop, on the same altitude, but they are distant now. It looks like they would violate the rules of separation, but air speed means they will clear by five miles. (Center, M)

Note how past experience and local knowledge shape this controller's moment-to-moment interpretive work, prediction, and planning. History, socialization, and local context matter. All controllers are trained to develop plans early and move traffic so no crisis develops. They describe their work as a series of "moves": a new player alters the plan, so controllers adjust by moving the other aircraft in the sector. An amount of preliminary planning goes into each action, but plans are reactive and adapted from moment to moment as conditions change.[2] At the center, the main techniques are adjusting altitude and vectoring. A vector is a magnetic compass heading assigned by a controller to change the direction of an airplane, as in "Fly heading 270," which tells the pilot to turn the plane to the west. The process of moving traffic and maintaining separation is one of continual revision, not one decision per aircraft, but many, even if a plane is in a sector for only five minutes. Although dead reckoning at the center is based on standardization and institutional rules—the airways and intersections that are the invisible lines in the sky—this controller's description of an unfolding situation shows how local and tacit knowledge figure into his interpretive work when a new player enters the sector. Pointing his pencil at the screen, he elaborates, supplementing the technological representation on the screen with the social context that guides his plan and moves. His interpretive work embeds the objects with a coherent meaning:

> This plane has twelve skydivers [in it]. I've got to keep everybody out from under him, or those guys are going to have a bad day. See these little dots here with no data? These are UFOs, don't know for sure but probably gliders. Now over here we have this big Lufthansa coming off the oceanic track, slowing to descend to Approach [the TRACON], which is over here. This guy has been

> flying all night, he's tired, so I've gotta keep everybody out of his way. Now I have to move this guy under this guy, take that one under this one, then get him back heading this way. (Center, M)

Recall their definition of common sense: the ability to avoid messes in the first place and get out of messes whatever way they can. One way they avoid messes is by giving a steady stream of traffic advisories, coordinating with pilots to enhance pilots' dead reckoning and thereby avoid error: "Traffic no factor"; "Additional traffic at 6,000 at your 12–2 o'clock position"; "Expect Sparta [intersection] shortly"; "Unfortunately, have a slow in front of you. Take a 360 to the right. Take your time about it." Controllers watch for warning signs of something going awry so they can prevent mistakes and avoid collision. Radar technology provides several useful aids. Controllers can project a diagonal line forward along a plane's route to help anticipate possible conflicts. They can reproduce the route history on the screen using hatched lines to duplicate it. When an aircraft approaches the boundary of an airspace, the data block flashes on and off, signaling the controller to contact the owner of the adjacent airspace, next on the route. As an aid, controllers can throw a circle around two aircraft whose routes will have them passing close to each other, drawing their attention to the five-mile separation limit necessary between them. These screen enhancements are aids that some controllers rely on more than others. However, if the rules of separation are broached, the computer system automatically projects a circle around the two airplanes that reads "CA-CA" (conflict alert), flashing on and off. This is a strong warning signal, never missed—despite mistrust of its accuracy due to occasional false alarms.

Memory is not just in the individual mind but a collective memory that comprises interactions between the people and devices in the room.[3] Flight progress strips are as important an actor in distributed cognition as the scope, and more reliable. Recall that a flight progress strip contains the aircraft's call sign, type of equipment, point of origin, destination, and the most recent information on route, airspeed, and altitude. But in addition, they are annotated (fig. 15). Controllers modify the strips, using a standardized, hieroglyphic-like language of air traffic control, as each controller along the route updates this information by marking changes in route, altitude, speed, or current direction on the strip, so it is the analogue of the plane's trajectory in the sky. A strip conveys so much information that in the year 2000, as the FAA experimented with a computerized replacement technology for the strips, it found that there were ninety-one total bits of information from the printed and handwritten

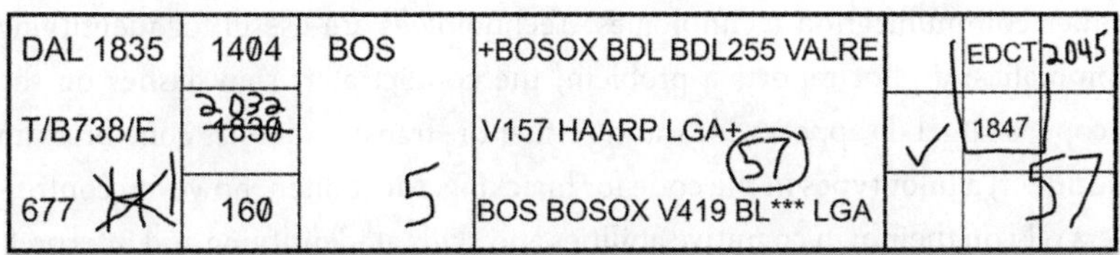

DAL 1835
1404
BOS
+BOSOX BDL BDL255 VALRE
EDCT
2045
T/B738/E
2032
1830
V157 HAARP LGA+
57
1847
677
160
5
BOS BOSOX V419 BL*** LGA
57

Figure 15: Flight progress strip with controller markings
Illustration adapted by Noah Arjomand

codes on a strip. At that time, the FAA couldn't replicate it all. The strips provide redundancy: if radar fails, controllers can still work traffic from the strips. Moreover, because they interact with the strips, touching, updating, and moving them, the memory stored in the strip is analogical with the memory in their heads and combines with what is on the scope, recognition, and experiential knowledge of the controller or author of the strip marking, thus producing a mental model.

Ethnocognition also comes into play as controllers give new meaning to the information on the strips, using them for memory enhancement and as a signaling system for others in the room. When the flight progress strip is printed and delivered by computer to the workstation, the radar associate slips it into a plastic holder and stacks it in the strip bay, which holds two columns of strips: one column holds the strips of aircraft en route to the sector, chronologically ordered, and the other column holds those already in the sector. The radar controller at a glance knows what's coming, but sometimes enhances his or her memory by running a pencil down the strips as a reminder of the positions in the sequence, sizes of aircraft, and destinations. They have a tactile relationship with the strips. Holding the strip enhances memory, as does even the physical act of writing itself, whether the controller looks back at the writing or not, having written something down alone is a memory aid. Controllers also give a strip symbolic meaning by setting it at an angle in the strip bay as a reminder of some pending activity of an aircraft. A quick glance at the strip bay enhances prediction; it not only is for the controller working an airspace sector, but also is a signal for the supervisor doing oversight from behind and the controller at the next workstation, who is alerted by the signal. In addition to hearing, situation awareness includes peripheral vision, so that as the radar controller locates the strip, picks it up again, places it on the desk near the radar assistant, or stands and holds it, other controllers are alerted to an ongoing situation.[4]

Controllers all are dependent upon radio, radar, flight progress strips, and other communication technologies. Technologies are useful for identifying anomalies: a pilot reports a problem; the conflict alert sign flashes on the scope; a target disappears, indicating either the transponder has gone off or an accident; a pilot types in the code for hijacking. Most often, however, controllers rely on their own cognitive abilities and skills at identifying and interpreting early warning signs. When I asked in interviews how they identified an anomaly that needs attention, controllers at all facilities respond by naming three conditions that are unambiguous: bad weather, air traffic control technical failure, and pilot or aircraft emergency. These are strong signals, immediate attention-getters. But controllers also mention more ambiguous signals. Instead of naming them, they invoke abstract principles to explain how they recognize them:

> You learn it from your trainer. I can't explain it, it's many things, but you learn it from your trainer. (Center, M)

> To great extent we know what to expect, recognize anomalies, and know what to do. (Center, F)

These controllers are referring to the elements of common sense: institutional rules, local knowledge, and tacit knowledge, gained from experience, so that actions are driven by habits of mind. They bring to each session a background grid of information from which controllers notice deviations from the expected. They don't watch everything equally. They give greater attention to the airplanes that stand out against the cognitive frame of reference that is their cultural system of knowledge. Although identifying anomalies is cognitive, some examples materialize in what they say and do. Radar controllers often made comments to themselves, pointed something out to their radar associate, or gave a shout-out to the controller across the room who would have the aircraft next, or if the deviation was egregious, announce to the supervisor. Room awareness alerts everyone in the area. All watch, listen, and act so that small anomalies don't turn into accidents. Watching them work, I made this list of warning signs, which I later found were the same at all four facilities.

- An aircraft is following a standard traffic pattern but uses a "strange approach" or is "on the wrong route."
- A pilot makes a procedural error.

- Rule violations: a pilot does not do what the controller asked or crosses one of these three boundaries without permission: separation limits, sector boundaries, or "busts altitude," meaning that the pilot, when descending, goes through or below the altitude the controller assigned.
- Hear-back or read-back error. Pilots give the wrong response. They repeat controller instructions to confirm, but read back the wrong altitude or heading. One difficulty is that so much is standardized that a controller may hear the expected response even when a wrong one is given.
- Pilots use the wrong phraseology or omit something or add something unnecessary. Controllers don't listen to each word; they listen to the whole, to the rhythm of it. They know what to expect so are able to recognize changes in the rhythm of speech when something is missing or added.
- Deviations in pilot tone of voice, speed of response, pausing, stammering, or accent that are inconsistent with the normal, rapid controlled tone of professional pilots that controllers take as signs of competence—these may indicate that the pilot is tired, distracted by something going wrong, or lacking in experience.

So the question that naturally came to me was, how do controllers who are trained to maintain the standard phraseology, professional tone of voice, inflection, and briskness to reassure pilots in a tenuous situation also manage to signal a sense of urgency and risk to a pilot when they need to? When I asked the radar associate next to me, she said, her voice feigning emotion and urgency, "We yell, 'Climb, you bastard, climb!'" Her radar controller partner corrected, in crisp, emotionless phraseology, demonstrating his frequency voice, "We say, 'Would you expedite that climb, sir.'" The official phraseology gives them a standardized vocabulary for just such occasions. They maintain the usual professional tone of voice and use phrases that incrementally increase the urgency of an action: "good rate," "expedite," or "immediately." Pilots understand and respond, as in this overheard exchange:

> *Controller*: Gardner Victory 431. Descend to 6,500 at Lobby intersection, and if you would give me a good rate of descent on the way down, sir.
> *Pilot*: OK, gonna hurry down.

Expedite (as in "expedite that turn") conveys more urgency than "good rate." *Immediately* is used only in extreme cases: for example, "heading 180 immediately" will have passengers spilling drinks and purses.

Constrained by the standardization, some controllers innovate, intentionally changing their speed and tone of voice to give a warning signal:

> I lower my voice. That's my way of indicating to them that I am a professional, I mean business. When traffic gets complex and I need them to do exactly what I say, I make what I say especially clear and become quieter. My way of telling them that this is serious. (Center, M)

Despite the mandate to maintain a professional tone, controllers also inadvertently convey warning signals to pilots. When controllers are dealing with severe weather conditions or a pilot emergency, or are abnormally busy, the recordings of their voices show that they, like pilots under stress, tend to talk extremely fast and elevate pitch. Controllers don't recognize these changes in themselves, but pilots do. So do their coworkers, who consequently modify the way they pass their airplanes to the next controller in order to adjust the flow of traffic to the controller's workload so the traffic flow moves more smoothly. Supervisors engage in moment-to-moment sense making, identifying signals to predict controller performance capability. Like controllers monitoring aircraft, supervisors recognize anomalies against a grid of the regular, expected pattern. They know their crews well. They can tell who needs a break before the official break time by unusual behavior: a normally soft, rhythmic "frequency voice" becomes louder or elevated; a usually laidback posture changes to leaning closer to the scope, a jiggling knee, a tapping foot. Supervisors recognize the early warning signs of controllers' fatigue or stress and avoid a mess by sending them to a less difficult position or on break.

Boundary Work: The Handoff and Turf Wars

To negotiate the trajectory of airplanes safely across the invisible boundaries in the sky, controllers must also negotiate the boundaries of people and organizations on the ground who own and use that airspace. Of the two kinds of boundary work, the latter is by far the more difficult at centers because they have more of these boundaries than do TRACONs and towers. Airplanes must be exchanged not only across sector boundaries within a center's area but also across social space to controllers in other areas in the same facility, to controllers in different facilities within the region, and to centers in other regions that abut, in this case, the New England Region. The characteristics of one airspace are not always compatible with another, leaving it to controllers to bridge the boundaries. To ease these transitions, an aircraft must meet certain formalized airspace "restrictions," which are rules about altitude, speed, and miles-in-trail (spacing between high-altitude aircraft) specifically designed to control traffic to fit into the peculiarities of an airspace. However, these restrictions also restrict the options of the controller who has to meet them: the airplanes must be handed off to the next controller a certain way. And the ownership principle always operates: the airspace owner has to be willing to accept the airplane.

The interdependence of the parts of the system and the variation of the airspace clash, leaving the problem of coordinating the movement of traffic across boundaries—and thus system resilience—to controllers. At all facilities, the "handoff" is the primary tool of boundary work.[5] As an airplane approaches an airspace boundary, the boundary crossing has to be coordinated with the owner of the next airspace on a plane's route, either verbally or by computer contact. The process, they say, is like "passing a baton in a relay race." One feeds, and the other receives. The owner's permission to cross into their airspace is often automatic, but not always. If the airspace is busy, the receiving controller may say "unable." It's a conflict of interest: one controller needs to get rid of an airplane; the receiver's needs are compromised if he or she accepts it. When a handoff is refused, a plane's forward progress is stalled. Traffic can move forward only when one controller compromises his or her traffic needs, yielding to those of another. It was at Boston Center that I first heard controllers express the sentiment repeated at all facilities, "It's not the airplanes that are the stress of this job; it's the people you work with."

Some situations are easier than others. For controllers in the same facility who work the same airspace with the same team or crew daily—a tower, a TRACON, or an area at the center—knowing the personalities of people in the room and having worked every position in the airspace makes it possible for one controller to automatically, and silently, alter strategies to accommodate to current traffic needs of a nearby controller. Controllers at all facilities are known for their remarkable situation awareness and skill at silent coordination. Work is typified by mutual monitoring and tacit task allocation rather than direct communication. Silent coordination is essential to the system because of the traffic speed and volume. It is as important to be able to coordinate without speaking as it is to be able to act without having to calculate.

However, controllers who coordinate with controllers in distant locations do not have the same advantages. They know neither the personality of the airspace sector nor that of the controller working it. An "unable" can be a lot of trouble for the feeding controller who must put the aircraft in a 360-degree holding pattern ("putting a guy on spin") and try again in a few minutes to see if the receiving controller will accept it. If not, it stays on hold or may be rerouted. This is when things get complicated. The boundaries separate the parts of the airspace, but they are interdependent: a change in one section of the airspace creates changes in others, creating a domino effect. An airplane circling in a holding pattern interferes with the existing traffic pattern in the feeding sector. Always, time matters. Depending on the length of the hold, the airplane on hold may overcrowd the airspace sector, causing the holding controller to refuse handoffs, too. Then traffic backs up into the adjacent airspace.

Here is where the trouble starts. If the receiving controller continues to be "unable," the feeding controller may have to hold additional planes in the airspace, and so builds a stack. Building a stack is routine for them but was a thing of beauty to me, the uninitiated. The controller stacks the airplanes by putting each "on spin," doing "360s" in the same area of airspace but at different altitudes. Imagine a downward-moving corkscrew. To build a stack, a controller puts a second airplane on spin above the one already doing a 360, descending the first one to spin at a lower altitude. For each additional plane on hold, the controller moves the others lower in altitude, each with the required thousand feet of separation above and below. When the adjacent airspace does open up, the feeding controller spirals the planes out of the stack one at a time, starting at the bottom. First in, first out and into the next sector.

Building—and working—a stack is both art and craft. It is a concrete illustration of the intersection of standardization, interpretive work, and improvisation. It calls for off-route three-dimensional thinking in an airspace where planes are moving on standard routes. Between the standard routes is a "safe" place for a stack. Without benefit of direct observation, the corkscrew is only in the controller's mind and actions, not obvious on the screen, where the targets (data blocks) show no pattern. Yet the rules of separation must be maintained. The origin of stacking was probably some previous controller who, in a jam, used common sense and innovated to "avoid a mess" or find a way out of one. Some of the more senior controllers, now administrators, told me it was of problem-solving origin, first an informal solution, then practiced and passed on from controller to controller. Once acknowledged as collectively useful, building a stack became a formalized procedure. It is possible only at radar facilities with sufficient airspace, professional pilots, and fast-moving jets. For controllers, a stack is a standard maneuver, irritating because it increases the complexity of a sector, workload, and delays. It is routine for them because they do it often, but it is still a difficult and delicate operation calling for improvisation in the moment, skill, and considerable concentration while they continue to work their other airplanes.

Physically distant from the social location of people in the same or other facilities, it is by their experience with handoffs—both giving and receiving—that controllers in different physical locations define the personality of an airspace or a facility. Areas D and E, located at opposite ends of the control room's center aisle, illustrate how the airspace affects the task and therefore the personality of a place (see fig. 14). The comparison also returns us to the question of how the past materializes in the present. It shows the system effects on the airspace, and how airspace determines personality; further, it explains why controllers say, "It's not the airplanes that are the stress of this job, it's the people you work with." Although every area in the center has a definitive personality and so could illustrate the boundary problem, Areas D and E are markedly different from the others and from each other, so they are the best examples. Further, Area E coordinates with New York Center, which creates similar problems for E, in turn affecting other center areas. Consequently, they illustrate the boundary problem within the same facility as well as between physically distant facilities.

Controllers in other center areas describe Area D's personality as laidback, "the country club." They observe that Area D controllers' physical posture and demeanor on position is relaxed. The relaxed attitude, they explain, is a function of a "less-challenging" airspace. Area D usually takes handoffs, which fuels the interpretation that they are less busy than the other areas. This perception of "less-challenging airspace" is reinforced by the number of graying heads they see working in the area. Although its crews are as diverse in age cohorts as the other areas, they make jokes about "the old guys" assigned to work in Area D. Center controllers are firm in their conviction because they know how historical events have shaped and reshaped the Area D airspace and its volume. Area D is the largest airspace at Boston Center, covering almost half of the center's total airspace. Area D overlies the eastern portions of Massachusetts and New Hampshire, most of Maine, and owns airspace extending around 150 miles east out from the coast. Prior to the end of the Cold War, its airspace included several active military facilities, which made it one of the busiest areas at the center. When the Cold War ended, military facilities were closed and the amount of traffic in Area D was significantly lower.

Although Area D is not as high volume as other center areas, it is not a simple airspace to work. It has its own boundary problems. Unlike other areas,

Area D owns large stretches of airspace down to the ground, so in addition to high-altitude traffic, deals with the "little guys"—smaller aircraft, less experienced pilots, general aviation—that need more assistance. In the summer Area D gets "hammered" by those little guys flying between the mainland and towers at Nantucket, Martha's Vineyard, and Cape Cod because they have to work those boundaries. Moreover, overseas arrivals and handoffs with Canadian centers across the border require time-consuming manual transmissions: that is, they have to talk to a controller to coordinate every handoff, whereas in the other areas, controllers can use the computer to "flash" the plane's call sign as it approaches a boundary to signal readiness to hand off to the next controller. It may have fewer operations annually, but it is the only area in which I saw four controllers working one sector during a summer rush hour, two standing and two sitting, to handle the congestion. However, controllers from other areas defined Area D's personality by its willingness to accept handoffs, knowledge of its lower traffic count, and observations of controllers when traffic is low. Location in the control room matters to how an area gets defined. The irony is that due to the walls between areas in the control room, other controllers never see Area D busy because during rush hour, thunderstorms, and other air traffic conditions, every area is busy and no one is walking the center aisle.

In contrast to Area D, about Area E center controllers say that "Area E thinks it's a TRACON." In confirmation, they point out that Area E set its lighting the lowest of any in the center, simulating the dark of a TRACON. They describe Area E controllers as "businesslike" and "professional" in demeanor while on position, quiet and intense regardless of traffic, always sitting straight and facing the radar, not talking much to each other, very regimented—descriptors not used to characterize the "personalities" of the other areas. Center controllers acknowledge the relationship between airspace and personality, again citing airspace history. The Area E airspace formerly belonged to New York Center. In 1988, during the final phase of the FAA's East Coast Plan to redesign airspace, intended to relieve congestion around New York, a section of New York Center airspace was reassigned to Boston Center.[6] Area E was created specifically to work it. Boston Center controllers went to New York to train, so they developed their expertise working with New York Center controllers who fed arrivals into the New York TRACON. The New York TRACON underlies most of Area E. Area E behaves like a TRACON because the airspace and traffic are analogical to that of the New York TRACON, guiding traffic into approach. The airspace shaped the techniques of New York

Center controllers, who taught Boston Center controllers to work it in New York Center before it switched to Boston. New York Center controllers were very regimented; Area E controllers worked the airspace the same way. The traffic and traffic pattern allowed for no alternative.

Area E works extremely heavy volumes of traffic landing and departing New York metropolitan airports. The airspace is small, with heavy traffic on routes crossing over many parallel routes descending into and ascending out of Newark, Kennedy, and LaGuardia airports. Area E's traffic volume and flow is tied to whatever is happening in the New York TRACON, which has a complex, layered airspace, is chronically understaffed, and demands that handoffs meet altitude restrictions or won't take them. "Unable" is a common response. In addition, Area E controllers report that New York Center also is hard to deal with: they expect people to do them favors but can't return them. So Area E has to put planes on spin, traffic backs up, and Area E says "unable" to other areas in the center. Area E is known—and resented—for its frequent refusals to take handoffs. To be accepted, handoffs have to meet Area E altitude and spacing restrictions. Center controllers describe Area E's personality in ways that I've heard applied to New York—difficult, elitist, and selfish—but at the same time Area E is respected for working a complex airspace with high traffic count.

In addition, the Boston Center's airspace divisions generate different personalities for each area, which translate into difficulties meshing air traffic across internal center boundaries. Controllers describe the five areas as "five little fiefdoms," the repeated conflicts between them frequently turning into continuing turf wars. Because routes are standardized, areas routinely coordinate more often with certain areas than others, and between certain sectors in those areas. Each center area has patterned refusals with sectors in other areas, based on incompatibility of airspace sectors and conflicting traffic demands, that tend to remain constant. This gets old. Repeating refusals create animosity, sometimes resulting in payback: "If you don't take mine, I won't take yours." They create frustration and ill will between controllers. Within the facility, the DSR monitor makes it possible to execute the "quick look": one controller can pull up on the screen the sector of a controller in another part of the control room to see just how busy it is. However, the interpretation of what the sending controller sees on the screen may not match the experience of the controller working the traffic. Because airspace traffic patterns vary, equipment varies, and pilots make errors, six airplanes may be a challenging workload for another controller.

Ultimately the supervisors of the two battling controllers try to negotiate a truce, but usually truce is temporary: the cause of the conflict is in the airspace design, so the problems repeat. Some get temporarily settled by "parking-lot justice" or result in an ongoing conflict. "Feed accordingly" takes on new meaning: the feeding controller conveys displeasure by "jamming" the other person with traffic instead of accommodating his or her needs. It is not always a conflict, though: some controllers get airplanes that do not meet restrictions and they'll take care of it and clean it up without many complaints and sometimes even a sense of pride that many people in their area will help out another area, for example, Area D. Alternatively, handoffs may also proceed smoothly because the two controllers involved have previously negotiated an informal bargain: "You take my airplane when I'm going down the shitter and when you are going down the shitter, I will take yours, even when I am getting hammered too."

Coordination between center areas is not the only "in-house" boundary problem center controllers have. They describe TMU as a sixth fiefdom: TMU is "the enemy" because it dictates how controllers work traffic and so challenges their ownership of their airspace. TMU's adjustments to regional traffic are announced in the areas by the supervisors. Controllers' own traffic situations become more complicated and their plans jeopardized. They don't know the national traffic situation or the logic of the changes, so some of the mandates don't make any sense. Similarly, TMU personnel do not know the immediate needs of areas. So the response of many center controllers to TMU directives for routing or spacing aircraft is "We're fucked." Complaints about TMU are a constant theme. As in the situation of Area D, physical distance and inability to know what's going on with the traffic is a problem. TMU is at the head of the control room's central aisle, separate from the areas. In the areas, physical distance and the lack of knowledge about the regional traffic conditions that generate the directives contribute to the antagonism. Despite the fact that TMU controllers are drawn from center controllers who worked in the areas, they are seen as "other,"—an extension of that other "other," the Command Center at Herndon. Airspace affects technology, physical layout, and architectural arrangements internally, all of which in turn shape not only the culture of an area and the work controllers do but also their emotional experience of doing it.

The center's in-house boundary problems over areas are a microcosm of boundary problems between the center, other facilities within the New England Region and at other regional centers. Traffic keeps flowing because the variation in this incompletely standardized system is bridged by controllers who are stuck between the system's need to avoid congestion and their personal need to control their own airspace to the best advantage of the aircraft in it. In having to compromise these needs to meet the needs of another controller, they experience a loss of control. Trained to control, the effect is anger, frustration, and resentment, but—ironically—most controllers don't fully grasp the systemic source of their problems. They blame individual personalities, or facility personalities—laidback, arrogant, selfish, big ego, bad attitude—not the way the airspace shapes the way the work is organized. The system effect at the local level is turf wars in the sky, experienced as stress with repeated refusals and smoldering resentments toward the airspace owner who is the obstacle to moving traffic along its designated route. The aggravation is not simply about the difficulty of keeping up with their own traffic during heavy traffic times or thunderstorms. For each controller and facility, the success of the handoff is essential to meeting the institutional mandate for the expeditious delivery of air traffic.

Bedford Tower

Airspace, Place, and Work Practice

Located about twenty miles northwest of Boston in Bedford, Massachusetts, Bedford Tower is located at Hanscom Field, an air force base and civilian airport located in the midst of a bucolic, rural area near the historic towns of Lexington and Concord, where the American Revolution began. Bedford is a small but busy tower with 120,000 operations annually. The traffic count is eight hundred on an average day, often reaching one thousand in good weather. Bedford controllers proudly point out that "it is the second busiest airport in the region," and "on a summer weekend, the workload per person is greater per day than at Boston Tower." A general aviation facility that also handles corporate jet traffic, Bedford offers major relief to Logan International Airport's congestion and also business traffic convenience. The fleet mix Bedford handles is both visual and instrumental flight rules, both military and civilian: corporate jets, private pilots, flight schools, commuter and commercial aviation, charters, and occasionally military transports from the colocated Hanscom Air Force Base, a defense-research facility. Bedford controllers are responsible for arrivals, departures, and ground control; in contrast to towers at larger airports where approach control is handled by a TRACON, Bedford do their own approach control, putting traffic into sequence for landing. Bedford "owns" the airspace up to 2,500 feet in a five-mile radius of the tower and watches the airspace between 2,500 and 3,500 feet. Boston TRACON owns the space above them. Bedford exchanges air traffic mainly with the TRACON. They "talk to" the center and the Massachusetts Port Authority (Massport), which operates public transportation facility lands and properties in the state, including Logan International Airport, Bedford, and others) when they need to report unusual circumstances, like closed runways, technical problems, weather shutdowns, or accidents, requesting assistance when needed.

Entry is through the high-security barriers of Hanscom Air Force Base. Military personnel in camouflage guard the gate. Once inside the barriers, those admitted drive through the quiet residential streets of the military base to the tower unaccompanied and without a badge. Barbed-wire fencing separates the tower from both the civilian terminal and the buildings belonging to the military. At 137 feet high, the small white tower more resembles an old lighthouse in need of paint than a modern air traffic control facility. It is surrounded on one side by low-lying, small buildings that house several flying schools, hangars for the schools' planes, and parking lots for privately owned aircraft. On the other side are Bedford's two runways. Attached to its base is a small rectangular frame building with two doors, one to the facility's air traffic manager, the other to his secretary and an office shared by the supervisors. Only an FAA decal on the office door window identifies it. There are no resident Air Facilities technical staff (they travel from Boston Center to the smaller regional facilities when necessary), no Traffic Management Unit, no weather specialist, no computer, no cafeteria, and no NATCA office, although the facility has active NATCA representatives. Directly behind the tower is a hole in the ground that marks the beginnings of the foundation for a long-anticipated new tower, requested in 1985. Construction began but stalled with lack of funding from the FAA. Controllers enter the tower by punching in a code at the ground-level tower door; visitors first must be identified and approved by the air traffic manager and then buzzed in and up by controllers.

The tower cab is up eight flights of metal stairs. On the floor immediately below it is a small kitchen with a table and television that serves as the break room. The single distinctive feature of that room is a red floor-model popcorn popper that belonged to a controller in training, who sometimes works as a DJ and takes it with him to parties. Opposite the kitchen is a tiny unheated, unisex restroom with a slit of a window overlooking parked airplanes and one of Bedford's two runways. Up one more flight is the tower cab. Halfway there, on the landing, is a low door leading to a narrow catwalk around the outside of the tower, where controllers sometimes escape for breaks—fresh air, a cigarette, privacy. The tower is high enough to work traffic but close enough to the ground to take in activity behind the airfield. Nearby, the pilot of a small private plane moves the plane into the compass rose, that early marine navigational aid continued into the present as a circle marked on the ground at most airports, with lines oriented toward magnetic north painted every thirty degrees. Assisted by a mechanic using hand signals, the pilot slowly performs the "compass swing" to check his magnetic compass for accuracy. The air force base, now mainly a defense-research facility with some military transport traffic, is usually quiet. Occasionally, military personnel in camouflage walk from one base building to another. Passengers cross the tarmac to board the shuttle to New York. An instructor from a flying school and a student get into a plane, sit for a while, then taxi toward the runway. Birds, feeding in the grass, are startled by the engine and take off.

As at the other facilities, the airspace governs the rhythm of their work at Bedford, but it creates a quite different rhythm from that of the other three facilities. Bedford is open from 7 am to 11 pm, staffed by two supervisors and three teams of five controllers each who divide two shifts a day, alternating weeks off. Because of their VFR traffic, there is a great deal of seasonal variation in number of operations, given inclement weather, and daylight also has an effect: after the sun sets, all such traffic stops and they have little to do. No midnight shift here: the tower closes at 11 pm. Traffic slows in winter, and as at all other facilities, summers are extremely busy, but again, Bedford's pattern is different. Traffic increases on weekends, rather than decreases, due to the "weekend warriors": student pilots and private planes. Their work includes both separation and sequencing: putting traffic in a line so that airplanes arrive and depart chronologically, one at a time. The rules of separation require aircraft to be three thousand feet apart during sequencing for landing and takeoff. But unlike the center, where controllers have total responsibility for separation, at Bedford the responsibility is shared with most of their pilots. The key concept for pilots using VFR is the same as in earliest days of dead reckoning before controllers and technology—"see and avoid."

Controllers are on position two hours, with fifteen-minute breaks between sessions and a one-hour lunch. Lunch breaks are in the kitchen, usually alone, paying bills, reading, TV, or video games. Sometimes, for lack of anything to do on break, controllers return to the tower early from lunch to be with the others, helping out if traffic is heavy. Each morning, the first in starts the coffee maker in the tower cab, turns on the tower equipment, and opens for traffic, guided by the "open watch checklist" Sometimes on Sunday, after morning traffic settles down, a senior controller uses his break to make omelets for everyone with ingredients brought from home. During slow periods, popcorn, the Sunday omelets, and dinner takeout from restaurants on the military base are often consumed in the tower. Controllers are not allowed to leave the base during a shift, but they could make food and coffee runs to base restaurants and Dunkin' Donuts. At night, someone on break might walk over to the terminal and pick up warm cookies for the group, donated to the cause by the concessionaire. Upon first arrival in the tower, a controller told me: "You're supposed to bring us donuts. That's what visitors do." After successful demonstration of competence, I was officially given "donut control," only to be stripped of it a few weeks later by the healthy eaters in the bunch. Although surrounded by barbed wire and governed by the same institutional rules and procedures as the center, the feel of the place was not of rules, regulations, confinement and surveillance, but personal and informal. After my first week there, one controller told me, "You will see better facilities, tighter operations, and more sophisticated equipment, but you will have more fun with us."

The connection between airspace and place is as vivid as it is at the center but stands in dramatic contrast to it. The tower cab is a hexagon, about twelve feet in diameter with a three-foot shelf doubling as an equipment and work desk attached to the wall extending around the tower's circumference to the stair well (fig. 16). Tower size is deceptive, however, because the elevation and the deep, circumferential windows expose controllers to sky, weather, and woods, giving the tower cab an expansive feel. The light and color was startling after the dark of Boston Center. Technology does not visually dominate, as it does there; people, runways, airplanes, and nature do. The elements of nature are not only visible to controllers; they experience them with their senses, with changes in foliage, weather conditions, temperature, day and night, the sound of wind and rain all visible from the tower windows. Subjected to nature's seasonal effects, they try to control them with sunglasses, sweatshirts, raising or lowering purple-tinted window shades, or adjusting temperature with a space heater or air conditioner. Whether the cab is too warm or too cold was an ongoing point of contention. Nature presents hazards for tower controllers at Bedford that Boston Center and TRACON controllers do not face. Most obvious are ice, snow, and fog, but the less obvious also can be harmful. Sunset is beautiful, but dusk is the most dangerous time of the evening because of the position of the sun and the transitional light. It is too light for pilots to see the lights on the runway; for controllers, either the sun is in their eyes or it is too dark to see planes easily.

During a shift, three controllers and a supervisor work in the tower cab, eyes on runways, alleyways, and the sky, and they are sitting, standing, or walking depending on traffic conditions and personal preference. When traffic is busy, the chairs are shoved to the other side of the cab as everyone stands for better movement and vision. The window glass blocks the sound of the planes, so the room was filled by controllers' conversations with one another and with other controllers, and occasionally, pilot voices were broadcast into the room. A radio is always on, tuned to Boston's WROR FM all-day seventies music (at the time) or ball games, playing so very low in the background that they are barely audible except to controllers who are trained and practiced in the sharp hearing skills of room awareness. Shifts are from 7 am to 3 pm and from 3 pm to 11 pm. In contrast to the center's shifts, with daily time changes, here each crew works two weeks of "days" (from 7 am to 3 pm) and two weeks of "nights" (from 3 pm to 11 pm), with five days off in each two-week period. Both crews overlap during a shift change, jamming the space, as the controllers on position go through the checklist for the position relief briefing with their replacements. In contrast to the center, women are a more obvious presence here: during my fieldwork, four of the fifteen controllers were women, divided between the two crews but still a visible presence in the small space. In contrast, at Boston Center, only 20 of 260 controllers were women, scattered among the five architecturally separated areas and between crews within those areas, so hidden among the majority.

In contrast to radar controllers who remain seated while they work, at Bedford the work was very physical. Controllers and supervisors are often on their feet and on the move, eyes scanning the sky in all directions for aircraft. Similarly, rather than adjusting technologies by touching panel buttons, keypads, and cursors, controllers manually operated lights on top of the tower and on the runways. Some of this entailed climbing and jumping. Suspended from the high ceiling is a light gun (or Aldis lamp) on a track. The light gun is a relic of the early history of tower devices for contacting pilots, now providing redundancy to radio communication at Bedford and other similar-sized towers. A handheld signal lamp, it is manually operated, like a large cylindrical flashlight. Leaping to retrieve it from its ceiling track, controllers use it to flash standardized red, white, or green signals to direct pilots lost in the weather or in event of radio failure. Planes can acknowledge the signals by rocking wings or flashing landing lights.

Whereas at the center, procedures and other information can be displayed on the new DSR radar scopes at the touch of the keypad, Bedford controllers search through books like the 7110.65, letters of agreement, and other manuals for quick reference. Rather than automated broadcasts for pilots, hourly, they personally updated their voice-recorded message on the Automatic Terminal Information System (ATIS), a device that informs pilots about current field conditions: weather, winds, runways, radio frequencies, construction, repairs, or birds on the airfield. At 11 pm, they work through the "closing watch checklist": adjust the runway lights and beacon on top of the tower, record an ATIS message that the tower is closed, collect and count the day's strips, log delays on the typewriter, notify the TRACON, and head home, their voices echoing and metal stairs clanging as they race down. At closing, the TRACON takes over Bedford's airspace, which after dark is all IFR.

The correspondence between the space, place, and work practice is not visible in the design of the interior physical layout, as it is at the center. There, if all controllers were absent, the connection would still be visible in the architectural arrangement of areas, workstations, and air sectors. At Bedford, this relationship is visible only when controllers are on position. The architectural arrangement at Bedford is basic to all towers. In contrast to the center, where each controller works an individual sector, the controllers in the tower cab all watch the same sky. The positions are interdependent: airplanes are passed from position to position, left to right for departing traffic, the movement changing from right to left for incoming traffic. When one controller in the line gets behind, it affects them all. Officially, Bedford has four main positions: flight data control (or data), clearance delivery, ground control, and local (or tower) control. These positions are the standard ones for towers throughout the system, with additional positions added when a facility fleet mix or traffic level requires it.

The architectural arrangements and technologies enable distributed cognition among all personnel and devices in the tower. Controllers on position each work every plane in the airspace. Although each controller controls a different part of the movement, they all hear what is going on in the air and ground. For departures, task interdependence works clockwise. Seated far left at the work desk, which is shaped to follow the hexagon tower architecture, and facing the runways is the data controller, who receives computer-

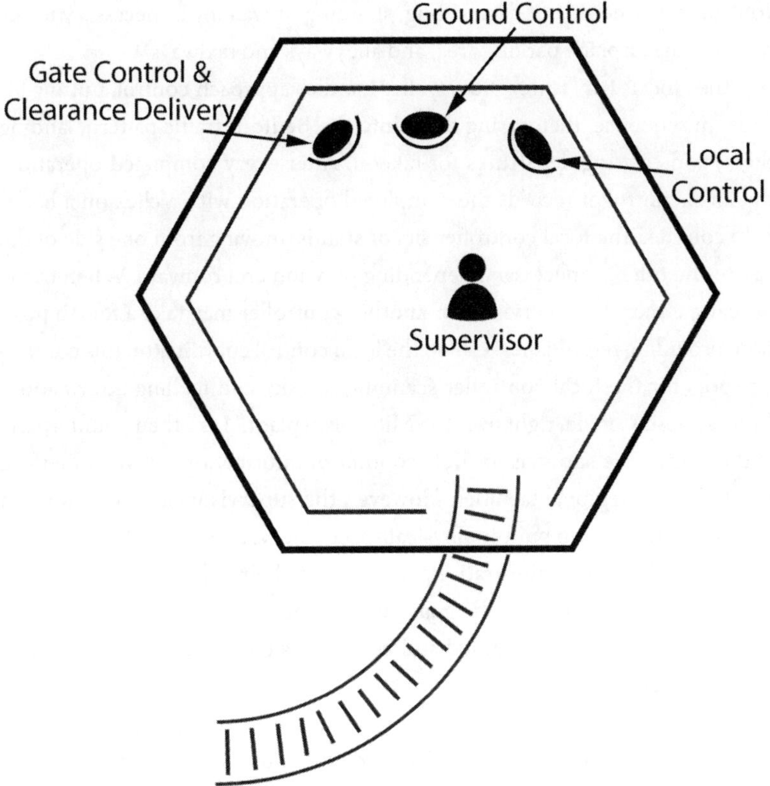

Figure 16: Bedford Tower architectural layout with positions, 2000
Illustration by Noah Arjomand

generated flight progress strips of aircraft coming into Bedford airspace from the printer. The outdated printer has the data controller "ripping strips": tearing off the information for one flight from a longer printout of strips and inserting it into a strip holder rather than having no-rip strips come one at a time, ready to go, like those generated by the center's updated system. The data controller then does the cleanup work of checking and correcting routes and arrival or departure times. Officially, Bedford has four positions, but in practice it has three, because the data position usually combines with gate control and clearance delivery (a separate position at larger airports, like Boston Tower). In this latter role, data gets departure clearance times from the TMU at the center, then coordinates Bedford departures with the TRACON. Next, the ground controller is responsible for the movement of all aircraft on the ground before takeoff and after exiting the runway upon landing. Ground control is on one side of data, sitting, standing or walking as necessary to see hangars, the airplane parking area, and alleyways and taxiways.

The "local" (or "tower") controller handles approach control, putting arrivals in sequence, then easing them into the Bedford traffic pattern, landing them, and clearing departures for takeoff. After every completed operation, the local controller records the completed operation with a click on a handheld counter. The local controller sits or stands, moving from one side of the cab to the other as necessary, depending on wind and runways. When traffic is heavy, either the supervisors or another controller may take a fourth position, providing redundancy. Called the local control coordinator, this controller spots for the local controller scanning the sky and feeding information: "There's your Cessna, right over the Mitsubishi plant." Like their counterparts at the center, the supervisors are a conduit of information between Bedford and the other regional facilities. However, the supervisors are not always in the tower. There is no built-in physical space reserved for them, like a desk or a console. They do administrative paperwork in their office downstairs, some of it "supporting" the air traffic manager's responsibilities. In their absence, a controller who has been trained and certified as controller-in-charge (CIC) fills the supervisor's role.

Bedford's technology was simple, minimal, and old: small airspace, small place, small share of the FAA budget. Ironically, the more "high-tech" of its technologies was actually low-tech and minimally helpful. A single radar screen—a monitor known as a D-BRITE—was suspended from the ceiling, showing approaching traffic and landmarks. It allows controllers to see radar images further than the eye can see. However, controllers used the D-BRITE only for rough approximations of airborne flight location because it was inaccurate. Images had to travel from a radar or host computer at Boston's Logan Airport through buildings and hills to the Bedford radar dish. The technical reproduction of images was poor. Also, the local radar dish was two sweeps behind the local traffic positions, so the data blocks on the monitor were "not even close to where the planes are." In one area on the screen, the data blocks completely disappeared because the tall buildings in Boston interfered with radar transmission. Most important technologies in the tower were the flight progress strips, radios, headsets, mics, and checklists. Technicians had to be called frequently to repair aging radio headsets and mics. Unlike the center, where mics and headsets were one piece, with the mics an extension from the headset to the mouth, at Bedford, they were two pieces; the mic was handheld, communication opened or closed manually. Controllers who instead chose to open and close their mics using a foot pedal periodically had to resort to duct tape to keep the pedal in workable position. The beauty of the strips, checklists, and binoculars was that they always worked.

As at the center, the flight progress strips were essential to distributed cognition at Bedford. A material artifact, they supply redundancy in case of a radar failure, but they also supply redundancy in case of human failure. Not only do they carry all the important information about an aircraft, including the hand markings and corrections the data controller makes to the strip, but also they are a physical symbol of ownership of an airplane. Controllers pass the strip from position to position as they hand off responsibility for the plane, in keeping with its movement. Memory is on the strip but also in the location and in touch: no one loses track of the aircraft because the responsible controller has the strip. Another essential device was less visible. My first day there, a controller turned to me and asked, "Has anyone shown you our pad management system?" (fig. 17). Consisting of rows of crossed-out numbers on a sheet of paper, nothing about it suggested an ordered sky. It is a pencil-and-paper list on scratch paper by the local controller to help keep track of all airplanes in the airspace during a session. A handmade technology of coordination and control, the pad management system also adds redundancy. Essentially a memory device, it is the handmade equivalent of a strip bay at Boston Center or Boston Tower.

The local controller (who initials at top left) records the position of all the aircraft as they move in Bedford airspace, as determined by contact with the pilot. The local controller makes three columns on a pad: "A" for approach (or inbound), "Seq" for sequence, and "D" for departure. Under approach, the local controller jots the call sign of each expected arrival in the order it enters the airspace. When the plane enters the pattern (i.e., the approach landing sequence), it is crossed off of approach and the call letters moved to the sequence column, marking its place in line. When the plane lands, the controller crosses it off. Departure is for sequencing planes for takeoff. The local controller enters the call sign to the sequence under "departure," and when it takes off, crosses it off the list. When the local controller is relieved, the paper holds the record of every operation during the session. After the number of operations is tallied, the list is thrown in the trash and the relief controller starts a new one.

Despite the national FAA hiring freeze that eliminated developmentals at Boston Center, training was a constant activity at Bedford. This, too, was a system effect: as a small airspace with diverse fleet mix and tower operations at 120,000 annually, Bedford trains controllers that are both upwardly and downwardly mobile. In the air traffic control system, controllers who want

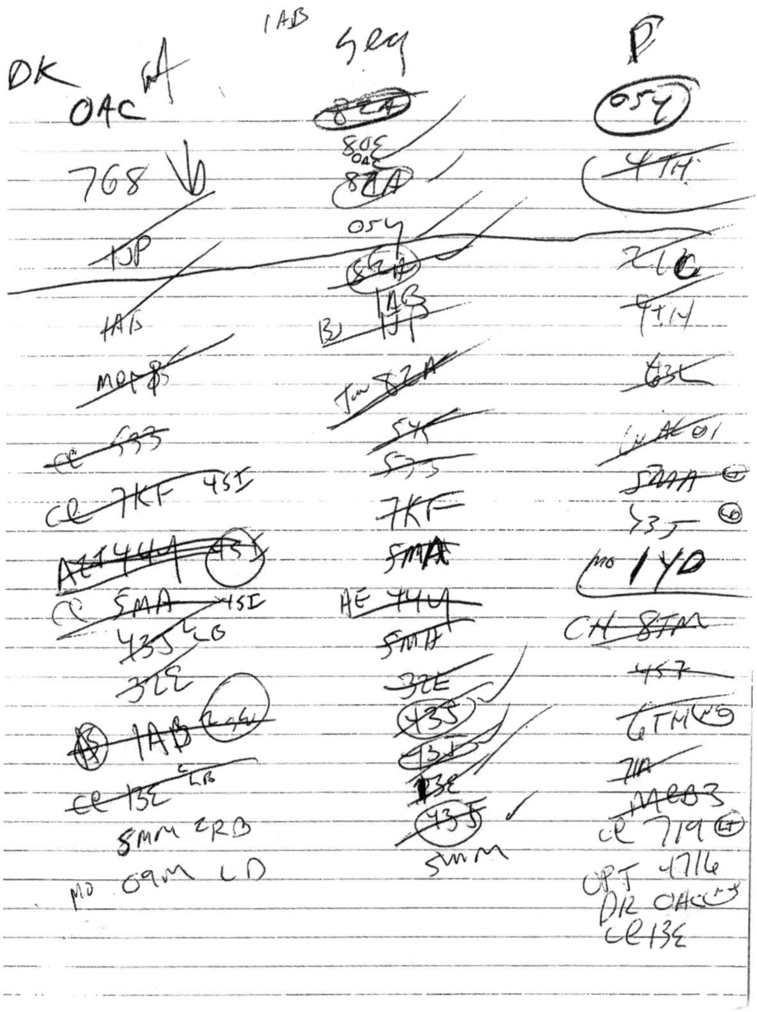

Figure 17: Bedford Tower pad management system
Photograph by author

to work at centers tend to be assigned to a center for training and if successful, tend to stay. If they wash out during training, they usually are reassigned to a small tower. Tower controllers have a different mobility pattern. They start at a small tower and try to move to larger ones, which increases their salary. Between my first (1998) preproposal visit to Bedford and my arrival August 2000 to start my fieldwork, three controllers had moved out and up, two to Boston Tower and one to Manchester TRACON. After retraining, all three were successfully checked out at their new facilities. Training to replace them at Bedford were a developmental who had washed out from another facility and two former PATCO controllers who had been at Boston Tower when Reagan fired the striking controllers in 1981. They had taken advantage of Clinton's window of opportunity to come back into the system. Their goal was to get back to Boston Tower.

Months later, when I arrived at Boston Tower, my last facility in the project, both of the PATCO controllers plus another Bedford controller were training there. Both PATCO controllers checked out, the other returning to Bedford. And so it normally went. The result was that Bedford had a high turnover rate with a core of controllers who remained. The origin of this core group partially defines Bedford's personality to controllers in other regional facilities. Bedford controllers complain resentfully that rather than having a reputation as a place where a small staff handles high volume traffic and a challenging fleet mix, or a place where excellent training is done, controllers at Boston Tower, TRACON, and Boston Center deride Bedford as an air traffic operation "run by a bunch of washouts." However, not all controllers want to be upwardly mobile. Despite the higher salary that mobility offers, some prefer to locate near their families rather than commute, as do most controllers at Boston Tower, many of whom cannot afford to live in or near the city. Others prefer the rhythm of the days and freewheeling air traffic control done at a smaller tower and would not want to work the fast pace of Boston Tower and TRACON or the size and regimentation of Boston Center.

I was at Bedford during the late summer and fall, when weather was good and the number of operations was high. This was the busiest time of year. The number of operations was a source of pride. The importance of the number of operations and delays was more in evidence here than among controllers at the center, where the number of operational errors was the major concern. As one controller joked upon taking position after hearing the latest radio story about congested skies and air traffic control delays, "Delays R Us." Every hour during the day, the flight data associate counted the strips for departures and arrivals, correlating them against the tally on a hand counter that the local controller systematically clicked after every completed operation, then tallied them on a sheet on a clipboard. Bedford's simple technology for tallying the count is a far cry from the computerized counting at Logan Tower and TRACON. Bedford was the only facility of the four without a computer for the daily log of activities, traffic counts, controller work schedule and sign-in and sign-out. Instead, Bedford controllers used a typewriter. Overall, the place had a general feeling of laidback ease, punctuated by moments of high stress. I was there during their busiest season, so missed the winter, but controllers described it as a quiet, even boring place because there was no VFR traffic. There was, nonetheless, a common complaint that was not seasonal: deprivation. Unlike the center, the source of deprivation was not rules, regulations, confinement, and surveillance. Controllers, supervisors, and the air traffic manager alike complained that despite being "the second busiest airport in the region," they were technology-poor: "The big centers get everything. We get the leftovers, if that."

Place, Technology, and Ethnocognition

Dead reckoning at Bedford is a throwback to another navigational era, when technology was simpler and used less, and human cognition and nature prevailed. Thus, place and technology call for a distinctive form of ethnocognition. The mix of standardization, interpretive work, and improvisation is different: working this airspace depends upon all three, but standardization is less than, and interpretive work and improvisation are greater than, at any of the other three facilities. Controllers watch the sky with the naked eye in a three-dimensional real world. Their airspace has no standardized airways or intersections indicating major traffic routes. Planes enter from both regulated and unregulated airspace, the latter adding another unpredictable element. Watching tower controllers work is dazzling and mystifying because you cannot tell what they are looking at (they are constantly scanning), and you can't identify the airplane they are talking to as you can when sitting at a radar scope that narrows your search with data blocks that state the call sign. My experience at the center gave me some important basics of air traffic control, but still I was ill prepared for air traffic control at Bedford Tower. Confused that first week, I said to no one in particular: "I can't see any patterns. How are you able to do this?" One immediately responded:

> Structure and routine, structure and routine: the rules and regulations (structure), and how we respond that becomes automatic (routine). If we have something new, we have at least had something like it before. If we have to improvise, we improvise from the base. (Bedford, M)

His comment was profound: a succinct, accurate statement of the fundamental principles by which all controllers, in every place, work. Despite the greater reliance on early navigational skills, local knowledge, and tacit knowledge to accomplish the work, Bedford air traffic control was still highly organized, but according to the needs of its airspace. Institutional rules and standardization serve the important function of orienting controllers to directional headings, destinations of aircraft departing and entering the Bedford airspace, and the coordinating of aircraft movement between one airspace and another, so controllers can coordinate with pilots and with facilities across great distances. And instead of the invisible lines in the sky that represent highways and intersections, towers have visible lines on the ground: runways. Runways are a major actor in distributed cognition between controllers and pilots. It is local winds that dictate the original and current runway configuration, and the runway configuration and winds dictate controller decisions about runways in use at a given moment, and consequently the airspace patterns of arriving and departing flights.

Placed to take advantage of local winds necessary to flight and harmonize with surrounding geography, Bedford's airfield layout consists of two crossing runways, one 7,011 feet by 200 feet, the other 5,107 feet by 200 feet (fig. 18). The standardized practice of all towers is for aircraft to take off and land into the prevailing wind, which reduces takeoff and landing distance. Controllers select the active runway or configuration of runways according to wind direction. The runway configuration then establishes the orderly flow of traffic in the airspace, known as "the pattern." The pattern comprises "legs." The terms controllers use to describe the legs are based on the position of the leg relative to the direction of the prevailing wind and the assigned runways. It works like this: When an airplane takes off into the wind, the leg is known as the upwind leg; a plane flying perpendicular to the wind on the departure side of the runway is on the crosswind leg; a plane flying parallel and toward the arrival end of the runway is on the downwind leg. The base leg is also perpendicular to the prevailing wind and is intended as a "base" as the plane begins its approach for landing. The last leg, when the plane is aligned with the runway for landing, is known as final approach.

The airfield itself is a technology of coordination and control. Runways, taxiways, and alleyways conform to international standards, so display numerous signs, numbers, markings and technologies such as patterns of lights and beacons, providing a common symbol system that is a major means of silent coordination between pilots and controllers. Runways are numbered and taxiways are named according to the phonetic alphabet (e.g., Alpha, Bravo, Charlie). Clearly marked signs identifying runways and taxiways are erected

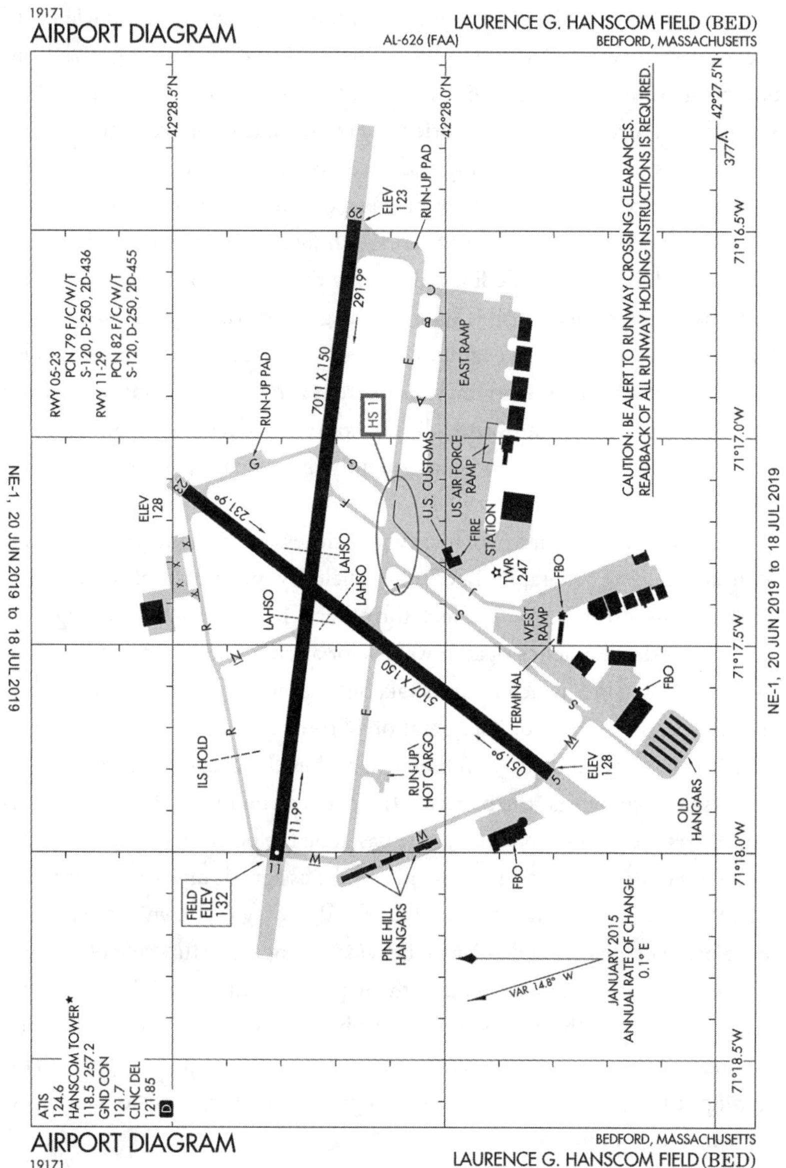

Figure 18: Bedford airport diagram

on the airfield, so passengers also can see them as their planes taxi by. However, passengers never see the complete layout that pilots see as they are on approach. Nor do they grasp the meaning of the signs and markers as they pass by.

My first week at Bedford, I noticed that local controllers frequently told arriving pilots to "extend beyond the numbers." Each time, I looked for but saw no numbers. Finally, I asked, "Where are these numbers that you tell them to go beyond? I don't see any numbers." The ground controller answered, "They're painted on the surface at the end of the runways." Then he followed with, "Do you know what the numbers mean?" I didn't. "They're headings," he told me. Safety and silent coordination were scripted into the runway symbol system: pilots were assigned the compass heading that matched the heading of the runway on which they were to land so they could see it and align with the runway for landing. Runways are identified by a number between 1 and 36, to correspond with a 360-degree circle on a magnetic compass. So if a plane is assigned to land runway 27, the pilot's assigned heading is 270 degrees, and the pilot lines up with the number 27 on the runway. Runway 27 points west, heading the aircraft toward 270 degrees on the compass. A runway can be used in both directions, and each end is named for each direction separately, so runway 27 becomes runway 9 when the winds are coming from the east. The two numbers always differ by 180 (180 degrees).

While the institutional rules governing radar facilities and towers are crucial to common sense at Bedford, the emphasis is different here. Their facility airspace has Bedford controllers relying more upon local and tacit knowledge in dead reckoning. As one observed,

> It is a small tower, but we have a complex, busy day. We handle a wide variety of equipment, both VFR and IFR, military and civilian, which increases the diversity of knowledge necessary to sequence and maintain separation. (Bedford, M)

In contrast to the center's standardized airspace and professional pilots, Bedford's airspace and air traffic is a hodge-podge of differences and unpredictable events. The traffic includes flight instructors and students from the local flying schools, aerobatics pilots practicing in customized planes, business and private jets, VIP flights from Washington, charters for college and pro sports teams, celebrity entertainers with their own planes (while I was there: John Travolta, Jon Bon Jovi, two local TV newscasters, Livingston Taylor), air shuttles, and occasionally *Air Force One*. Controllers handle special functions, like arrivals for the Blue Angels show in the summer (the Angels bring their own controllers who manage the air show while Bedford controllers watch from the tower), military operations, F-18s for a "flyby" to produce a sonic boom for a special event, and idiosyncratic pilot requests, such as circling for a photo shoot of the military base. The blimp (i.e., airship) that floats over Fenway Park during a Red Sox game takes off, lands, and parks at Bedford all summer. The changing air parade is colorful and ever entertaining. Controllers are "plane spotters": when an airplane comes in that is a type they have not seen before—old military plane, new aerobatic model—always there is admiration and discussion. However, the consequence of this variety is unpredictable traffic and aircraft piloted by people with a range of competency.

Signals and Interpretive Work

Working traffic across the boundaries in the sky amid this diversity is the supreme example of distributed cognition as practiced in the earlier navigational form of dead reckoning, when standardization and technology were minimal. Controllers' own interpretive work and improvisation are highly visible in every operation. Whereas at Boston Center, traffic is guided along formal standardized routes, at Bedford the many general aviation pilots using VFR fly into their airspace unannounced and from all directions. The local controller has to work these new arrivals into "the pattern," an invisible but standardized flight path that includes incoming and outgoing traffic, in sequence and aligned with the runways, with two additional loops of traffic consisting of "the touch-and-gos": student pilots and teachers repeatedly practicing take-offs and landings. As the wind changes, the runways change, and the local controller alters the pattern. Timing is crucial. The planes already on final descent need to land, and those in the air, on approach, need to be repositioned for the new pattern. The same strategic moves are necessary at all towers when the wind changes.

At Bedford a runway change is subtle operation, coordinated silently. Everyone in the tower watches the same airspace and hears the same conversations. Trained at and rotated through all positions, they are able to coordinate activities without articulation. Rather than depend upon material technologies like radar to monitor movement of aircraft in the sky and on the ground, controllers notice when the windsocks on the field indicate a change in the direction of the wind and so begin a runway change. The local controller starts incrementally redirecting airplanes in a new direction, taking off and landing from the other end of the main runway. The only visible evidence of the change in the tower is that the local controller moves the position equipment—the chair and small table on wheels that holds strips and the pad management system—to the opposite side of the tower cab when the change is a fait accompli. The first few runway changes I missed completely. Noticing I was still looking the other direction, a controller said, "We're over here now, Diane."

Once planes enter the pattern, the local controller keeps them in sequence, maintaining separation by issuing "traffic advisories" to pilots, such as "Follow the downwind"; "You should be turning your left base near the mall, so look out for it"; "Follow the Mooney"; "Watch out for the Cessna off your left wing now." Whereas at Boston Center the controller's plan results in the controller imposing an order on the screen, Bedford controllers are ordering aircraft in the sky and on the ground, putting them in line. Separation is estimated by sight and experience. Unlike center controllers, who visualize in three dimensions by envisioning a block moving through space, Bedford controllers visualize "the pattern," with planes in sequence with three thousand feet between each one. Controllers plans are dominated by one-at-a-time moves, not a series of them. New players "get in line," or the controller has the newcomer "loop around for a while" until it can be fit into the sequence.

Controllers' interpretive work is analogical to the dead reckoning of early marine navigation, but it is based on more than wind, drift, and position of sun and stars. Keeping separation is based on local knowledge. They are able to predict the position of aircraft and organize spacing between them because they recognize equipment by sight and know from direct observation its performance capability in all kinds of weather. They don't calculate airspeed and wind, for example, but know equipment and mentally adjust for wind effects. Changes in the weather—and therefore the wind—are predicted by observation: "You can tell the weather is breaking up because the sky was all one color, now we're beginning to get cloud definition, light and dark." They notice the behavior of the flocks of birds on the field, cloud movement and formations, and their knowledge of the different effects of wind when a plane moves between the sea breeze at six hundred feet to the tailwinds above. They watch the windsocks on the field. In the summer, the blimp that flies over Fenway Park during baseball games is tied down at Bedford, unintentionally serving as a giant windsock: like a boat at anchor, it heads into the wind. These objects convey symbolic meanings that are part of collective tacit knowledge.

Whereas at Boston Center, landmarks are technical objects, like beacons, represented by symbols on a scope, at Bedford they exist in nature. Controllers use them in dead reckoning to assess aircraft location and distance: a water-processing plant, three towers, a trailer park, the shape of surrounding hills, the skyline and terrain, building tops, shopping malls, highways, and rivers. Controllers guide incoming traffic by giving "visual reporting points" that give their location, such as Sandy Pond or Minuteman Memorial Park. Also, controllers use these markers to guide pilots into the pattern: "Southbound Route 128, turn right at the mall." If an aircraft is in a particular position, controllers know how much time they have to get it down and off the runway before the next comes in. Like center controllers, they have an experience-based system of known positions, distances, and timing in their heads and bodies, so they predict without calculating the distance. Into this equation goes local knowledge of pilot capabilities, which is based on direct observation in real time. Some pilots are in and out of the airport regularly, so controllers recognize their voices and/or their aircraft and know what to expect: who they can count on, who they need to watch. They know some of them personally. Pilots who work at the flying schools routinely brought their students into the tower as part of their training, and controllers sometimes fly with local pilots as part of a safety program at the field.

Bedford airspace boundary abuts only the TRACON airspace, so unlike the center, the major source of stress is not coordinating handoffs with other airspaces and facilities. The boundary challenge at Bedford is unruly pilots and delivering on the "orderly" institutional mandate along with safety and expeditious delivery of air traffic. Pilots are of varying strengths, needs, and capabilities. Many pilots are students and "infrequent fliers." At the center, pilots are primarily experienced professionals who give a predictable performance as they traverse a highly organized high-altitude system of standardized routes and intersections. Formality and institutional rules govern controller-pilot exchanges. At Bedford, standardization and institutional rules are still the basis of their interaction with pilots—phraseology, runway layouts and markings, headings, letters of agreement, procedures, frequencies, modes of communication, equipment requirements—but airspace and place leave room for the personal and informal. Phraseology remains consistent with institutional and local rules, but experience dictates circumstances in which informality and thus deviance from the formal rules are the accepted norm. A pilot, an excellent flier who is in and out of the Bedford airspace regularly, drops the standard phraseology and announces his presence with "Shazam—I'm here." They know him. If they are having an easy day, "We don't do much with him because he knows what to do. If we are busy, we respond formally, and he snaps to."

Informal courtesies are often exchanged between pilots and controllers. For example:

- A pilot, who was the only flight in position for departure, notified the tower that he had left his American Express card on a counter in the flying school. Controllers contacted the school, which sent a van out with the card.
- To a pilot's request for approach, a controller responded, "Your choice, sir. Be aware of border traffic." He said, "Thank you, I appreciate that."
- Controllers think of Bedford as "a teaching airport." They may give instruction to a student pilot practicing solo ("Start your turn now") or pay a compliment.
- To another pilot's request for approach, a controller responded formally with call letters, followed by "Lots of people are talking to me right now." The pilot looped around until traffic settled down. When the controller was free to get back to him, the controller said, "Thanks for your help," then to all in the tower said, "Nice guy."

These exchanges are part of the fun controllers get from the job. Yet these moments of courtesy and informality are interspersed with difficult exchanges that provoke other emotions and deviations from the professional presentation of self they were taught. Bedford controllers are trained to control traffic but their airspace and fleet mix makes traffic harder to control. At Boston Center, I never heard a pilot contest an instruction from a controller. At Bedford, professional pilots want priority in sequencing over general aviation. General aviation pilots protest when priority is given to commercial flights. All pilots contest delays, often caused by conditions elsewhere unknown to them, so the data position becomes "a hand-holding operation." Weekend warriors get lost, have less reliable equipment, forget basic phraseology, or mumble into the mic. Student pilots make beginners' mistakes, turning on the wrong taxiway, or wander into or out of the airspace without notifying controllers.

Identifying anomalies and correcting them is a major activity in this airspace. Unpredictability is the norm. Controllers refer to their work as "seat-of-the-pants air traffic control." Planes come from every direction. Some get lost. A pilot said he was coming in from the east, so the controller watched with binoculars but didn't see him. Noting the pilot's tone of voice and hesitation, the controller checked another direction. He was there. Many student pilots come from other countries because they can get better training in the United States at less cost. Often, their knowledge of English, the international language of air traffic control, is poor. Controllers pay extra attention to them. Deviations happen quickly and are bold, undeniable signals of incompetence and danger. Many pilot errors became part of facility lore. A student pilot taxied to the hold position for takeoff but faced away from the departure runway. To correct, the controller asked for a 180 turn; the student did a 360. In another instance, a student pilot entered the tower airspace without permission. The local controller gave the call letters and said, "State intentions." Getting no response, the controller repeated. After a pause, the student slowly said, "I want to be airline pilot." Having a firm control of English is no guarantee of experience. Controllers at the other facilities who have worked at VFR towers and moved to IFR said that being a controller at a visual tower is like being "a ringmaster at a circus."

At the center, standardization and institutional rules are the ground against which the deviations stand out; with professional pilots, conformity is the norm. At Bedford, the inexperienced pilots that populate the airspace make anomalies the rule, not the exception. This changes the nature of the controllers' work, accounting for the increased salience of interpretive work and improvisation. Moreover, their traffic calls for greater incorporation of social control. Controllers in every type of facility have the responsibility to sanction pilots who jeopardize safety by disobeying a controller directive, by having poor or squealing radios, by lacking or misusing equipment, by miscommunicating, by not following the prescribed route or rules, by being inattentive or difficult, or by acting in other ways that indicate a risk to themselves or others. However, towers discharge this sanctioning responsibility more often than en route centers or TRACONs because tower controllers have access to pilots when they are on the ground. Because of the variation in pilot competency, far more frequent at Bedford than at the other three facilities are pilot actions that indicate inattention, inexperience, bravado, or technical problems that create risks. Responsibility for social control adds to the work. "Pilots are the enemy," said one frustrated controller, "You can tell at the first transmission whether you can trust them or not." A controller can have an operational error because of a pilot mistake. Worse is the potential for tragedy, which some of them had witnessed on the airfield.

Frequently, an inexperienced pilot has a "stuck mic": after use, the mic button sticks down, and the mic is left open without the pilot realizing it. Conversation in that cockpit is heard by all pilots and controllers on that frequency. Meanwhile, other transmissions are blocked. Because pilots and controllers know the same international rules, they know how to coordinate activities without being told. An example of standardization, silent coordination and the resiliency of the system, when one pilot has a stuck mic everyone else in the air automatically switches to an alternative frequency. But the pilot with the stuck mic cannot be contacted, and sometimes will violate separation or a procedural rule, like landing without a clearance. A pilot not in communication can cause an accident. After asking a pilot three times to turn on her transponder, a controller said, "Full stop. Cleared to land." He said, "When I ask three times and she can't hear what I'm saying, that's danger. I've got to get her out of the sky." Controllers have to be alert for surprises. An airborne pilot takes an unannounced 360. Or a pilot turns left into the pattern so has a two-mile final approach rather than the mandated five miles. Another is slow to get off the runway and the controller has to send the next pilot in the sequence on a "go-around": to preserve separation, the pilot is ordered to abort the landing and circle away from traffic, then gets directed back in line.

Anomalies, nonconformity, and responsibility for social control take an emotional toll. They increase the complexity of controllers' task and make the job more difficult. Sometimes controllers' frustration breaks through the inhibitions of professional training, escaping in language and tone to pilots who deviate from the mandated phraseology. It is also the case that being a ringmaster in a difficult-to-control circus with an overabundance of performers in the air and on the ground means that sometimes controllers miss something. However, the VFR principle of "see and avoid," gives the pilot responsibility for separation. I got to experience firsthand this reversal, where the controller misses something and the pilot corrects.

The Bedford controllers had the idea that I should see the field from the sky and hear them working from the cockpit, as pilots hear them. I was enthusiastic, so they made an appointment with one of the instructors at a flying school on the field to take me up for a demonstration. It began as would a flying lesson: an explanation of the principles of flight and aerodynamics, complete with blackboard diagrams. Then he took me to a Cessna 178 and walked me through the preflight safety check, exterior and interior. Inside, he taught me to start the engine and to use the mic, and he dictated each of my contacts with the controllers, simultaneously teaching me to work the pedals to navigate the plane from its parking place to the nearest alleyway. Maintaining a straight line with the pedals was not easy, and I was meandering side to side, depending on which foot I was using. Meanwhile, I noticed that my pilot recognized the controllers' voices. I finally reached the alleyway and was weaving back and forth down it toward the "taxi into position and hold" point, with great embarrassment, because I figured the controllers were all hooting at my incompetence. I was wrong.

The instructor suddenly told me to hold short, pointing out a much larger plane (it looked huge) coming off the taxiway right at us, to enter the alleyway. Apparently, ground control hadn't been watching us at all. My instructor had seen and avoided. He said, "It's important to pay attention here. Ground control should have stopped us, but sometimes they are very busy and get distracted. I know her voice. Usually, she is good—very clear, very helpful, very careful." He attributed the incident not to individual carelessness or error but to a system effect. He said Bedford didn't have enough controllers for the amount and variety of traffic it had. He compared it to other towers of similar size, complaining that it was hard to find a small tower with sufficient traffic, equipment, and controller staffing to train students. Then my pilot launched into a story of a student pilot of his who was practicing solo touch-and-gos at Bedford. He was in sequence in the pattern. The local controller neglected to turn him from the downwind onto base, and lost and waiting for instruction, he continued straight to the ocean, which gave him the idea he should go back.

I will not give the details of my flight experience, except to say that my instructor thought the arrangement was for a flying lesson while I thought it was a free demonstration. I was surprised by being given responsibilities ("OK, now pull out the throttle"). The pilot was, of course, in control, with me doing some things poorly. As a learning experience, it was invaluable. I experienced the principles of flight, saw the field from above, saw "the numbers" painted at the ends of the runways, heard the voices of all the controllers and the pilots' responses on the frequency. At one point in the descent, about fifty feet from the ground, he said, "Cut the engine," and as we floated down in the silence, he took over, landing with no incident. I returned to the tower cab a hero. "Tell us war stories," they said first. Then, "What did he say about us?" I told them.

Boundary Work: When Technology Fails

In emergencies, both kinds of boundary work have to happen: controllers must manage traffic across the invisible boundaries in the sky and negotiate the boundaries of people and organizations on the ground that also use and own the airspace. Interdependence, coordination, redundancy, and resilience come to the foreground. In contrast to Boston Center, Bedford controllers all watch the same airspace and have fewer external organizational boundaries. Although they regularly coordinate traffic with the TRACON, less often do they coordinate with Boston Center, Massport at Logan, and less frequently still with other relevant FAA units. In emergencies, these parts of the system must be brought together. Bedford controllers have less material technology than larger facilities; they rely less on what they have and more on common sense: their acquired cultural system of knowledge. Nonetheless, they are fully dependent upon radio. The day the radios failed at Bedford, so did technological redundancy: the backup receivers and transmitters also failed.

Pilots and controllers could not hear each other. Controllers improvised: they tried alternative techniques, increased their reliance on one another, and took actions to bring the necessary parts of the system together. Nearly deviceless, the tower team went from its normal state of interdependence to become a more tightly coupled system. The division of labor changed. Formal positions abandoned except for the local controller, the others scrambled to solve the problem and provide backup. With technical specialists distant from the tower rather than part of it, as at the other facilities, they were left to their own resources during the crisis before help arrived. The reaction of the controllers to the unexpected breakdown shows what happens in a technical system failure and how distributed cognition changes in a crisis. The local controller reinterpreted the situation, reconstructing communication and coordination with pilots based on a silent signaling system that was a reversion to an earlier form of air traffic control. Collectively, all controllers were working toward resolving the problem, but the division of labor changed, so with one exception—the local controller—the other controllers were not doing the work of their assigned positions. They took of different support activities, investigating alternative technical solutions. The reaction of controllers to the technical system failure demonstrates most fully the human component in dead reckoning, the resilience and redundancy within a single socio-technical system, and boundary work.

It was a clear, bright Sunday morning in September 2000. The tower opened at 7 am, as usual. Oddly, traffic was low. They later remarked that it was "weird" that there wasn't more traffic on Sunday morning, when people usually are departing for day trips and students from the flight schools are in the practice area or doing touch-and-gos. The supervisor was not coming in until later, so Rebecca was working as controller in charge (CIC). Denise was working local, Jim at ground, and John at data. At about 8:30 am, they heard an unusual sound on the radio frequency. At first it was a clicking, then a loud squeal, then settled into a constant static. The speakers were on, so everyone in the room could hear it. Rebecca thought it sounded like an electronic locator transmitter, or ELT, used to locate planes if they crash. She was about to call Boston Center to alert them because the center originated rescue operations for the region. Then Jim noticed that the ground control radio was not working—he had a garbled message from a Beechcraft Bonanza—so the team concluded it was more likely a stuck mic.

But then Denise announced that her local control radio wasn't working. She had cleared somebody for takeoff, repeated herself two or three times, and still got no response. She had a departure in the air that wasn't responding either. Concluding that the main receivers and transmitters were out, they pushed the button to switch to backup receivers and transmitters. They got the same sound of interference on the line. John muttered, "The antennas are putting out this interference so it's like somebody's got their mic keyed the whole time, on every frequency that we have." With no computer in the tower, Rebecca, as CIC, reverted to the telephone. She called the TRACON to keep them appraised of the situation, then called Air Facilities to request a technical specialist, then Massport: if it were a stuck mic causing the situation, Massport would need to make contact with the aircraft. Finally, as a precaution she telephoned Boston Center in case it was, after all, an ELT they were hearing.

Other devices essential to the prediction essential to dead reckoning for both pilots and controllers failed. Every element providing redundancy, both human and technical was not available. The printer that produces the flight progress strips wasn't working, so controllers had no idea what was coming into their airspace. Neither was the ATIS, which pilots check for hourly updates on weather, runways, and airport conditions. Controllers couldn't update to warn them of the situation in advance. Jim and Denise decided to try the handheld, battery-operated, portable radios kept in the tower. Jim tuned his to the ground control frequency of 121.7, and Denise tuned hers to the local control frequency of 118.5. John left the data position, useless because no strips were printing. He ran down the stairs to get as many backup handheld radios as he could find so they would have extras. However, the handhelds didn't work either; they had no external antennas. Jim and Denise were standing close together, soon concluding that each radio was blocking the other's transmissions. Air Facilities phoned back to say that the telephone company Verizon had reported a chip out but didn't know when they would be out to fix it. Air Facilities contacted an AF specialist to come out "immediately," in the air traffic control phraseology sense of the word. But it was Sunday, Bedford had no Air Facilities representative locally, and it would be a long wait for help.

By 8:40, ten minutes had passed. Listening on the speakers, everyone concluded that transmissions on the ground control frequency 121.7 seemed to be coming in more strongly than on the local control frequency, so they switched both ground and local to the ground frequency. John said, "Pilots were able to hear us a little better [on 121.7] and we were able to hear pilots, but it was very difficult, you had to have them talk real slow, and we had to talk to them real slow and basically told them, we tried to get across to them what the problem was. You would have to do it about four or five times, it was very difficult for anybody to understand anybody."

At 9:10, Rebecca told Denise to take her break. She had been working local for two hours and they didn't want someone working the position tired, especially under the circumstances. Jim was due to switch to local, but would work local and ground combined because he already was working ground and both positions were now on the ground frequency. Combining positions was a strategy used at all facilities when traffic was low. Because the most challenging position at Bedford was local, it made sense that Jim was more rested and should remain. However, Denise stayed in the tower and both she and John provided support. She went out on the catwalk, experimenting with using one of the handheld radios to see if they could get better reception out there. She tried making blind transmissions, telling pilots who had been calling in on 118.5 to switch because they could be heard more clearly on 121.7. Jim suggested Rebecca call Massport to see if they had some more powerful handheld radios they could bring over. They did, but these worked only a little better than the tower handhelds.

There was little discussion: they coordinated action with minimal exchange of information and ideas. Distributed cognition continued in a different form: situation awareness and collective knowledge persisted despite being out of formal positions, absent the normal routine, deviating from prescribed phraseology, by improvising with different devices and finding new means to contact each other and support the local controller. Rebecca called the supervisor to tell him he should come in. She telephoned the flight schools on the field to tell them that they were stopping all departures. John rushed to get a set of walkie-talkies he kept in his car, heading out to the runway to inform planes that were holding short for takeoff. Denise took one, to stay in touch with John from the catwalk, as John got in his car and drove out to the approach, just in case he could help somehow, directing others coming in. Several pilots had the idea to call in on the tower telephone line to find out what was happening because they could not get through on the radio. Rebecca was taking these calls. There was one emergency. The pilot of a "Lifeguard," the call sign for a small plane, capable of IFR, that transports serious medical patients to hospitals, telephoned: "Hey, I've got to get out of here." On the phone, she advised as she would have on the mic: depart VFR, then contact Boston Center frequency to pick up instrument rules when he was able. Jim heard the call, but Rebecca conformed to standard procedure, telling Jim to expect him for departure and calling Boston Center to give them a heads-up that the Lifeguard was on the way.

It was a triple technical failure: the main transmitters and receivers, the backup transmitters and receivers, and the handhelds. For everyone, the situation was unprecedented: neither their training nor their work experience had prepared them. Later, Rebecca said, "I just remember thinking 'What else can we do?' Our radios don't work, that's the worst I've ever seen. It's not so bad losing your main radios, but when you lose your backups and your portables too . . . they worked but we couldn't hear well, it was like they didn't work. I mean it would have been better if they didn't work. If it had been a total loss of radios the tower could have been shut down and no services provided."

John said, "Closest thing I had to this was one time they were doing some construction over here and they cut some phone lines and we lost all our frequencies, all our electricity, but the handhelds worked. We could use hand sets and that was no problem."

Lacking training and precedent for this kind of technical failure, Jim, now working both local and ground, innovated, drawing upon a related experience:

> I figured I'm going to run this like an air show. I mean, what else have I got? I'm just going to bring them in to the downwind, I'll see them visually, you know call them out by their color if I have to and have them respond. I didn't try to make it like an air show but I figured that I knew—I had used that type of technique before, working an air show. It's a lot different because you don't have a structured training program like you do here. You go in, you get a half-hour briefing, and they sign you off to work traffic. When I did it we were sitting on a flatbed truck and working it from there with binoculars and handheld radios. And again, due to the equipment you can't talk to planes up as far out as we do here every day. Even though we don't have the greatest equipment, it's still better than a handheld. So really, they're in a lot closer to the

> field when you start talking to them. Then you just kind of have to pick people out on the downwind and you choose who you want to talk to. Instead of having them report at a certain point. And a lot of it is done visually, saying, "Blue-and-white Cherokee, what're your numbers, what's your intention?" Stuff like that, versus having the full call sign, having the standard pattern entry, telling them the weather report.

In switching to air-show technique, Jim also switched to full-time use of binoculars and an air-show scan. Instead of the local controller's normal eye movement between the pad, the D-BRITE, and out the window, he concentrated only on what he could see out the window. He jotted notes on a pad but did not structure information in its usual columns as they did for the pad management system. Instead, Jim just jotted down call signs, but not in any particular order. He relied mainly on his own memory of what was moving in the air and on the ground. However, as he improvised to meet the situation, he did not realize that the air-show technique that he knew had itself originated in the early days of air traffic control, when minimal technology was available and controllers' dead reckoning relied on jotted notes and silent signals of acknowledgment by pilots.

Planes approaching the airfield for landing began to hear him when they got three to five miles out, but how many of them could and how clearly he did not know. A lot of times there was interference, so he might pick up a call sign but not hear where they were because it was garbled by static. Jim cleared three flights using IFR for takeoff, one Cherokee and two jets, a Gulf Stream and a Challenger. Incoming traffic was all small planes, single engines, trainers—all VFR traffic accustomed to the "see and avoid" rule for entering the pattern and landing. Jim instructed them to use the standard procedure: squawk emergency "7600," the code for "no radio" to indicate they were aware of the situation, and enter the pattern. Only one pilot conformed: he squawked 7600, stayed above the traffic pattern to overfly the field and monitor the traffic, then entered the pattern. But they had planes come in that never talked to them, never used any of the official "lost radio communication" procedures. With VFR, they just entered the airspace, got in the pattern, and landed. Jim didn't know if they couldn't hear him or if they were ill prepared for the procedure. For those pilots who were not responding, John, back from the runway, grabbed the light gun from the tower ceiling to communicate with planes on final approach. Pilots are supposed to acknowledge light-gun signals by flashing the landing lights or rocking their wings. No one responded.

At approximately 10 am the Air Facilities specialist arrived, having "burned up the highway" during the hour trip from his home. He had the radios back within half an hour. The problem was fixed by changing the telephone lines, which had caused interference on all the frequencies—main and stand-by—controlled by the antennas on the field. Everything was back to normal by 10:30 am, at which time the supervisor arrived. Verizon never showed up. There were no further problems with radio failure that afternoon, at least, as John said, "no more than we normally have, the system being so old." In interviews, the team in the tower cab described themselves as elated in the aftermath. Everyone felt the same. They agreed it had been a rough situation and everyone had stayed calm and worked together and the situation had worked out.

Each person had performed a separate duty to assist everyone else, without discussing who would do what and when. They reflected that there was no yelling; indeed, everyone had remained outwardly calm. When issues came up that had to be discussed, everyone's opinions were taken into consideration. They had operated as a team. They said the situation was not stressful because the weather was clear and traffic was low: "nobody was in danger the whole time," "the adrenalin was not pumping because it was not busy traffic," and "busy traffic or sequencing that doesn't work is much more stressful." They commented that it was frustrating in the beginning as they were trying to figure out what had happened, but it was exciting and fun because it was not routine: they had to innovate and experiment. Proud of their team effort, they were angry at the air traffic control system, however. The radios had long been a problem, controllers had complained in the past, and the solution was always a temporary technological fix. John commented, "If the employer doesn't provide proper equipment, coordination between controllers goes up."

Because of the unique aspects of space and place at Bedford, the radio failure the controllers there experienced provides a clear example of what people bring to the job that technology can't replace. However, the capacity to innovate is a requirement of the job in all facilities. Recall this comment made by a controller at Boston Center:

> You need neat and orderly, but you also need flexibility, and you also need the ability to just pull something out of the air. And you have to because there are going to be times in your career when neat and orderly is going to go out the window and you're going to have to pull an idea out of your head. And it had better be there. (Center, F)

But also recall also this controller's insight:

> Structure and routine, structure and routine: the rules and regulations [structure], and how we respond that becomes automatic [routine]. If we have something new, we have at least had something like it before. If we have to improvise, we improvise from the base. (Bedford, M)

Their response to the technical failure was an improvisation, but ordered by common sense: the embodiment of institutional rules, local knowledge, and tacit knowledge. Notice how, in this unexpected situation, they enacted distributed cognition and deployed symbolic meaning and interpretive work while moving to different locations and technologies, so they were not tied to the tower cab and their accustomed material objects and devices. Moreover, they followed all the institutional and local rules, did all the required coordination with other facilities, innovated with other technologies immediately at hand, and borrowed technique from air shows. Structure and routine, structure and routine—yes, and they did then "improvise from the base." Tacit knowledge is experiential: much of what they did reflects their own previous experience with technical failures at Bedford and with alternative strategies of working traffic learned elsewhere. Ethnocognition intersected with distributed cognition. As for redundancy and resilience, they were their own best resource. As John said, "If the employer doesn't provide proper equipment, coordination between controllers goes up." The radio failure shows the relation of space, place, and dead reckoning, in particular the relation between technologies and human cognition, and how that varies by facility. It also points to how airspace locates a facility within the air traffic control system in terms of the distribution of resources and how that impacts the work of controllers. The construction of the tower requested in 1985 was stopped after the foundation: "The big centers get everything. We get the leftovers, if that."

6

The Terminal

Boston TRACON and Boston Tower

As startling as the light at Bedford after the dark of Boston Center is the contrast between the secluded rural locale of both those facilities with the urban, public environment of Boston TRACON and Boston Tower. The two are located at Logan International Airport, the nineteenth-busiest airport in the United States. The tower is a twenty-two-story, 285-foot structure of two cylindrical pylons, with six stories of offices near the top forming a cross between them that supports the tower cab. An iconic architectural structure, it rises out of a parking garage adjacent to one of the main terminal buildings. At every level of the garage, parked cars abut the lower portion of the two supporting pylons. What could be more public and urban? The TRACON is in one of the terminals, not far from the tower. The public location and its suggestion of accessibility are deceptive, however. Security is as high at the TRACON and tower as it is at the other two facilities; it is just not as obvious as barbed-wire fencing.

The terminal is like a small city within the larger one of Boston, populated by some two thousand employees and visited by thousands of transient travelers daily. There is even a Catholic chapel. It regularly serves its "city," holding mass for airport employees, people from East Boston, and interfaith travelers. Beautifully appointed and quiet, it is inconspicuously located down two flights of stairs off the walkway between two terminals. Just inside, an alcove holds a stylized Madonna atop a brass globe with airplanes circling it. Its discovery—and its name, Our Lady of the Airways—brings a jolt of recognition that it is a ready haven for families awaiting news in the case of an aircraft emergency. For controllers, the terminal offers bright, brief diversion from working air traffic. Invisible in the crowds, controllers go to Dunkin' Donuts for coffee to take back to their team or for energy-regenerating aerobic walks. Controllers are never fully away, however. They are tied to their supervisors by pagers or cell phones, and the crowds of travelers are a constant reminder that people are riding in the airplanes that they are controlling. The rhythm of their workdays can be measured not only by the number of airplanes in the sky or on the radar but also by the lines of people and piles of baggage in the terminal.

At Logan, as at all airport terminals, the airfield and the runway layout determine the airspace for the tower and TRACON, and consequently the kind of work controllers do at each place. The Logan airfield is located on a small, irregular piece of land almost entirely surrounded by the Atlantic Ocean and Boston Harbor and bounded by the city growing behind it. Because all airplanes must take off and land into the wind, all airport runways are aligned with prevailing winds and fit to the local geography. When the Logan airfield opened in 1923, it was two cinder runways on an 189-acre tract of mud flats, surrounded by marshes. The expectation was that, if needed, the area could be expanded by landfill. It was, but as traffic at Logan increased over the years, the city actively blocked further airport expansion, so runway additions were jammed into the small surface area. Compare, for example, Boston Tower's layout of crossed runways with that of Dallas–Fort Worth, which is located on a flat, open plane with parallel runways to handle high volume and the very different challenges for controllers' at each place (figs. 19 and 20).

Historically, the colocation of the Boston TRACON and Boston Tower at the terminal originated from their contiguous airspace. The tower is responsible for ground to two thousand feet; the TRACON owns the airspace from two thousand to fourteen thousand square feet for twenty-five square miles. Although the facilities are colocated, the differences in their airspace created differences in each place and in the social organization of work that are as opposite as those between the Boston Center and Bedford. All flights arriving and departing Boston Tower must cross the airspace boundaries between the two facilities. The tower "launches" departures, and the TRACON "catches" them, handing them off either to smaller towers in its airspace (e.g., Bedford or Providence) or to Boston Center, depending upon the airplane's destination. In turn, the TRACON receives incoming flights from the center and other facilities and sequences them, putting them in line to hand off to the tower for landing. This latter function is emphasized by the TRACON's full name—Terminal Radar Approach Control, known to controllers as "Approach" and

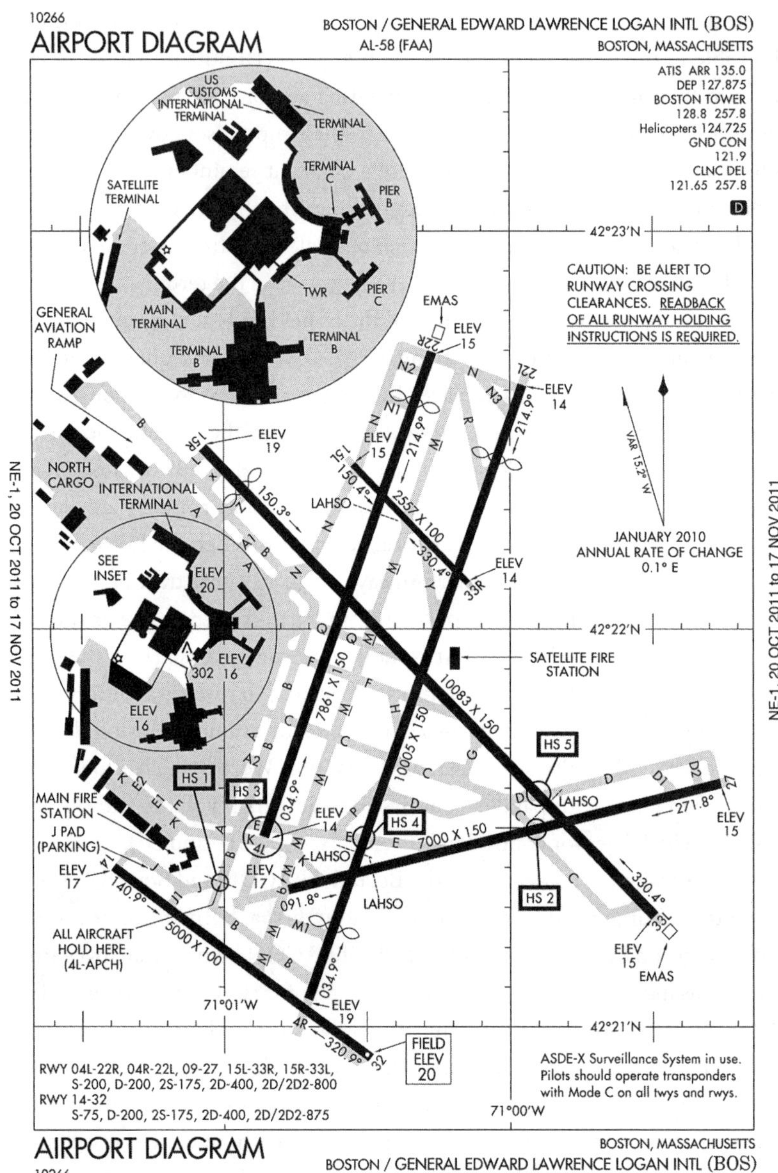

Figure 19: Boston Tower airport diagram

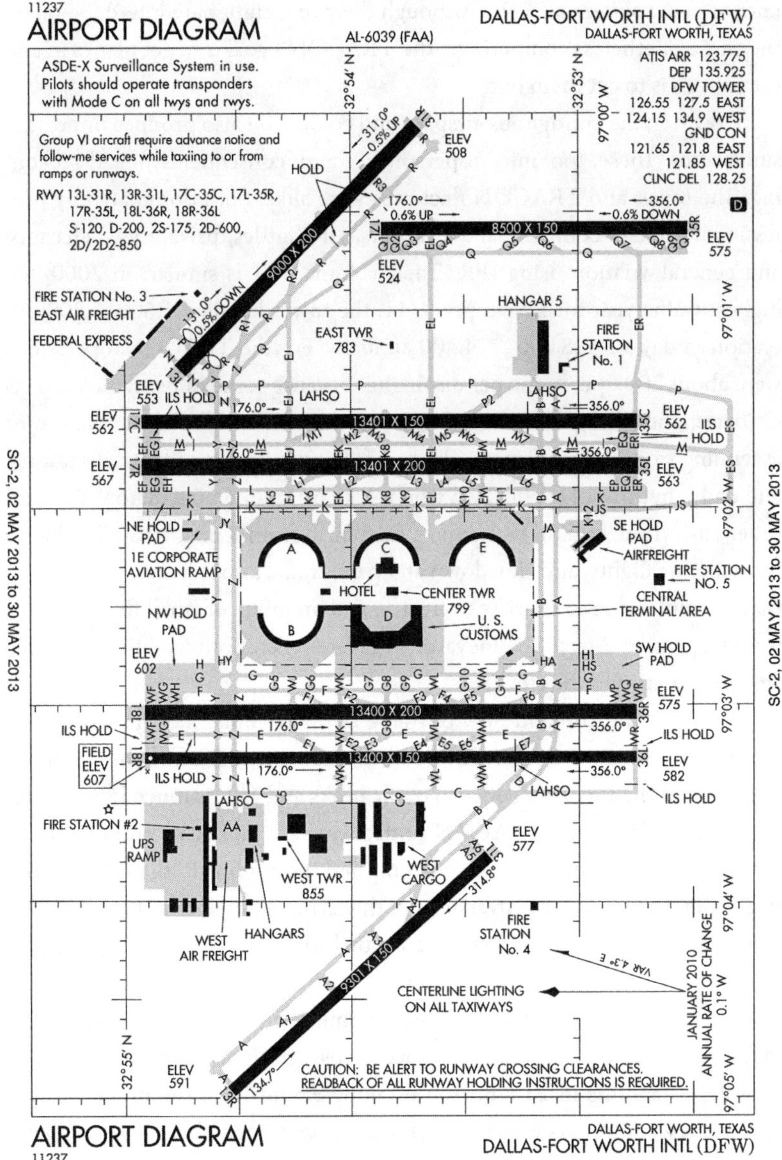

Figure 20: Dallas-Fort Worth airport diagram

known in the system as A-90. Although the two facilities are interdependent, they are nonetheless competitive: the TRACON's goal is to get planes in and the tower's is to get them out.

However, the contiguous airspace and colocation also produce important similarities. These, too, inform personality and, consequently, dead reckoning. The tower and TRACON fleet mixes are alike, both dominated by professional pilots—commercial air carriers, air shuttles, private jets, charters, and general aviation using IFR. Their volume, too, is similar: in 2000, the highest-traffic year for Logan pre-9/11, the tower handled 1,500–2,200 operations a day, or 528,000–774,400 annually. For the TRACON, operations were about 200 operations per day higher because some of its traffic consists of "overflights"—flights in TRACON airspace that are not descending to or ascending from Boston Tower. The two facilities also have identical seasonal and daily rhythms of traffic: heavy on weekdays from 6:30 a.m. to 9 a.m., another rush from 11 a.m. to 1 p.m., a final push from 3 p.m. to 8 p.m. due to transatlantic flights, and slowdown at 9 p.m.; traffic also is heavier in summer than in winter due to tourists to Boston and an influx of small flights to and from Cape Cod, Martha's Vineyard, and Nantucket. Both facilities run three shifts a day, around the clock.

Although the ebb and flow of traffic at the terminal has the same basic rhythm as at Boston Center and Bedford, the high volume in the small airspace makes the variation between traffic peaks and valleys much less. As a result, Boston Tower and TRACON controllers have more concentrated traffic activity, less downtime while on position, and shorter breaks than at the other two facilities. Despite the diversions of the terminal, their trips to it are brief. In the evenings, both facilities have a ritual that permits a short escape for one controller. Unlike Boston Center and Bedford, where controllers are not allowed to go outside the barbed wire, terminal controllers can take advantage of their urban location. Each place keeps a stash of takeout menus from favorite restaurants in East Boston or along Route 1. As the dinner "hour" approaches, controllers who make a thousand decisions a day about airplanes struggle to choose a restaurant and an order. Their indecisiveness about this is a standing joke among them. Once it's called in, one person goes on the "chow run." They eat in the break room as the break schedule and traffic allow.

Two additional similarities between Boston Tower and TRACON distinguish them from Boston Center and Bedford. Both facilities have a heightened preoccupation with speed, time, and schedule. These production concerns did not displace safety but rather coexisted with safety concerns, forming an aspect of personality at both tower and TRACON that was much more visible than at the other two facilities. This, too, was a system effect. As air travel increased in the late twentieth century and at the beginning of the twenty-first, quantitative measures of "the safe, orderly, and expeditious delivery of air traffic" became indicators for how the air traffic system was faring in the eyes of the public, the airlines, the government, and the FAA. All facilities track operational errors—the measure of safety—but the numbers for centers receive the heaviest scrutiny. However, the statistics for expeditious delivery of air traffic—the number of on-time arrivals and departures, and of delays—are counted only at terminals. A flight is considered delayed if it arrives at or departs the gate fifteen minutes or more after the scheduled arrival or departure.

Logan's limited physical capacity—its runways and taxiways are cramped on this small, angular jut of land—is a systematic delay producer, limiting the arrival rate to a maximum of 68 an hour. Compare Logan to Dallas–Fort Worth, built for volume with an arrival rate of 108 per hour. In addition, Logan's location on the water makes it vulnerable to weather conditions in the Northeast, another impediment to on-time arrivals and departures. For these reasons, concerns with speed, time, and schedule were a preoccupation at Logan, driven by the sheer volume of traffic and the need to get planes off and on the runways to make room for more. In addition, however, in 2000 Logan ranked fifth nationally in delays, which resulted in an ambitious plan to improve operational efficiency. Designed cooperatively between the FAA and the Massachusetts Port Authority (Massport, the public agency that owns and operates Logan Airport), the plan included a new short runway, terminal construction to reduce gate delays, additional taxiways, and technology improvements and procedural changes for controllers.

Contiguous airspace and colocation had an impact on Boston Tower and TRACON controllers' dead reckoning in another way as well: controllers at both places understood the airspace and how the other place did its work, so coordinating across facility boundaries was easier for them than at other facilities. The explanation for this begins with history and system effects. Before 1987, controllers were assigned to one facility: the terminal, which was a type known in the FAA as an "up-and-down." The tower and TRACON were joined under a single trainer and training program that taught local knowledge. When moving to the apprentice stage, developmentals had to train and qualify in both places. Those who qualified worked a week "up" in the tower, then a week "down" in the TRACON.[1] The result was that administrators, supervisors, and controllers all had local knowledge of both airspaces. Consequently, the boundary problem that plagued controllers at Boston Center was not duplicated between the tower and TRACON: boundary work between the two was a usually smooth operation.

This easy boundary relationship remained the same even after the FAA system change in 1987 that officially ended the practice of working up and down at major facilities. Boston Tower and TRACON physically remained at Logan, but each operated as a separate facility with its own supervisors and teams. Nonetheless, local knowledge of both airspaces persisted. In terms of structure and organization, all administrators remained colocated in their offices on the floors immediately below the tower cab, a situation conducive to exchange of information. Second, the tower training retained its collective element. Prior to the facility split, the training program taught all controllers to work both the tower and TRACON; but even after the split, TRACON and tower developmentals still learned both airspaces because they were trained as a single group, learning about both airspaces for many weeks before they divided into their respective specialty classes.

Propinquity also mattered. Knowledge about work at both places was learned informally. During my fieldwork, many of the administrators and several controllers who had worked there when it was an up-and-down were still there. Moreover, although no longer officially an up-and-down facility, controllers still went up and down, to visit friends, to hang out during a break, or to watch changes in procedure or technology. Finally, some controllers who trained, certified, and worked in either the TRACON or the tower sometimes decided years later to switch, so they retrained and became certified in the other place. At the turn of the century, though, a few controllers still had the experience of working up and down. Unlike other major facilities, the tower and TRACON had in-depth knowledge of the personality of the facility with which they most frequently exchanged airplanes. However, the differences in the airspace and the tasks at each place still had to be bridged. The result was the complex relationship between the tower and the TRACON: the two interdependent facilities are, on the one hand, competitors, and on the other hand, capable of extraordinary coordination. Whereas at Boston Center, distributed cognition across system boundaries produced conflict and stress to the point it was a routine emotional component of daily work, at Boston Tower and TRACON, boundary work, although more technically challenging, was a relatively smooth operation. When the two had conflicts, which they often did, controllers settled them face-to-face.

This chapter continues our inquiry into how controllers who have trained to work in a standardized system engage in dead reckoning in a system fraught with variation. Having described what Boston Tower and TRACON have in common and the source of their complex relationship, next I separate them to show the difference and the distinctive cultures that institutional, social, and technical arrangements produce in each facility. For each, I begin with "Space, Place, and Work Practice," followed by "Space, Technology, and Ethnocognition." Then I bring together the tower and TRACON again for "Boundary Work: The Routine Drama of the Runway Change," showing them engaged in dead reckoning during runway change. It is a complex maneuver requiring a complete swap of direction of all moving planes both air and ground, requiring close coordination between the two facilities. To accomplish it, they go from being loosely coupled to tightly coupled, dependent on each other at all times. This one maneuver is a close-up of distributed cognition, interpretive work, interaction, and negotiation as separately and together Boston Tower and TRACON move traffic across the invisible lines in the sky and ground.

This chapter continues the theme of part III, showing the distinctive challenges to dead reckoning at each of the four facilities and the unique forms of expertise required to meet them. It demonstrates how interdependence, resilience, and redundancy are built into everyday work practice. It foregrounds controller's expertise and what they bring to the job that technology can't replace. In a small, local socio-technical system, it reveals the effects of factors outside the immediate environment—history, institutional, and organizational factors—on ethnocognition, boundary work, the form of distributed cognition, and the culture of each place. Like the previous chapter, these two cases affirm the variation in the system, making clear the uniqueness of each facility in order to show how the parts of the air traffic control system work, separately and together.

The TRACON: Boston Terminal Radar Approach Control

Airspace, Place, and Work Practice

Boston Tower juts out of the terminal parking garage, visible for miles. The TRACON is difficult to find, even with careful attention to directions. Down a long public hallway joining two airport terminals, after shops and between some offices, a short corridor runs off to one side and turns past an elevator. Near the elevator is a wall telephone. I dial in the number I've been given; a supervisor answers; I give identification; the supervisor checks the image on the video cam trained on the phone; a controller comes down the elevator to escort me up. The elevator doors open onto a narrow hall with four doors, all with punch-in combination locks. Two are to the restrooms, one to the supervisors' office, and directly opposite the elevator is the entry to the TRACON.

Immediately inside is a small square room with a sign-in book, coat rack, and metal lockers. A lot is jammed into close quarters. Around the corner to the left are more lockers and cubbies for controller headsets and mail. Next on the left is the inside entry to the supervisors' office, where eight supervisors, or "sups," share two desks. Used mainly for paperwork duties, the office is rarely occupied because the supervisors are always in the TRACON. All my interviews were conducted in that office. Opposite it is the NATCA office, barely big enough for a desk, a phone, and an extra chair. The desk is bare, and the office usually empty. The union is active and influential here, but the activity does not have the visibility of Boston Center, with its larger numbers and larger NATCA office space. A small sofa across from the NATCA office marks the entrance to the break room, a kitchen with all the basics and a television in the corner. In the middle is a round table that seats six. NATCA keeps a supply of candy, snacks, oatmeal packets, soft drinks, and the like in the cupboard and refrigerator. Regularly replenished by a controller, they are for sale, honor system. From a picture window, the ledge strewn with magazines and newspapers, one looks out on airplanes moving on concrete and in the sky. Next door, construction has begun for an expanded break room that will make room for La-Z-Boy chairs, a bigger sofa, and a bigger TV. Although far from luxurious, the words of the Bedford supervisor seemed to ring true: "We get the leftovers, if that."

At the opposite end of the hall is the door to the TRACON. Inside, the relation between airspace, place, and dead reckoning take material form (fig. 21). The room is eleven feet by thirty-eight feet. Because of its dedicated airspace, radar and radio are the primary technologies. It is dark, even darker than the old control room at the center. Just inside the door is the supervisors' console. The only light in the room is focused here: a small lamp and a few ceiling track lights. Seven feet long, the console is shaped much like an island in a restaurant kitchen and about that height. Behind the supervisors' "desk" is a wall full of state-of-the-art, computer-generated information displays. Incongruously, a vintage 1970s lava lamp with red and pink moving globs sits between the high-tech wall displays and a computer workstation that serves as the facility log: daily sign-in and sign-out; break schedule; team shift schedules; days off; and the daily record of unusual conditions, incidents,

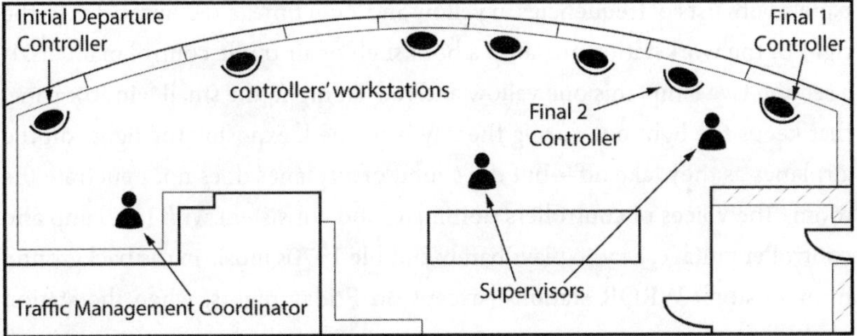

Figure 21: Boston TRACON architectural layout with positions, 2000
Illustration by Noah Arjomand

and emergencies. In contrast to Bedford, which has no Traffic Management Unit, and Boston Center, where TMU has a large separate unit located down the aisle from where controllers work, the TRACON's TMU is right in the midst of things, next to the supervisors' area and only a few feet behind where controllers sit. Backed by a wall of high-tech TMU information displays, a single TMU representative—known as the traffic management coordinator, or TMC—on each shift aids in the prediction necessary to dead reckoning by keeping everyone abreast of moment-to-moment changes in routing, speeds, and spacing changes coming from the center's TMU, as well as weather and traffic conditions in other areas.

Just steps away on the opposite wall are eight controller workstations with radar scopes, side by side, with the end positions angled to make a gentle arc. More in keeping with the lava lamp than the walls of state-of-the-art technology behind them, controllers were still working on monitors from the 1960s, the same ones that had been replaced that year in the big DSR installation at the center. Despite a level of darkness that makes it hard for the unaccustomed eye to identify the controller who is sitting at the most distant positions, the many bits of colored light emanating from TRACON technology give the small room a cozy, almost festive, air. The wall of computer displays is an array of bright colors. At the workstations, the old keyboards have keys lighted in soft orange, green, blue, and red. Controllers' scopes are circular and green, the outside of the circle edged with tiny red lights around the compass rose that rings the edge of every scope. Unlike the Boston Center radar, here the sweep of the radar is visible on the scope, a sparkling rotating line of yellow that skitters over objects and land formations as it rotates, a radius pivoting around its axis in the center of the circle. Above the workstation are black panels with lists of frequencies in yellow and Zulu time in red numbers. To the right of the workstation arc, atop a bookshelf of air traffic control manuals, is a second lava lamp, this one yellow and red. At night, the small window panel that keeps the light out during the day is opened, exposing the lights on the airplanes as they take off—but the sound of airplanes does not penetrate the room. The voices of controllers dominate, and consistent with lava lamp and controller vintage, a radio plays barely audible 1970s music in the background from Boston's WROR station—except on Friday nights, when the station plays all disco.

After observing at Bedford, where technology was subservient to human cognition in dead reckoning, I have gone again to the dark of radar, where the interdependence of technology and cognition was a striking contrast. Despite the 1960s monitors, the radar is more accurate here than at Bedford or Boston Center because the radar at Logan terminal is the major one for the region. Controllers get the position of aircraft in real time, whereas Boston Center gets a composite picture from many locations, constructing a radar image of an aircraft twelve seconds behind its actual location. The absence of flight progress strips is notable. The TRACON did away with them. Because of its high volume and small airspace, the TRACON didn't have time to do the strip marking every time it changed altitudes and speeds of an aircraft. Also, and in contrast to the center, the TRACON has an airplane in its airspace only for four or five minutes, so its controllers can rely on memory: "Sometimes it gets so busy we can't look away from the scope. We have to remember what we give for an altitude. We use a different part of our memory than at the center, where they rely on the strips." No strips, no radar associate. Each sector is worked by a radar controller alone.

Although flight progress strips are out, other technologies convey the essential information faster, and speed is essential. First, the wall-mounted "strip camera" at one end of the room displays a live video feed from Boston Tower showing the departure strips in the tower's strip bay, organized chronologically by the local controller. The strip camera is crucial technology for silent coordination between the two facilities and for dead reckoning; all TRACON controllers down the line can see it, so they know what's coming off the runway. The strip camera, and the strips on it, are crucial to distributed cognition and the prediction that is the core of dead reckoning. When an aircraft takes off, controllers can see the strip removed from the tower strip bay. Immediately, the aircraft appears as a data block on all TRACON monitors. Second, once on the radar, the radar technology tells controllers where each aircraft is going. The "exit fix" is indicated by the initial preceding the data block on the scope (e.g., if *R*, TRACON's Rockport sector takes it; if *P*, Pease sector takes it; if *C*, Boston Center takes it). By this initial, controllers know the route the plane will take through and out of TRACON airspace and into the next one. Aircraft entering the airspace to land at the terminal will have the letter *F*, for "final approach," which precedes the data block.

The connection between space, place, and work practice shows in the physical layout, as well. Airspace traffic patterns—arrivals and departures—shape the architectural arrangement of the TRACON workstations. They are ordered so that traffic flows from left to right in the room. Just below the strip camera, the position at the far left in the line of workstations is known as initial departure, which "catches" planes "launched" by the tower, then hands them off to the right, to the controller in the room working the next sector of air through which a plane is routed. At the far right end of the line are final 1 and final 2, the two positions that turn arrivals toward the appropriate runway, line them up for the final approach, sequence and space them for landing, and hand them off to the tower. In between are six positions that work crossing traffic: incoming flights on approach heading to final, and outgoing flights from initial departure that are leaving TRACON airspace. Each of these sectors has different altitudes and unique traffic patterns. Controllers change positions for each work session during the day; as the airspace sector varies, the task varies.

Fifty-two people work here: eight supervisors, two TMU staff, and forty-two controllers, five of them women. They are divided into eight teams, such that one supervisor heads a team of four to seven controllers. They work a shift system affectionately known as "The Rattler": each team's week of shifts has a two-one-two pattern, so two consecutive shifts starting the same or similar time (e.g., 2 pm–10 pm, 3 pm–11 pm), then the one, known as the "quick turn" because they work late and must report for an early shift next day (e.g., 6 am–2 pm or 7 am–3 pm), followed by two days of shifts at the same or similar time (e.g., 10 am–6 pm, 11 am–7 pm). Once a month, this last shift of a controller's "week" alternates with the midshift, either 11 pm–7 am or 12 am–8 am. Then two days off. Two teams and their supervisors work a shift, so anywhere from eight to fourteen people may be on hand. As at all facilities, redundancy of personnel is essential. Enough people have to be available to handle the heaviest traffic situations. Full teams report, except on Saturdays, Sunday mornings, and for the midshift.

As at Boston Center, the radar technology accommodates these traffic and personnel fluctuations, consolidating or expanding the traffic between sectors as needed. On the TRACON midshift, one scope—initial departure—has it all. One supervisor and one controller work. More than at Boston Center or Bedford, supervisors are actively involved in traffic from minute to minute. Standing at the console a few feet away from controllers, they are mobile and on top of the sectors, sometimes sitting at an unoccupied scope to watch the TRACON airspace. The supervisors stay in close touch with the TRACON's traffic management coordinator, the tower supervisors, and Boston Center, and keep controllers informed about what's coming next. Like the supervisors, the TMC is in the midst of things and interacting with controllers: unlike Boston Center, where the Traffic Management Unit is far up the aisle, TMU is defined not as the enemy here but as part of the group.

The pace of work is noticeably faster at TRACON than at Bedford and Boston Center. The volume of traffic has controllers busy talking to pilots with little downtime. The level of concentration in evidence at Bedford and Boston Center during peak traffic hours is required here more hours every day. Emergencies happen more often, requiring additional attention. The TRACON has two or three emergencies daily, some due to aircraft technical problems, some from instrument-rules pilots in private planes who need help, some flights carrying medical emergencies. These emergencies are handled so quietly that they could pass unnoticed to the casual observer. In one instance, a pilot flying by instruments got lost in the clouds. Immediately the supervisor transferred control of the aircraft to a controller with a pilot's license who gave the pilot flight instructions that guided every move through the clouds to the nearest airport. Had I not been plugged in with the controller taking over, I would have missed it. Usually the only visible indicator that something nonroutine is going on is the supervisors' activity. One goes to watch the sector with the problem, the other to the phone to notify the tower or to alert Massport to send fire equipment to the runway.

Emergencies happen so often here that they are taken for granted and normal, even when circumstances are unusual and outside controllers' experience. One evening, during the busiest traffic period of the day, the alarm in the hall just outside the TRACON rang for over fifteen minutes. "Astonishing" is the only word to describe the level of concentration controllers maintained. The alarm was a penetrating, whooping siren, accompanied by loudly repeating recorded instructions about exits and rapidly flashing white lights that transformed all controller movement into something resembling an early silent movie. So loud was the siren that pilots on the TRACON frequency heard it. I was sitting at a scope, plugged into the frequency. Several pilots asked if they had a fire. One said, "You guys using the radar to warm up those TV dinners again?" Not one controller broke stride in their rapid exchanges with pilots. Neither did they ask about the alarm. They didn't have to. Their room awareness let them hear everything the supervisors were saying on the phone and to technical people who came in. The cause of the alarm, it turned out, was a water pipe broken in the ongoing break room construction.[2]

Now at my third facility, I noticed that the TRACON had a cohesiveness and a sense of collaboration and collective responsibility that permeated the place, both when controllers were on position and when they were off, and this distinguished it from the others. Teamwork and collective responsibility typify all air traffic control facilities—they teach it, they claim it, they practice it—but here it was more extensive, going beyond the idea of helping each other out. The same instruments of social control, both bureaucratic and technological, existed here as at Boston Center and Bedford. Controllers' workdays were governed by standardized schedules, procedures, and routines. Every move was recorded: they signed in and signed out for shifts and breaks; their conversations with pilots and on ground lines were recorded and regularly reviewed; and the aircraft activity at each sector could be reproduced in case of error or accident investigation.

However, the feeling of the place was not of rules, regulations, confinement, and surveillance. Not in two months of observations or in one-on-one interviews, during which controllers at all the other facilities were quite frank about their irritations, difficulties, and resentments, did I hear complaints about deprivation or gripes or derogatory comments about the facility ATM, supervisors, or coworkers, all of which were the norm at Boston Center and Bedford. The traffic volume and density were greater than at the two other facilities I had visited, so the work was challenging and the pressure great. Yet animosities between controllers and between controllers and management that were present at Bedford and Boston Center were missing. Indeed, good spirits, congeniality, and comfort with one another and the place prevailed. TRACON controllers and administrators alike talked with pride about having a team spirit that other facilities didn't have. There were fewer sick leaves, fewer grievances filed, both of which the FAA uses to index the quality of a facility in annual evaluations. This unusual level of collaboration and collective responsibility is a cultural difference: a system effect resulting from the relation between airspace and place.

Place, Technology, and Ethnocognition

The TRACON airspace is a transitional zone for pilots, making dead reckoning a sort of hybrid: not exactly like a center or a tower, but having some characteristics found in both. By virtue of the altitude of their airspace and the need to rely on radar, TRACON controllers' dead reckoning has much in common with that of controllers at Boston Center. However, the peculiarities of TRACON airspace create subtle differences in ethnocognition. Whereas many Boston Center controllers told me they visualize a data block in three dimensions as a result of learning and working with the rules of separation, TRACON controllers said that their three-dimensional abilities came from learning their airspace "shelves":

> We learn shelving. It's complicated, but it makes sense because we are learning why shelves are there by understanding how traffic flows. For example, I own surface to 5 [i.e., five thousand feet] here. Reason is because Rockport [sector] is bringing their arrivals in at 6. Out here, I own to 8 because Rockport is bringing their arrivals in on top. It's like geological strata. We learn in three dimensions. We don't require visualization strategies. (TRACON, M)

When it comes to separation, however, the airspace has them operating more like a tower. The rules of separation are the same: three miles, one thousand feet. An airplane is in a TRACON sector for only four or five minutes, compared to ten to twenty at Boston Center, which leaves controllers with little time to calculate separation. Rather than rely on information in the data block about route, equipment, and speed to identify a potential conflict and then calculate distance to maintain separation, they draw on experience and tacit knowledge:

> We don't have time to calculate. We just see it. You've got to know when you've got one here [far left on the scope] and here [pointing to the far right] that they will be in conflict. (TRACON, F)

However, the quality that most distinguished the TRACON from all three of the other facilities was the level of collaboration and collective responsibility in dead reckoning. On my first day of fieldwork there, as I plugged my headset into a controller's sector frequency and sat down, the controller turned to me and said, "We are self-regulating." He went on to explain that they looked out for each other. In the following few days, I saw what he meant. TRACON controllers talk about and enact a collective responsibility for the entire airspace, not just their own sector. The people working the sectors all acted as a check on one another, to monitor and correct for mistakes in the making. It was a form of dead reckoning dedicated to safety through prevention and staving off errors, mistakes, and harmful outcomes *and* accomplishing the maximum arrivals that their restricted runway layout would allow, not for individual accomplishment—although that was part of it—but the collaborative effort was for the collective accomplishment "of the room" (in their own words).

This distinctive collaborative, collective quality was not solely due to teamwork, a skill all controllers have, but it was a system effect: a result of the effects of airspace, of space, and thus of the social organization of work. First, the small airspace allows the radar technology to reproduce the entire twenty-five square miles of TRACON airspace on each scope, which aids both collective prediction and collaborative efforts. The technology enhances interdependence between controllers, allowing them to be more tightly coupled than the other facilities, and also redundancy, enabling them to back one another up to an extent otherwise not possible. Each controller not only sees and works his or her own sector but also always sees the full airspace displayed. Second, the arc of the workstations, with initial departure and final slightly angled out, marking both ends of the arc, allows controllers to see one another and tell from body language at a given moment how other controllers are doing at their sectors. Third, all are doing the same task: working airplanes already in space. The result is that everyone's attention is on the whole airspace, all the time. Consequently, they are able to act in a collective way that is not possible in other facilities where the radar technology limits their vision to one sector, as at Boston Center, or where controllers are more dependent on room awareness to alert them to problems in other sectors.

Signals and Interpretive Work

At the TRACON, the identification of early warning signs—anomalies—and their interpretive work stands out because collaboration and collective responsibility are an audible aspect of everyday work. Several times a day, someone will say, “Hey, you all right with that guy?” to make sure the controller who has an airplane sees some anomaly. It’s such a frequent refrain and so much in the caretaker mode that controllers who transfer from radar facilities where competition runs rampant or where setting the other guy up to fail is practiced have a hard time adjusting. If two planes at four thousand feet are heading to final, there’s a chorus of “Pair of 4s, pair of 4s” to alert everyone that two planes are assigned to the same altitude: “Are you watching that Eagle flight?”; “Why hasn’t this guy turned?”; “Go after that USAir. Head him over here”; “Take care of him because he’s a first timer [at Logan]”; “See how he’s doggin’ it?”; “Are you watching that 5250?” At Bedford, Boston Center, and Boston Tower, heads-up statements also occur between controllers, but the difference at the TRACON is that controllers are calling out about traffic in other people’s sectors because they can see and concentrate full-time on the entire airspace. In these comments, made to me while I was plugged in with working controllers, TRACON controllers’ express the pervasive sense of collective responsibility:

> Here it is less about boundaries and more about separating airplanes. Everybody approaches operational errors as everybody in the room screwing up. (TRACON, M)

> One little thing can mess you up. Here people anticipate problems and help correct. Everyone knows what it’s like to be at the bottom trying to get to the top. We sink or swim together. (TRACON, M)

> We are used to seeing patterns. This guy is coming in, we are used to seeing him turning southbound. Someone sees it not happening, “Hey, is that American coming around?” Or, if an aircraft has a C [in the data block, *C* is the exit fix, identifying Boston Center as the next facility to take it] and the center is not taking the handoff, someone will say, “Hey, see this guy, should we be calling the center?” (TRACON, F)

Their collaborative self-regulation extends to every position. However, the focal points are initial departure and final. Both initial departure and final are time critical, with sixty arrivals and sixty departures an hour, and separation is crucial. For both positions, the institutional goals of expediting traffic and safety most visibly come together. These work arrangements show the interdependence of technology and human cognition in dead reckoning as well as provide a close look at how controllers supply the redundancy of the system in their everyday work.

Initial departure is located on the far left of the workstation arc. Departure is the first position out of the tower airspace. When a departure is handed off, the receiving TRACON controller has an airplane for about one minute before handing it off to another controller in the room. The "strip camera" that displays Boston Tower's strip bay is mounted above and next to departure's position. Everyone can see it. The tower's strips tell the order of departures, but the human component supplies additional essential insight. TRACON controllers can predict how planes will come off the runway by the "handprint." The hand that the strip camera shows stacking and removing strips from the strip bay is the hand of the tower's local controller. From their experiential knowledge, TRACON controllers all recognize whose hand it is and know the person's techniques with departures on all runway configurations, so they can predict not only *when* but also *how* departures will be coming into TRACON airspace.

The TRACON controllers are providing redundancy, not only for one another but also for the tower's local controller handling departures. The TRACON's initial departure controller has the responsibility for catching errors on the strips that the tower controllers may have missed. The departure controller checks the strip camera to see which plane is departing next, the runway and the heading assigned, the first departure fix, and the transponder code. A code is a number transmitted by a plane's transponder. It appears in the plane's data block on the screen and identifies the airplane. The error of greatest significance is no code or a wrong code, because then an aircraft cannot be tracked by the radar. If the pilot doesn't turn the transponder on or types in the wrong code and the tower has taken the strip down (which they do immediately at takeoff), the TRACON doesn't know which plane it is. A code error is a small thing, but in a high-volume, time-critical position like initial departure, an error can start a chain of events that pile up and make bigger trouble. Everyone watches the strip camera, comparing the strip to the data block on their scope. If departure is busy, hasn't time to correct a code, or misses it, and an aircraft comes off the runway showing no code or a bad code—self-regulation kicks in. Several times a day, someone shouts "untracked, untracked" or "unacquired." As one controller put it, "We are the last line of defense against bad codes."

The other focal point of collaboration and collective responsibility is the final position. Controllers say it is the most desirable and exciting position, the one that attracts most controllers to TRACON work. By putting airplanes in sequence with even spacing between them, the final controller manages the arrival rate. The runway system is an obstacle to be conquered. The challenge is how to get the most out of it so the TRACON can maximize the number of arrivals per hour to avoid delays. As aircraft move through the TRACON airspace and approach the runways, the final controller puts them into sequence for landing. As new planes come into the airspace, final makes holes in the established sequence to work them in. As one controller told me:

> We have to be fast and expeditious. We make holes and want to get them in the holes. We want them to come, then want to get rid of them as soon as possible. No one has to tell you to move a lot of airplanes. You do it because you see them coming. It's the fear factor. You have to get rid of them or you'll go someplace you don't want to go [a euphemism for "down the shitter"]. (TRACON, F)

Dead reckoning on this position is an art; it takes finesse and what controllers call "precision vectoring." The final controller is receiving planes from all directions. Planes must be merged in a tidy line, evenly spaced with others on the downwind, then turned on the base, then turned into the wind onto the localizer indicating the runway designated for landing. Part of the Instrument Landing System (ILS), the localizer generates and radiates signals to the cockpit to provide information that guides the aircraft to the "glide slope" and the runway lighting system. Timing is everything. If the final controller needs to make a hole, it has to happen fast enough to insert the plane in the line. These two controllers articulate the experiential knowledge necessary to predicting the positions of airplanes in space:

> It's geometry. Turn him to intercept the localizer. Headings are all boxes and angles. This is the real world. In the [training] lab, you don't have wind. Here it is wind and equipment variation. You have to know when to cross the localizer, where to cross, and the speed. We run planes as close to the localizer as possible before turning them into it. Sometimes run them closer, to turn tighter, so assign each 230 [compass] heading. (TRACON, F)

> To turn them to get them in the line of traffic on final, this one on the downwind, if I extend an imaginary line 1 mile ahead of the last plane [on final] and extend an imaginary perpendicular to the plane on the downwind—if I start

> the turn then, he will slip right in behind him. Gets 3.31 miles [separation]. With a headwind on final that will compress a little bit. End up at 2.5, which is OK inside the marker. (TRACON, M)

The trick is to run them close, but not too close. To sequence them for Boston Tower and landing, the final controller takes into account aircraft speed, equipment capability, headings, wind, and wake turbulence. Separation is maintained by manipulating speed and altitude. When controllers talk about how they do it, we can see how the three elements of common sense—institutional rules and cultural understandings, local knowledge, and tacit knowledge—combine when working final:

> We try to establish a consistent final, with speeds 170-170-170. At the fixes, we reduce speeds gradually, so [the spacing between them] looks like a pretty picture. If we slow them down too much, it backs everything up. (TRACON, M)

> It's in the book [the 7110.5] that we have to have a heavy [aircraft] in front, a small behind, with four miles in between to allow for wake turbulence. Varies by size. Five miles between two heavies. You have to build an eye to be able to gauge it. You don't have time to measure it out. (TRACON, M)

> You have to know your wind. Depending on the direction of their approach, the one on the left may have a strong tail wind, the one on the right struggling against the wind, you have to know immediately which one is first. You have to know wake turbulence and space for it. I'm watching how the aircraft maneuvers, as I'm waiting for the turn. I know, depending on the aircraft weight and destination, when and how long a maneuver will take. I'm constantly watching to see if they do as expected. (TRACON, F)

> I always use altitudes [i.e., separate planes by assigning each a different altitude] on a turn on the downwind. It keeps you safe. You can make a nice tight turn. If it's a favorable wind, you can turn them close, have them right up the tailpipe coming around that turn. (TRACON, M)

The TRACON's handiwork is visible at any large airport. If you can find a window to look out at arrivals, you will see the airplanes on the horizon "all lined up like Rockettes," coming in to land. In this tightly spaced, one-at-a-time sequencing, final needs pilots to perform as they are directed, but they do not always. Failure to conform to controller instructions creates safety hazards and extra work. For the nonconformist pilot, the final controller reserves the "penalty vector":

> If you're trying to fit them into the pattern and they don't perform as you need them to, we have [in the 7110.65] the threat of going to the end of the line. I gave this Northwest several chances: "Have you started that turn yet sir?" "Affirmative." Nothing. This is not working. "Take a 180 right." This puts him at the end of the line. He's typical Northwest, slow on turns. I'm trying to help him out. If he's not going to help me out then I'm going to have to put him at the end of the line. (TRACON, M)

"Feeding final," too, is regarded by controllers as a collaborative effort and collective responsibility. The final controller is dependent upon the controllers working the TRACON's three feeder sectors. They begin early putting planes in sequence for the final controller. They are responsible for ensuring that aircraft are at the designated altitude for handing off and also at an appropriate rate and speed for landing. If the feeders are not doing a good job, then the final controller struggles, flights are delayed, and the arrival rate goes down. Helping and protecting the controller working final is the common goal. A controller in a feeding sector asks the final controller, "Would you like this guy at 190?" or "Do you want them straight in?" Feeders begin setting up the spacing for final, watching to see the other aircraft on approach, their speeds, type of equipment, and predicting position from wind and the behavior of other airplanes already in the final controller's airspace. A feeder with a plane headed to final predicts the position of others to anticipate how his or her plane is going to fit into "the pattern." For example:

> If final has a fully packed final and lots of airplanes on the downwind, then something is falling apart in the room. It's nonverbal communication. The ability to know what's happening in the room. Getting a feel for what the plan is, what is working. Not really communication, it's being on the same page, one hand knowing what the other hand is doing. I watch the final plan, then either stretch my guys out or speed them up. It decreases the need for communication because everything is so standard—standard routes, standard procedures. (TRACON, F)

> You have to learn to feed the final controller. It's an intangible to teach because it's different every time. You get a feeling for it. It's one of the things that makes the job enjoyable here. If you don't keep the feed going, the final will dry up. You're not maximizing your potential. At the [Boston] Center, when they're not busy, they kick back and relax. Here, when final dries up, controllers start shouting for more airplanes. (TRACON, F)

The fact that they all watch the entire airspace means that no performance goes unnoted. At the TRACON, as at all facilities, praise is never given publicly. Good work is expected. But the group collectively sanctions the final controller who doesn't achieve both goals of safe and expeditious delivery of traffic. A polished session on final has perfect spacing, the final controller running them tight and at the same time maintaining the rules of separation: close, but not too close, and no wasted space. If not, self-regulation goes negative. A poor session, or work that negatively affects the group effort, produces verbal sanctions: "Hey, you gave that AAL too long a ride." The feedback can be humorous, but the humor is biting. To a controller getting relieved from final, someone shouted, "Hey, Jack, nice try." Another time, "You do pretty good work for a blind guy," both meaning the opposite—a lackluster performance, below group standards.

The personality of the TRACON is an ongoing display. Controllers' collaborative, collective sensibility is cultural, pervading other aspects of life on the job. In all facilities, ongoing joking and fun were a mainstay of daily activity and one of the main attractions of the job. Although in many ways the TRACON humor was like that of other places, it also was distinguished by a collaborative, collective style. Occasionally, at the evening rush hour—everybody flat out busy, with both final positions open—someone would begin whistling "Sabre Dance" and everyone immediately joined in. The rapid beat of the well-known music exactly fit the pace of the work.[3] All multitaskers and masters of the interrupted conversation, the unison whistling carried above the flow of talk with pilots while the nervous jiggling of legs, feet tapping, and moving elbows kept time. Moreover, under certain traffic circumstances, the group response was to collectively make what they mysteriously referred to as "animal sounds."

Even the TRACON break room conveyed the TRACON's collaborative, collective orientation. Bedford's small break room was usually empty or occupied by one controller at a time, since only a few (usually three) controllers worked a shift. At Boston Center, too, the small official break rooms were usually for individual activity, like reading, sewing, bill paying, or conversation. The cafeteria was a center of social activity, but even when the rows of tables held numbers of controllers—some alone, others in small groups—they were dispersed across the large room. They engaged in small-group activity, like playing cards or lunching together, and in addition, NATCA held many social functions at outside venues. However, I saw or heard nothing indicating that controllers working in the same center area collectively engage in the same activity while at the center. They often took a break from each other, getting together with friends from other areas. At the tower, as the next chapter will show, controllers were more likely to separate from the others, either mentally or physically.

In contrast, TRACON controllers behaved like a collectivity when off position. Although the terminal had its lures, TRACON controllers tended to take their breaks in the break room, socializing with whoever was there. They not only stayed with the group but also frequently participated in group activity: a running card game, Pitch. They played partners, game ending at twenty-one points, which cannot be reached on one twenty-minute break. The game is fast and doesn't require a lot of strategizing or remembering of what has been played before, like Hearts does, so the game lends itself to substitution and perpetual play. When a controller was relieved for break, he or she left the TRACON and relieved a controller in the Pitch game. When one person's break was over, that controller relieved the next person on the break list, who picked up the hand and played. Supervisors, considered part of the teams rather than administrators, also played.

Finally, confirmation of collaboration and collective responsibility comes not just from what was visible to me but also from what was missing. The boundary problem created by clashing individual personalities and air-sector personalities at Boston Center was nonexistent at the TRACON. Absent were anger and resentment due to conflicts of individual controllers' personalities and air-sector personalities while moving traffic across airspace boundaries. Their cooperative work ethic extended to airspace boundaries outside the TRACON, to their relations with the other facilities—Bedford, the center, the tower—that they have to coordinate with in order to move traffic into and out of their airspace. At these other places, controllers remarked about TRACON, "No matter how busy they are, they'll always help you out." Recall, now, the Boston Center controller who couldn't figure out why the TRACON controllers, always busy, nonetheless would always go out of their way to take handoffs.

Changing Space, Changing Place, Changing Culture

In contrast to the cooperative culture I witnessed in 2000–2001, the more senior controllers and supervisors told me that, historically, the TRACON had been a place of animosity, grudges, and trying to best or bust the other guy. Then a major airspace realignment transformed the conflict-ridden TRACON culture. The aim of the FAA's Expanded East Coast Plan was to reduce congestion in the New York metropolitan area and associated airspace, which included the New England Region.[4] In 1988, that same initiative took airspace from New York Center and gave it to Boston Center, constituting Area E, as described in chapter 5. Separating Boston Tower and TRACON in order to eliminate the "up-and-down" practice was part of the Expanded East Coast Plan, which also called for a "realignment" of the TRACON airspace. The realignment changed the arrangement of workstations, the number of positions, and how the airspace was worked. A senior controller told me:

> It was a huge transformation of everything we did here. So, going through that wasn't fun. The people who were used to it didn't want to change; that's normal, nobody wants to change. The new people, which I was still a fairly new guy then, we all tried to just figure out what was going on . . . But it was all done at once. They split the tower and TRACON and then redesigned the airspace and the TRACON all right at the same time. It was a huge project. (TRACON, M)

The system effect was also huge. The ramifications for the TRACON were the same as they had been for Boston Center's Area E: the airspace change altered place, both technology and architecture, and thus also the social organization of work, hence altering the personality and simultaneously the culture of the facility.

Prior to the change, the TRACON owned the same altitude of 3,500–14,000 feet and twenty-five square miles, but the airspace divisions within it were different. The original airspace was divided into six positions (compared to nine after the plan): satellite north, satellite south, departure high, departure low, approach A and approach B. Controllers were all on different scopes, but no one owned his or her own airspace because the airspace had no internal boundaries. The airspace was shared. Each person owned a plane and worked it through the airspace from beginning to end, handing it off to either the tower or Boston Center. Consequently, the airspace bred competition and conflict within the TRACON. Controllers had to compete with one another to get their planes through the airspace and toward their destinations. This TRACON controller, who worked before and after the airspace realignment, described it as follows:

> If you were working departure high or departure low, as soon as that airplane told you "airborne," you worked him, no matter the altitude, you climbed him through everything on his merry way and handed him off to the center. We used to share the final. That wasn't even there, that scope. We opened up that last scope, that final vector, when we realigned the airspace. We had approach A and approach B. So, what you had was two people basically sharing final. Approach A was working the arrivals coming from the south, and approach B arrivals from the north and overseas.
>
> But approach A was calling the sequence. So, say approach A was working all the Providence arrivals, they would have to make the holes for the B controller who had traffic on the downwind. Then you'd just have to hope that the B controller was where he was supposed to be when he was supposed to be there to get the guy in the slot. So, it was constant coordination, constant.
>
> There was a lot of yelling. Because you were yelling across the room, and it wasn't like yelling in an anger type thing necessarily, although there was a lot of that too. Departure high, it was like eminent domain, if you were there first, you would work the airplane. I mean, there were certain guidelines you would use, obviously. You were expected to climb the airplane as soon as you could, and get him above all the arrivals. Which is fine, as long as the person working the arrivals worked them down as soon as they could, it all worked out very nicely.
>
> Now [in 2000, post–airspace realignment] you can work within the confines of your own little hunk of airspace, and if you don't want to coordinate anything with anybody, you don't have to. You can just go by the book and work the airplanes and no airplanes get anything special. Nobody is being expedited and climbed through something. Before, it was impossible. (TRACON, M)

A supervisor said the old airspace design made the work more like a game of chicken:

> It used to be we did a lot of what we called "look and go," where you look at the other person's traffic and you miss it—you go over them, under them, whatever it takes. And it was always—the two were at odds with each other, because one person might be thinking you're going over the top, maybe the other person was thinking, "I'm going to go under you," and it was a lot of butting heads going on.
>
> We had two people that would vector to the final at the same time (approach A and approach B), with very little talking between each other. One person, if they had a dominant personality, would turn in, let their plane go in, leaving the other one nowhere to go. The other person had no place to put their airplane. So, it would be almost a game of chicken, but that's a—I don't want to make it sound like it was dangerous, but it was a "I'm going in, get out of my way," that kind of thing. Everybody wanted to be top dog, better than the other one. And one way of doing that is if you put down the guy next to you it will make you look better, or something.
>
> Now [post–airspace alignment] you own your own airspace, and you do your own vectoring. It takes a strain out of wondering what the other person is doing. The other people now give you the airplanes at a specific place, and you turn them. You even watch that they come in from the same flow now. There's not planes shooting all over the place.
>
> Now it's terrific. It's quiet in there. There are no arguments. And if there are, they're minor. Occasionally there are some outbursts, you've got to deal with some guy yelling and screaming and swearing, calm them down, they're fine. You haven't seen it. I haven't seen it for quite a while. (TRACON, M)

Other system conditions reinforced the former conflicts. The old airspace existed at the time of the 1981 strike and after, when new controllers were replacing the ones Reagan fired. Those controllers who had stayed on during and after the strike resented the replacement hires. The replacements were "still wet behind the ears," and besides, they were taking their buddies' jobs away from them by being there:

> Senior people would ignore junior people. They would hog final—you couldn't get a space. The old-timers stuck together and they'd help out their

> buddy, but not us. Most developmentals failed—hardest place in the country to certify and to work. If you couldn't vector [a plane] into a six-mile hole and take care of all this other stuff, they got rid of you. It was a very punishing job. Sometimes no breaks, or else two hours on. Horrible. Nobody wanted to come here because of the reputation it had. (TRACON, M)

A second factor was that prior to the late 1980s, traffic management—with its spacing programs, metering, holding orders, and reroute plans that organized the traffic flow—had not yet come to the New England Region. A supervisor said:

> We didn't have the traffic management system. So, they [the Boston Center] would just unload on us. We'd get tons of planes. The first thing out of your mouth is slow up as slow as you can go. We were doing our own holding. If you're doing 360s all over the place, plus turning the final and managing that final—it was insane, absolutely insane. People were yelling at each other. I mean, I saw people go out in the parking lot and start fighting. Almost come to blows. Now it's very calm and nice. (TRACON, M)

The 1988 airspace realignment immediately decreased conflicts among controllers. The TRACON airspace was organized differently, creating a new traffic flow. The individual sectors, renamed to reflect the changed airspace, would be governed by the principle of ownership. The radar technology, number of workstations, and organization of work changed accordingly. The role of technology as a representation of the TRACON airspace was crucial. It was not only the airspace that was changed during the realignment; how those changes were built into the TRACON's radar-room technology also mattered. With changes in the organization of work, system-bred competition gave way to the collaborative, cooperative form of dead reckoning. This marked the beginning of a cultural transformation.

Over time, other system changes completed and then institutionalized the transformation. First, the traffic management system came to New England. The regional TMU at Boston Center and an internal TMU at both the TRACON and Boston Tower regulated the traffic flow into and out of Logan. Second, at about the same time, the cohort of "old-timer" controllers who had been there during the strike began moving into administrative positions and/or retiring, until very few remained working traffic. The generational conflicts that had affected dead reckoning stopped. Third, after realignment, and because of it, the "team concept" was emphasized in the reorganized training for the TRACON and Boston Tower. Fourth, at the TRACON (but not at the tower) the teams (eight teams and eight supervisors, four to seven controllers each) were mixed together in different combinations on shifts and so were not distinguishable from one another by practices and techniques. As a result of the mixing of team combinations, TRACON controllers tended to work alike, eliminating idiosyncrasies that would give them individual team personalities that created conflicting work styles. So alike were they in work practice that only the supervisors on duty were a clue to which teams were working a shift. Fifth, personnel hiring practices changed to perpetuate the cooperative culture that had developed. One operations manager told me:

> Once we saw what we had, we changed our hiring practices. We made personal phone calls to find out what the controller was like. We didn't necessarily hire the hot-shot controller; we went for the one that would fit in with the people we had. (TRACON, M)

Developmentals who did not conform to the TRACON's self-regulating, collaborative form of dead reckoning did not make it through the training. "Some people who set you up for a problem have a hard time transferring here. The teamwork—'Are you all right with that guy?'—gets a defensive reaction." The TRACON's self-regulating system perpetuated the TRACON's changed culture: controllers doing the training made sure that teamwork was taught and that those people who were training fit in with the collaborative way of working the airspace. Otherwise, goodbye.

Boston Tower

Airspace, Place, and Work Practice

A Boston landmark, the cab of Boston Tower sits like a teacup perched atop the large, H-shaped structure that supports it. Reaching it is incremental, in stages reached by gaining access at higher and higher levels of elevation. Controllers use ID cards and punch in door-lock combinations at each level. Visitors, as at all facilities, must have administrative clearance. Then, under scrutiny of a surveillance camera at each checkpoint, they use a local wall phone to identify themselves and be buzzed in—and up. The elevator reaching the lowest of the six stories of offices near the top that are trussed between the tower pylons is the first indication of getting close. Exiting the elevator, one enters a carpeted hallway with conference rooms and executive offices on the one side and secretarial offices and the break room on the other. Of all four of my fieldwork facilities, at Boston Tower the offices of both the air traffic manager and assistant air traffic manager are the only ones that are spacious and handsomely furnished. And there is that view, of course. From the corner office of the ATM, almost the entire runway system is the backdrop, with ocean beyond. Judging from the architecture and furnishings, managing one of the busiest terminals in the country is a high-status position in the FAA.

Across the hall, the break room is dominated by a long sofa against the back wall and a three-section lounger perpendicular to it, under the window. Its contents are like the TRACON's: a large TV in one corner, lockers, coat racks, individual cubbies, and kitchen table, microwave, refrigerator, toaster oven, and dishes. One item sets this break room apart from all the others: a new iMac desktop computer on a table at the end of the lounger. In contrast with the TRACON break room's collective activity, tower controllers on break are quieter and often opt for isolation from the group. During peak traffic periods when runway use is maximized, they may have "the 120 hour"—or sixty departures and sixty arrivals, two planes a minute. On break, some are visibly tired, "zoned out," slumping into the lounger to watch TV, or have lunch at the table alone or with another controller. Some take off—again, alone—for aerobic walks in the terminal. On weekends, when traffic is low, there is a visible increase in energy and interaction in the break room. As at the other three facilities, Boston Tower offers little privacy. As in the TRACON, the union is strong at the tower, but NATCA has no office space, using instead a conference room on the administrative floor below. With carpet and a long polished table, the conference room is also where a trainer is likely to take a developmental for feedback after a session if the break room is occupied.

Around the corner from the break room is a short set of black iron steps leading to another door with a telephone access system. A call is picked up in the tower by the supervisor or traffic management coordinator, the door is buzzed open, and yet another set of stairs spirals to the cab. Low lighting is embedded in the wall at each step to guide the way. At the landing are the first sounds of the cab: the clicking of flight progress strips in their plastic holders as they are moved from position to position and stacked in strip bays. Upon approaching the top, you hear the collective murmur of controller voices, and then the stairwell becomes drenched in light from the windows. Looking up, you see the sky then suddenly you are in the eleven-sided cab.

The view is spectacular. On a clear day, you can see for miles in every direction: Boston Harbor, the Harbor Islands, and the Charles River, with sailboats, brilliant-colored spinnakers flying even in December, tall ship fleets in the summer, and oil tankers and small boats year round. To the left are the surrounding waterfront communities situated on the flight path that for years have aggressively protested airport expansion, noise, and dirt. To the right and rear is the city skyline, with visible landmarks like the financial district, the North End, the Prudential building, the cable-stretched Zakim–Bunker Hill Bridge, and Fenway Park. The spectacle is not just the landscape; it's the airfield: a fascinating, ongoing display of airfield and aircraft technologies; the international, national, and local demand for air travel; and the reality of controllers' complaints of "too many airplanes, too little concrete." Forget the moon on the water; the lighted runways and aircraft at night are beautiful. Beginning at dusk and into the night nothing can quite match the visual impact of the TRACON's work: the perfect line of planes on final approach, lights on, equal spacing between them, forming a graceful arc over the earth's horizon as TRACON controllers systematically lower the altitudes at each fix, heading them toward Boston Tower airspace and the designated runway. Or the lumbering yellow army of airport snow equipment hauled out during blizzard conditions to clear first one runway then the next and then back to the first while off to the side passengers sit in airplanes being de-iced then de-iced again as they wait for a runway to open up for departure.

Upon arrival, it is not the analogies but the differences that are striking. Once again, I am initially overwhelmed by the visual complexity of place, technology, and work compared to my other fieldwork sites (fig. 22). Boston Tower's cab is larger than Bedford's, but it is more cramped, packed with twice as many people and more technology. Usually eight people work in the tower at a time. In total, thirty-seven controllers work there, seven of whom are women. Controllers are divided into three teams of twelve to thirteen controllers each, with two supervisors per team, and there are four TMCs, two per shift. In contrast to the TRACON's "Rattler" shift, Boston Tower controllers work a week of days, a week of nights, and a week of mids. During my stay, the cab was exceptionally crowded. Five of the thirty-seven controllers were developmentals, assigned to different teams. Despite the FAA hiring freeze, a continuing source of developmentals is controllers already in the system who have aspired and applied to be at Boston Tower, having worked their way up through the system toward greater complexity and salary. When training, their trainers added to the number and sound level at the tower.

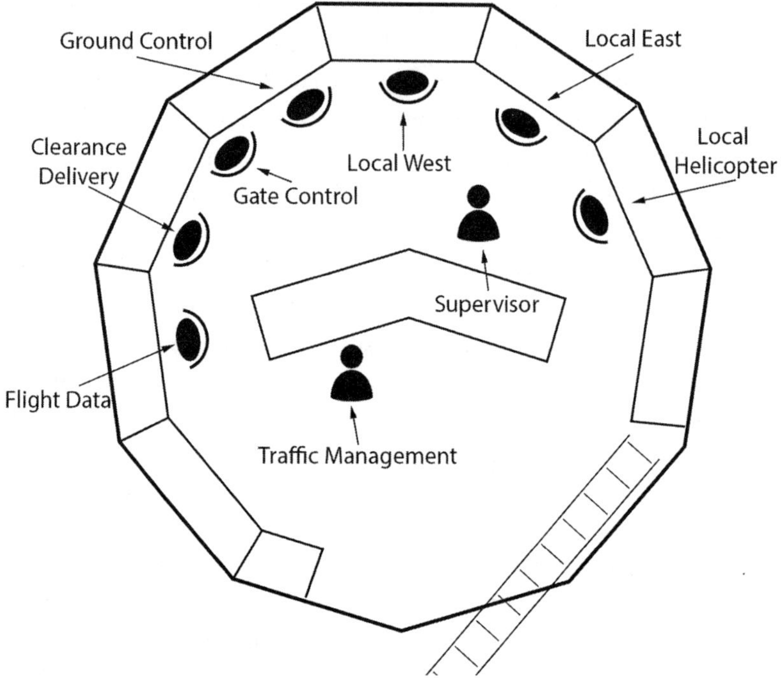

Figure 22: Boston Tower architectural layout with positions, 2000
Illustration by Noah Arjomand

In addition, in a change designed to decrease delays and increase runway safety, the tower was experimenting with dividing the ground control position into two positions when traffic was heavy. Ground 1 and ground 2 each worked a different side of the airport: lots of talk and yelling went on between them as they worked out the division of labor. During shift change, the volume increased. The sounds of airplanes were merely background noise. The cab's construction and thick glass windows protected controllers' ability to communicate with one another and pilots. All towers have certain fundamentals in common. Like Bedford, a work counter runs all the way around the tower cab perimeter with controller positions and equipment arrayed around it. Also like Bedford, next to the stairwell a short extension of the work counter is cluttered with coffee pot, mugs, purses, a stash of aspirin and Tums, and, like every facility except Boston Center, a radio playing the barely audible 1970s music of Boston's WROR. Following the tower's eleven-sided structure, the work counter's interior angles roughly mark where a position and its technology stop and start. Instead of Bedford's four positions—flight data, clearance delivery, ground control, and local control—Boston Tower has five normally open positions, adding a local controller to that configuration, and as traffic demands increase, it may open as many as seven.

Informally, the controller positions are known as the front row and the back row, remnants of a previous architectural arrangement in which all tower controllers faced southeast, toward the runways, in two rows, the front row standing, the rear seated directly behind. Although physically two rows no longer exist, the referent persists. Ground control, local control west, local control east are the front row. They stand. The two local controllers face the runways (southeast). Ground control paces, direction changing from one moment to the next. Helipad, controlling an area designated for the helicopters is rarely open, and although it is facing due south, away from the runways and toward the harbor, it is still considered the front row. Flight data, clearance delivery, and gate historically were the back row and remain so called, despite the fact that there is no back row, and the positions now face northeast, away from the runways and the front row. They sit.

The installation of a console was the architectural change that rearranged controller positions. A metal support post marks the center of the tower cab and extends to the top of the structure. Angled around the front of the post like a boomerang is the console. The top is chest-high and serves as a desk for the supervisors and the TMCs, who stand behind it. The console is loaded with technology. The front—the side facing the runways—contains all the radio frequencies, backups, and tapes that record all controller radio communications. The front-row controllers plug their headsets into the frequency designated for their position. As in the TRACON, the supervisors and TMCs are right in the midst of things and part of the team. Their physical location, the architecture, computer technologies, and their observations and interpretive work are actants in distributed cognition in the tower.

At each shift, a TMC works behind the console's left side. The console holds the TMC's computer displays and phones that connect to TMU at Boston Center, Central Flow, and other regional facilities. The team's supervisor occupies the console's right side, the back of the console holding ground lines, computer displays of weather, arrival and departure rates, and a direct line to the TRACON, Massport, and emergency phones. Behind the console is an iMac for easy retrieval of air traffic information that doubles as the tower's facility log—shifts, scheduled days off, controller sign-in and sign-out, and incident reports. As at Bedford, my first obligation at Boston Tower is to keep out of everyone's way. I stand behind the console, beside the post, between the TMC and the supervisor—both important sources of action, play-by-play interpretations, and other information. From here, with my headset plugged into the frequencies located on the front of the console, I can listen to controller-pilot conversations for all positions. The pace of work was such that my conversations with controllers about what was going on were restricted to breaks and interviews, but when they had time, the TMC and supervisor filled me in.

The tower's primary technologies are flight progress strips, radios, computers and computerized information displays, and—suspended from the ceiling above the windows facing the runways—three radar scopes and two ground monitors called ASDEs (pronounced "as-dees"): Airport Surface Detection Equipment. Designed to reduce runway incursions, ASDE is a surveillance system being tested at the time of my fieldwork in a few towers. It allows controllers to track surface movement of aircraft and vehicles, especially at night and when visibility is poor. In contrast to the TRACON, strips at the tower are a mainstay of the technology. Because of traffic volume, even more than at Bedford or Boston Center, the strips are the essential mechanism of distributed cognition and silent coordination: they save time by reducing the need for conversation in a place where every second counts. As at Bedford, the strips reduce error: possession of the strip indicates who is working the airplane, and the strip markings allow for the next controller to see what the last controller has done or forgotten. Equally important are the TMU technologies that predict and regulate traffic flows; unregulated traffic flows produce delays and congestion, or even shut down the airport. The TMC at Boston Tower is in constant communication with the center's TMU, coordinating departure spacing programs and miles-in-trail to regulate the flow of traffic arrivals and departures.

The pace of work at Boston Tower is fast and unrelenting. Its rhythm is distinguished from that of the other facilities by lengthy periods of heightened physical movement of strips and bodies. The interdependence of controllers in the other facilities is here reproduced at high speed. As at Bedford and at all towers, the sweep of the work for departures goes clockwise around the tower perimeter: flight data to clearance delivery to gate, to ground control, to local control west to local control east. Simultaneously, arrivals are going counterclockwise—from local east, to local west, to ground control. All positions are marking strips and passing strips. The work does not have the same rhythm for each position; each has its own. To a great extent, the airline companies—the customers—control both the rhythm of the room and individual positions, because most flights are scheduled on the hour. Clearance delivery has a complete turnaround of airplanes every thirty minutes, when aircraft need to push back from the gate to meet the departure time. Ground is busiest on the hour. At 3 pm, on an average weekday with normal weather conditions, ground is working about fifty airplanes. Taking into account the different demands of each Boston Tower position, the supervisors limit a controller to working no more than two hours at a stretch (sometimes less), rotate each controller between standing and seated positions, and make sure that no one works the same position twice during a shift.

To follow traffic movement from position to position, we can follow the movement of the strips. Recall from Bedford that the departure process begins when a pilot files a flight plan. From a computer, flight data receives flight progress strips for each aircraft scheduled to depart. Flight data updates the flight plan, then slips the strip into a plastic holder and puts it into the strip bay in front of clearance delivery. Clearance delivery communicates to the waiting pilot the route that the aircraft is to fly and the expected departure clearance time. A third position, gate control, tells pilots when it is OK to push back from the gate, so officially starts the plane moving, but gate is open only when there are a lot of delays. Usually, clearance delivery handles gate. When the plane pushes back, clearance delivery shoves the strip down the "strip slide" to ground control. The strip slide is two rudimentary tracks at the edge of the work counter between clearance and ground. It is a simple yet essential technology, each track telling ground which runway the airplane goes to for departure and which local controller gets the strip—and the airplane—next. Keeping the strip from overshooting or undershooting its destination, or from falling off on the floor, is an acquired skill.

Ground control owns all maneuvering areas: any place where aircraft have left the runways or departure gates. Unlike most other major airports, where the airlines handle their own "ramps" or "aprons"—the area for loading or unloading passengers, cargo, refueling, parking, maintenance—Boston Tower works its own ramps. Snow plows, fire trucks, runway safety checks—all are part of ground control's domain. Next, ground passes a departure strip to one of the two local controllers. Local controllers are the equivalent of the tower controller at Bedford: two instead of one, they divide the sky and the runways. The local controllers' main responsibility is to clear aircraft for takeoff or landing and to ensure that the runway is clear for them. They regulate all runway crossings. If either local controller identifies any unsafe conditions, he or she can send a landing aircraft on a "go-around": the plane quickly aborts the landing, instead climbing and turning away from the runway to be resequenced into the landing pattern by the TRACON. A pilot also may initiate a go-around. This is part of the daily routine for controllers—a pilot misses the approach, or another plane is slow to get off the runway or into the air—but the sudden shift from gradual descent to sharp climb and turn has passengers white-knuckling their armrests. When a departure is airborne and transferred to the TRACON, local west removes the departure strip from the strip bay and drops it in the "strip bucket."

Arrivals go in the reverse direction. The TRACON's final controller hands off to the appropriate local controller, who transfers the arrival to ground. Ground takes the airplane as it turns off the runway and guides the plane through the maze of taxiways and alleyways to the gate. Then ground drops the strip in the bucket. When the bucket is full, flight data gets it, sorts the strips to compile statistical data on arrivals and departures, and records the total number of operations per hour. The TMC tallies the delays per hour and send the figures at midnight to the TRACON supervisor, who adds the TRACON's and forwards the day's total delays for the terminal to FAA headquarters in Washington. Although the process seems more complicated because of the number of positions, traffic volume, and the tallying, in principle the movement of airplanes from controller to controller follows the same pattern as at Bedford, where the day's count is taken from the local controller's notes on the scratch pad that constitutes the pad management system and the number of clicks the local controller records on handheld counter.

Training is an industry here, as it is at the TRACON, because traffic complexity gives controllers who work here higher salary and higher status. Some come because they have always wanted to work in a major tower, or this particular tower, or because they like challenging traffic, but they don't come because it has a reputation as a kinder place. Like the TRACON, Boston Tower is a hardworking place, a place where teamwork within a team is highly effective, but it does not offer the compensating comfort of the TRACON's collaborative and collective culture. Of the four facilities, Boston Tower is the most competitive. It begins with the airspace. At the tower, even more than the TRACON, controllers are competing against time as they are pushed by traffic demands into racing against airplanes and airline schedules on their small airfield. In addition, however, the team concept operates differently in the tower, reinforcing competition.

In the TRACON, controllers all have similar work styles because of their training and because shift assignments regularly mix them, which results in all teams having the same personality—that is, the social organization of work is identical, which minimizes conflict and enabling cooperation. In contrast, Boston Tower airspace results in a division of labor in which the individual contributes differently to the team effort, and this allows for team differences to develop. Also, Boston Tower teams do not mix together on shifts. Team 1 never works with team 3, for example, with the result that each team has its own personality, its own way of doing the work. Should a member of one team need to work a different shift, the differences in technique cause conflicts, breeding competition between the teams, not cooperation. Making it as a developmental is hard: a developmental has to be able to work competently at the necessary pace and fit with the personality of the team to which he or she is assigned. It's a package deal, and the two don't always come in the same package. Many controllers, successful at other facilities, come to Boston Tower and wash out because the speed and volume of traffic call for competencies they do not have, both personal and professional. Stories about developmentals who just couldn't handle the pace are part of the facility's lore. A developmental is assigned to a team, and the team decides on the fit and trains accordingly. Boston Tower controllers, like the controllers at the other three facilities, are a special breed, survivors of training who are shaped by their location in the air traffic system and the peculiar demands of their airspace, which produces a place with a distinctive personality.

Place, Technology, and Ethnocognition

Relying on the skills that characterized early navigational forms of dead reckoning is a quality that all towers have in common. Like Bedford, dead reckoning at Boston Tower is rooted in the direct observation of events as they happen in nature, seen through the eleven-sided cab windows. Thus, the importance of the view at Boston Tower is that it provides markers by which controllers judge separation, speed, altitude, and distance. The position of the monument to the northeast, the Citgo sign in Kenmore Square, the Tobin Bridge over the Charles, the Boston Harbor channel, the islands, the gas tanks at Everett, bird activities, wind socks, cloud formations, and the darkness or smoothness of the water are all aids in predicting position, wind, weather, and how long it will take each kind of aircraft equipment to execute a maneuver. Controllers' ability to calculate distance and separation based on these markers and knowledge of airplane and pilot capabilities requires local and tacit knowledge.

But nature also creates obstructions to dead reckoning. I remarked to the supervisor once on a beautiful sunset. Because of their talents at room awareness, no comment goes unnoticed: the controller working local says to me over his shoulder, "The sunrises and sunsets are beautiful, but we never get to see them." They occur during rush hours, not exactly the most opportune moments to reflect upon the glories of nature. Moreover, the sunrise is blinding, even with sunglasses; it is directly in their line of vision of the runways in the hours between 6 am and 9 am. The sun sets behind their backs during the evening rush, yet because they scan to follow the traffic west, the sun remains an obstacle to the focus of their attention. But weather is an even greater hindrance to dead reckoning. In all seasons, the water and temperature conditions can suddenly produce a fog so thick that airfield and airplanes completely disappear, shutting down all VFR traffic. Conditions vary—wind gusts, temperatures—making plane performance less predictable. Snow and sleet stick to the windows. Bad weather not only poses safety issues—skidding airplanes, icing, obstructed vision—but also reduces the number of operations per hour. Separation between aircraft is increased, speeds slowed, runways closed. Extreme weather conditions cause cancellations and delays, sometimes cutting arrival and departure rates to nearly zero, or closing the airport. Bad weather always has implications for the number of operations any tower can handle and increases the number of delays.

In combination, weather, airspace complexity, and high volume at Boston creates greater need for technology than at Bedford and more of it is available. Technologies provide redundancy, affirmation, or correctives to what the human eye can see. As weather conditions worsen at Boston Tower, controllers also rely on the technology of Boston Center's TMU to regulate the flow of traffic or ground-stop it, to avoid the dreaded airport gridlock, when it becomes so jammed that nothing more can get in or out. Although Boston Center controllers say that "the best place to be during a thunderstorm is a tower," that is not Boston Tower controllers' experience. During a ground stop, local controllers have a lull, but not flight data or clearance delivery. Flight data scrambles to enter the many reroutes and changed departure times on flight progress strips and get the changes to the other positions. Because weather alters the original flight plan, creating ground stops or slot delays, clearance delivery is hard hit by calls from anxious pilots in a hurry to depart: "Could I have an update on the ground stop?"; "What number am I for pushback?"; or "What's the delay for?" When the weather finally begins to clear and the traffic flow opens up again, both Boston Tower and TRACON controllers have to work extended shifts, sometimes on into the midshift, to shovel out from under the jam up of delayed and canceled flights.

Signals and Interpretive Work

Signals and interpretive work—the ability to spot anomalies in time to prevent errors, mistakes, and accidents—is not the collective activity that it is in the TRACON, where everyone working watches the full airspace simultaneously. Like the TRACON, Boston Tower gets ratings of excellence annually for teamwork, but Boston Tower airspace creates differences in architecture, division of labor, position placement, and time, so that teamwork depends more on individual (rather than collective) noticing and alerting. As at Bedford, Boston Tower controllers work positions with different tasks, so all eyes are not on the full airspace all the time. However, the greater size and complexity of the runways and traffic create a different challenge at Boston Tower. All controllers have their own job to do and tend to see best what is in their control. Redundancy exists, but it's a different version of redundancy. The controllers in the front row—local control west, east, and ground—all watch moving planes in different locations. They see anomalies in their individual domains but also because their scans are continuously moving between air, sky, ground, radar, and ASDE, as well as what the other two in the front row are doing as they monitor each other.

The back row—flight data and clearance delivery—face in a different direction, so often can see things coming unraveled before the others do. The supervisors and TMCs are extra pairs of eyes on everything—sky, runways, radar. They are still occupied with their own jobs but have time to take in more of the scene. The positions are still interdependent, but teamwork relies more on individual rather than collective alerts. Time also works against the kind of collaborative coordination that distinguishes the TRACON. A Boston Tower controller pointed out this difference: "If the volume is high, you don't have time to look out for one another. We only have them two minutes, they have four or five to decide what to do." The Boston Tower volume is high, and the airspace is smaller than the TRACON's. Consequently, controllers there are more reactive. They have planes for only a short time and don't have as much time for preplanning. Like controllers at Bedford, Boston Tower controllers all said that the most challenging part of the job is pilots—I found this surprising, given the high proportion of professional pilots who fly in and out of Logan. For pilots, both departure and arrivals call for heightened activities and multitasking, which lend themselves to the routine production of anomalies. One controller described it:

> Never assume. You can't ever take anything for granted in this job. You can never be sure they're going to do what you tell them. Something as simple as a plane waiting to cross a runway—it's unpredictable how long between the time you tell them to cross and the time they actually start moving. (Tower, F)

Common sense—the combined package of institutionalized rules and cultural understandings, local knowledge gained from training and experience, and tacit knowledge—is the stable background against which anomalies stand out. For example:

> If a plane varies from the norm, we see it immediately. We know because from experience we know what it is supposed to look like. Like last night, that Beech turned the wrong way, into the pattern. Must have gotten confused or not familiar with the airport, turned right into the departure path. Things happen so quick here. No time to plan. (Tower, F)

> You have to watch the turn, the timing of the turn, which airlines turn fast or slow. Delta shuttle is a fast turn. They're in and out of here all the time, so really crank it up, but there's refinement. If it's going to DC, it's heavier because of the fuel. If to LaGuardia, lighter and so turns faster. Also [you] have to watch their speed on the runway, keep them moving. Then the crossings—MD-80s, DC-9s, they're old, they have to spool up the engine, it takes five to seven seconds—like an eternity—for a runway crossing. The newer ones just power up and cut across. The Cessna is like a car. Shoots across. You have to power up the big ones before you give instructions to the Cessnas so they are timed to cross together. (Tower, M)

For ground control, where planes typically are moving slowly, a pilot in a hurry is an anomaly that leaves little time to correct:

> I had an Olympic Air screaming in one night on Bravo [taxiway]. I wasn't talking to him yet and I had planes lined up front of him, going across the runway. I said [to local] switch it to me now [on ground's frequency], he's got to stop. And he was barreling inbound. I mean it was a 747 and he was taxiing fast and it was at night and we had this 727 with not many lights on it in front of

> him and they were going to meet and I finally just picked up and said on the mic "Olympic Air, STOP!" And he just smoked his brakes, he locked them up. Instead of [my saying] "Give way" or "Look out, traffic," it was too late for that. At that point I'm just saying "Stop, I don't want you to move." Don't move and he did that. Actually he flat-spotted his tires, he locked up so much. They asked later, why did you tell him to stop? I said because he had an airplane in front of him and he was doing about thirty miles an hour taxiing and you're only supposed to do five. So the guy was hauling, you know? . . . He was late, wanted to get to his gate . . . Most of the time you have a situation like that it's like a comedy of errors. Things happening that aren't noticed or recognized or foreseen and all of a sudden the situation is there. (Tower, M)

> Like the TRACON, emergencies happen several times a day. Planes take off and either Boston Tower or the pilot discovers a problem immediately and the pilot returns to land. It could be failed equipment or a medical emergency. Controllers clear the airspace and runway to make way. The supervisors call Massport to send the crash trucks, fire trucks, or whatever the pilot asks for. If the pilot wants the crash trucks, the supervisor asks the pilot the required, always-sobering questions: number of pounds of fuel on board? Number of souls on board? Time is always of the essence. Two or three times a day, I witnessed "saves": someone spotted a potentially dangerous anomaly and quickly corrected or gave a shout-out. Incidents ranged from the routine—clearance delivery notices the just-launched United still had its gear down—to the not so routine. The supervisor shouted, "Send him! Send him!" when he noticed a plane told to wait at a hold point had instead wandered onto the runway in the path of an arrival. In an occupation where emergencies are handled quietly, the shouted order for a go-around was startling. The arrival immediately nosed up, then circled to get back in line. A minute or so later, without turning his head, the local controller yelled to me, "Hey, Diane, you buying stock in Amtrak?"

A fast reaction is essential. However, Boston Tower is far from the seat-of-the-pants air traffic control that Bedford's airspace and traffic demands. Boston Tower work is standardized and formulaic. Work practice is strictly regimented to meet the demands of crossing runways, jet traffic, and the fast pace necessary to do the work and still meet goals of the system for the safety and the efficient, expedient delivery of traffic, which produces a distinctive culture. As one Boston Tower controller put it, "Everyone (here) says we are multitasked, and we do a lot of things, but we have one goal. We're a production line. Departures are our product." Experiencing work as a production line is a consequence of the airspace—the high volume, small geographical area of the airfield, crossing runways—and how place has been organized to deal with it. The division of labor and standardization of work give controllers the feeling of being on a production line, each person is one link in a process where the routes are standardized, work is systematized, tightly organized, the "products" passed rapidly from one controller to another on a high-speed schedule driven by an airspace with traffic demands that leave little room for deviation or downtime. Another controller continued the production line analogy, describing Boston Tower as a "machine" that is a tightly coupled system of interdependent parts. Notice, in the following, how he indicates the connection between airspace, place, and the social organization of work:

> It's possible—and I'm speculating—that the difference being in a center is you still have sectors and in each sector there's individuality in the way they run it and that is where it's left. Whereas in the Tower, it's that one airport, that is the sector, the sector is the entire airport and now you're involved with several other people and everything determines the outcome so your consequential impact is more vivid and is much more easily seen. What I mean by consequential impact is whatever a controller at ground does profoundly affects local and vice versa or clearance delivery, data—how they work back there.
>
> When the [Boston] Tower goes down, it starts at clearance delivery. So if the clearance delivery controller—it sounds absurd but especially there—when the clearance delivery starts to go down, that's the beginning of the end, then you see ground go and it's like dominoes and there goes local. I mean it's like if you have this clearance delivery smoking, this guy's on his game, he's really got everything up to date and everything taken care of, all the times are right, nothing's timed out, all the routes are right, all the altitudes are right, all the releases are there, then ground, all ground has to do is taxi and sequence, taxi and sequence, all local has to do is say "cleared for takeoff" and the tower is just a machine, it's just beautiful, it's amazing, and when you do that it's, like, this is the best.
>
> But the profound effect is that you have that one airport, that's your whole sector so that interaction is I guess more greatly specific, and maybe that influences the way that the team ends up working. The potential is greater for the individual to affect the group. Whereas [Boston Center] radar, it's just them, right there, that one sector. (Tower, M)

More than at any of the other three facilities, even the TRACON, at Boston Tower I heard production talk: awareness of and concerns about time and timing were expressed in interviews, conversations about techniques, talk between them while working, and instructions to pilots. Tower and TRACON controllers all know the maximum amount of planes that can be landed and departed on each runway configuration. Logan has six runways of varying length. They are aligned in three directions, with runway ends pointing toward six distinct compass headings (see fig. 19). They combine in various configurations, providing flexibility in a setting often buffeted by highly variable wind conditions. A *configuration* is a particular combination of runways that is optimal for arrivals and departures during a particular wind direction. Certain configurations can either constrain or maximize the number of operations that will be produced that day. At Boston Tower, how they are doing on "the count" is a frequent topic of conversation. Production and output clearly matter. On a Friday, the busiest traffic day, the TMC told me, "We had 92 ops between 2 and 3, 103 between 3 and 4. The optimum is when we're on the [runways] 4s and 9. We've had eight "hundred hours" in a row. On a good day, we can handle that without breaking a sweat." A controller coming back from break stops by the computer to check out the traffic count for the last hour. "Forty-three? Can't be right. It seemed so busy. Who was on?" The importance of production breeds competition. More is definitely better, for both the status of the facility and the controller who was working local.

The Production Line: Filling Holes and Shooting Gaps

Production and competition go together. Boston Tower's dilemma is how to maximize departures when they have to be worked between the TRACON's arrivals. As a supervisor put it:

> The goal of the TRACON is to run a tight final. We need it to be predictable and tight. The challenge is to work in the Departures. We have opposite goals. Theirs is to get as many airplanes in as possible. Ours is to get as many out as possible. (Tower, M)

The crossing runways on the small airfield generate arrivals and departures in crossing patterns. The system effect is a built-in competition between them, felt mainly at Boston Tower, due to the volume of traffic being funneled from the TRACON into the tower's smaller airspace, and the volume of traffic being forced through the funnel out to the larger TRACON airspace. Although the TRACON also demonstrated a keen preoccupation with expediency and safety in getting arrivals in and departures out, the TRACON has more airspace—and therefore more time—in which to deal with both arrivals and departures. The squeeze is on the tower.

The preoccupation with time and timing shows at every position but especially for the front row. Watching local and the ground controllers work is like watching two (or three, if two locals are working) conductors leading separate sections of a giant, sky-filling orchestra. Their instructions to pilots sometimes are unconsciously accompanied by arm motions signaling urgency—to hurry up, move in a particular direction, pointing with emphasis at the aircraft being addressed—common sense and emotion made visible in the body. Instilled during training, the voice is steady, with no change in inflection: the body vents the emotion and displays the urgency. Stripped to its essence, the job of local and ground controllers is filling holes and shooting gaps. "Filling holes" refers to local getting departures out through the spacing the TRACON controller on final creates between one arrival and the next. Shooting the gap is ground getting departure and arrival traffic moving across active runways and taxiways. Hold points are like stop signs: departing aircraft wait for the local controller's "cleared for takeoff" to depart through a hole between two arrivals; arrivals wait for a clearance to dash across a runway, known as "shooting the gap." The need to attend to time and timing is generated by the institutional concerns about delay and the press of daily traffic for local and ground controllers. Speed is not in conflict with safety; speed is essential to safety.

For arrivals, local controllers "call speeds" to adjust the spacing distance between them to make the holes. For departures, they "prep" pilots to make sure departures fill those holes: "Wheels up time is about ten minutes away"; "Traffic on four mile final, be ready to go right out sir"; "You have three, no four, arrivals before departure, last one is American 1807"; or "Taxi into position and hold. You will have 10 seconds to take off, be ready to move immediately." Similarly on runway crossings, ground control preps pilots to shoot the gap: "Give me a good rate," "without delay," and "expedite" are a routine part of instructions to pilots during rush periods. From experience, controllers have time-sensitive techniques for working local:

> I give an instruction, I know I've got fourteen seconds before I have to deal with him again. I can do several things between now and then. That's an eternity in this business. If he contacts me again, I lose the fourteen seconds. Now

> I'm behind. Takes double calls to get him out. If anything comes up needing special attention, I'm really in trouble. (Tower, M)

> It's important to keep control of the frequency. I call them [pilots] first. I can tell from the strips where they are and what they want. I don't want extra transmissions, so control of the frequency is important. It's important in all configurations but especially when we're on the 4s. There's a bottleneck at Bravo [hold point], so it's especially important to work the bottleneck or we can't get anybody out. On the 4s and 9, they bunch up at Echo, Whiskey, and Sierra [taxiways]. We can't afford to let them accumulate because they will string across the runway. (Tower, F)

> Finesse is getting a "two-fer" or "three-fer" at a busy time. They can cross [a runway] faster if you keep them rolling. He just crossed two in almost the time it takes to cross one when they come to a complete stop. Have to pay sharp attention to arrivals. Got to get them off the runway so arrivals can continue. Sometimes local west passes them in bunches of four—have to make sure to talk to them first. There's not much time to move them. Don't want someone stopping on the runway. (Tower, M)

If they don't move them fast, safety is jeopardized. The need to be both safe and expedient at Boston Tower is, to a great extent, "programmed in": institutionalized in formal international and local rules and experiential knowledge that govern practices. For example, the go-around may look like improvisation, but it is not: it is an officially sanctioned, standardized practice:

> In a go-around, you give a heading to turn them out of the pattern. Based on the runway configuration, I know what the traffic pattern is, and don't have to calculate, I automatically know the heading to assign for them to circle and get back in line safely. I learned this in training—which headings go with which configurations, under different conditions. It has to happen fast, but it's an SOP [standard operating procedure]. (Tower, F)

The go-around is a tool that allows an arriving airplane to dodge the departure and arrival patterns and get out of the way when the pilot has a wrong approach, separation is violated, or other some unexpected incident—either in the sky or on the runways—dictates. It has long been standard practice, but originated as a natural response to an urgent situation.

As we saw in chapter 2, system development was historically marked by improvised solutions to events and local problems that were first informal, then—when they worked—became formalized and codified nationally. As the system became more complex in response to changing external conditions, many local solutions that were originally deviations from rules became the norm and then over time became the rules.[5] Several Boston Tower controllers and supervisors told me about practices that originated as local problem solving to speed up the work, cope with congestion, and avoid delays, and then as the air traffic control system became more congested, and as a consequence, more standardized and safety minded, the informal procedures became formalized. At that point, the routines could be taught and their use could be regulated, to be used or not depending on individual pilot approval. One example at Boston Tower is "15 to circle"—also known as "the slam dunk."

The approach is particularly suited to Boston's complex airspace. It was developed as an informal procedure to ease inbound traffic congestion on runway 4R and increase landing capacity, then it became codified in the 7110.65. Adapted specifically for Boston Tower airspace, the approach allows controllers to transition aircraft from the expected, usually assigned runway to another one (see fig. 19). The 15-to-circle approach is used primarily by small commuter aircraft that normally land on the shorter runway (15L) instead making a visual transition to the longer instrument landing runway, 4L. This strategy speeds the arrival of commuter traffic when traffic on 4L is low and arriving traffic on 4R is high by moving the smaller aircraft to the runway with less traffic. During approach, the pilots of the smaller planes "dunk" under the arrival stream on 4R and slip into the pattern for the parallel runway. ATCs can request, but it is voluntary. The maneuver is as beautiful to see as it is expedient. Instead of coming straight in over the Tobin Bridge on the normal approach to 15L for this configuration, after the bridge, the plane circles, starting the downwind. It follows the Boston Harbor channel, keeping the city to the right. It then circles low—no lower than six hundred feet mean sea level—in front of and under the planes in line to land 4R and sets up for landing on 4L. A controller's ability to execute this procedure is one of the more difficult maneuvers at Boston Tower, requiring expertise at assessing conditions that affect the predictability of airplane equipment to perform it, as well as the ability to watch it circle behind the controller's back:

> We can land twice as many that way. It's more work on our part, involves more contacts with the pilot to control it. Also, we provide visual separation. Also it

> takes more skills. We need radar skills to work [the airplane] as it comes in and we need tower skills to complete it and bring them in. A heavy won't do it, they are too slow to make the change, but the props, some general aviation, they will do it. They don't have to hold short, so they land and taxi direct to the gate, plus they get to fly the airplane, take it off the auto pilot. They love it. So do the people downstairs [in the administrative offices]. It increases the arrival rate [by] six per hour. If conditions are right to do it and we aren't, they will call up and ask why. It's not always an option we offer. Pilots have to be willing. Low [visibility] ceilings limit arrival rates because VFR pilots refuse. (Tower, M)

In the Fire

Of all Boston Tower positions, the ground control position has the greatest potential for congestion and chaos. As planes come from different parts of the airport for departure, ground control merges them on the taxiways. Ground establishes the departure sequence—who precedes whom—and builds in separation. Simultaneously with the departures, ground is taking arrivals when they turn off the runways onto the taxiways and has to quickly cross other active runways to get them to a gate. Because of the irregular layout of the terminal buildings and crossing runways, ground is the most physically demanding position. Ground is all over the place, pacing back and forth behind the other controller positions, looking all directions, at the radar and ASDE, and importantly, at the strip bays of the local controllers. The demands of this job are further complicated by the antiquated headsets in use while I was there. To see all movement areas, ground has to get quickly from one side of the tower to the other by walking between local controllers and the console, where the headsets are plugged into the frequencies. It's an obstacle course. Ground has to step over—but at peak traffic times usually suppresses by stepping *on*—the swaying cords of local controllers' headsets. "Why is it," a ground controller yelled during a hectic session, "that McDonalds employees have wireless headsets and we have these?" Why, indeed?

In describing their techniques for working ground, controllers exhibit the same concern with integrating safety, time, timing, and schedule as do local controllers. In the following interview quotes, as ground controllers talk about their work, note that ground, too, is organized by common sense: institutional and local formal rules, experiential knowledge, and tacit knowledge. Note also the conjunction of safety and expediency that go into dead reckoning. On departures:

> It is safer to decrease the congestion in the airport. Efficiency allows us to get more planes out. Even at push back [from the gate], I'm trying to think about the departure sequence. I alternate heavies and props, and alternative west bound and east bound. When sequencing on Ground, I focus on the exit fix. I try not to give the same fix to back-to-back planes. If they have the same fix, I vary the altitude. I try to group heavies and 7-5s [757s]. If I have a little guy following a heavy, I need six miles separation, if two heavies, only four miles separation, so I try to group heavies together. We can launch more that way. (Tower, F)

> Alleys are hazardous because carriers are competitive to push back. They complain about each other—"He pushed back into my turn." I watch where the tail is. If push back in the alley, no problem, if push back onto the taxiway, that is a problem I have to pay attention to. You have to visually make sure a plane has pushed back before you start them moving, so you know exactly where to look—each gate—we have some eighty gates. Know the exact location. Then, we know what they are headed for, plus how long it takes them to get there, so you can look right at the spot and check progress, make sure the plane is where it is supposed to be. But they don't taxi at the same speed. The Airline Manual says a "brisk walk." They don't do that and we don't want them to. It would close the airport—the increased delays. They go faster than that. Professional pilots are pretty responsive. (Tower, M)

The volume of arrivals and departures on the small airfield of crossing runways and taxiways make communication between ground control and the two local controllers a second-to-second necessity. Their room awareness keeps them alert to each other's conversations with pilots even though they are talking nonstop themselves. However, they could not work together as effectively without the technology of the strip bays. Local controllers use the strips as a silent signaling system—a communication short cut that allows each to predict what the other is doing or about to do, and plan his or her own moves in response. As at Boston Center and Bedford, the strips themselves are actively used as memory, physical symbols that, by their positioning in a strip bay, remind them of an aircraft's position in line, an instruction just given or due to be given, or an action completed or yet to be done. Each controller has a set of techniques for positioning the strips, as reminders for themselves and to alert the others.

The local controllers can create their own, using a blank strip and mark in red the call sign of an aircraft doing something special, like waiting for a gate, or waiting in front of the tower, or short on fuel. But some of strips of special significance are permanent—wooden or plastic, painted in bright colors, placed in the strip bay as memory aids—"Runway closed—22L"; "Airborne"; "Bravo Hold Point," below which are all strips for planes holding where the taxiways come together, known as "that mess over there." Some of these permanent strips are symbolic reminders of past errors. Each has its own story of an accident or close call resulting from something some previous local controller once forgot. So, for example, the strip "vehicle on the runway" commemorates the time an Air Canada departed over a "field check" vehicle that speeds down the runways between arrivals and departures every few hours to check for debris, bits of rubber, and such. A controller gave the field-check vehicle permission to proceed, then forgot it was there and cleared the Air Canada to take off down the same runway. The vehicle driver saw it at the same time as the controller, swerving off the runway as the plane pulled up, the wing barely missing it. A catastrophe was avoided, but the lesson learned from the near miss was institutionalized in the commemorative "vehicle on the runway" strip.

It falls to local controllers to ensure that the TRACON's spacing creates the holes that allow departures to get out and gaps that allow taxiing traffic to cross runways. The local controllers can call the TRACON and yell about the spacing. Well aware of Boston Tower's traffic needs, the TRACON sometimes is so pressed by the onslaught of traffic that "running them close" is the only answer. A Boston Tower supervisor explained:

> The plane was coming in on [runway] 2-7, almost at touchdown and went around. The TRACON was busy, had lots on final, clear out to the edge of the airspace [twenty miles]. When they send them in so tight, we can't get our departures out. We had one still on the runway, so he [the arrival] had to go around. (Tower, M)

In response, the local east controller joked he was going to "stick it to the TRACON." He grabbed the 7110.65, opened it to the page stating rules of separation for towers (I noticed he did not have to search), and for a moment laid it open under the strip camera, displaying the separation rule to the TRACON. The TRACON controller running final then gave them three holes in a row to catch up departures, but then had to "jam them" with two and a half miles of separation (the allowable minimum, instead of three) on final to catch up arrivals. These competitive tensions between Boston Tower and TRACON goals are continually and daily negotiated between the controllers working final in the TRACON and local in Boston Tower. To keep things running without major antagonisms between controllers, the Boston Tower supervisors will call the TRACON supervisors downstairs to negotiate boundaries, asking for "a five-mile hole" so they can get departures out.

The contiguous airspace of Boston Tower and TRACON creates interdependence, which leads to a complex boundary relationship between the two facilities. The competing goals of each facility lead to system-bred competition between them that calls for daily negotiation and adjustment that could create lasting animosity and stress were it not for the TRACON's collaborative personality and willingness to compromise in face of the obvious—that with less time and traffic density Boston Tower controllers have no room for flexibility. As in the other facilities, compromise between controllers and facilities is at the core of all boundary work. However, several times during a day, nature forces them to set the competition aside, and the tower and TRACON controllers meld efforts in a beautifully coordinated activity that shows the complexity of this interdependent relationship and the resilience of the system.

The Terminal: Boston TRACON and Tower

Boundary Work: The Routine Drama of a Runway Change

Many dramatic incidents played out daily at each facility with an almost eerie calm as controllers—trained to hold their emotions in check and whose common sense was so ingrained—were able to act without thinking: an action was embodied to the extent that it came automatically, without pausing to calculate, thus freeing them up to concentrate on anomalies. The runway change at the terminal was one of them. It calls for reversing the direction of all moving aircraft in both Boston Tower and TRACON airspaces. Runways are changed every day, sometimes several times a day, sometimes several times a shift. Passengers are completely oblivious to the maneuver, so it was dramatic for me while routine for controllers, still supremely aware of its perils. As the wind changes, runways change because planes must take off and land into the wind. The problem is that all planes approaching the airfield are on headings based on current winds and the designated runway configuration for that wind direction. Departures are lined up waiting for takeoff. Planes on taxiways and runways are sequenced and moving to and from the existing configuration. A wind change requires controllers to change the direction of airplanes moving on the ground and in the sky. Timing is everything. Departures on the old configuration have to be gotten out and arrivals on final have to get down as all other moving aircraft in the sky and on the ground are changing directions to align with the new runway configuration.

Logan has several primary runway configurations in use, with over eighty operating subsets. The configuration in use affects the number of operations and the number of delays. The "operational capability" of a configuration ranges from approximately 120 operations per hour when the weather is good and three runways are available to fewer than 60 operations per hour for a single runway. So it is always an important decision, and one that calls for controllers to take into account all three institutional mandates of the system: safe, orderly, and expeditious delivery of air traffic. With the high speed, high traffic volume, and crossing runways at Boston, a runway change becomes a major—and visible—operation, fascinating in its execution and result. It requires close coordination not only within and between the two facilities; it also has a ripple effect on all the facilities with traffic heading into or out of the airspace, requiring them to alter the directional movement of aircraft in their airspace as well. Consequently, at the TRACON I heard the planning; heard the coordination with Boston Tower, Boston Center, and satellite airports; and saw the radical change in the traffic pattern on the radar as controllers shifted the planes landing at Logan to the new configuration. When I moved from the TRACON to Boston Tower, I saw the initiation of the change, the other side of coordination, and how planes in line for departure got out and how planes on the ground reversed direction.

The drama occurs during the moments leading up to the transition: the precision timing, the reversal of the cognitive map, the concentration, the attention to separation, and the boundary work involved. As remarkable as the maneuver itself was the smoothness of coordination within and between these facilities—especially in light of the conflicting goals of the TRACON and Boston Tower about arrivals and departures. All that is put on the back burner for a runway change. Together, they respond as nature dictates the runway configuration they must use.

The execution of a runway change hearkens back to the dead reckoning used in early marine navigation: human cognition versus the elements. Each shift starts with the supervisors and TMCs from Boston Tower and TRACON conferring on telephone to develop a "plan of the day" based on all the weather information, delay information, runway closures, and field conditions, to decide whether they will stay with the runway configuration in use or whether they can anticipate a runway change and, if so, at what time. Technically, it's the tower supervisor's decision, but in actuality, it's collaborative. Technology contributes importantly. The TMCs at both places collect all the necessary data to anticipate the new configuration: weather information, volume, field conditions, and forecast. But the plan is dynamic, always subject to change. They predict when a wind change is coming. They see it in the forecasts and in the wind speeds and direction indicators on their computer that are generated by devices on the airfield. The wind sock is supplemented by a computer that measures the wind and displays moment-to-moment changes digitally for local west, east, and ground control. However, these measures are often not the best indicators. Direct observation of nature is the key: like Bedford Tower, the Boston Tower supervisors and TMCs look out the window. One TMC told me:

> Weather is our bread and butter up here. You get to know the patterns. You have to be able to predict in advance to try to stay ahead of it. In the summer, winds are southwest, then by midday there's an easterly flow, then southwest again. They usually shift aloft before they shift on the surface, so we can anticipate. (Tower, M)

Supervisors described the subtleties of the craft:

> I watch the smoke, I watch the water. You can tell a wind shift by looking at the water because the ripples will start reversing out on the ocean so you can see a sea breeze coming in by just watching what color the water is. Especially if you have a southwest wind and the wind coming back in the water will reverse ripple and you can see it coming. While the smokestacks down here to the southwest, the smoke will be going the opposite direction of what the airport weather information is showing. There's a lot of telltale information out the window. Also we watch what the satellite facilities [small towers] are doing [runway changes]. Because they're scattered geographically, you can tell the movement of the winds. A lot of it is just looking at the forecast and knowing from being here so long with seasonal changes what to expect. Sometimes we get lucky, sometimes we don't. Sometimes we make runway changes and all of a sudden, shit, the wind goes the other way so you've got to change again. (Tower, M)

> We want to be landing into the forecast. If a cold front is coming, we can predict a wind change, it's fairly accurate. Sometimes the wind is coming around like a sailboat. We can tell by watching the planes. Look at the ground speed in knots, how they turn, are they slowing or speeding. The heavies will land long on the runway. For pilots in planes built after 1990, ask the pilot. He can punch in and tell you the wind speed at eleven thousand, one thousand [feet]. With a warm front, it's harder to predict. (Tower, M)

The change has to happen very quickly. Those arrivals on final have to land and departures all lined up to go have to be gotten out. The preference is to wait for a "natural hole" in the traffic, but depending on traffic volume, that may not be possible. Most often, the wind dictates the timing. No natural holes appear and they have to "do it on the fly": execute the maneuver despite heavy traffic. Alternatively, they may opt to make an artificial hole in the traffic by asking Boston Center to put their airplanes on hold during the runway change. Some TRACON supervisors always prefer to have Boston Center hold rather than "doing it on the fly." A TRACON controller commented:

> Some runway changes are a lot easier than others. I mean, if you go from landing 4s to landing 22 Left, that's messy. Some runway configurations, the transition can be very smooth with very little disruption to the room because the traffic flow is not that dissimilar from the original one. But we still have some supervisors who always want to keep it simple. I don't need anything remarkable to happen on my shift, you know? Let's get this cleaned up. Stop the take offs. Stop the inbounds. Shut them off at [Boston] Center. (TRACON, M)

The pivotal decision is which airplane will be the last arrival and which the last departure on the old configuration. On this the supervisors at both facilities consult their controllers. The crucial part is the coordination between the tower supervisor and the TRACON supervisor—how well they're able to communicate and work with each other to get the change done at the same time. Tower controllers describe the maneuver's dangerous "in-your-face" potential, the cognitive challenges of a runway change, and how much coordination matters when they are working departures in the local west position:

> The actual procedure of changing runways can be complex because there are airplanes that are still arriving in the opposite direction. For example, let's say we're departing runway 2-2 Right, so all of our jets are going out this way. Now

all of a sudden a runway change so now we're going to land on runway 4 Left and 4 Right. My last few departures are going out on that 140 heading at the same time the TRACON is now feeding me airplanes in from this same direction. Hopefully everything has been coordinated such that I'll be done with my departures by the time they get their first arrival inside of, say, ten miles. That's the perfect world and most of the time it happens that way, but there is a transition period in there where you need to be extremely vigilant of where, how close the first arrivals are because I'm still sending departures out basically in the face of those arrivals.

Your working speed is critical at that point because now you're trying to get rid of all your departures as much as you can so you're changing your priorities a little bit and it does take a few minutes for me to then get my thinking turned around so that now I have a new heading that I'm going to issue. I'm just looking at the airport from a different perspective. I'm trying to flip my own cognitive map and I'm absolutely relying on the supervisor, the traffic management coordinator, the other local controller because I expect them to know who's going to be that first arrival in the new configuration because my job is to miss them. And sometimes there is some confusion there, when the TRACON may have an airplane that's better situated for the old runway configuration versus the new and they may decide to send him into the old configuration. I'm expecting not to see that airplane and I'm absolutely relying then on the other Local controller to not only coordinate for me if that's what it takes but for the supervisor and traffic management person to be feeding me some correct information. (Tower, M)

Remember, it is the wind coming around that creates the urgency, and it's extra work for everybody in Boston Tower. Flight data and clearance delivery hold airplanes at the gate until ground control gets traffic movement on the ground either to a gate or moving toward the new departure runways. The two locals and ground use the land lines to transfer control of the runways, which amounts to two calls each, plus local west calls initial departure in the TRACON, making three. That is a lot of phone calls, given the time pressure and everything else they are doing. The supervisor feeds them the wind information from the weather radar system during the change, then quickly, at the change, switches runway lights to the new configuration. Local controllers change the maps of runway configurations displayed on the ASDE and radars hanging from the ceiling. The supervisor then informs Massport of the change. Tower controllers said a runway change is no big deal, they are "used to it," although they admitted its tensions and hazards. However, they unanimously declared that it's much more difficult for the TRACON than for the tower. A supervisor said:

> Our guys are on the ground. They're separated. Their aircraft are in the air, they've got to move, and they're all going in one direction. Sometimes they have to change 180 degrees to go in the other direction. They have to make a decision to stop doing it this way and take all these guys around, which is much more difficult for them. And their airspace is all changing. Little pieces of airspace, shelving, wedding cake, all that stuff it all moves around so they have a lot more difficulty. We just change. All we need to do is find out when their last airplane is and then we can judge when we need to finish up our departures. If the two sups haven't communicated well, we'll end up with a whole bunch of airplanes left to depart off the old configuration and they're all ready to come into the new one. And it's either they have to hold the airplanes while we get the departures out, which means it's going to build up out there, or else we have to taxi all our airplanes to another runway. (Tower, M)

TRACON controllers agreed, saying for them the runway change is probably one of the most difficult things to learn and do. One captured the general sentiment in the TRACON, noting that when the supervisor announces a runway change, "the hair on everybody's neck goes up. It's like saying 'fire.'"

The TRACON experience is different, and more difficult, because of the added complexity of the TRACON airspace. Boston Tower controllers view themselves as watching one sector—the airport—whereas the TRACON airspace is divided into different sectors. And although all controllers are watching the entire airspace on the radar, people are working different sectors, with different altitudes, shelving, and traffic patterns. TRACON controllers describe the many different cognitive challenges a runway change presents for them:

> When we're changing runways, the airspace is up for grabs. There's no set airspace. We're landing on 4 Right, now we're going to turn around and go to 27-22, so right where that last one ends and the new one starts, in the transition , all the airspaces are different. You pick an altitude where you know you're safe, and you go there and you just wait and let things happen, until we start to get things turned around. Like if we're going to 33-27, a safe altitude for me with my arrivals would be seven thousand because departures only climb to six. If I stay at seven, I'm going to be out of everybody's way. (TRACON, F)

> The entire airspace changes completely. It's like you're having to memorize a completely new set of airspace boundaries. You definitely have to do some more thinking. Sometimes, if traffic is heavy enough, I'll do one map overlaying another map—I'll keep on the original map, but I'll overlay the new map [on the radar scope]. If you can see the departures still coming off the airport, you can see how that is going to interfere with your operation as an approach controller now getting the airplanes in. That helps. (TRACON, M)

> When we change, everybody owns a different piece of the airspace. You have to think about your traffic, who's going to have it next, your airspace and the airspace you're going to have next—the changes in altitude, traffic patterns, plus the person on position. We could be in the midst of a runway change and something happens so we change again. I've seen all configurations on the same shift. You need people who really have it together. A sudden wind change adds complexity—doing it on the fly, nobody knows who's working what airplanes. (TRACON, F)

For the TRACON, change happens for each sector at different times. The goal is to keep the flow even for outgoing initial departures and arrivals into final. Change starts at final but is not all in order, like a wave. It depends on the configuration. After conferring with his own TMC, and collectively with Boston Tower's supervisor and TMC, the TRACON supervisor goes to the break room window to check weather to compare with the computer. He confers with controllers to decide the last plane in on the present configuration. "It's kind of an abstract judgment call." Anybody can be made the last arrival on the old configuration, and anybody can be made first arrival on the new. They try not to miss a beat between "last" and "first" so they don't lose a space. So the supervisor will ask controllers, "Who do you want to make last for this configuration? Can you make this guy last?" The question of who's last really is a question about a controller's skill—can he or she maneuver a particular aircraft from its present position to be the first arrival on the new configuration without losing the flow. A supervisor commented:

> Then we go through that who's last process, and some people are pretty good. If you feel that within your ability and your confidence level that you can actually take this airplane and get him to the next configuration without causing a problem or losing a space, then it's not too bad. (TRACON, M)

When both final 1 and final 2 positions are open, the two final controllers have to negotiate the division of labor, deciding who will do what and which airplane will follow which on final. When it's settled, Boston Tower and TRACON supervisors announce the plan to controllers. All pilots are advised of the cutoff—who will be last on the present runway configuration, who will be first on the new runway configuration. They begin. Boston Tower moves its waiting departures out and ground starts reversing the traffic in the taxiways. In the TRACON, the final 1 controller announces and initiates the runway change: "I'm gonna do the configuration everybody." "OK" they answer in chorus. Final 1 changes the runway configuration map on everyone's radar scopes to the new configuration map: "It's done. Grab your stuff [airplanes]." And occasionally, in the excitement of moving traffic into the new pattern someone shouts, "Release the hounds!" Whereupon the TRACON fills with the sound of controllers barking.

PART IV

Emotional Labor, Emotion Work

The stress of the job is part of the cultural image of controllers. By virtue of their responsibility for the safety of human life, air traffic control also qualifies as risky work. The risk inherent in the job is announced to the public in media coverage of accidents, close calls, and near misses, sure stress producers. Moreover, controllers work within an institutional and organization context that often appears to increase the risks. First is the problem of congestion: "too many airplanes, not enough concrete." Scarce resources and local politics limit the FAA's ability to build new airports and expand existing airport runways when air traffic increases dictate. Second is the problem of not enough controllers. Throughout my fieldwork, the concern that controllers most frequently voiced about their employer was that the FAA was not hiring. They complained that many facilities were already understaffed due to a long-standing FAA hiring freeze. Further, controllers were alarmed by the coming wave of retirements. Most air traffic controllers working in 2000 were replacements hired after the 1981 strike. Mandatory retirement age is fifty-six, but rules varied over the years, so depending on when controllers began, retirement could begin as early as fifty. It takes years to produce a controller with the knowledge, skills, and experience necessary to do the job. Why wasn't the FAA doing something about the staffing crisis that would be fully upon the system in five years? Controllers reasoned that fewer controllers working more traffic would mean more overtime and shorter, less frequent breaks, thereby increasing fatigue, stress, and risk.

This part of the book takes another look at system effects, exploring air traffic control as emotional labor. In particular, what is the emotional experience of working air traffic on an everyday basis and when things go wrong—mistakes, errors, accidents? How do controllers experience and manage risk and stress? These more social psychological aspects of the job are particularly interesting in occupations like air traffic control, where the job requires employees to control their emotions while dealing with various publics. Arlie Russell Hochschild, in her definitive work on emotions in the workplace, describes emotional labor as work for which emotions are a regular and natural part of the job.[1] Emotion work is the extra effort employees must exert to conform to the feeling rules that indicate what emotions are appropriate to the organization's goals. It is not just that they must produce a certain impression of emotion that disguises true feelings; emotion work is the work that goes into actually feeling as the occupation prescribes. Air traffic controllers are trained to maintain a calm tone of voice when talking to pilots—the front stage of their occupation that the public sees—but what is happening backstage, in the privacy of each facility?

On this issue, too, I entered the field full of the cultural imagery about the job. I did not know what to expect, but I certainly did not expect what I saw. Every day, at every facility, the work was conducted in a systematic, routine, orderly, and calm way. I was usually there ten hours a day, seldom was out of the room (except for interviews, which were interspersed with observations and began only after observations had been under way for some time), and rarely took a day off. Most of the time, controllers appeared relaxed and at ease. One day while observing in Boston Tower, I started a conversation with a supervisor about stress, overheard by everyone, courtesy of room awareness. A controller busy working local said (over the shoulder, eyes on sky), "Hey, Diane. Do I look stressed to you?" It was the noon rush, and indeed, he was the epitome of cool. They all seemed to be. I was with controllers during all kinds of challenging work in the sky—thunderstorms, snow, heavy traffic, emergencies, technical failures, lost pilots—situations guaranteed to increase risk and therefore stress. Daily, in each facility, I saw controllers engaged in the challenging and often vexing situations arising from the need to negotiate the boundaries of people and organizations on the ground in order to keep planes moving: pilots, other controllers, the Traffic Management Unit, management. I saw what controllers meant when they stated: "The stress of this job is not the airplanes. It's the people you work with." Yet these challenging situations were handled smoothly, with great concentration and all due diligence.

I saw controllers in many situations that I thought surely must be stressful.

Occasionally I saw people get mad. When they did, most often their displays of feelings were short-lived. Controllers are trained to be direct, and they are direct in releasing anger. Then it's over. I never once saw anyone freak out, although I heard stories about a few controllers at Boston Center ("the older guys") who had a reputation for "snapping"—yelling, throwing down their headsets, and storming away from position. Controllers love to gossip. I got the latest news about things going wrong and things going right, who did what to whom, gripes and conflicts, but I neither heard about nor saw any such incidents while I was in the field. In situations that I perceived as stressful, I saw no visible display of emotions. I was awed by what they had to go through in on-the-job training. I understood, from them, how stressful training was for both the developmental and the trainer. I saw many controllers being trained and saw the effort and concentration required to talk to airplanes, do manual tasks, and coordinate with others while also taking in the instructor's rapid comments. Yet even during training I saw very little overt display of emotion. In interviews I learned that trainers occasionally lost it, yanking off the developmental's headset and throwing it across the room. The replacement hires told me this tactic was "old school" from the military days, used on them but seldomly anymore. I never saw it, although at Boston Tower there was a late-night incident, which I missed but heard about the next day: "Hey, Diane, you went home too early."

I had seen controllers under fire stay very cool indeed. They were trained to control emotions with pilots and "get out of messes whatever way they can." They did both very effectively. But I had to wonder—could it be that people trained to control their emotions could at all times control their emotions around me? It made sense. They were trained to suppress emotions in conversations not only with pilots; those who survived the hazards of training came out of it with a strong sense of their own competence and well-developed egos. To show emotion in front of coworkers would be taken as a sign of weakness, which controllers were loath to show. However, I had read research in which observers rode in patrol cars with on-duty police officers for seven weeks to observe how they dealt with suspects, under which circumstances they used force, and how frequently the use of force was excessive.[2] Undeterred by the physical presence of observers in the back seat, the officers regularly used excessive force with suspects. The researchers' explanation was that the police automatically responded to the demands of a situation as they normally would, observers forgotten in the urgent need to act and the adrenaline rush.

With controllers, I could be seeing the real thing. Sometimes, sitting in the darkened radar rooms, controllers got so busy that they forgot I was there, and even in the crowded towers where it was impossible for me to disappear into the dark, air traffic, coworker, and pilot difficulties often trumped observer presence, as the police study would have predicted. I got glimpses of "real air traffic control," including verbal emotional outbursts associated with the traffic, pilots, and coworkers. ("Well, suck me!" shouted one in response to a crack from a coworker, followed by "Visitor in the aisle, visitor in the aisle," calmly announced by the controller I was sitting with). Still, the verbal display of such emotions was the exception. I still wondered not if my presence was changing their behavior but to what extent.

Nonetheless, their bodies betrayed their feelings. Supervisors say, "A quiet controller is a busy controller." As the pace and concentration necessary to the work picks up, any non-traffic-related talk between them shuts down. Everything they do happens faster—talking, marking, coordinating, typing. Some voices become louder, some quieter. Physical postures go from relaxed to tense, as radar controllers leaned forward into their scopes, both elbows or even both forearms on the workstation during rush hours, bad weather, or emergencies. I saw legs twitching, feet bouncing, pencils tapping, fingers tapping clickers nonstop. Tower controllers who usually sat, stood instead. Those who usually stood covered more distance faster, either with their scan, binoculars, or on foot, especially local and ground, but also the TMCs. Some controllers at every facility walked away on break with circles of perspiration under their arms or barely perceptible drops of moisture in their hairlines. These physical indicators occurred during the busiest times. Were they simply indicators of increased physical activity and concentration, or indicators of stress? How to evaluate what I was seeing?

I also saw symbolic indicators of deep concerns about risk. My attention was first drawn to these concerns when, after spending three weeks in Area B at Boston Center, I moved on to Area C. As typical throughout my fieldwork, my first day in the area began at the supervisor's desk. He gave me an overview of Area C's traffic patterns, described the overall airspace, and then described the characteristics of each sector airspace, any unusual activities that day, and weather and traffic conditions. After about an hour of instruction, I stood beside his desk to watch for a while before I walked into the darkened area to plug in with a controller. The controller at the workstation next to me looked up and, as if to correct his supervisor's explanation of the area, said, "We are mainly known for TWA 800."

Over the next weeks, controllers and the supervisors in Area C filled me in. TWA Flight 800 took off from John F. Kennedy International Airport in New York about 8:20 pm on July 17, 1996, heading to Rome with a planned stopover in Paris. About eleven minutes later, the Boeing 747 suddenly exploded midair, killing all 230 people on board, the debris falling into the Atlantic Ocean near East Moriches, New York. At the moment of the explosion and for about eight minutes before, it had been in Area C's airspace. TWA 800 was one of their "regulars": it flew the same route, New York–Paris–Rome, through Area C every night at about the same time. This night, it flew into the Erick/Sardi sector, as usual. Lori was radar controller working the sector. No radar associate was working with her. She had another plane—an air shuttle—at fourteen thousand feet, so when TWA 800 came into the sector, she instructed the TWA pilot "to expedite" through up to nineteen thousand feet to get above the air shuttle. Then Dave came back from break to relieve her at the sector. She explained the expedited climb during the position relief briefing. Dave had about eight airplanes in the sector when he took over. To slow things down for the few minutes it usually takes to reset the scope to suit personal preferences and get comfortable with the traffic at the start of a session, he stopped the TWA below the other plane rather than allowing it to continue to nineteen thousand feet. The pilot slowed the climb and had not yet reached the newly assigned lower altitude when the plane's radar transponder return stopped and the plane disappeared from the scope.

Dave had no indication from the pilot that anything was wrong. Other pilots in the sector airspace began calling Dave on the sector frequency to report seeing an explosion with flaming debris falling into the ocean. Pilots flying in Hampton, Area C's high sector above Erick/Sardi, contacted that controller, reporting a fireball. Pilots in two sectors of Area E, about fifty miles away, saw it and called Area E controllers. Air traffic facilities in the New York and Long Island area also received pilot calls reporting a flash of light. Lori, who had headed to the cafeteria and missed it, was called back to relieve Dave, who, according to Lori, was "pretty wigged out." Being the last person to talk to a pilot is traumatic. Dave had no idea what had happened. The uncertainty, knowing what happened but not knowing why, or sometimes not even knowing what happened adds another layer of anguish to that trauma: "Did I miss something? Did I do something wrong?" The controller is immediately removed and replaced, for his or her own sake and for the sake of the other airplanes in the sector. They have no time to grieve or come to grips with what has happened because they must immediately participate in the required internal investigation. They are debriefed about the incident, then asked to help interpret the data with others by listening to the tapes of the pilot-controller conversation and watching what they just saw on their scope replayed on the radar computer backup tape. They relive the incident and the trauma. The FAA recognizes the emotional impact of accidents in its Critical Incidents Teams of controllers who have taken training in posttraumatic stress disorder and counseling, making help available to those affected.

TWA 800 was the first time Boston Center had ever had a plane blow up while on radio frequency. In addition to Dave, additional controllers in the building were emotionally affected. Lori also had talked to the pilot. She recalled:

> If Dave had not gotten me out, I would've been the one, and it would have been worse. What I was thinking was that if Dave hadn't sat down and if that plane [TWA 800] actually had climbed above the other plane and blown up, he probably would have taken the other plane out also. That's all I kept thinking for days. He would have blown up on top of him and come down. I kept thinking I am really glad he got me out of there. That air shuttle probably never knew the difference. I mean he saw it, but . . .

The supervisor in Area C was emotionally affected. The other controllers in Areas C and E who were called by alarmed pilots were affected. Other Area C controllers who had worked TWA 800 other nights it had flown through their airspace wondered if they had ever talked to that pilot. But Dave was the primary person. The trauma may subside, but the last person who talks to a pilot never forgets the exchange. Also, being the controller associated with an accident at the moment of a tragedy has an impact on their identity. To the other controllers in the building, Dave would always be known as the controller who was talking to TWA 800 when it blew up. The accident is part of his history. A symbolic social reminder, this identity would follow him even if he transferred to another facility.

More than just a public identity, the controller internalizes it, accepting it as his or her own. When I was at Boston Tower, several controllers told me that I should be sure to talk to Ken, a controller on duty in the tower at Los Angeles International Airport on February 1, 1991, when there was a runway collision. A 737 landed on top of a Metroliner, a commuter plane that was taking off on the same runway. The local controller was immediately taken off position and Ken took over, doing both her tasks and his. He had since transferred to Boston Tower, and his identity as the controller who not only witnessed but also took over at LAX after the runway collision had followed him to Boston. "Have you talked to Ken yet? He's the guy who . . ." He was on leave when I arrived, so I had not met him. I knew he had returned because he had signed up for an interview. At the appointed time, he walked into the office I was assigned for interviewing, sat down, leaned back, and said, "Do you know who I am?" He waited. "Well, I think I do, why don't you remind me." "I'm the controller who was working at LAX when the USAir landed on top of the SkyWest." He did not mention his name. His experience at LAX was his work identity.

Accident history lives on in the present in the memory and identity of the controller, but also in the memory and identity of the area and the facility. TWA 800 became part of the identity of Boston Center's Area C: "We are mainly known for TWA 800." So did Area C's unusual experience of having many historically famous accidents while aircraft were flying in its airspace. On December 22, 2001, the day of the failed "shoe bomb plot" on American Airlines Flight 63 en route from Miami to Paris, an Area C controller sent me this e-mail:

> Just thinking. ZBW [Boston Center] is one of 23 ARTCCs [en route centers] in the NAS and Area C is one of five areas of specialization [in it]. TWA 800, SWR 111 [Swissair], AAL 11 [September 11], UAL 175 [September 11] all flew through this airspace and first encountered problems here on the days they entered aviation history books. EgyptAir 990 went thru New York missing Sardi-Erick-Hampton by twenty miles, although if it climbed well enough out of JFK, Hampton might have gotten a point out on it from [New York Center]. JFK Jr. flew under our airspace, missing it by a few thousand feet. Pan Am 103 was flight planned thru the airspace that December day but of course never made it. AAL 63 in today's news [personally I'm shocked the passengers didn't kill the exploding shoe guy and I bet there's a great story there] was also flight planned thru the airspace. Of course the American flight was karmically protected by the fact I was working. (Boston Center, F)

An accident becomes part of the identity of the facility, part of the lore that is passed on to generations of controllers. As a symbolic indicator of deeper concerns about risk, there are additional repercussions, however. When an accident happens, controllers in that facility and elsewhere recognize the risk and the real possibility—like Lori—that it could easily have been them. They need to know why it happened. They study it; discuss it. Was it a technical failure? Was it the controller's fault? Could it have been different? How can we make sure this doesn't happen on our watch? Accidents are, of course, the epitome of controller's experience of risk and stress. However, I also saw that controllers' response to mistakes and errors—lesser events but of great significance and meaning in controllers' lives as well as organizationally and institutionally—also seemed to be symbolic indicators of deeper concerns about risk. A mistake differs from an error in their view: an error being a specific kind of mistake that is officially identified as an operational error—a violation of the rules of separation—that calls for an official investigation. I saw the same controller preoccupation with mistake and error as with accidents and the same response pattern—emotional trauma, identity, the search for lessons so there won't be a "next time."

Their search to understand how things go wrong is part of the culture. Safety is one of the three mandates of the system. It is central to their training, in which learning from mistake is institutionalized practice, perpetuated after training in mandatory refresher training sessions using films, computer-based simulations of accidents, and written materials. Their mistakes during training become part of their identity, memorialized with nicknames, like "Cement Head," "Butcher," "Jigsaw." After training, when mistakes and errors happened, controllers, on their own, were asking one another what, why, and how. As with accidents, the details of mistakes and errors rapidly fly through a facility:

> You pass on stories all the time, constantly, not just to developmentals but to coworkers. You should have seen what happened to me last week. I did this, I did that. If you listen to it you say, wow, I've got to watch that and make sure it doesn't happen to me. (Center, M)

Accidents are rare. However, mistakes and errors happen more frequently, resulting in close calls that have the potential for serious harm. Controllers are trained to fix them so that catastrophes do not happen, but what about controllers' experiences and their emotional responses to mistakes and errors? And what about events such as emergencies, bad weather, thunderstorms, and technical problems, which I had seen handled very professionally, quietly, and calmly, with no visible indicators of risk awareness or stress? And what is their experience of those everyday busy periods when traffic becomes heavy and they are getting hammered?

Stress, tension, and related emotions such as fear, anger, anxiety, and depression are not always visible at the time they are experienced, especially when an occupation requires—and trains for—their suppression. Further, the effects of the work on individuals will vary—with experience, disposition, age, and competence. How could I understand their emotional experience of working air traffic and the meaning of mistake, error, risk, and stress to them? Interviews were the only way to get at the interior of their experience, but what would they be willing to tell me? One possibility was that in the privacy of an interview with a stranger they might feel free to say things he or she would never say to another controller. However, I had several times heard the comment, "Once a controller, always a controller." Trained to suppress emotions, they may have been unwilling to reveal them to me.

I developed an interview strategy based on my field observations. I had learned that learning from mistake was a tenet of their training. They had to be able to recover from a situation when things went wrong. I also learned that controllers experienced mistakes and errors emotionally. As one said to me, standing in the lunch line at Boston Center early in my stay, "We come from all walks of life, but we are all bound together by that little frisson of terror we feel when we think we might have missed something." Mistakes are usually caught and corrected; nonetheless, they produce emotionally searing, unforgettable experiences. In conversations with controllers, I saw that explaining what happened regenerated the memory of the incident and consequently a recall of and in many cases even a reexperiencing of some of the adrenaline or emotion they had felt when those incidents occurred. They talked about the emotions they felt during specific incidents as readily as they described the incident itself. Consequently, to get at their experience of risk and stress, I revised my interview schedule to evoke their recall of mistakes and errors they had made.

Each interview started with questions about how they came to the occupation, which naturally led to their training. After my questions about their training and learning from their trainers, I asked about learning from mistake. Everyone said that mistake was an important way they learned. Then I asked them if they could recall some mistake they had made either during their training or as a certified controller, what had happened, and what they had learned from it. Everyone had an incident, details still stark and clear in their minds. These experiences had scared them, and as I expected, describing what happened in detail put them into the past, reviving the emotions they had at that moment. I did not ask about emotions: they volunteered their feelings as part of the natural chronology of the incidents they described.

Describing an incident, controllers became more animated, talking louder or faster. When they talk about mistakes and errors they have made that might have serious consequences, they become very physical. From across a room, you can tell when they are talking to each other about airplanes because of the way they use their arms and hands: fingers tight together, arms and hands move to show relative position and direction in the sky, "This guy was descending 32 and this guy . . ." In their line of work, a lot can happen in seconds, and voice and hands have limitations when it comes to conveying the full dynamic of an incident when the experience starts tumbling out of memory. Having begun their description to me using their hand movements, often they would grab paper and pen so they could draw the position and movements of airplanes to continue the explanation more quickly. My field notebooks contain the rapidly drawn diagrams they drew on the pages and on the sticky notes I started carrying.

As I expected, explaining how a mistake or error occurred generated a detailed account of the event and, simultaneously, emotion and adrenaline. Their narratives were filled with descriptions of their moment-to-moment feelings. Then, having put them back in touch with events that had scared them, I asked questions about risk and stress per se, in this sequence:

> When you are working air traffic, do you experience your work as risky? Why or why not?
>
> Are there any other kinds of activities or occupations you think are more risky than yours?
>
> Do you experience your work as stressful? Why or why not?
>
> Are there any other kinds of activities or occupations that you think are more stressful than yours?

Prior to these questions, I had not mentioned risk or stress in the interview or asked people informally about their personal experiences of them.

The following two chapters are drawn from both my field observations and controllers' answers to these questions. They go behind the cultural imagery and myth of air traffic control to expose the emotional labor and emotion work that are integral to the job. Chapter 7 investigates air traffic control as emotional labor: an occupation for which emotions are a natural and regular part of the job. It shows how they define a mistake, how that differs from an error, and the hierarchy of mistakes and errors they make. Their compelling descriptions of these incidents demonstrate just how deeply emotional the job is and the meaning of mistakes and errors to them. Although controllers interpret mistakes as individual failures on their part, their accounts reveal how mistake and error—and the emotion they generate—are system effects, natural products of the design of the air traffic control system itself.

Chapter 8 explores emotion work: the work that goes into not just managing emotions but actually producing the feelings that the system requires to accomplish system goals. In startling contradiction to the deeply emotional experiences that are splayed open in the previous chapter, controllers uniformly insist the work is not risky. Further, only a few said it was stressful. Most said it was not stressful at all or was stressful only under limited circumstances. The chapter concludes with an explanation of this contradiction, showing controllers' agency in response to the emotional labor built into the job. Affirming Hochschild's work but going beyond her analysis, the chapter shows how institutional, organizational, and cultural factors combine to generate meanings that redefine the experience of work itself. Controllers respond to the exigencies of their boundary work by constructing cultural understandings that effect a social transformation of their risky work. They redefine emotions and transform the meaning of work itself in ways that protect them. Van Maanen and Barley observed that culture is the solutions that people develop to resolve the problems they face in common.[3] Integral to ethnocognition and boundary work, controllers' emotion work is a collaborative effort: both individual controllers and the group initiate cultural solutions that enable them to continue doing the work day after day, mediating the emotional costs.

7

Mistake and Error

Emotional Labor

In controllers' line of work, making mistakes is taken for granted. They make plenty of them, and talk about mistakes openly:

> Of course we make mistakes; we make six thousand decisions a day. We've all made them. I mean, even a CPC [certified professional controller] will make mistakes. If you can learn from your mistakes, you're going to make it. If you can't learn from your mistakes, you're going to keep doing the same thing over and over again and you're not going to be able to do the job, not after a while, anyway. That's one thing about the training process. You want the developmentals to make the mistakes because that's how they're going to learn a job better. . . . You can only talk so much about well, this happens, this happens, you know? This is the outcome. But if they see it right in front of them, it tends to leave a lasting impression. (TRACON, M)

Controllers describe a hierarchy of mistakes—mistakes that have various degrees of seriousness, depending on the outcome. Their credo is "the safe, orderly, and expeditious delivery of air traffic." In working to carry out operations at the lowest level of potential harm, they describe actions or inactions that jeopardize the orderly and expeditious aspects of the credo. They count as mistakes the actions for which the outcome is wasted time and increased operational complexity, even if not harm to passengers:

> I'll tell an aircraft "Contact Manchester approach, 125.05." And then not a minute goes by, did I tell him? That's a mistake in my mind because now I'm wasting time on something I've already done. Did I climb that aircraft to the proper altitude? When I'm busy, to replicate transmissions to do the same task two or three times possibly, isn't serving me, or serving the customer. Because now I'm making myself that much busier, and making the operation more complex. For me, that's a mistake. Anything that's going to make the operation more complex, or make it more complex for somebody, say next to me, or up in another room, or wherever that next controller is. (TRACON, F)

Another type of mistake in this category is the inability to "make something work":

> If I personally make a mistake, I feel bad about it, not even so much for the airplanes. It's the impression I'm making in front of all my coworkers. Because there are so many variables that go into things. What the pilot's doing, what another controller did to you, what the read-back is, the situation of the equipment, the radios, timing. So you might not have been able to make something work. No matter how perfect your plan might have been, all it takes is one pilot to say, "Say again?" and your plan is out the window. So sometimes that's all it is. You've built the perfect space on final to put an airplane in there, and when that airplane reached the turn point, you turned him, and the transmission got blocked. Now you've got to turn even harder [read: give the pilot a heading requiring a sharper angle]. You turn again, and he says, "Say again? Was that for me?" Now you've missed the hole. Now you've got this final going, stretched out for forty miles, and you've got this big eight-mile gap in the final where there should have been an airplane and there's nobody to put in there. Not a thing you can do about it. (TRACON, M)

Both of these controllers are describing situations in which they were not able to be as orderly and expeditious as possible, which is a mistake in their eyes. The outcome of concern to them is not passenger safety, but the implications for schedule and for their coworkers—they made the job harder for someone else or lost status by making a mistake that affected the group goals.

However, controllers also describe situations when mistakes have the potential for more serious consequences, actions taken or not taken that threaten to violate the safety component of their credo. At this level, one frequently mentioned mistake is the "hear-back or read-back error." Controllers talk about these as mistakes, but at the administrative level, they are official errors: a mistake that is institutionally recognized, named, and tracked, with interventions administered to reduce them. In a hear-back or read-back error, a controller gives a pilot an altitude instruction and the pilot mishears it. The pilot repeated the instruction (the read-back) to confirm, but repeated the wrong altitude and the controller didn't catch it (hear-back). The pilot goes to the wrong altitude, leading to a controller "operational deviation":

> [So I say,] "Descend and maintain one-one-thousand [eleven thousand] and the read-back is one-zero-thousand [ten thousand]." That aircraft ends up going through 11 to 10 into the path of an aircraft that is at 10, and then you have to do basic maneuvers. Or he descends into somebody else's airspace. This would count as an operational deviation against you, where one of your planes intrudes on another's [controller] airspace without your getting their permission. No loss of separation. It's like you're a controller at the center, I'm a controller in Boston, and one of my airplanes gets in your airspace without a handoff or any coordination. (Center, F)

Hear-back or read-back errors happen so often that controllers describe them as a routine problem, and just as routinely, controllers spot and correct them. If they miss the read-back, they save the situation by their scan: they see the pilot doing the wrong thing and correct it. One controller told me he had already caught three that day. But sometimes they catch it late or don't catch it at all, which results in a close call or an official error: either an operational deviation or the dreaded operational error, a violation of the rules of separation. Informally, this is known as an "OE" or a "deal." Controllers call it a deal because personally, organizationally, and institutionally, it is a big deal. Close calls, operational deviations, and operational errors are the mistakes that challenge the safety component of their mission credo. They can arise from hear-back or read-back errors and other transgressions, acts of omission or commission, made by pilots and controllers alike.

Although accidents are avoided, close calls, operational deviations, and operational errors have an emotional impact. These are the mistakes that leave that lasting impression. When controllers describe them, they talk about "two planes coming together," "going to get together," "this guy descended into him." They do not talk about collision or accidents. They are talking about airplanes coming to the point at which the rules of separation might be violated. When publishing statistics on operational errors, the press tends to equate operational errors with close calls and near misses, which is not always the case. You can have a deal without having a close call, and you can have a close call without having a deal. For controllers, both trigger a deep emotional response.

Close Calls

In conversation about mistakes and errors with possible serious consequences, controllers use common phrases that indicate an emotional reaction: "You surprise yourself"; "You scare yourself"; "Your eyes play tricks on you—you think you see something or you see nothing but you're wrong." The impact stays with them:

> We think in seconds. You look away and look back—how long was it, was it ten seconds? There is always the fear of things going wrong. Looking, seeing something you hadn't noticed or some change in the few seconds your eyes were on something else. You never forget. The shock. (Center, F)

> We are all bound together by that little frisson of terror we feel when we think we might have missed something. [*Raises his left forearm and moves his right hand above it, wiggling his fingers a little to suggest a physical sensation moving up and down the arm.*] (Center, M)

It's physical. They talk about "the pucker factor" and "the scare you don't get over." For example, "I stood up from my seat," "I was scared shitless," "I was shaking," "I couldn't speak." They frame it is a matter of the heart: "Your heart goes ping," "My heart was in my throat," "My heart stopped," "It was heart-wrenching." They know the feeling well. In training, everyone gets to have the experience:

> I scared the crap out of myself when Kevin [the trainer] and I were training on ID [initial departure, the first training position in the TRACON]. I turned a guy too soon and he [Kevin] had to step in. It wouldn't have hit, but it was not a smart thing to do. And that's like a bucket of cold water in your face. But then something like that happens and your awareness just grows exponentially. It's hard to explain. It just—your vision, your scope, everything just broadens. Or it should. Your awareness from then on—that's why now I look at the wind, and I consider the wind so much when I do things. It affects the turn rate, who it is. You scare the hell out of yourself and it's like something happening in cement. It's always there because you never want it to happen again. (TRACON, F)

Serious mistakes and errors are forever embedded in controllers' memory. They remember the call signs of the involved airplanes, what the pilots said and did, altitudes, directions, weather, time of day, the controllers working on either side of them, what they did, what the supervisor did. This next example I reproduce in its entirety because it is typical of the detailed recall when they miss something—and the lesson learned. A former Boston Tower controller who moved to Boston Center described his experience:

> [When I was] working in the tower, an Eastern Airbus, Eastern 1030, I remember the call sign, landing on 4 Right, called on a five-mile final. I had a TWA 747 departing 15 Right, which crosses runway 4 Right. Eastern checked in. I gave him my best "Roger." I did not tell him about the traffic that I needed to depart on 15 Right. I cleared TWA for departure on 15 Right, and then was busy with other aircraft doing other things. Eastern was on about a mile and a half final and I heard a loud roar off my left side in the tower, which was the TWA jet starting his roll on 15 Right. He had hesitated, didn't go when I told him to. And I, not knowingly, didn't pay attention to it. Eastern is coming in on 4 Right. TWA is departing on 15 Right and they're going to meet at the intersection. That's the only thing I could think of.
>
> I looked at the Eastern jet and calmly said, "Eastern 1030, [gives heading] go around." He didn't answer me the first time. So the second time, thinking that it was going to help, I pointed to him—like he was going to see me point to him in the tower?—and very calmly said "Eastern 1030, go around." And I saw the nose of the airplane coming up. TWA kept rolling. TWA went through the intersection. Eastern wasn't even close to him because he didn't even hit the threshold before TWA went through the intersection. So it was going to work, but it just was very tight.
>
> Eastern went around. The supervisor was looking the other direction. The controller in at ground looked at me and he said, "You OK?" I went, "No." I said "I need to be relieved" because it scared the hell out of me. The supervisor came over and he said, "What's the problem?" I said "Eastern went around, TWA took off." And he looked at me and he said, "So?" And I said, "I got close, that's why I sent the Eastern around." He said, "Well, you did your job. That's what we pay you for. That's what makes you an air traffic controller. Continue working." And I said, "No, I want to get out of position." And he took me out and to the back of the tower and explained to me how he thought I had done the job that I was paid to do and I shouldn't have a problem with it. And this was the first time that I thought I actually got airplanes close together where

I thought I was going to kill somebody. To this day, I remember that, and I can still see it in my brain, still see it. If I shut my eyes, I can still see those two airplanes, and I learned a lot from that.

Q: Wow. [Silence, both of us clearly affected.] So—what did you learn from that?

Scan. Always keep looking. And if you tell somebody to do something, watch them. Make sure they do it. And if I had only told—now Eastern called after he went around, we had Eastern call. And Eastern, the pilot told me the reason he didn't answer me the first time I told him to go is, he said "When you said 'Go' I hit the throttle and started my climb procedure. I didn't have time to hit the mic. You tell me to go, I'm going." He said, "If you had told me about the TWA," he said, "I could have slowed down." He was coming in faster than normal.

So lesson learned—paint the picture. An old supervisor once told me that. Paint the picture. Had I told Eastern that I had a TWA jet to depart on 15, he would have slowed up. Had I told TWA that Eastern was on a four-mile final, no delay, take off, he would have moved a little faster. So hence neither pilot had the picture that I had. So had I only given them more information, that situation would have never happened. And now when I train somebody, or when I talk to somebody who is a new developmental, I always tell them, paint the picture.

That was the first time in my air traffic career that I scared myself. I saw two airplanes, two very large airplanes, and it was all my fault, and that bothered me a lot, that it was my fault, that I had screwed up a lot of people's day, so to speak. I could have screwed up a lot more than their day. Once you make a mistake like that you never forget it. I have never known anybody to really make the same mistake twice if you follow. I mean you do make other mistakes, but our rules [of separation] are set up to correct those mistakes so that you have enough buffer zone to fix them, especially at the [Boston] Center, some down in the TRACON, not necessarily in the tower.

But you learn. You don't forget. I can shut my eyes and still see that to this day and that happened probably fifteen, sixteen years ago. But if I shut my eyes I can still: I can see who's in the tower, I can see all the people working with me. And I can see that Eastern and that TWA jet. (TRACON, M)

He took an action that saved the situation. The perhaps surprising response of the supervisor, "Well, you did your job," was not surprising to controllers. The important thing is to save the situation—"to get out of a mess whatever way you can." Their job is to separate airplanes. And he did that. Thus, the supervisor's comment.

He didn't see it coming. He missed the TWA's delay in starting his roll. Missing something is hard to get over:

> It's more likely that you'll replay something that you've missed. Two airplanes coming together, and then take some really serious action to separate them, expeditious turns, fast descends, something that you just didn't spot in time is what it is, and you just missed something. You go, "Oh my God." And whether you have an operational error or not, or maybe somebody else spots it and saves you from it, things like that are when you go, "Oh my God, I can't believe I didn't see that." Stuff like that is what you tend to replay over and over again. (Center, M)

Missing something is a fundamental blow because it is about not being in control, which shakes their belief in their ability to do the job. This controller talks about the experience of missing something versus having a deal:

> If you make a really bad screw-up, that stays with you.
>
> *Q: . . . How do you define a really bad screw-up?*
>
> Not seeing traffic.
>
> *Q: You mean having a deal?*
>
> It's not so much having a deal. I mean, you can have a deal and not feel bad about it because you tried something. I mean, you could have a deal that you never even saw, those are the ones that really bug you. Never saw. That means you missed it completely. That means there is nothing—it's one thing if you have a deal where you're trying to do something. You know these two are going to get together, but you tried to do something and it didn't work out, OK, that's fine. You had a deal. It was a bad day. That doesn't bug me. But when you don't see something, that's the one that sticks with you. (Center, F)

Having a Deal

Objectively, controllers are sanguine about the meaning of having a deal:

> We all make mistakes. I could have a deal tomorrow, for all I know. We all learn from it, and no one gets hurt from it. Because with anything in this business,

> a deal is a compromising of limits that have been placed upon us. At the [Boston] Center, you could get them as close as 4.995, like big deal. If you actually think what five miles [separation] is, you start at Boston and then you go five miles north of here, then you're like four cities away. That's not that bad. Here [the TRACON] the rules of air traffic, three miles, one thousand feet, on the runways and stuff. There is a margin of safety built into those. And if you're looking out the window, you look outside one day and see a couple of planes go by each other. They're probably only four hundred or five hundred feet apart, the little Cessnas and stuff. In a half mile, they look like they're days apart from each other and there is no way they could hit. But on a radarscope, they're this close. [*Laughter.*] But they never were really that close. So I do also realize that if a plane comes within two miles of each other and six hundred feet, I had a deal but nobody was going to get hurt. They weren't pointed at each other, they just laterally were close. So it was a safety violation, but there wasn't danger. So just learn and don't do it again. (TRACON, F)

However, subjectively and experientially, the operational error is the most serious mistake in the hierarchy of mistakes, in their view, because it increases the risk of harm to passengers. It is an emotional moment:

> The first thing you do if you have a deal here is you feel you hate yourself. How stupid were you? 'Cause it puts that fear [in you] of "wow, what if this happened, what if that happened?" It takes you to that step that we don't think of until it happens. It takes us to that step of fearing the accident. And that's a scary thing. (Tower, F)

Almost everybody I spoke with had a deal. In interviews, controllers talked openly about how they felt emotionally when they had deals. For most, operational errors have a lasting emotional impact. The emotional jolt is the same even when two airplanes come close but a deal is avoided. Controllers describing these incidents talk about "almost a deal," "so-called deals," and "unofficial deals"—the latter is when the controller believes he or she has had a deal, but no deal was officially identified. They describe feeling fear or panic or their heart pounding in the moment. They also talk about being afraid after the deal that they will have another deal or even an accident. They describe feeling bad about themselves and about having made a mistake. A few described the experience as embarrassing and humiliating.

In each of the following incidents, the controller acted to avoid a collision.

Note the elements of surprise or shock, and how integral to the retelling the controller's emotional experience is in every case. This, from a Boston Tower controller, where one pilot's slow roll-out caused a deal and started a spiraling-out-of-control traffic situation:

> Just one [deal] occurred, when I first got checked out. We were on 27/22 [runway configuration: arrivals 27, departures 22]. I had a 22 Left departure that I had an operational error on. Cleared the guy for takeoff and by the time he finally got rolling it was, you know, too late. I let him keep rolling, a 27 guy landed and this guy [departing 22] was just getting airborne, so it was an error. It kind of went downhill from there 'cause 27 was busy, kind of, you know, I needed help, 'cause then the next guy I had to send around on 27 too because the first 27 arrival hadn't got off the runway yet. So now I've got this other 22 Left departing. I knew I had this operational error, so that kind of shook me up a little bit but then I had to send the guy, the next guy on 27, around because the one that had landed had stopped short of 22 Left on the runway and I had to work him in around the other traffic. After that I was real nervous about 22 Left for about a year afterwards. I can remember getting really scared when I had to depart 22 Left. It took me about a year to get over that. That's probably the biggest incident that stands out in my mind that had an impact on me. (Tower, M)

And at Boston Center:

> I've had a couple of deals myself that were close. Once you . . . put your heart back down in your chest, you just, you know—it's pretty scary, you know, it's pretty scary. When I had my first error I was newly checked out and I had an error. I descended a guy into somebody else's airspace without coordinating with him. And the aircraft got together with a refueling track. And it was very, very close. And it stunned me. I was actually speechless. And when we later went back and listened to the tape of the incident, on that tape all this stuff takes place, and then there's this Lufthansa who's calling on the frequency. And says, "Lufthansa, Boston Center. Lufthansa 471 at 3-9-0." And there's nothing because I am like stunned now. I've just had the error. And I'm just, I'm in shock. And this Lufthansa 471, "Boston Center, Lufthansa 471, do you read?" Nothing. Finally, he says, "Does anybody read Lufthansa 471?" So then somebody says, "Uh, yeah, Lufthansa we hear." He goes, "Well, I called Boston again and again. I don't think he likes me." I was just gaga. (Center, M)

The newly checked out controller having a first operational deviation or error can be stunned into inaction, but after the first deal controllers learn to work through it until they are relieved—the training kicks in and they "get out of a mess whatever way we can"—because, meanwhile, the sky is full of other airplanes needing attention. The emotional impact doesn't diminish with time on the job, however. An experienced TRACON controller described the following incident, in which a hear-back, read-back error—an official error—led to a second official error, an operational deviation:

> I had a situation myself. It happened mid-September. I was working Lincoln at 7 [o'clock]in the evening. I wasn't busy at all. And I had a guy who was going westbound, and his request route was ten thousand. And I climbed the guy to ten thousand, but he read [back] twelve thousand. So he ended up going into somebody else's [a controller's] airspace who he shouldn't have gone into. So that constitutes an operational deviation. So there was a big—you know they had to file a report, do the whole thing. And I'll tell you something, after I had that airspace thing, for like the next three or four weeks I was feeling pretty bad. I was going into work and it was like, you know, anything can happen here. Because I was always of the posture that I'm checked out. Everything is going good. I feel good about it. And then boom, something happened when I'm not busy. So now I'm thinking, well, Jesus, if that can happen when I've only got three airplanes, then what can happen when I've got twelve airplanes, right? So that experience changed my stress level, I guess. (TRACON, M)

The deal itself has an emotional impact, not only in its own right but also by reminding controllers of the possibility of an accident. But a second emotional jolt comes from the administrative response to a deal. Institutionally, deals get more attention than other kinds of mistakes because loss of separation increases risk of collision. In addition, deals get more attention because the rules of separation establish spatial boundaries that allow for the identification and counting of an operational error, and for administrative action to be deployed to reduce the occurrence of such errors.

A Big Deal: The Degradation Ceremony

The pressure on every facility to reduce errors is a fact of life. The number of operational errors is the FAA's official measure of safety. Collisions happen rarely, yet the FAA does not use number of collisions as the official measure of safety because they are rare and also because accidents happen for many reasons that have nothing to do with controller capabilities: technical failures, pilot disorientation, error, or weather, for example. For each facility, the FAA records and counts the number of operational errors in relation to the number of operations for that facility, and thus identifies trends for each. It can then compare facilities of the same type and rank them. The error count weighs importantly in the annual evaluation of FAA facilities. Further, the numbers are taken very seriously as a measure of safety by the external political bodies responsible for the agency's funding. Journalists jump on the trends published in the *FAA Administrator's Fact Book*, which records monthly and annual error counts by facility. Local headlines blaze with accounts of the number of "near misses" and "close calls," but controllers object because not all operational errors are "near misses" or "close calls" (at Boston Center, for example, 4.9 miles apart is officially an operational error, but controllers point out that planes are still quite far apart).

For both safety's sake and political reasons, the FAA engages in a system-wide effort to reduce operational errors by investigating every incident of alleged violation of the rules of separation to understand what happened and why, and to take steps to remedy the error. The incident itself is already a big deal emotionally for the controller. The administrative investigation makes it a big deal institutionally. The investigation is experienced as a degradation ceremony by controllers, stigmatizing and temporarily stripping them of their profession their professional identity, thus extending the emotional impact of the original incident.[1]

The facility response is immediate and public. At all centers, major towers, and at some TRACONs, radar technology is equipped with a computer detection system that measures distance between airplanes. When the computer detects that two airplanes have reached the separation limit for a facility, a "conflict alert," a computer-generated ring, suddenly appears on the radar around the airplanes that are close. The letters "CA-CA" flash above the ring to warn the controller working that airspace. The controller then initiates "avoidance strategies"—that is, tries to get out of the mess whichever way he or she can. Next, the controller is immediately removed and a replacement sits down to work traffic (ideally—if staffing levels provide a controller to fill in, but a replacement is not always immediately available, in which case a traumatized controller must keep on working).

An investigation is held immediately to review evidence and come to a conclusion about the cause and the controller's responsibility. The controller fills out a written report. Next is an adversarial hearing. The controller is accompanied by a NATCA representative; and the supervisor, facility ATM, and quality assurance representative (if the facility has one) all interview the controller. Together, they review the physical evidence: recordings of controller-pilot conversations, controller-controller tapes from the ground lines, and, if a radar facility, pull the computer tapes reproducing the positions and actions of all the involved airplanes, exactly as they appeared on the radar.

Several outcomes are possible. The controller may be charged with an operational error; the pilot may be found responsible for a pilot deviation (the pilot didn't do what the controller instructed, thus violating the rules of separation), or (alas) if related to radar, it may have resulted from a computer measurement error. If the controller is charged with an operational error, the controller is decertified. This means that the hard-won title of certified professional controller, earned after successfully training on every position in their facility airspace, is temporarily taken away. Controllers cannot work traffic until they spend a designated time—one to several days—retraining, usually doing problems on the DYSIM simulator or computer-based instruction system. When their performance is deemed acceptable, they can be recertified. The consequences can be serious: two or three deals in a year depending on the facility, and controllers will lose their job.

After the emotional impact of having a deal, the controller is immediately hit by the administrative response, which adds the trauma of the moment:[2]

> Initially, it's just like [*sighs*] all of a sudden it's there, and it's like "Oh God," you know? I mean it's more of a surprise, because you didn't know it was there and you didn't know it was happening. All of a sudden it's flashing [i.e., the computer signaling conflict alert on the scope], and you're like—it's a little heart wrenching. Initially, when you see it, it's like you're turning everybody out and panicking, then decertify, get off position, and everybody says, "What happened? What happened?" Your heart goes ping. (Center, F)

The next steps in the administrative response to a deal also can generate emotions—anger and resentment. Controllers can find the investigation and its outcome insulting and unfair:

> They take a decision you made in a fraction of a second with five other distractions going on, then they leisurely review it, have a cup of coffee, then tell you

> what you should have done. If we all only had twenty minutes to hit a Roger Clemens fast ball. (Center, F)

Furthermore, although retraining is pushed as an important and necessary skills upgrade by the FAA, controllers were unanimous that they don't learn anything from the retraining while decertified. Indeed, they scoffed at the whole idea of it. They insisted that assignments to work DYSIM problems on the simulator and a few days of computer based-instruction did not teach them anything helpful in reducing errors. Instead, they experienced the entire decertification and retraining process only as punishment—and unnecessary and undeserved punishment at that:

> I wouldn't say it's humiliating. It's embarrassing. *Humiliating* is, you know, a big word. They happen to so many people, which is why it's embarrassing not humiliating. You know, you can't work. And all of the sudden you are not good enough to work. You have to prove yourself again or whatever. You know it's kind of like riding a horse. If you fall off, what's the first thing you do? Get right back on the horse. What do they do here? You fall off the horse, they take you away from the horse for three or four days and do this meaningless stuff that you know doesn't add up to anything. It's ridiculous. I could do the job up until this moment and now you are telling me I can't do the job until I go up and run two DYSIM problems [simulations] and do a bunch of CBIs [computer-based instructions], you know. (Center, M)

The emotional impact has several sources. For controllers, their job safety is on the line and being responsible for an accident is unthinkable, but also they are trained that errors are caused by individual failure and to take responsibility for their errors. They assume that personal responsibility immediately and often feel it deeply:

> There is nothing they could do to me to compare with what I do to myself after I have a deal, or I come close, beat myself up for days. Then I think, OK, you've really got to slow down. You've got to be mindful. You've got to watch what you're doing. And I'm willing to bet that's true for just about everybody. It's humiliating. There is nothing they could do to you that compares to what we do to ourselves and to each other. You're just so humiliated. You can't look people in the eye. You feel like such an asshole. It's the worst feeling in the

> world. You just want it all back. I suppose, I've never been in a car accident where I've been responsible, but I suppose it's like that. You just want it back. You just, you know, have you ever just wanted something back? I just want to go back in time. Let me start over. I'll do it right this time. I promise. It sucks. (Center, F)

Moreover, they want to avoid being the one responsible for raising the facility's count, which, under normal circumstances, results in a downgrading of the facility's status among other facilities of the same type. Sometimes, one person's deal has a direct impact on his or her crew. Under one ATM at Boston Center, days without a deal were counted to give an area points toward days off, much like the prison principle of "good time." When someone in the area had a deal, the area lost all accumulated good time and had to start over. Everyone feels punished. Finally, it is also that the deal itself has personal implications about how, as professionals, they are doing the job:

> Part of the pride of the job is that you are expected to be perfect all of the time, and having a deal is being less than perfect. I've always, I've come to work, and if I'm not perfect, not that I like pressure myself, but I just, it's my mode of operation, I strive for perfection. I don't go crazy if I didn't get my exactly five miles, you know, banging my head. But I will make sure that my operation is clean, and I won't leave any room for a matter of semantics to get us in trouble. I'm very precise and clear when I tell somebody to do something. If I'm responding to a question, I don't want to leave any open-ended answers. Very matter of fact. No interpretation errors. (TRACON, M)

Coworkers, usually ready with a witty zap when someone has a sloppy session or makes mistakes of lesser consequence, tend to be silent when someone has a deal. They know from their own deals that the person feels badly enough already. But the controller who has the deal often experiences that silence as isolation and a negative sanction. And sometimes it does reflect a condemnation. Depending on the circumstances surrounding the error or the person's history of deals, having a deal can affect a controller's reputation as competent and also his or her status among coworkers.

The administrative response is ritualistically performed after each reported incident. Much lore surrounds it; each controller's imprinted memory of the operational error is embellished by the memory of the investigation experience, which adds to the personal and facility history of mistakes and errors that cycle among controllers. However, although the emotional impact of the incident itself seems to affect all controllers similarly, the administrative response to the errors varies across facilities.

Space, Place, and Boundaries: When Is a Deal Not a Deal?

The variation in administrative response to errors is consistent with what we already know about how airspace builds variations into this standardized system. The across-facility variation adds another level of subjectivity to a ritual already perceived as biased and unfair. The true number of operational errors at any facility is unknown. There is a gap between the number of operational errors actually made, the number identified or reported, and those eventually designated officially as operational errors. It is much like other forms of statistical measurement of performance and efforts at accurate counts: the results have interpretive flexibility. From some hypothetical true number, after internal processing, not all end up in the official count. What gets counted as an operational error varies by type of facility, airspace, rules of separation, technology, and norms of the place. These factors make some facilities predisposed to having more operational errors than others. Moreover, some facilities are predisposed to having more operational errors *counted* than others—this is what controllers mean when they talk about unofficial deals. The theme of system effects repeats: airspace affects place—its technology, architectural arrangements, and the social organization of work. Consequently, the meaning of a deal and the emotional impact on controllers varies by facility.

Boston Center

Boston Center competes with other centers to reduce the number of operational errors, which makes having a deal at Boston Center a bigger deal than at Boston Tower, TRACON, or Bedford. Boston Center has the largest airspace and the greatest number of annual operations, so more operational errors than the other three facilities. However, it's complicated. The rules of separation there give the greatest distance between airplanes—five miles between, one thousand feet above and below—and thus what would appear to be the greatest opportunity to prevent operational errors and, if an error occurs, the greatest amount of time to maneuver to prevent a collision than at the other facilities. Planes travel very fast, though, and Boston Center has many sectors with complex airspaces and sector boundary issues, so operational errors are not necessarily reduced simply because of the rules of separation. Perhaps the most important factor to take into account at Boston Center, and at all centers, is that radar technology makes operational errors detectable.

The "snitch machine" is the not-so-affectionate name controllers have for the technology that lets the administration know when a controller has violated the rules of separation. The snitch machine is a surveillance and alarm system tied to each individual workstation and airspace sector to monitor violations of the rules of separation. At the same moment that the conflict alert flashes on the scope, an alarm rings at the watch desk in the control room. The operations manager at the watch desk notifies the area supervisor, who removes the controller from the sector. The snitch machine increases the probability that errors are identified, thus increasing the error count. However, the snitch machine is not always a reliable measurement instrument. It can falsely signal a separation violation when none has occurred. The radar sweep indicating the position of the airplane lags some seconds behind the plane's current position. The snitch machine can be set off, indicating planes approaching a separation limit, but during the time it takes for the radar signal to transfer to the scope, the pilot has already changed direction or altitude, following either the original flight plan or the controller's previous instruction:

> If you're working a session, it's either intense or it's complex, and all of a sudden that damn machine goes off. And you go, *yiiiii*. Your heart just jumps because you think you got this whole thing under control and, all of a sudden, the machine goes off, and it's a guy at thirty thousand feet flashing with a guy at thirty-seven thousand feet [actually, the rule is one thousand feet above and below] because the program's all screwed up. You know, it doesn't work perfectly. That's when I get scared. You know, it's that instant—what did I miss—and then you yell at the machine, "That damn machine, you know, got me again." I'm surprised people don't have heart attacks over that. (Center, M)

In this situation, then, the scope shows the conflict alert, the trauma for the controller is the same, and the investigation proceeds, but it is not a deal because one plane already turned away. The snitch machine has also been known to completely malfunction, failing to signal at all when there is a violation of the rules of separation—thus the language of the unofficial deal. Pilot errors are usually caught and corrected; nonetheless, they can produce emotional trauma, deal or no deal. This controller explains an unofficial deal:

> I've only been nervous once, when I was training. I almost had a near midair, that got me a little shaken, you know? The guy missed his corner. And he was like right here, got 11-700 and climbing out, dead-fuckin' wired. And I'm sitting there and I asked for help. I looked around and my trainer, he was looking at the situation, like he has this friggin' blank stare. And I said, "Oh fuck, I'm alone." And I dug my ass out of it, you know. But I mean the guy missed his corner by two [radar] hits. It's twelve seconds a hit; twenty-four seconds. First hit could be bad [false positive] because we have false hits all over the place. Second hit, I knew he missed the turn. I said, "Are you making your turn?" He says, "Oh, yeah, we're going to turn now to Rover." You know, going to Kennedy. He waited too long, turned too wide. Guy right smack in his kisser. And, I mean, these guys were merging.
>
> *Q. Oh no. And what did you say?*
>
> I said—I couldn't believe how calm I was, I couldn't friggin' fathom at how calm I was, I said, "Traffic safety alert, traffic twelve o'clock. One half mile, opposite direction, out of 11-7, climbing to 1–2,000. Make an immediate right turn to a ninety-degree heading." And he says, "Roger, center." This was like six years ago. And I said, "Sir, do you have that traffic?" He said, "Yeah, we have it in sight." I says, "Can you turn your airplane any faster?" and he says, "We got this sucker on a forty-degree bend. We can't do any more. " We had one [radar] hit in the [conflict alert] circle. The guy was turning so fast in a 747 [that] the first hit was like a half mile, second hit was five miles. But we didn't have a . . . I didn't have a deal. I was shaking. (Center, M)

Notice that he did not remark upon avoiding a collision. He was relieved that he didn't have a deal. It wasn't an official deal, yet the controller was just as traumatized. Several Boston Center controllers complained that having had a deal, they had to continue working the position for as long as an hour: "It was the worst, the worst situation I had ever been in." Everyone's working; not enough controllers were scheduled. Inadequate staffing then adds to the emotional impact of the moment for the controller.

Also, the center is large—a bureaucracy within the FAA bureaucracy—and the identification and investigation process is more formalized than at the other three facilities. It is quasi-legal adversarial process that feeds into the ongoing internal "us versus them," "Big Brother is watching you," management-worker bureaucratic conflict. In the investigation, interpretive flexibility kicks in. The snitch machine's occasional inaccuracies extend to identification of a violation of the rules of separation at 4.95 miles or at 5.1 miles. Center controllers complained vigorously about the 5.1 because of the serious implication of the small percentage difference for their career. Also, there can be complex situations in which many things happen cumulatively—these are not necessarily mistakes on controllers' part, but they happened in a controller's airspace and culminated in an operational error. Does that mean that an error should be charged to the controller? How do these more complex situations get decided in the internal hearing, given the pressure to reduce errors and other subjective elements that might come into play? How often is a complex situation settled as a pilot deviation rather than an operational error?

Boston Center controllers expressed never feeling that these uncertainties would be resolved in their favor, even with a NATCA representative participating in the investigation as advocate. Instead, they experienced the administrative response as punishment designed to place blame and reduce error counts, not guided by any effort to understand the complexities of their airspace and traffic situations: "If we all only had twenty minutes to hit a Roger Clemens fast ball." Center controllers more often expressed alienation and resentment about internal investigations and the processing of errors than did controllers at any of the other three facilities. They pointed out—with some envy and more than a little antagonism—that TRACON and Boston Tower controllers "have more outs than we do," meaning they have rules that allow their airplanes to fly very close but those operations do not get counted as operational errors.

The TRACON

Although the TRACON is also a radar facility, it had no snitch machine. It wouldn't work in the TRACON's airspace because controllers there must work planes that are close together, in crossing patterns of arrivals and departures. Because of this requirement of the airspace, the TRACON has different rules about the distances required for separation—three miles apart (increasing to five miles as the plane approaches Boston Center airspace), or altitude separation of one thousand feet. On final, the rules say three miles apart, but depending on the controller and conditions, this might just as often be two and a half miles. What counts as an error is not strictly determined by these numbers, but rather by additional rules created out of the necessarily greater proximity of the airplanes. Safety—and forgiveness—is built into the system of rules governing Boston TRACON, and all TRACONs for that matter. These are the rules of divergence and visual separation. A TRACON controller explains:

> The snitch machine. I'm sure you've probably heard of that at the [Boston] Center. We don't have that breathing down our back here. If we did, we'd have operational errors all over the place—the thing would be going off every two minutes because we run the final so tight, you know? First of all, we can apply rules that they can't apply up there, like diverging courses and visual separation, you know, pilots see and avoid. All of these things allow us to give less than three miles of separation.
>
> *Q: So less than three miles is really legal?*
>
> Yes. [Using divergence,] I can have two hundred airplanes flying along like this, and if I can say that I know that this [one's] course will pass behind this aircraft, that they are not on a collision course, I can now give him the same altitude. They may, he may only pass a mile behind him, but it's legal, all day long. And here we can use visual separation [see and avoid]. And it's amazing how close pilots will get to each other. You know, if we were doing that [running them that close], they'd be screaming at us and calling in on the telephone when they landed. But he sees this guy, he climbs right up in front of him. They miss by a few hundred feet, but it's fine because they see each other. They're doing what they need to do, you know, whatever, to fly the airplane. (TRACON, M)

Not only are the rules of separation and what counts as an error different, but recall the TRACON airspace reorganization reduced conflict and competition, instead producing a cooperative team environment that has reduced the real rate of errors. The airspace design and division of labor has all eyes watching the entire airspace, so everyone is in a position to help by watching and warning on behalf of the others. They collaborate, and so more effectively prevent operational errors. The arrangement of airspace on individual scopes prevents Boston Center controllers from such a cooperative effort. At the TRACON, though, the responsibility for deals is shared, as a supervisor told me:

> If someone has an error in here today, everyone kind of takes a little part of it. They don't like it. It's "What could we have done to stop it" even though I'm four scopes down working on a different sector. Most people think of it as our error, not somebody else's error. (TRACON, M)

Indeed, in 2000–2001, the Boston TRACON had the lowest rates of operational errors nationally. The airspace gives controllers there less time to correct errors than at Boston Center, but simultaneously, the architecture—the social organization of space and place—make them more effective at catching them. Also a factor, the TRACON is smaller, not bureaucratically organized like the center, but more informal and personal. Although the administration at Boston Tower and TRACON also has a keen interest in reducing errors, a difference is that the top administrators know the airspace and its challenges well because the airspace is smaller. This understanding breeds a more sympathetic approach toward a deal. TRACON controllers experience the investigation of operational errors as nonpunitive and built on the administration's understanding of the complexities of their crossing traffic patterns and the requirement to bring planes close together. This does not mean that the emotional impact on the individual controller of having a deal is lessened, however.

Boston Tower

Towers are another case entirely. Judging what is and is not an operational error by eyeballing distance between planes in the sky is an imprecise art, to say the least. Moreover, like the TRACON, the rules of divergence and visual separation give Boston Tower controllers "outs" that center controllers covet but do not have. Also, pilots own the responsibility for separation by the see and avoid tenet. Tower controllers can still have operational errors, however, so are not totally off the hook. This controller, who formerly worked in towers before his assignment to the center, compares the two:

> It's a very formalized process at the [Boston] Center because they have the snitch machine. In the towers, we don't have that happening, but most of the time I worked in towers, we would give somebody a little extra room to make sure that they didn't have any problems like that. Sometimes you had to run them close, and nobody can say, looking out the window, if they're a little too close or not. Unless they get on the runway, then you can. That's the only place you're responsible for separation in a tower is on the runway. People in the air, no. (Center, M)

The controller points out the operation of a norm of forgiveness, As at the TRACON, the tower process is less formal, less punitive, guided by the understanding that the airspace creates situations in which a deal is impossible to avoid and eyeballing distance in the air is insufficient. The greater problem of operational errors at towers is from runway incursions. Runway incursions are incidents that have a negative impact on runway safety. Runway incursions may or may not involve a violation of the rules of separation. Either way, they are counted as errors against the facility. A system effect, runway incursions are more likely to happen at facilities with crossing runway configurations, like Logan, than on airfields with parallel runways:

> An incursion is basically an airplane or a vehicle on a runway that's not supposed to be there at a time when they're not authorized to be there, so it doesn't matter whether the controller caused it, the pilot caused it, the driver of the vehicle caused it. If they're on that runway and they're not supposed to be there that's a runway incursion. It doesn't matter whether there's a loss of separation or it doesn't matter if a vehicle is down on the runway when somebody is trying to land.
>
> So, you told the pilot vacating the runway, hold short of runway 22 Right and the pilot acknowledges that he's going to hold short. And what does he do? He forgets and he crosses the runway by mistake. Doesn't have to be any other traffic around. Could be a vehicle—snow plow, some idiot that got lost that walked out onto the runway. Most frequently a vehicle, 50 percent of the time. So when you hear the press talking about the increase in runway incursions take it with a grain of salt, even though that is the hot topic, the runway incursions. Because that doesn't mean a near miss or anything. If a plane is landing and there's a runway incursion on that runway, that's an operational error. (Tower, F)

Boston Tower has its own version of a snitch machine for runway incursions: The Airport Service Detection System, or ASDE, tracks the location and movement of all vehicles on the runways and signals conflict alerts. Also, as at Boston Center and TRACON, Boston Tower has the computerized voice recording system for pilot-controller and controller-controller exchanges. Add to this traffic, so fast and congested that avoidance strategies must happen on the fly. Then add to that the division of labor, so very different from at the TRACON, thus limiting the ability of the team to work together to reduce errors. These conditions produce both a higher incidence of real errors and a higher error count than at the TRACON. A supervisor shared:

> We used to keep a tally of error-free days, [Boston] Tower and TRACON. And the TRACON was always so much higher. You might have two thousand error-free days in TRACON, and the tower had twelve, and it didn't go over well with tower. It's like apples and oranges. It's generally, at least here, the tower has more, not because of lack of talent but because the way the tower operation is, one mistake out there can happen so fast you can't fix it, whereas down here [in the TRACON] you have some time to fix it, usually. It's just so fast that the environment's different. (TRACON, M)

Bedford

Bedford is a terminal facility without a TRACON, so controllers have the additional responsibility of setting up their own approaches. The risk of violating the rules of separation for both planes in flight and runway incursions is increased by the inexperience of many of their VFR pilots. Taking into account those airspace conditions, the number of operational errors is surprisingly low. A supervisor told me:

> This facility went almost fifteen years without an operational error. And then we just, in one year we had several of them. One was an equipment problem. We have a blind spot on the runway. An aircraft was in that blind spot and didn't acknowledge the instruction and couldn't get out of the way in time and they just didn't hear it. And we haven't had any, you know, for a while now, thank heavens. And we haven't had any this year. (Bedford, M)

How to account for this low number with its inexperienced pilots? Bedford has the same "outs" as Boston Tower and TRACON: the rules for divergence and visual separation. Recall also that during my fieldwork Bedford was so technology-poor that controllers had no effective radar detection system for planes in flight or runway incursions:

> In the traffic pattern, especially in the VFR environment, there's a policy of "see and be seen" between pilots. Now we would inform one pilot of the other, who to follow in sequencing, and it's really the pilots' responsibility to ensure that their aircraft don't collide. There are no distance minimums between VFR aircraft as long as one aircraft has the other in sight and avoids him at a reasonably safe distance.
>
> But where we get into problems, or we will get into an operation error,

> is if there are two aircraft on the runway with less than their legal separation [a runway incursion]. Between light aircraft, we have a 3,000-foot minimum, and between larger aircraft, intermediate size, there's 4,500 feet. In the larger aircraft, it's 6,000 feet or more. At times we have to have at least one aircraft off the runway before another aircraft can land or start takeoff roll. And that's where we would get into operational errors, is on the runways themselves. As far as the separation goes that's a judgment call based on experience. We know what the distances are, and if we observe it and we lose that separation, then there is an operational error. (Bedford, F)

Absent technology, investigations were hampered by the inability to completely reproduce the incident to determine what happened. When it came to counting and investigating errors, Bedford controllers, like controllers at Boston Tower, operated on a shared understanding of exactly what they were up against with their traffic: the inability to predict and control the actions of their pilots' movements in the air and on the ground. The supervisor on duty was often downstairs doing administrative work, and even when a supervisor was in the tower, the error had to be egregious before it was recorded and reported. On supervisors' days off, it usually was left to the controllers to report on one another's errors. In the absence of a supervisor, one controller acted as controller in charge. A controller explained to me how operational errors are handled:

> We feel bad for the person who has the deal. We do everything we can to share the blame, or deflect the blame, or whatever, we don't want to see each other going down. If I'm the controller in charge up there, unless there's an accident there will not be a deal. If you know what I'm saying.
>
> *Q: Is it the controller in charge whose responsibility it is to recognize it?*
>
> To pursue it. [*Laughs.*] At [Boston] Center it's almost impossible to have a deal without alerting the world because they have the snitch machine. Here, unless it's an airspace violation of some sort, or a flagrant violation of procedures, for which another facility gets involved, or unless a pilot makes a formal report, it's all subjective. (Bedford, M)

Other factors enter into a decision about calling a deal a deal, officially. Carrying out the required administrative response hits hard at a small facility. The facility has only two supervisors, one on duty at a time. If they see an error, they report it, but for the supervisor, the realities of a small facility can mean only that those operational errors that are clearly attributable to a controller get turned in. For the supervisor, a deal is time consuming and requires a mound of paperwork, including what the other controllers saw and heard, and there's a deadline to file a report. There is a shortage of supervision. If the supervisor is at home, getting there can take an hour. Moreover, in a small facility, the decertification and retraining puts an extra burden on every other person. A supervisor explained:

> Sometimes the staffing makes it really hard to deal with. Staffing permitting, you get the controller off position, then make your assessment. The problem with an operational error, say for example there's only three controllers on. One's a CIC [controller in charge], they're all in position. One of them has a deal. The one that had the deal is supposed to be removed from the position, which means the other two controllers are going to be working their little fannies off. (Bedford, M)

According to some of the Bedford personnel, the effect on the facility and the institutional response doesn't prevent errors; it prevents reporting them.

Mistake and Error as System Effects: Crossing Boundaries of Time and Social Space

Controllers at all facilities talked about the causes of deals in a contradictory way. On the one hand, controllers experience having a deal as personal failure. They blame themselves. They accept responsibility. They try to learn from it and do better. On the other hand, they talk about having a deal as a normal part of the job. Their view is that everybody makes mistakes. An operational error can happen to anyone on any given day. They've all had deals even if they haven't been identified and recorded as part of their record. Weak controllers and strong controllers alike have deals. Controllers believe that deals are inevitable. They understand the deal as part of the routine and as a normal part of their occupation. They do not go so far as to say outright that having deals is a system effect, but indicate as much when they acknowledge (which they all do) that both strong and weak controllers have deals:

> This job is all about survival. Because it will pinch anybody. It doesn't matter who you are, you can have an operational error. It does not differentiate between you and the next guy. Anybody can have one. I don't care how good

> of a controller you are, or how bad of a controller you are. But the sharper you are, and the quicker you are to think, and thinking on your feet, you survive. (Center, M)

Controllers agree that operational errors go with the job. A few controllers had ideas about the causes. Deals are a natural by-product of attempting to provide service and move traffic ("If someone hasn't had a deal it's either because they have been very lucky or are running them so far apart they aren't providing any service, you know?") as well as human factors ("Whenever human beings are involved you are going to have mistakes"). There's also the "too many variables" theory: "There's so much stuff out there that happens that's out of your control, it could happen to you at any time, I don't care how good you are." No doubt each of these explains some portion of operational errors.

When are mistakes and errors most likely to happen? If something systemic is producing them, we would expect patterns. The FAA search for patterns in the causes of operational errors is handicapped by the small number of deals that occur in relation to the total number of operations. It is amazing that all twenty-one regional centers combined produce only about 411 operational errors out of forty million operations in a given year. All the major facilities have operational error review boards that investigate the circumstances surrounding each operational error. The members—both NATCA representatives and FAA quality assurance personnel—meet once or twice a year to examine the number of operational errors and look for common causes. Their data are the reports and other documentation gathered for each operational error during the postincident investigation, and they analyze these semiannually. The average number of operational errors at Boston Center is about fifteen a year, which is about one error for every one hundred thousand operations (a rate comparable to the national rate), or one every twenty days. The sample is so small that finding patterns is next to impossible. One year Boston Center went the entire year with only six operational errors. Then, in one week in January the following year, it had five errors. It went from six in 365 days to five in 6 days. As one controller put it: "Oh my God, errors are up 400,000 percent, or something like that. And nothing changed. The traffic didn't get any busier, the controllers didn't get any less vigilant, the weather wasn't any worse."

One controller, a regular member of Boston Center's operational error review board, described unsuccessful efforts to link the cause to airspace characteristics. When the board looked at investigations of close calls rather than deals, it found that more close calls seemed to happen at specific sector airspaces, indicating that something systematic was going on, but examining deals alone just pointed to the randomness of it all:

> Operational errors are very tough, because, statistically, they're insignificant and they happen for countless reasons. You can't prevent errors, you can't stop errors, that's a human nature thing anyway. But you try and make an environment where they're less likely. And if you find you have a lot of close calls but not necessarily errors, a lot of times that'll spur a change in either airspace or procedures. It can happen that there's a sector where controllers might not even be having errors, they might just be having some really tough sessions, with some really close calls. They might have to take some very abrupt maneuvers, which, you know, we prefer not to do with airplanes because, you know, there's people in the airplanes. On the other hand, you might have sectors where you do tend to have operational errors. They happen, but they're different people sometimes making the same mistake. They're different people making different mistakes. They're the same person making different mistakes. And, again, they're so few that sometimes it's like hitting the lottery, you know? (Center, F)

In interviews, I asked controllers when they felt most vulnerable to making mistakes and errors. In answering they drew upon the totality of their experiences—their history of mistakes, close calls, operational deviations, and deals—and described circumstances in which they felt vulnerable to getting it wrong. These moments of vulnerability, too, are emotion generators. As they anticipate a challenging situation that reminded them of a situation they had faced before, they felt vulnerable to making a mistake or having a deal, whether nothing untoward happened, they had a close call, had saved a situation, had an "almost deal," or actually had a deal. These are situations where their perceptions of risk were shaped by the past and where they felt at risk for another mistake or error—or worse, the possibility of an accident. Risk is in the eyes of the beholder: the person who has been cut by the knife will be leery of the knife the next time. Recall the incident described by the Boston Tower controller:

> After that I was real nervous about 22 Left for about a year afterwards. I can remember getting really scared when I had to depart 22 Left. It took me about a year to get over that.

Drawing from their history of mistakes, close calls, operational deviations, and deals produced a bigger sample than the operational error review board had by looking only at deals annually or semiannually. In total, controllers named eighteen different situations during which they felt particularly susceptible to making mistakes and errors. On the surface, the number of situations seemed to support the findings of the board on operational errors: no pattern.

A number of controllers said they felt vulnerable during complex traffic situations or when working a complex airspace. What constitutes a complex airspace or situation varies by facility. Controllers said that if either a pilot or a developmental made a mistake at a busy time or in a complex airspace—or worse, both—it was difficult to act quickly enough to fix things. But feeling vulnerable in these complex situations was only one of eighteen situations they named. Moreover, each of the other seventeen situations was mentioned by just as many controllers. No pattern was immediately apparent. Even controllers talked about them as if they were singular, with no relation to one another, which itself suggested a regularity in the way that controllers think and talk about certain kinds of mistakes.

However, when I looked closer at what they said, I saw that all the other "vulnerable moments," taken together, did have a striking pattern. Each of the conditions they described was a *transition*: either controllers or airplanes—or both—were crossing boundaries of time or space. Significantly, the transitions they described were all part of the daily routine, not some rare or unusual circumstance. They were a normal part of the job, built into the organization of the air traffic control system—thus, they were system effects. Controllers identified three transitional moments when they feel vulnerable to mistakes and errors: when the controller is crossing boundaries of time and space, when airplanes are making airspace boundary transitions, and when controllers and airplanes are making boundary transitions at the same time.

Controllers and the Boundaries of Time and Space

Controllers recalled their training as an extended period of vulnerability to mistake and error. Training was a long period of transition across boundaries of time and space. They moved physically from the academy to the facility, where they transitioned from one position to the next until they had mastered each position they would work in the facility. As novices, they not only could not predict what the airplanes were going to do; they did not have a repertoire of responses down pat to deal with new situations. Controllers also said they felt vulnerable when they first get checked out. The trainer isn't there to direct them, to save a situation by taking control, and they continuing learning by experience. The feeling of vulnerability they expressed was borne out by review board investigations showing that most operational errors were by controllers who had been certified professional controllers for fewer than three years.

Experienced controllers at all four facilities talked about feeling vulnerable during a position transition. Whether moving into or out of position, controllers were crossing boundaries of both time, at the end or beginning of a session, and of space, both physical space and airspace. These transitions were times when they did not feel they were up to speed: they did not feel up to the speed required, mentally and physically, to competently and confidently meet the demands of the traffic. One such moment of vulnerability was coming in at the beginning of a shift, returning to work after a break or lunch or even vacation. When controllers replace another controller at a position, they are making a physical transition from nonwork to working traffic in an official position, working a specific airspace. Mistake and error were expected start-up costs to being away and returning. Uniformly, they talked about feeling vulnerable to mistake and error during the first few minutes on position.

To ease the transition into a position, controllers rely on formal procedures to reduce the probability of mistakes: checklists for the position relief briefing; the requirement that the replacement must watch the action for a few minutes prior to sitting down; the departing controller's standing behind to watch a while before departing. They pace themselves at the beginning until they have a grasp on what is happening. Despite these strategies, taking over a position is, in the short run, an emotion generator:

> You know that they always say it's the first ten minutes when you first take a position. You come back from break and you get Local West and it's busy. First ten minutes it's not really your baby; it's not your thing cause you're taking over what the other person had and you haven't established your way of doing things or your routine, what's going on. . . . Your first ten minutes in position are the scariest. No matter how many times you review the status boards and listen and watch all that stuff, your first ten minutes in there is when you're getting, I was going to say sowing your wild oats but, you're getting the feel of it. You're settling into the position mentally, getting your head into the picture. After that, basically everything that's happening is happening because of what you've done so now you're into the flow of things. (Tower, M)

Controllers named additional kinds of position transitions where they felt vulnerable to mistake and error. These, too, were between boundaries of time and space. Different from those described already, controllers emphasized that these were all times when they were tired: after a quick turnaround, those two back-to-back days in their shift schedule when they have the least number of hours to get home and rest before returning to work; the midshift, when they have few airplanes to keep them alert, the facility is quiet, and their circadian rhythms are out of whack; the last session before they leave for the day and the last day before they have their days off. In these situations controllers are mentally fatigued, have a hard time concentrating, and begin making a mental transition to what they will do when their shift is over. They have strategies to correct for these moments of vulnerability, too. For the quick turnaround or midshift, some "prepare": go for a run or work out before they come in for the shift. For end-of-the-day fatigue, controllers rely on caffeine, or go for aerobic walks on breaks, play video games, foosball, cards—anything to keep them mentally and physically stimulated during working hours.

Time-of-day transitions also can make controllers feel vulnerable to making mistakes and errors. For Boston Tower controllers, the sunrise and sunset are visual trouble. A supervisor described the effects:

> Right at sunset it's hard to see the aircraft and it's hard for aircraft to see each other. And the lights—it's not dark enough yet for the lights to become visible so they can be picked out. And that's just a very critical time, that's the time when I will staff a coordinator position—someone to assist the local control, just another set of eyes. It's just a very dangerous time for flying, whether from the controller's standpoint or the pilot's standpoint. (Bedford, M)

Time transitions also make them feel vulnerable because the rhythm of traffic varies by time of day and day of the week. Controllers talked about being more vulnerable to making mistakes or errors when traffic transitioned from a busy period to a slow period. No one said they felt more vulnerable when they were busy; in fact, they said the opposite. Heavy traffic has all their attention. Adrenaline is pumping. Vigilance is automatic. When traffic subsides and they have only a few airplanes, they relax their focus:

> I love it when it's busy. I really like it. When it's busy, people are all focused. They're working. And most things don't happen then—things happen when it's slow, that's when you really have to watch and you have to force yourself to watch. People relax, don't concentrate as much. Seriously, people have deals with only two or three airplanes. (Center, F)

They also feel vulnerable any time when traffic normally is slow: Sunday mornings, Saturday afternoons, the midshift, some holidays, bad weather when planes are grounded. During slow times, they said, they feel vulnerable to mistakes and errors. Uniformly across the four facilities, controllers said that busy is good, busy is safe, slow is risky:

> Operational errors occur when your guard is down. They don't occur when you're working a hundred hour, they occur when you're working a thirty hour and your awareness isn't as acute.
>
> *Q: So you think that the busier you are, the safer it is?*
>
> Correct. (Tower, M)

Why is it that the busier they are, the safer it is? Trained to multitask, when the sky holds many planes controllers deploy all their embodied skills—the scan, room awareness—and they feel the rhythm of their own techniques and procedures. Slow times are a different matter. One controller made this analogy:

> When it's busy you're in a better state of readiness, I think. You know, I think of myself when I'm driving at home on the freeway in the car at night, and it's dark after a long day. I mean not necessarily working here, but any time. With the radio off, nobody in the car, you know, you think you'd be much more focused on the drive but actually, if the radio's on I have something else to pay attention to, and I think I pay much more attention to the traffic. Of if someone else were in the car talking to me, much more likely to fall asleep if I'm sitting there by myself concentrating. You can forget about people sometimes. When there's not a lot going on, because, yeah, you're not focused out the window, you're not sequencing you're not giving traffic calls, you're not landing people, giving landing clearances, giving take off clearances. So you're not focused [*snaps fingers*] out the window. (Bedford, M)

Airplanes and Airspace Boundary Transitions

The second major category of transitions during which controllers described feeling vulnerable to mistake and error was when airplanes were crossing airspace boundaries. As the relation between airspace and place would suggest, the situations they described varied by facility. Boston Center controllers talked about feeling vulnerable when airplanes were making a transition across the boundaries between their airspace and that owned by another controller: the handoff . Discussing the handoff, the controller below nails the source of the problem exactly: system effects—"a built-in frailty in the system is that boundaries are there." He explained:

> Sector boundaries is where there is problem; it's probably where you are most likely going to have a problem, a mistake. I mean, it really varies, because some sectors have built in confliction points in the middle of the sector, and that's where mistakes happen. . . . But a lot of times at the sector boundaries, you might get an aircraft that's just, it's a late handoff to you. And when it's a late handoff, you don't have much time to react to it. You don't have as much time to call the previous controller and issue him a control instruction to separate the traffic, or whatever. You have that, that's a built-in frailty in the system is that boundaries are there.
>
> If you're not working an aircraft, you can't control when the other aircraft comes in, or when the other controller initiates a handoff to you, or when he takes yours. When you're handing an aircraft off to another controller, you're trying to get them to take their handoff. But as he's charging the boundary, you're waiting for them to take their handoff; you're waiting and waiting. When he gets a certain distance from the boundary, you call up by going "Hey, handoff on this guy." But most of the time by the time you make that phone call, it's a little too late to even try and contain them inside your airspace. Because if you can give them a 360-degree turn, he's probably going to clip into that other sector's airspace. (Center, M)

Center controllers also talked about feeling vulnerable when airplanes made transitions into and out of holding stacks. The holding stack happens only at centers: no other facility has sufficient airspace to pull off a holding stack. A holding stack results when the boundary to the next airspace on a plane's route is temporarily closed because of a complex traffic situation or more airplanes than the controller can safely handle. Meanwhile, airplanes keep coming in that direction. At the closed airspace boundary, the controller still controlling the airplanes holds them in his or her airspace by having each make 360-degree turns; as they accumulate, the controller builds a corkscrew-shaped stack:

> The holding stack. It's the easiest time I think to get in trouble. Because it is very easy to say a wrong altitude. I used to think, how can somebody clear two people into the hold at the same altitude? How stupid is that? Until I did it one day. I almost had a heart attack. Like if you have a string, five planes com-

ing in like this, the first guy is going to get twelve thousand, the second guy is going to get thirteen thousand, the third is going to get 14, 15, and so on.

Q: So they are a thousand feet apart?

Right. But I cleared two guys years and years ago, I think it was to 15. Never even dawned on me I said 15. Because what I'm doing is, I'm looking at the 15 guy, wanting to keep this guy at 16, but because I'm looking at 15, 15 comes out of my mouth instead of 16. And I see this guy make the turn, and this guy is going through 16-2, because that's what I have in the data block, to 15. I'm like, holy shit. And I turned him the other way so I can at least keep them apart before I need to climb them back up. But oh yeah, scared the shit out of myself. But that's when I realized how easy it was to make a mistake. (Center, M)

Controllers, Airplanes, and Runway Transitions

Finally was the situation when both airplanes and controllers were making boundary transitions simultaneously: the runway change. This is a transition peculiar to terminals. Bedford Tower, Boston Tower, and TRACON controllers talked about the runway change as a time when they felt especially vulnerable to mistake or error. Reacting to a coming shift in wind direction, they change arrival and departure runways, which amounts to turning the airport around. As the wind is in transition, controllers engage in four different transitions simultaneously. They are changing the airspace configuration, changing their own cognitive maps, switching control of runways from one controller to another, and changing directions of airplanes on the airfield and in the sky. Bedford has only two runways, but the inexperienced pilots who frequent the airport complicate the maneuver:

> When you change runways, you're so used to working things with one runway that once you turn it around and start working on another runway, that's a very critical point. Especially if you're going from one end of the airport to the other. Everything is totally turned around. So it's very easy to make a mistake. A lot of the times it goes back to inexperienced pilots that usually create a mistake during that time. (Bedford, M)

At the terminal at Logan, both Boston Tower and TRACON are involved, adding complexity. Whereas at Bedford the execution of a runway change fits in with its "seat of the pants air traffic control" personality; at Logan a runway change is more regimented and governed by procedure:

> A big one for us is a runway change. We, you know, we've had some problems during runway changes, transitioning to the different configurations, sometimes something falls through the cracks, but usually, you know, we have certain safeguards . . . do it the right way, get all your releases, get control of this runway and that runway. I try never to assume that the person next to me knows what I'm thinking. . . . And you know sometimes things like that will happen [something falls through the cracks] and it's usually because . . . everything is so regimented in what we do, on this configuration you do this, on that configuration you do that, well in the transition you're doing a little bit of this, a little bit of that. (Tower, M)

Time is scarce at Boston Tower, so a runway change in heavy traffic can leave less time for teamwork as airplanes cross the boundaries of each position. Specialization inhibits coordinating with individual controllers and inhibits each person from looking out for the others. Contrast this situation with that of the TRACON. Tower controllers emphasized that the TRACON had it much tougher than the tower during a runway change—"Our planes are all on the ground. Theirs are in the air." However, TRACON airspace and division of labor create ease of coordination and cooperative self-regulation that reduce error and mistake:

> [A runway change] certainly leaves the most room for someone to make an error, definitely. I think that the chances of it occurring are extremely high. No, what's the word I'm looking for—it's like the opportunity is certainly there, if ever there was one. I think because everybody is so focused on everybody else's traffic and trying to pay attention to go through this cleanly that for something to actually happen, no, it's minimal. We have not had any deals or operational errors during runway changes that I can remember. But I think that's due to the increased vigilance and increased emphasis that everybody puts on it. Everybody says, "You know what? This is a dangerous time" and we look out for each other. (TRACON, F)

Emotional Labor as a System Effect

Controllers had no hesitation in talking about mistakes and errors that they had experienced and the emotions they felt. Their emotional reactions were a natural part of the retelling. The story's chronology brought out every detail of the incident, including its impact on them at that moment and even after, sometimes long after. They spoke of shock, fear, nearly having a heart attack, rising up out of their seat, and physically shaking. They talk about mistakes and errors as having a hierarchy of seriousness, with the most serious outcome being an operational error. The removal from position, decertification, and official investigation that follow a suspected deal extended the trauma of the incident, causing a variety of additional emotions: embarrassment, feelings of personal and professional failure, concern about loss of status among coworkers, anger at the administrative response, indignation at not being able to work, feelings of being punished unfairly and possibly having put their very job in jeopardy.

The meaning of mistakes and errors for controllers is double. On the one hand, they view their failure as individual—"I couldn't believe I missed it". On the other hand, they recognized that something systemic was going on. They said that having a deal was a normal, routine part of work that was built into the job. As evidence, they pointed out that everyone has deals, even the best controllers—and some said especially the best controllers. No factor distinguishes controllers who do and who do not have details. As they talked about the inevitability of a deal, many talked about having deals as a normal cost of providing good service. They were well aware that this system goal affected their own outcomes in this way.

However, when I asked them to talk about moments when they felt particularly vulnerable to mistakes and errors, the variety of their experiences revealed another system effect that was not as obvious to them. Drawing on their career history, the undeniable, dominant pattern was that they felt most vulnerable to things going wrong at moments of transition, when either airplanes or controllers—and sometimes both—were crossing the boundaries of time and space. One Boston Center controller acknowledged "the built-in frailty of the system is that boundaries are there," yet those boundaries extend far beyond the airspace boundaries to which he was referring. The transitions controllers described included crossing the boundaries of the social organization of work, time, and physical space in the workplace. These incidents revealed stress in anticipation of a mistake, based on past experience, that a situation in which things had gone wrong before might reproduce itself in a similar situation. They often spoke of the emotional impact of surprise; however, the routine, everyday aspects of the work also are also emotion generators.

The emotions controllers experience are a system effect. When I asked controllers when they felt most vulnerable to mistake and error, they revealed that the causes of close calls, mistakes, and errors are at the heart of the system.

The external pressure on the system for on-time arrivals and departures and the FAA attempts to meet the expeditious service challenge is one they readily acknowledge. Complex airspaces and complex traffic situations are emotion generators. Still others, unrecognized by controllers but visible in what they say about vulnerable moments, have their source in the social organization of work and the multiple transitions generated: training, shift work, the need for controllers to change positions during the day, to have breaks, the going home and returning, the time-of-day patterns associated with the ebb and flow of traffic. Airspace boundaries also are crucial. They affect the division of labor in each facility (the structure of positions) as well as the need to hand off airplanes from one controller to another. We see that mistakes and errors arise from the unanticipated consequences of standardization in the division of labor as well as time and space requirements that originate from characteristics of system designed to make it safe, and from standardized airline flight scheduling to meet passenger demands and maintenance requirements. Standardization has created a system of boundaries, not only in the sky but also in the social organization of work on the ground. As we have seen, controllers' task is to keep airplanes moving across the sky by bridging these many different boundaries.

Systemic causes of failures are widespread in organizations, but the more typical response when things go wrong is to target individuals responsible for errors. In many instances, the systemic causes are never identified because those investigating some harmful outcome are not trained to do so. Instead, they are trained to concentrate on human factors: the pilot who fell asleep while autopilot was on, the operator who misread the instrument, the manager who made the wrong call. Further, in the United States, the broader culture is oriented toward attributing responsibility to individuals for their success or failure. Finally, it is more expeditious to blame individuals when things go wrong rather than searching out the organizational and institutional sources of a failure. It is easier to change personnel (fire, transfer, or reeducate) than it is to identify systemic causes and change the organization. Also, this tactic has political benefits: having dealt with the individuals identified as responsible creates the public impression that the mess has been fixed and thus the organization has a clean slate and can go on.

The FAA, however, does try to search for systemic causes when things go wrong. When accidents happen, the circumstances surrounding the accident sometimes lead to changes in procedures to try to ensure the elimination of the accident-triggering factors. Also, operational error review boards nationwide examine errors to search for patterns. Samples are small, but with overtime and from their own work experience, board members do become aware that specific airspaces within a facility have more errors than others. Thus, they may alter their own airspace or procedures. Efforts to change and improve the system are ongoing. Even so, the FAA's central target for reducing and preventing mistakes and errors is the controller. As we saw in chapter 2, the secret to safety in air traffic control is controlling the controller:

> There's a mentality that the buck has to stop somewhere. There always has to be some human being that could have kept these two airplanes from getting too close, because if there isn't, then you question that's a possibility that two aircraft could hit despite our best efforts, right? So the FAA does it with aircraft accidents, too. And the [National Transportation Safety Board] does it, right? No matter what events happen, no matter nearly all aircraft accidents are from minor equipment failures, it's always expected that the chain of events is going to end with that guy flying the airplane. He's going to somehow overcome it. Same thing with controllers. They don't like to chalk it up to anything other than a human being, you know, solely responsible in the very end. (Center, M)

The knowledge that the system has built in frailties tied to the boundaries of time, space, and the social organization of work that make mistakes and errors inevitable would hardly reassure the flying public. It is nonetheless very reassuring that despite the built-in frailties, the number of errors in comparison to the number of operations is very low. It is even more reassuring that accidents are rare. Now, knowing that the system factors behind the production of mistake and errors are built into the social organization of work, we must ask how this level of achievement is possible. The answer is yet another system effect.

Perhaps the most salient fact to remember about air traffic controllers in relation to mistakes, errors, and accidents is this: They are not trained to fear accidents. They are trained to fear a violation of the rules of separation. It is this emphasis that is crucial to understanding their ability to avoid collisions. The rules of separation focus their attention prior to a collision point. They strive to avoid an operational error and, in doing so, avoid the accident. Even when they have a deal they've still got a cushion of time and space to take an avoidance action before an accident happens, although the amount of buffer zone varies by facility airspace.

The deal itself—the realization that two airplanes have come close—and the emotional response it generates at the moment of occurrence, plus the immediate administrative response and its repercussions, lead controllers to fear violation of the rules of separation. Ritualistically, after an error, they go back and listen to the tapes, talk to other controllers, try to figure out what went wrong. How can I make sure that doesn't happen again? And coworkers, for the same reasons, try to learn from the other person's errors to avoid these same consequences—thus serving institutional goals—decreasing the risk of collision by focusing on the rules of separation. Air traffic control is a proactive system of social control, one in which accidents are unaffordable; therefore, the system is designed to prevent accidents before they happen by actions and standards that minimize the possibility of their occurrence.

That is the logic behind standardization. It provides redundancy, the basis for coordination, common language, and common understandings that forestall trouble. However, the air traffic control system has one strong reactive mechanism of social control: punishment after the fact for operational errors. But controllers are punished not for the accident but for the violation of the rules of separation—this is an odd combination that is nonetheless still preventive, consistent with a proactive system, because it does not wait until the worst has happened to punish. It is a system meant to deter individuals from making mistakes. It is a system that works, but controllers bear the cost as they are the ones who experience the social, physical, and emotional costs of system-generated boundaries and the mistakes and errors they produce.

8

Risk and Stress

Emotion Work

Controllers' accounts of specific incidents of mistakes and errors, their physical and emotional states, and the memory of them that endures even years later would appear to be solid confirmation that theirs is a highly stressful, risky occupation. But here's the conundrum: that is not how controllers describe their experience of the work. After asking them to describe some mistake they had made and what they had learned from it, I then asked, "When you are working traffic, do you experience your work as risky?" Almost all controllers said no. Some even seemed surprised that I asked the question—"Risky? What's risky about this?" or "That never occurred to me, that it was risky. Why is it risky to you?" To them, either the work was not risky or they just never thought of the work that way and didn't experience it that way. They were matter of fact, unambiguous:

> No. You don't even think about it. When you're working . . . no, I'm just totally concentrated on what I'm doing. I'm not thinking about risk or danger or stress. Or anything like that. See, I don't even think that this is risky. I just think of this as a job. (Bedford, M)

> It doesn't even cross my mind that it's risky because I've learned my limitations. What I should and shouldn't do. I think I have a fair handle, operationally, on my technique, and it's developed, obviously, over years. So you know what you can and can't do and you know the consequences. That's the other thing you have to know. You know the consequences of doing something

> wrong. And you have to respect that. You have to have respect for your limitations. It's like swimming. You have to respect the ocean, but you don't think of swimming as risky. (TRACON, M)

Almost all controllers said flat out that they did not experience the work as risky. However, a few acknowledged that an awareness of risk was in the back of their mind or in their subconscious:

> Yeah, you always know the millions of dollars of equipment and the lives are there, but you're not really thinking about it. (Tower, F)

> No . . . I mean rarely. Every once in a while I will have a dream. It's always something bizarre. It's not like a real vision. It's always like I've only got two airplanes and I can't think of what to do to make them, you know, they're coming together, and I just can't think of how to fix it. Here all I've got is two. Something ridiculous like that. I guess in the back of our minds, we all kind of fear that. (TRACON, M)

Of those who said it was not risky, a few added that only rarely or infrequently had they experienced it as risky. These occasions are the emotional events they described when I asked about mistakes and errors:

> Normally it seems safe and that you're good at it, but during weather, emergencies, and bad days when you struggle with complex situations all day, I see it as a risky day. I do feel it physically and sometimes leave the sector shaking. (Center, F)

> You're in there working and something happens. . . . I've had experiences where I've had airplanes coming closer than I ever intended them to be, and you definitely feel it. For me it's a physical thing where your heart starts pumping. You kind of get this bad feeling in your gut. You kind of all sit up in a chair like this and you're thinking, "Jesus, here we go" kind of thing, you know? And therein lies the risk as far as knowing what the risk is and knowing that there is a risk. You know this ain't no game, you know? This is the real deal here. (Center, M)

Unexpectedly, many controllers responded to my question by immediately talking about the risk of having a deal. These controllers were clear about where the true risk lay:

> Your biggest fear is getting planes within five miles, not crashing, ever. That never crosses your mind. (Center, M)

> There's a certain amount of risk involved, sure. 'Cause every day you come to work there's that specter of a deal. There's also that minute specter, I would imagine somewhere in the subconscious that people probably don't think about a whole lot, of serious accidents where lives can be lost. I mean, you know, that's endemic to the job but nobody I don't think really thinks about that because it's less likely. But there is that specter of a deal because that happens. (TRACON, F)

Ironically, more controllers associated risk with having a deal than the number who talked about rare and infrequent incidents that made them aware of risk. Note also that the meaning they are giving to *risk* here is not risk of collisions or harm to passengers, but the risk of an outcome that harms themselves. They talked about the risk of having to go back to training and recertify before they can work again and the risk of losing their careers. Thus accidents become a "minute specter," in their view. When they mentioned accidents at all, it was to discount them as unlikely. The low probability of accidents appeared to be definitive in their understandings of the risk of the job.

Only five controllers responded that, yes, they experienced working traffic as risky. One controller said it was "absolutely risky" because he can kill people; another said that the likelihood of death as a result of his actions was "slim to none," but the risk was always there because he was human and could miss something, especially when a sector had too many airplanes. Two described experiencing risk when they had a close call or when safety got compromised and separation got lost: the emotion-generating incidents totally defined risk for them. One mentioned that he knew his feelings were different from those of his coworkers, "It bothers me and I know it doesn't affect the others as it affects me." Another said:

> Risky, yes. Dangerous? Probably not. It's not like you are going to be hit by anything. But risky, yes. Because the work takes a subtle wear and tear on your body. It's not, it may not be a physical risk, but it definitely is a mental risk. (TRACON, M)

Comparison is a means by which all of us understand our experiences and our place in the world. When I asked controllers what kinds of other activities or occupations they felt were more risky than theirs, they all named occupations in which the risk is to oneself, with possible harmful physical consequences, including loss of life: test pilot, police officer, construction worker, firefighter, astronaut, cab driver, convenience store clerk ("those guys get shot and killed all the time"), ocean oil-rig worker, road construction worker, or machine worker. Several specifically pointed out that, in contrast, air traffic controllers' lives are not in danger and they have no risk of being hurt or killed.

The dominant pattern was clear. Almost all controllers said they did not experience the work as risky. There were no gender differences. There were no facility differences. This was a surprise, because I had imagined that perceptions of risk would vary by place: higher in towers, where controllers see planes and pilots in real time, lower in radar facilities where the technological representation of airplanes distances controllers from the human element.

What they said when they talked about stress was equally surprising. When I asked the follow-up question "When you are working traffic, do you experience your work as stressful?" many laughed in recognition, saying they got this question a lot: "If I only had a nickel for every time someone asked me that question." When flying as passengers on airplanes, some controllers said they never tell their seatmate their true occupation in order to avoid this very question and the others that would follow. They know the public thinks of the job as stressful, and they imagine that the public thinks the source of the stress is the possibility of airplanes hitting each other, but controllers define their experience differently.

Although many controllers said that working traffic was never stressful and some said it was always stressful, most controllers said that working air traffic normally was not stressful or was stressful some of the time.[1] Then they went on to describe how they felt when working and the specific conditions in which they experienced the job as stressful. Much of what they said shatters stereotypes of their job. Perhaps most surprising to me were the controllers who found it a relief compared to the stress in the rest of their lives:

> No, personally, no. I mean I come to work to relieve stress. . . . I've got a thirteen-year-old . . . handicapped-challenged child at home, and to me, I get to come to work and I get to relieve my stress cause I deal with enough at home that you want to come here and you want to do your job and you don't want to be bothered, you know. And to me this isn't stressful. The worst part about my job is my drive. You know, I drive ninety miles to work so I'm in the car for an hour and twenty minutes. Now that's stressful. (Tower, M)

> In the outside world, I control nothing. My wife is divorcing me, lots of stuff going on. I come here, I control everything. I love to come to work. It's peaceful. (Center, M)

A consistent theme was that the problem was not stress but boredom. At Boston Center, Boston Tower, and TRACON, controllers remarked that everything is so standardized that they know what to expect: planes with the same call sign, coming in or departing in the same direction, flying the same route day after day, the same standardized procedures and routines to deal with it. Some controllers at Boston Tower and Boston Center compared their work to an assembly or production line. At the center, they look forward to working difficult positions or air sectors because it breaks the routine and avoids boredom. In winter at Bedford, once it gets dark, they are desperately wishing for airplanes. Summing up their experience of work, the accepted aphorism among controllers is "Work is 99 percent boredom, 1 percent sheer terror."

Controllers think that the public perception of the job is that high volume is a stress producer. However, most controllers do not define being busy as stressful. Being busy is routine; the traffic rush at specific times every day is normal for them. Further, the more things they are doing at once, the better they like it. The busier they are, the less likely they are to make mistakes, they say. Because high volume increases attentiveness, they equate being busy with safety. Busy is tiring, and some days work is exhausting because higher volume means more boundary work: communications with pilots, coworkers, and controllers in other facilities. They acknowledge having a physical experience when they are busy. And they see recognize it in each other: the voice either goes quiet or high, the body leans into the screen, foot tapping, sweating. However, in trying to explain what it's like, they don't label it as stress. It's fun, or an adrenaline rush, a high, being pumped, in the zone, being wired, or it's exciting and thrilling. Others describe it as a "heightened awareness" or "feeling a little antsy" that enhances their performance. They believe they work best under those conditions. If they mention stress associated with being busy, it's "good stress" or "positive stress."

Some insights about their experience of stress come from the kinds of conditions they describe that are stress producing and those that are not. In the case of stress as well as risk, controllers compared other life experiences and activities that they know well to explain the stress of their job. To distinguish their experience from stress or from "bad stress," they often compared working air traffic with other occupations they thought were stressful. Notice, in these comparisons, how often the low frequency and short duration of stressful incidents is meaningful to their understanding of what makes a job stressful:

> My brother has a Carvel—a Carvel ice cream store, people come in. . . . People are a pain in the ass. Can you make that more chocolate, can you make . . . I'd be like, could you shut up and get a life. When I am keyed to mic, I'll make an editorial comment. I'll say [who is that jerk out there]. But to have to deal with retail, forget it, I just couldn't do it. My brother says me too, the stress, the this, the that. I say I could not last a week at the Carvel. I would kill people. I would literally kill somebody because I could not take the constant whining and complaining that goes with those people that come into your store. And he says, how do you deal with the stress of air traffic? I said it's nothing compared to the stress of a Carvel. It really isn't. (Center, F)

> There is some kind of stress, but it is not bad stress. My dad was an engineer, became a corporate executive bidding on multimillion dollar deals . . . and he couldn't sleep at night, he had chronic eating problems because his stomach was stressed out. He lived with stress twenty-four hours a day until he finally died of a heart attack. But the stress that my dad felt was every day, every waking moment, and apparently every sleeping moment. He would have nightmares, you know? I go home every night and it was fun. (TRACON, M)

> I think it has its moments of stress certainly, just the same way any other job does. I think most of the time it's a very limited stress. When I think about a stressful job, I think about a cop. A cop gets a radio call, he doesn't know if it's going to be a mad man with a submachine gun or a kid stuck in a tree. At least I know basically what it's going to be. There is only so much that can happen to me. I think there are so many more stressful jobs than what I'm doing out there. (TRACON, M)

In answering the question about experiencing working traffic as stressful, controllers explained themselves by reference to other occupations, without any prompting from me. In contrast, when I next asked specifically about other activities or occupations that were more stressful than their own, they drew upon their own everyday experience: "Getting my four kids dressed and off to school is more stressful than this"; "Driving the freeway to get to work is more stressful"; "Getting Thanksgiving dinner on the table so everything is done at once"; "You know that guy at McDonalds with ten people in line shouting orders? That's more stressful than this"; "Working in a cubicle. I could never do that. That would be really stressful"; "I could never do what you do. Teaching. Talking in front a group. To me, that would be really stressful." One person said participating in the interview was more stressful than working air traffic.

Some compared working heavy traffic to their participation in athletics, where in anticipation they are nervous before, but once they start, they find the pace and then feel exhilarated. Their experience of the same feelings in another familiar activity that they loved had normalized their experience, leading to an interpretation of work as fun, not stressful:

> I never find it stressful. I find it fun . . . I love sports. And every time before a game, I get my adrenaline, nervous excitement, and stuff. Same thing with this. It would make my stomach so bad I'd have to run to the bathroom, you know? And then once in the first couple of minutes, I would feel totally weak sometimes in a game, like I couldn't do anything. But once I snapped into it, it was the most, best thing in the world, you know? (Bedford, F)

They also acknowledge experiencing physical effects even after they leave the facility. Many have long commutes (an hour one way is the norm; the longest is a controller who commutes two hours from home to Boston Tower). They use their commute to chill out, making a transition to home by listening to music or reworking events of the day over and over in their mind so that by the time they arrive they don't have them so much on their mind. They talk about going home drained. But here also they make distinctions: you can be busy and experience stress and be tired, or you can be busy and tired without experiencing stress. Many talk about how being busy has left them exhausted, but nonetheless they define it as good stress or they avoid the word *stress* altogether: instead, they describe it as a physiological and psychological state of alertness, of being pumped or wired.

Losing Control: Stress-Producing Conditions

So if the work generally is not stressful, but it is some of the time, then under which circumstances does controlling air traffic become stressful? Talking this through, one controller arrives at two important insights:

> Yes, it's stressful at times but then again, anything that somebody does is going to be stressful. Everything comes with stress . . . but if you like and really enjoy doing something, then your enjoyment will displace the stress.
>
> *Q: Do you equate being busy with stress?*
>
> Not always.
>
> *Q: As it gets busier does it become more stressful?*
>
> No. I guess physiologically it does, but I find the more . . . stress, the most stressful situations are the ones that are not necessarily the busiest but the most difficult.
>
> So I've had busy sessions where I've worked 120 airplanes, 120 out of here without stopping. I got out of here, boy that was fun, that was a blast and then I've worked sixty hours [sixty airplanes in an hour] where I've gotten off and I'm like, my head's spinning, I'm like, "Oh shit I need to go for a walk." Every single little thing was just a federal project. So the stress—not necessarily stress—isn't necessarily tied to the amount of traffic; it's tied to the complexity. (Tower, M)

That controller's comment contains a principle about when being busy *becomes* being stressful. Being busy is a routine, normal part of the day. Busy becomes stressful when another condition or conditions are added, *increasing complexity* so that the work becomes *more difficult*. At that point, normal rules that make air traffic predictable no longer apply, thus producing stress. A Bedford controller pinpoints just what it is about the complex and difficult that gets to her: feeling a loss of control. When asked about experiencing working traffic as stressful, she said:

> Seldom. It is stressful sometimes; most of the time it is fun. Most of the time you have a good time with everybody. When people [pilots] aren't listening to your instructions, that's when it gets stressful, and sometimes you get four or five of those guys on the frequency at the same time and you don't have control over it anymore. When you lose control is when you start getting stressed out. (Bedford, F)

Trained to control, the job becomes stressful for controllers when they lose control. The controllers who said that normally working traffic wasn't stressful, only sometimes, named several different conditions in which they experienced the work as stressful. Their responses contradict what anyone with a different job would expect.[2] The top four stress-producing conditions, in declining order, were weather, their training, coworkers, and institutional and organizational conditions that affect their work. These cluster together with the highest ranks, receiving an almost identical number of mentions from controllers. The four next-most-common conditions were related to everyday work and traffic: unusual traffic situations, pilots, close calls, and mechanical problems—the very conditions the rest of us might imagine controllers would name first. All of these conditions had one thing in common: they increased task complexity and difficulty, so that the controller experienced a loss of control. The following brief sections are presented in declining order of importance, as indicated by the number of controllers who talked about them as major stress producers.

Weather

Complexity and difficulty combine with high volume in summer: commercial air travel is up and "you get pounded from the second you walk in the door." Controllers deal with high volume all day, day after day. When weather is bad, the normal rules that make traffic predictable no longer apply. Summer thunderstorms bringing delays, reroutes, and TMU restrictions on departure times and spacing programs. Pilots want to avoid flying through thunderstorms and turbulence. They request route or altitude changes. Weather conditions also can create airport gridlock: delays and flight cancellations result in congestion on the ground. Winter has its own extremes. At Boston Tower, snow and delays pile up. During icing or snowstorms, the runways become a site of major vehicle activity, adding work for controllers even though traffic may be reduced because of flight cancellations and delays. Time for de-icing airplanes, extra safety checks, and snowplows has to be filtered into decisions.

Training

For some controllers, training was the first thing that came to mind, even years later. They recalled it as an intensive learning or hazing experience when being out of control captured the entire experience for them. At the FAA Academy, they lacked control over their future. They lived with uncertainty about the outcome. Assigned to facilities after graduating, they became apprentices under the supervision of a trainer who could make or break their career. Working live traffic for the first time, they experienced working traffic as being out of control until they gained sufficient competence to master one position, then they moved on to learn the next. The experience of being out of control, losing control, was repeated even during the first few years after checking out. They talked about retraining when changing facilities. Being controlled and simultaneously feeling loss of control over traffic were continuing themes. One controller said, "Retraining is stressful for most people, even for the really great naturals because they have to do it as they are told."

Other Controllers

As one controller said, "It's not the airplanes that are the stress of this job; it's the people you work with." They complain about others "giving you a lousy interval and too many airplanes"; "people you talk to on the phone giving you a hard time or delaying in answering"; and "getting screwed by another facility." Individual differences in work technique, character, and personality become a major source of stress:

> I don't feel the job is stressful, no. It's fun. Honestly, the most aggravating part of the job is watching a controller that misses departure holes or he doesn't, you know, provide the best service he can. I sit there and I watch that, and it really, it gets on my nerves. (TRACON, M)

> We had a guy, played piano, thought he was a comedian. When traffic was slow would type on the keyboard raising his hands like he was doing a concert. Talked to foreign pilots in a foreign accent. Drove us nuts. (Center, F)

Controllers complained about the whiners, who blame others for their own mistakes, and also about working with people who are less competent and compelling coworkers to pick up tasks to keep the traffic moving. Some controllers identified other controllers as the only thing that was stressful about the job.

Institutional and Organizational Conditions

Controllers spoke passionately about production pressures due to the airlines' practice of scheduling planes to arrive at the same times, which resulted in pressure on them from the FAA to reduce delays. For example, "They [the airlines] want everybody to get there at the same time and that's not just not feasible" because of the rules of separation and airport ground congestion. At all four facilities, controllers said that pressure to handle more operations faster was more stressful than having a deal, a close call, or the unlikely possibility of a collision. At the agency level, FAA policy plans to ease congestion and the way they are implemented are a major source of stress—even those changes designed to make the work easier frequently make it more difficult and complex initially. At the facility level, management rules and requirements is a stress producer for some. One said, "If they would just make fewer rules and let us work things out among ourselves, it would be smoother."

In contrast, the conditions that they mentioned less frequently were tied directly to the daily experience of working traffic: in declining order, a loss of control due to unusual traffic situations, pilots, close calls, and mechanical problems. In common, these situations all happened unexpectedly so had an element of surprise and were of short duration, in contrast to those in the previous group. Unusual traffic situations included planes flying off route, "routine emergencies" (which occur regularly, as opposed to "real emergencies"), or several factors coming together to create a one-off situation a controller had never had before. Controllers talked about experiencing stress when pilots don't do what they have been cleared to do or are slow to do it, and about pilot errors, such as busting an assigned altitude or wandering out on a runway when told to hold at an intersection. Communication problems, like a pilot not answering or too many people talking at once, jamming the frequency, were stressful moments. Least mentioned were mechanical problems, such as an airplane technical malfunction or failure of controllers' own equipment.

In combination, all these conditions and events increased task complexity and difficulty, contributing to controllers' sense of having lost control. So for most controllers—those who said the work was stressful only some of the time—these were the moments that produced stress and had an emotional impact. Consequently, we have to take seriously the common physical and emotional reactions of the controllers in training, the newly certified, and the experienced controllers not only during experiences of mistake and error—those searing, unforgettable moments of fear—but the other stressful experiences as well.

The Social and Cultural Transformation of Risky Work

Now the conundrum has been fully exposed. The public imagery of the emotional aspects of the job is contradicted by controllers' experiences and their interpretation of those experiences. The last chapter showed that when talking about mistakes and errors, controllers drew from their career history, describing incidents in total recall, including recall of their emotional reactions. Although the passage of time mediated original feelings, the emotional impact was as much a part of the memory as were the actions of airplanes, pilots, and other people in the room. The recollections were filled with emotion words such as *fear, scary,* and *heart-wrenching,* and with details of their physical response. For a moment, time and action stood still: "I'm in shock now," "I was stunned," "My heart stopped," "I couldn't speak," or "I was paralyzed."

Then, this chapter reveals the surprising contradiction. In spite of being formally responsible for thousands of lives a day, work that often leaves them feeling mentally depleted and exhausted, and their very long list of stress-producing situations, all of which are emotion generators—the dominant pattern is that controllers do not define air traffic control as risky. Rather, they view it as stressful only infrequently, under certain conditions when they feel they are losing control. How can we understand this contradiction between their emotional responses to mistake and error on the job and how they talk about their experience of risk and stress?

In *How Emotions Work,* Jack Katz understands emotions and feelings as the same thing. If people act and experience emotionally, these processes are always connected with transformations of the body. At those moments, people always take into account the social expectations of others and so can try to modify their emotions, but the emotional experience sets limits to these adjustments.[3] Given the intensity of controllers' training to suppress emotions and the expectations of coworkers, we must wonder at the intensity of the deep fear they experience with mistakes and errors that cause them to break with the normative constraints of their usual emotion work. Moreover, we must wonder how they manage their emotions in front of others under even the more routine circumstances in which they feel out of control. The answer is emotion work that transforms the experience of risk and stress in the workplace. Controllers' emotions are managed from above by the system they inhabit. Moreover, controllers themselves enact individual and group strategies that are cultural, normalizing and mediating the daily experiences of risk and stress: "We are self-regulating."

Arlie Hochschild famously has written about airline attendants and other commercial jobs that, like air traffic control, produce emotional labor.[4] She points out that our emotions act on us by sending signals to us that tell us about how we are experiencing our situations. This signaling process is shaped by social factors that influence what we expect and thus what those feelings mean. All social activities and settings are governed by "feeling rules." For jobs that involve emotional labor, the employer, through such things as training and supervision, exercises control over the emotional activities of employees through such things as training and supervision.[5] As a result, employees do "emotion work": the work that goes into not just managing emotions but also actually producing the feelings that the system requires to accomplish the job by conforming to the feeling rules of the workplace.[6]

Hochschild links the institutions we inhabit to cognition: In jobs that require regular and extensive emotional labor, the institutional management of emotions is layered. Emotion management is learned from directors of formal training, both at the academy and in on-the-job training during the apprentice period in a facility. In a standardized system, controllers are expected to execute their responsibilities in a standardized way. Trainers stress accuracy and adherence to rules. In particular, they urge precise ways of talking in keeping with international standards as stated in the book of phraseology. Feeling rules are actively taught and practiced. Controllers are taught to suppress emotion in conversation with pilots because remaining calm and maintaining a methodical, consistent style of talk is associated with safety. They develop a "frequency voice." Like other aspects of their work, emotion management becomes embodied. As one controller said about the required phraseology, "A way of talking becomes a way of being."

It makes sense that so many controllers said they did not experience working traffic as risky and stressful. Those who survive the training are people who have a high tolerance for both. On training at the FAA Academy, controllers recognized at the time and after that "the screen" screened for many qualities necessary to the occupation, tolerance for stress among them. They saw that tolerance was as important to making it as a controller as was gaining skill proficiency and mastery of information. Those who couldn't handle stress either developed a tolerance, dropped out, or failed because succumbing to emotions interfered with the successful performance of their tasks. Once a controller is assigned to a facility and apprenticing with live traffic, the training screens for those who can tolerate not only stress but also risk. Mistakes and close calls bring them face-to-face with their own lack of experience and the reality of possibly causing an accident. Risk, which at Oklahoma City was risk of failure, in the facility becomes associated with the possibility of fatal mistakes. Several controllers said that the ones who saw it as risky had been "weeded out a long time ago."

To what extent individuals had these qualities when they came to the occupation is unknowable; neither my observations nor my surveys showed anything that would suggest it. However, the training is designed to create and increase that tolerance. So, for example, consider the somewhat surprising fact that almost all controllers said that being busy was not stressful. But recall that at Oklahoma City, controllers began academy training by working a small number of airplanes, represented by flight progress strips, the number of which increase over time. Once assigned to a facility, they work on simulators until they can handle far more traffic than they would expect to ever occur in that airspace before apprenticing with live traffic. As the training builds up and increases their tolerance for volume, it simultaneously changes their experience of what it means to be busy. What was "busy" at the training academy had become now routine. Controllers distinguish between a routine busy session and a "really, really busy" session.

This insight forces our attention in a direction that goes beyond Hochschild's work. My interviews and observations reveal that cultural mechanisms also affect controllers' interpretation of their experiences and their capacity to do the work. They reflect the emotional component of ethnocognition.

Culture, Cognition, and the Normalization of Risk and Stress

The gradual development of tolerance for the difficult-to-tolerate and the nonroutine is an example of the normalization of deviance.[7] Actions that outsiders see as deviant or unusual, insiders—who may have started from that same outsider perception—gradually come to define as normal, routine, and acceptable. The process is an incremental one: a gradual expansion of activities, so that some initial tolerance level is increased, building a new base, and then increased again, creating and re-creating a new normal over time. In training, interaction, and on the job experience, controllers' definitions of what is normal and acceptable change. They talk about how, as they gain experience, fewer things surprise them because they develop techniques to handle almost everything. As a result of training and experience, being busy becomes a taken-for-granted aspect of the job. The emotions associated with it also become normalized, and consequently are not defined or experienced by them as reflections of risky or stressful work. The normalization of deviance helps explain why nearly all controllers say the work is not risky and most say it is stressful only in certain circumstances, when they feel they are losing control.[8]

I want to emphasize that the normalization of deviance is not simply a matter of time and an individual "getting used to it," as expressed in the old saying, "You can get used to anything, even hanging, if you hang long enough." Instead, the normalization of deviance is a complex layered process in which institutional and organizational factors affect individual thought, meaning, and action.[9] For air traffic controllers, cultural understandings about the work—in this case, risk and stress—come into being and are stabilized by three factors: institutional and organizational factors; the characteristics of the workplace (i.e., the architecture, technologies, and social organization of work in a facility); and controllers' response to the task and experience of work. As a result, they redefine risk and stress, producing collective cultural understandings that reinforce conformity to feeling rules.

Standardization, Formalization, Rules, and Routine

When asked to explain why they did not experience the work as risky and stressful, few controllers talked about the rarity of accidents, although I sensed that was important in their thinking. Instead, they talked about international standards and formal rules as the backbone of safety because they facilitate coordination, provide the rules of separation, and tell them what to do in every circumstance. At all four facilities, controllers evidenced a cultural belief that conforming to rules mitigates risk. Uniformly, they responded that if they were to follow all the rules, the work is not risky. Even if they lose separation, they expressed confidence ("I know") that they would never lose an aircraft through any fault of their own. The belief that following rules mitigates risk is not unique to air traffic controllers; it also has been found in other occupations that outsiders would define as risky work.[10] Second, controllers reported that because of standardization and formalization, work practice is experienced as routine rather than stressful because "much of it is rote—done without thinking." Standardization and institutional routines preprogram what they expect to occur and how they will handle it. The training and subsequent on-the-job experience transforms formal rules learned at the academy and local knowledge from the facility into tacit knowledge, constituting a cognitive package that they talk about as common sense. Thus, routine traffic situations can be handled without thinking, taking on a taken-for-granted character that mediates the experience of risk and stress, as when they told me, "Risky? What's risky about this?" or "The problem is boredom, not stress."

Formalization, standardization, and routines act in a third way. An emergency or difficult traffic situation shrinks the time controllers have to experience and express emotion and keeps them focused on procedures and routines necessary to their tasks. Any possible fear and awareness of risk are suppressed or, as controllers described it, remain in the back of the mind or in the subconscious, because they have to focus the immediate present and the traffic keeps coming:

> In emergencies, we have procedures, things to do. We can't focus on what is happening because we have to get emergency vehicles out there, get other traffic out of the way. We have to concentrate on the procedures because emergencies increase the workload and we have these other airplanes. (Tower, F)

> Even if we see an accident it's more what you're required to do to help fix it more than that person's probably bleeding, possibly hurt. You might think about that a little bit, but I don't know. (Bedford, F)

Finally, formalization, standardization, and routines alter controllers' interpretation of the experience itself—how they think about the work. They transform the very experience of emotion, thus normalizing their experience of risk and stress:

> People get accustomed to, you know, handling sixty arrivals an hour or, you know, fifty departures an hour, and we're not really focused on that possible catastrophic accident. I think everybody—sometimes I think of it as just production work, just one airplane after another, just doing the same turns, same runways, same separation, just one airplane after another. It's like a factory job. (Tower, M)

Standardization can strip situations of emotional meaning. Even in "real" emergencies, the standard phraseology mutes the seriousness of events. To prepare for the possible worst-case scenario, controllers must ask pilots: "Number of souls on board? Pounds of fuel on board?" The first time I heard these questions asked, they took me to a different place mentally. I had become inured to risk by seeing the effects of their training and watching daily the calm execution of standardized routines and coordination. The question to the pilot suddenly reminded me of the awful possibilities presented by combining people and fuel in an emergency. However, the question has a different meaning to controllers. For them, these questions are standard operating procedure, one of many deployed in a real emergency. Like the rest of phraseology, asked routinely, most often with no resulting accident, it is one among many standardized utterances that become devoid of emotion content:

> It never crosses my mind that it is risky, even when we ask how many people on board, it just never, it is just a number. It doesn't . . . we have to ask. Like, twenty-seven plus three SOBs. Souls on board. (TRACON, M)

Like other bureaucratic language, the phraseology loses its association with risk by repetition.[11] To controllers, the answers are just numbers to be forwarded fast to other involved facilities and emergency services.

Airspace and Place: Architecture, Technologies, and Intimate Space

Logically, it makes sense that the greater the proximity and sense awareness of a controller to some close call or harmful outcome, the greater the experience and perception of risk. Following that logic, I expected that tower controllers would be more likely than radar controllers to say the work was risky because tower controllers see airplanes, close calls, and accidents in three dimensions and in real time. In contrast, radar controllers are distanced from events by radar technology and data blocks as representations of airplanes, so I thought they would be less likely to experience the work as risky. However, regardless of facility, nearly everyone said no, the work was not risky. This result was most surprising at the two towers, where controllers see people in the terminals and on the tarmac boarding planes and deplaning after landing, thus making the human element more salient for them than for radar controllers.

Architecture and technologies of place had an effect, but the effects were complex and in unexpected directions. At radar facilities, the visual representation of planes on the radar distance controllers from the reality that people are in the airplanes, muting the experience of risk during routine work. However, proximity still exists, technologically: the sensory experience of risk that is kept distant by the radar can be brought close by the radio. In a "real" emergency, radio and speaker systems change the experience. When the speaker switch is flipped on to promote coordination, the pilot and controller voices are broadcast into the room. Everyone hears the pilot's voice and, during the final moments of flight before a crash, hears the pilot's last words—and cries.

The pilot is typically heard only by the controller who is talking to the pilot; but in the case of a crash, the broadcast voices have impact on everyone in the room. Radar controllers talked about the helplessness they feel and how emotionally shattering those experiences had been for them.

Tower architecture and design mutes the experience of risk in normal conditions. Tower elevation reduces the size of airplanes and specially designed, soundproof tower windows reduce the noise. Airplanes appear smaller, less powerful, more controllable. The transformative powers of architecture and technology on controllers' experience and perception of risk were dramatically brought home to me during my fieldwork at Boston Tower. I was awed the first few days by the contrast with Bedford traffic—the larger airplanes, the greater speed, the volume. But then, as my fieldwork at Boston continued, I adjusted, concentrating more on the controllers' work and my own, and the planes, always fascinating to me, were the objects they worked with. I did not think about risk or danger. In other words, their work situation had become normalized for me.

An incident changed this, allowing me to personally experience how tower architecture and technology affect the senses, and thus perceptions of risk. A controller going on break asked if I had been out on the catwalk. We stooped to get through the small door in the tower stairwell, stepped out on the catwalk, and stood up. It was the evening traffic rush, December, and already dark. It was exciting. The noise produced by all the activity was enormous: the engines of many airplanes as they backed away from gates, rumbled along alleyways and taxiways, pierced by the roar of takeoffs and landings during the rush hours. Tugs were towing planes to parking places, shuttle buses ferrying passengers out to board planes parked on the ramps, and other vehicles moving in the alleyways and taxiways. The noise was penetrating, making conversation difficult and at moments impossible. The smell of jet fuel was another blast to the senses. The exposure to the sounds and smells of the airfield was a sudden reminder of the power, speed, and danger of aircraft technology.

Back inside, risk and danger again seemed a remote possibility. The noise of the airfield was muted by the tower height and the windows' technology-reducing sound, all amid now-familiar routines, standardization, internal arrangements, technologies of place, and the immediate need in the present to do my work. Equally important were the intimacy of controllers' small workspace and the comfort of the familiar: people accustomed to working with one another on a shared task, the need to coordinate activities, and the prominence of the sounds inside the tower. The movement of flight progress strips, the daily routine, the joking, the conversation—these experiences of place dilute the experience of risk. Recall this controller's comment about risk:

> No. You don't even think about it. When you're working, no, I'm just totally concentrated on what I'm doing. I'm not thinking about risk or danger or stress. Or anything like that. See, I don't even think that this is risky. I just think of this as a job. (Bedford, M)

There was no denying the analogy: I understood that what had worked on me was what worked on them.

The Individual, the Group, and Cultural Devices

The final contribution to the normalization of risk and stress comes from strategies initiated and deployed by the controllers themselves. Controllers' acknowledgment of deep emotional experiences and the simultaneous denial of risk and stress indicates a social psychological strain. Geertz has written about institutionally induced strain, observing that such strains are often expressed in symbolic form through cultural devices.[12] Deep cultural understandings often cannot be articulated by those whose lives are shaped by them. To surface these deeper meanings, we have to look to cultural devices that reveal taken-for-granted understandings associated with the occupation.

As individuals and as a group, controllers respond to the job with actions that both indicate the existence of this strain and at the same time offer them relief from it. In combination, these individual and group actions or reactions are emotion work that help controllers conform to feeling rules and do the tasks necessary to the system. To conclude our discussion of the normalization of risk and stress, we return to the intersection of culture and cognition in emotion work, examining both individual strategies and group responses that rechannel emotions to conform to feeling rules.

The Individual: Risk Strategies

Individual controllers actively construct cognitive strategies that alter their perceptions of risk by distancing them from the reality of passengers being on board the airplanes:

> If every time I had to make a decision, I thought of the three hundred people on board, I would never be able to do the work. I don't think any of us think of people [while we're working]. If we did, we couldn't do the job. (Center, F)

The personal necessity creates a response to the job that becomes cultural: a common way of thinking and acting, passed on through generations of controllers, that conforms to feeling rules. Controllers talk about how they do not think about the risks of the job when they are working. Instead, and regardless of gender or facility, they actively deploy what I call "risk strategies": they reimagine the work, effectively depersonalizing it.

Rules become a mechanism for deflecting attention from possible risk of harm to passengers. They work in several ways. We have heard controllers describe how following accident procedures in emergencies narrows the opportunity to experience emotions. We also have seen that training and experience inculcate a cultural belief that if controllers follow all the rules, the work is not risky. Now we see how they actively use the rules of separation to depersonalize their work:

> I don't think about the lives, actually. I focus on separation standards and the airplane itself, and not thinking about the people in the back actually, you don't do that. (Tower, M)

> In our built-in system of five miles, it's very safe. I mean, they're not running into each other. And where we get actually, I get into it, and I'm sure a lot of others do, that you're actually trying to save the five miles and you don't want to get involved in an OE. That's more of a concern that anything else put together. So you're working very hard not to have the OE and not thinking about risk or accidents. (Center, F)

Another risk strategy they deploy that depersonalizes the work is thinking in terms of airplanes ("this guy"), call signs (AA 2973), data blocks, or "talking to a pilot." In training, they learn to think and talk in terms of aircraft call signs. Call signs become a way of thinking that neutralizes the human element. Call signs, seen on the sides of planes, on the scopes, read off of flight progress strips, exchanged in communication with pilots, simultaneously stand in for and distance controllers from the lives of passengers and crew while on the job.

Place also matters in shaping risk strategies, and therefore perceptions and feelings. Facility differences in architecture and technology provide different opportunities for controllers to depersonalize the work, reduce the size and power of airplane technology, and thus minimize thoughts of risk and harm so they can do the job. At radar facilities, controllers talked about these risk strategies, which include reimagining radio communication as well as the representation of airplanes on the scope. Note in the following comments the extensive use of analogy:

> I don't think about people in the airplanes. If I was thinking of it as people in the airplanes, I probably wouldn't like the job as much. It's too personal. To me, it's about disembodied voices that come out of the wires and dots on the scope that I'm supposed to keep apart. (Center, F)

> I compare it to a video game, whether it's a ping-pong ball going back and forth across a screen, or you are controlling a laser shooting this or that, it's just an object on the screen and you have a set of rules. And the rules say that they can't get any closer than this. The object is to work for an hour, or however long it is, and bring all of these things in the air without letting them hit each other. That's really the way I look at it. (TRACON, M)

> It's a video game. One quarter, no extra men. (Center, M)

> I've never looked at it as a plane full of people. It's sort of like when you're working a construction job and you have to put the concrete foundation first, then we have to do two by fours. It's sort of like you're working a job to get to your final task. And you know, it's just constantly building those blocks, and getting where you want to be, to your final goal, which for us is getting data blocks to the final. (TRACON, M)

Some tower controllers depersonalize the work by using tower height to perceive a reduced aircraft size and power as a strategy to redefine risk:

> You're like a kid with a train set. I mean you've got an airport and a microphone. What could be better? You're not even a grown-up. I don't even have a responsible job. You know, I just sit and talk to airplanes all day. Working ground control, I make believe they're ants and that's my ant farm and they're all tunneling around. One at a time, one behind the other. Different places in the ant farm. (Tower, M)

> I'm not looking at three hundred people on this airplane, one hundred on this, thirty-five on this, you know. I think it is like, you know, a Matchbox game where I put this guy here and he goes and this guy goes here . . . you know, the little cars, as a kid you'd play cars and you'd move them through the streets and navigate them. All right, this truck's got to go behind him like this, you know, and look at it as just a game. It takes an element of stress out of it if you don't worry about that. And I mean it's not that I'm not worried about it. I mean, of course we're all concerned for safety, but if you start getting into it, oh, man—just do the job and go, you know. (Tower, M)

Controllers frequently used the game analogy: a chess game, puzzle solving, or video games, Matchbox cars, ping-pong—but no one talks about losing, and the costs in these games are never loss of life.

The Group: Rechanneling Feelings in Acceptable Directions

What happens when the institutional mechanisms fail, when individual risk strategies also fail, and inappropriate emotions are displayed in a setting where the management of emotions is institutionally valued and reinforced? The group makes efforts to bring the violator back into conformity with culturally legitimate behavior.

The affective deviant is the person whose behavior doesn't fit in with the norms of their group.[13] In air traffic control, the affective deviant doesn't conform to the feeling rules of the group. This controller has the wrong feeling for a situation, and for the controller, the right feeling would be a burden, so he or she displays the wrong one, failing to demonstrate solidarity with the group. The responses from coworkers revealed unarticulated cultural understandings and feeling rules guiding how air traffic controllers are to behave as professionals. These examples show the group effect on the management of emotions. I give two examples, one from a man and another from a woman, to make clear that the affective deviance is gender neutral.

In the first situation, a Boston Center controller in the radar controller position was working in moderate traffic conditions. He and his partner, the radar associate sitting beside him, were relaxed, having a casual conversation between transmissions. Both were certified professional controllers, one no more senior than the other. Occasionally, the radar controller would consult with his associate, saying, "This guy is headed here, I'm thinking about doing X. What do you think?" or "He wants direct, these guys are all going the same place—do I send them?" They would talk the situation over, then the radar controller would take some action. Consultation was interspersed between transmissions and other conversation and had gone on for some twenty minutes when the controller working radar on the opposite wall suddenly swiveled around toward the radar controller and yelled, sarcastically, "That's it, that's it. Run everything by your d-side," a reference to the radar associate, informally known as the "data side," or "d-side."

The shouter was objecting to the display of uncertainty by the radar controller, in the position of key decision maker at the scope. Controllers are trained to be decisive and confident. The shouter was condemning an example of inappropriate professional behavior that deviated from group norms. Also, the shouter was affirming group feeling rules about showing weakness (self-doubt) and trying to pull the radar controller to conformity by shutting down the consultative conversation. I imagined I had something to do with the strength of the reaction because I was a visitor observing how they worked traffic, that is, their professional work. Moreover, everyone else was silent, which was not typical of controllers: even when they are very busy, they always make time to get off a zinger about each other's performance. I learned early that silence was the harshest sanction they could give. At a deeper level, the angry reaction was not only about professional competence and the inappropriate display of uncertainty but also about the deficient masculinity of the offending controller that was taken by the others as an outward manifestation of lesser competence.

Another time, in a different area, a shift change brought a new radar controller to the position next to the one where I was observing. Soon after assuming the post, she said, "Lord have mercy." She was hunched over the keyboard, giving audible sighs, and soon she said it again. Controllers at positions on both sides of her now stopped chatting with each other and were paying attention. When her radar associate pointed to something on the scope, she said, "I didn't even see . . ." However, her supervisor did not come over and stand behind her (an indicator of supervisor concern), and her radar associate soon announced to no one in particular, "We had so many airplanes, we actually put ourselves on hold," meaning they closed the airspace to other airplanes until they got it cleaned up. She had an anomalous situation immediately after taking up the position. Traffic had been put on hold in sectors feeding into Newark approach for some reason, and other aircraft were being rerouted into her sector. She was not forewarned by the Traffic Management Unit. She saw her sector filling up with excess traffic, far beyond the norm for the sector.

When she initiated a self-imposed hold on her traffic and got all her planes in a holding pattern, her voice matched the others in tone. Even though her traffic was excessive, her body posture relaxed, and her vocal expression of feelings stopped. But during the crisis, she was affectively deviant, uttering words and sighs showing emotions that signaled her feelings to all within earshot. Later, her violation of feeling rules appropriate to professional controllers was affirmed by her voluntary disclaimers to me when our paths crossed a few hours later ("[Traffic] is never like that"; "That was highly unusual"). In addition, her radar associate's announcement to the room and her supervisor, who went out of his way to say to me, "She always thinks she is busier than she is," suggest that her affective deviance had occurred before, contradicting her own attribution of her display of feelings to the unusual circumstance. Like the previous example, the issue was competence and gender.

Her violation of feeling rules was considered inappropriate because professional controllers were supposed to be sufficiently competent that they were at ease in complex traffic situations. Also, they are trained to be aggressive, tough, and in control. In this case, expressions of weakness violate expectations about what is legitimate professional behavior. These feeling rules hold for both women and men. Together, the two examples of affective deviance and the response of other controllers to them show that the institutional constraints that repress expressions related to risk and stress are supported by group norms about which feelings are appropriate. In each facility, I witnessed moments when controllers worked in circumstances in which they experienced a loss of control, yet I saw very few incidents of affective deviance. The rarity of such displays of affective deviance shows the existence and power of the group on the other controllers, who work without such displays. No one wants to be known as the "weak stick."

A second group response actively rechannels feelings of stress and risk in other emotional directions that are culturally acceptable to the group, offering relief from it. There are several ways this happens. Recall how controllers described the physiological effects of being busy. They did not define being busy as stressful; instead, they recognized the adrenaline rush they experienced when traffic volume went up as being pumped, wired, high, a rush, fun, or in the zone. If they described the feeling as stress, they defined it as "good stress" or "positive stress" to distinguish it from bad or harmful stress.

It is likely that the meaning of the adrenaline surge that controllers feel during busy sessions is based on interaction with the group, with the result that feelings of fear or stress are socially transformed, during training and after, into an experience of pleasure, not hardship or bad stress. The intensive training transforms people who come to this occupation. However, logically the transformation of feelings about risk and stress comes about not so much by learning from direct teaching as by the more indirect means of observation and conforming to the cultural meanings held by other controllers.[14] Defining busy as an athletic high, as being wired, or as fun results from experiencing the physiological effects of the adrenaline rush and the recognition and connection of these feelings with the meanings coworkers express.

Anger is a legitimate expression of emotion when controllers feel they are losing control—anger about pilot actions taken or not taken, about difficult situations, or at other controllers—"They keep refusing our airplanes and they know we are getting hammered." Anger expressed to the people in the room is culturally acceptable because it reflects professional frustration with situations that threaten safety and/or common enemies that are obstacles to doing the job, such as other controllers or pilots. Occasionally, even anger expressed over the frequency to pilots (an FAA no-no) is also acceptable when a pilot disregards instructions or errs in ways that make more work or jeopardize safety when a controller is already in a difficult situation. However, controllers are affectively deviant who are routinely angry at pilots, expressing it both over the frequency and in the room to the extent that it is their professional style. The inability to control anger is a signal that the controller too often feels out of control. He or she violates the feeling rules scripted into the mandated phraseology and calm tone of voice—the prescribed professional criteria—and thus signals incompetence and incurs the disrespect of the group. Towers are the places where *pilots* are most likely to err, not controllers:

> Some people get all peeved and they're pissed off and yelling and yelling at pilots, you know, somebody makes a wrong turn and they get all upset and then chew this guy out. There's a lot of people that are good people in the tower, but there's a lot of people that are just ridiculous, shouldn't be allowed to speak on a radio the way they do. [Shouting,] "I told you this and that and you know and you didn't . . ." It's just stupid because you get paid to be a professional and do the job and you should deal with your stress in a professional manner and not freak out on these pilots, you know. So what if they make a mistake? That's the reason you're there, is to deal with these things and make things happen. You're not there to yell at people for doing something wrong. (Tower, M)

As with the deployment of individual risk strategies, revealing anger is a symbolic indicator of and emotional release from the social psychological strain. Humor is another.

The Hilarious Backstage: Emotion Work Can Be Fun

At every air traffic control facility, humor effectively channels feelings of risk and stress in ways that release emotion in acceptable directions and affirm appropriate group feeling rules. Laughter is the emotional release, the safety valve, that "keeps us sane." This controller talks about the inappropriateness of showing emotions related to risk and stress and the group's use of humor to mediate these feelings:

> If you think risk is in there, they're not going to let you think long about it. First of all, nobody gives an outward sign. "Oh man, what am I going to do," you know? "I can't handle this," you know? They weeded them out a long time ago. But we'll start laughing, you know? I don't know, like if, you know, too many planes are jumping on the frequencies, somebody will throw their pen and say, "Oh, we're done," you know. And everybody laughs. (Center, M)

Rather than express fear, uncertainty, or other weakness, controllers make jokes to make light of difficult maneuvers or situations with potentially serious implications and at the same time to signal others that they are dealing with something difficult. Certain phrases become iconic and part of the lexicon of everyday talk:

> I had a guy on the frequency one day, he just starts babbling. He just comes on the frequency and says, "I'm a VFR student pilot and I'm flying a yellow and red Cessna 172." Like I'm on radar and can see colors? The guy tells me everything in the world except where he is. And I said, "What is your position?" He comes back and says, "I'm a bank president." It left me speechless. I just finally keyed the mic and said, "I was speaking more in a geographic sense, sir." And he tells me where he is. I thought, "What's your position? I'm a bank president." A bank president? And really, it took me a half minute to even focus, what the hell did this guy just say to me? "I'm a bank president." So now when a pilot acts like a moron, we say, "Must be a bank president." (Bedford, M)

"This is a cluster fuck" is sometimes muttered at Boston Center when a controller is working a small airspace with crossing traffic and it becomes jammed full of airplanes. Or, they might be dealing with a "shit-box route":

> Sometimes when you have one big plane like a 727, a big Boeing, you have a bunch of little planes down there, that's bad. You have the big fish in the little pond and you're trying to work the big fish through the little pond with the little fish surviving. But when you have all of those little fish swimming around down there, usually I call that a shit-box route. I was talking to a FedEx one day and he [the pilot] said, "We seem to be having some trouble getting lower," and I said, "Yes, well, you've got a shit-box route going on down there." I said "shit-box route" on the frequency! And he said, "OK, at least now we know what's going on." I said, "I shouldn't have said that." It didn't come back to haunt me so it was OK. (Center, F)

Controllers say that, in common, they have a "whacked out" sense of humor. Controllers who had been at several facilities over the years said that air traffic controller humor "is a universal": it has the same basic contours everywhere. The styles of humor vary slightly between types of facilities, depending on amount and rhythms of traffic. However, humor is integral to the work even when controllers are extremely busy, but more of it occurs during downtime, either after a busy period or while waiting for the next wave of airplanes. People talk about the tension in the room at those moments and how humor diffuses it.

One form of humor is play. Boston Center and Bedford have the most pronounced lulls between traffic. At Boston Center, some of the areas, but not all, are into playful pranks during traffic lulls. Anticipating the return of a controller from a day off, a crew member put the returnee's cubby in upside down so when he came in for his shift and opened the drawer to get his headset, everything fell on the floor. A controller was standing talking to the supervisor with his headset cord hanging down to the floor and another controller came around the corner of the supervisor's desk, looked down, saw the dangling cord, and without even hesitating, bent over and tied it to a nearby table leg.

Sometimes the play is collective. At Bedford and other small facilities, one told me, "You sit there waiting for the airplanes—where are they, where are they? You sit there forever. It gets so tense, so tense." At Bedford, rather than playing tricks on one another, sometimes they sang while they waited for airplanes to come. Gossip was another collective preoccupation during downtime. Boston Tower and TRACON did not have the same significant quantity of downtime between waves of traffic as did the other two. However, TRACON controllers did play collectively during busy sessions, whistling "Sabre Dance," singing songs, or making animal sounds in unison. And on break, they had the running card game Pitch as ongoing play. Boston Tower's traffic and division of labor left little opportunity for these kinds of physical play, although controllers there did manage it occasionally. Once, forgetting I was there, the controllers working the front row (ground, local east and local west)—on that day, all men—began joking about Kegel exercises and performing their version in unison as they talked to airplanes.

Everywhere, joking is constant. A lot of the fun is good natured and innocent. The jokes primarily are inside jokes: you have to know the people, the local history, and the situation to get them. Controllers make up nicknames from people's official initials—the initials they use to sign in and out and when talking on the ground lines to other controllers. And the nicknames stick because they are based on something from the person's history, some personality quirk or incident that is associated with that person. Nicknames are used to a person's face, even the ones that are derogatory, because controllers, like a lot of groups, have norms about teasing. A supervisor explained, "If I'm giving you shit to your face, it's because I respect you, but if I wait until you're gone, that's because I really don't like you or I really don't think that you can do the job."

Everyday incidents become grist for the humor mill. Early in my stay at Boston Center, a supervisor said, "See that guy over there who seems to be sleeping? He's a natural." He explained that a natural is a controller who is exceptionally talented ("born to it") and always appears relaxed, even to the point of not paying attention, uses fewer moves, and can handle lots of traffic without getting stressed out. One day, this same controller—the natural—was coming back to the area from break. He stumbled over a small low table in the middle that holds reference books. The supervisor said, "John, what's the matter?" John says, "I was looking the other way." The supervisor: "You're always looking the other way." Laughter all around.

Sexual jokes have historically been part of air traffic control culture. I rarely saw and heard sexual humor, except accidentally when controllers forgot I was there, although people openly talked to me about it, both in interviews and in casual conversation. Several supervisors explained that generally what I was seeing was controllers acting normally, with the language cleaned up some. Also, supervisors talked about how different sexual humor had become with the arrival of sexual harassment and discrimination laws. So naturally, in front of a (female) outsider, the sexual humor would be restricted. Nonetheless, one supervisor told me:

> A vulgar sense of humor seems to be an important part of fitting into this group. We talk trash, we get really bad. (Tower, M)

As with the other comments made about humor, no gender differences showed up during interviews in relation to sexual humor, as one might expect. Like risk and stress, controllers said that those who couldn't tolerate it—both males and females—were weeded out. Others tolerated it but didn't participate. For many, however, it was as important as the other kinds of hilarity, a normal, acceptable part of the day. Supervisors complained about being unable to control sexual humor when it crossed the line, in part, they said, because most people thought it was all part of the fun, including the women. One woman controller viewed sexual humor as play, but she also indicated that, like the constant harassment and teasing about technique during training, appreciation for sexual humor required the development of a thick skin:

> Our area has a lot of sexual innuendoes. That goes on a lot. Our area is extremely—I think we are the sex area. I don't know if the other areas are like that. Our area is definitely very sexual. 'Cause it makes it easier than trying to hurt somebody else. It's more of a fun, happy, playful way to be. It's more playful, yeah, it gets easier—it gets nicer. I think it has gotten nicer and nicer every year in the area but by now you are hard. (Center, F)

Only a few women complained. One resented the terminology used while working radar at Boston Center. Normally, if one person is working a sector as radar controller without a radar associate, the position is referred to as a "one-holer." However, if that single controller is a woman, it is called a "two-holer" and if both the radar controller and the radar associates are women, it is a "three-holer." One said, without describing what was said, that while she was pregnant she was the target of remarks that were very hard to tolerate. Another said the jokes were fine, but she couldn't stand the stories: "Take it to a bar." However, these were the exceptions. It could be that those who objected would not talk about it to me—"once a controller, always a controller"—the priority was loyalty to the group. An observation expressed by both men and women was that generally the women gave as good as they got, and then some. One supervisor said:

> If you get just all guys in the room, let's face it, the language and the things they say are going to be different than if there are a few women in the room. But some of the women don't care. Some of the women could care less, and some of the women, not so much in the TRACON but there are a couple in [Boston] Tower and they'll embarrass the men. (TRACON, M)

Another form of controller humor is the ongoing banter, or verbal fencing—competing to get in the best dig of the day. The banter goes on throughout a shift, but when traffic is normal or low—downtime—controllers go into high gear:

> Get a clearance, bang. Turn around and give a shot to somebody else. It's a tension reliever. It sometimes gets carried away, but it can be fun. When you are busy, you kind of are put into, "This is my job, I can do this part." Then when you do have the time to relax is when you get a chance to just break the tension barrier. You can actually feel that in the whole area. It's very quiet when people are working. And then all of a sudden the planes go away and then it starts to be joke time. If you have a chance to look around, you'll see body postures and things that give you a clue that somebody is busier than you are at the time, then you'll see the difference in people as the tension diffuses. (Center, F)

They also recognize that joking mediates the experience of stress:

> People assume that air traffic controllers are more tense than they are on a daily basis. It isn't the most stressful job in the world. Somedays it is but for the most part it's not. But there are days when the world goes to hell you know when the delays are really bad, the weather's really bad and situations might happen that are really, really stressful and even then people are joking but there's probably an undercurrent of tension which we're relieving by joking all the time even when we're not that tense, you know. Humor is really important to us. We crack up, we really have a good time. Jokes are wild and I wish that I could write it. I wish that I could write an accurate portrayal of what really goes on on a day-to-day basis in the tower because I think that it's a fuckin' riot what goes on up there. (Tower, M)

> I mean, we're down there laughing all the time. You've seen that, just joking and ribbing each other. It keeps you on your toes. It keeps us sane. It's just a big joke in here. We're all very serious. We're all very professional but that's what relieves most people's stress is the way we talk to each other. It's just joke after joke ripping on each other. (Center, M)

It's not like a typical office. Controllers have a unique work environment in which the same people sit so close together day after day together that an integral part of the job is being able to vent, whether it's venting through humor or putting people down for being a "weak stick" or "you missed a hole." They can't go off to their own office or cubicle to escape. Everyone knows each other's weaknesses and strengths, embarrassing moments, quirks, and they know which buttons to push to set each other off. If they want to annoy someone, they know how to do it. Everybody gets to be the target:

> If things are going slow and you know you feel like stirring the pot a little bit and then it starts, throwing all the little barbs at one guy, it's a great tension reliever in the aisle when it happens. Usually at the time it happens, we've been busy, and we come to kind of like a lull and we are kind of waiting for the supervisor to make some moves and it isn't happening, there's like a slow simmer in the aisle. You can just almost feel it when the pressure's up in the aisle. And it kind of breaks the tension and everybody's like turning and talking, we all have a common thing to go on. We all sit around and laugh and it really lifts the spirits of the whole aisle. (Center, M)

They are masters of the zinger and the hilarious remark:

> You have got to have quickness. You know, there are some very, very sharp people here, and some of the conversations go so fast it's just amazing. The quickness, the little subtle thing that just cracks you up, pure simple funny, but also the hurts can be very quick. You have to have a thick skin. If you don't, you're dead. (Center, M)

Not everyone participates in firing the zingers. But controllers must be able to tolerate it. Indeed, being the target of brutal criticism and jokes is integral to the training, and the survivors are used to it. A TRACON controller who had lousy spacing while working planes on final approach got a sarcastic "Nice try, Jack" from someone down the line as he left the position for break. Everyone can have a bad day, even the best controllers. When they do, they are subject to caustic comments and all-day ribbing. While controllers generally sympathize with someone who has a deal, sometimes the circumstance of the deal makes a controller the target of a punch line that becomes part of the facility's lore:

> This controller had the longest deal in air traffic history. He had like three minutes of the snitch machine going off. They came down and they said something to him. "George, you're having a deal." "I've got these guys on parallel headings." "Yes, but they're four miles apart." (Center, F)

"Yes, but they're four miles apart"—One phrase summarizes a whole incident and breaks everyone up whenever someone throws it out. Of course, timing is everything. At Bedford a controller shared:

> You've heard "Well, how else was he going to get there?" right? We had a controller check out on ground control even though during checkout she taxied someone across the closed taxiway, which is a big taboo. And it's a no-no, against the rules. And then she made the comment, "Well, how else was he going to get there?" And Mike, in his infinite, you know, he's terrible at ragging people, keeps going with that. It's going to go down as the greatest phrase of all. "Well, how else was he going to get there?" (Bedford, M)

A lot of it is good-natured fun, but not all of it is. Some controllers are consistent targets. One controller said, "It's like every facility needs a football to kick around." Those people who are targets have attributes that set them off from the group in some way: they don't fit in. It could be personal attributes or work competence or some combination of both. The controller who doesn't take criticism well, who bristles at the least suggestion that there might have been a different way of handling something, becomes the target of jokes ("You suck," "No, you *really* suck") then gets madder and madder until he snaps. Snapping is not due to the stress of the airplanes. Controllers can be merciless. Someone gets singled out and, as one supervisor described it:

> It's like a piranha feeding sometimes. There's a lot of teasing that goes on and joking that's light hearted and good and some of it even funny, but it can turn into an attack and then it becomes like a piranha feeding. The rule is never let them see your weak point because they will just go for it like piranhas and especially if they know that they can get to a person in a certain way they will

> stick with that vulnerability until the person snaps. Fuck this. I need this like I need a hole in the head. I'm out of here. And no matter where you are, as long as you're away from the scope, it's better than here. You can be at the damn dump. (Center, M)

A controller at Boston Tower said, "Nothing's too cruel if it's funny enough." Joking provokes laughter and works as an effective emotional release from tension and stress, but it has a second serious function related to risk. For strong and weak controllers alike, the daily ragging is aimed at attributes that don't measure up to professional standards. Controllers use humor as a mechanism of social control. They target poor technique or specific weakness in an attempt to bring the controller into conformity with group norms. "Nice try" to the TRACON controller running sloppy spacing on the final is a statement about falling short of the mark, which is running them tight and evenly spaced—the institutional goal of expeditious delivery of traffic. For the group, having a common target generates feelings of camaraderie, and for some of the recipients as well, the experience ultimately makes them feel that they are a part of the group.

For others, not so much. Although controllers appear to be indiscriminate in who they poke fun at, humor is a weapon wielded in a specific way, making status distinctions among them. On a daily basis, the public jabbing doled out about performance while working traffic gets at everyone—the natural, the competent controller having a deal or having on off day, or the "less competent" controller—controllers' define less competent as the one who is not confident handling a large number of airplanes, or plays it safe and keeps, say, six or seven miles of separation rather than five, or who cannot run them close and fast, who hands other controllers a mess, or who can go by the book but cannot innovate when it is necessary, who misses things, or shows inappropriate emotions. Ragging on or ripping someone can be a sign of either respect or disrespect, which is based on competence and trustworthiness as a coworker. But ripping the less competent controller takes on a different character. This supervisor explained the distinction:

> They just know this person can't be trusted. He's also not the strongest controller out there, and so yeah, they joke and fool around, but it's a whole different level of things, and it's done a whole different way and every so often there's some bite to what's said. You take somebody like Sam, Sam gets picked

> on. He gets harassed all the time about his lack of success with women, but they pick on him because they love him. A lot of it starts with a person's competence as a controller. The person who is the weaker controller to some degree takes a lot more shots, especially when they are not around. It [competence] is probably a lot closer to the truth when I think about it. When I think about the stronger controllers they are almost universally liked and respected. And when you think about the weak controllers, the respect level, and the whole attitude and discussion when they're around or not around is very different and that's probably the big criteria. (Center, M)

Often the barbed comments specifically target failure to live up to group standards for working traffic. The overarching point of all the bantering and digs is about increasing the quality of the work, the group performance, and it is risk averse and safety related:

> It seems like that's in your job a little bit to get under each other's skin, almost to push each other to be better, to be the best. Because I guess one of two things. One, you want your area, you want pride in what you're doing. I certainly take pride in what I do. You want everybody to be top of the line. The other thing is you don't want that guy giving you a load of crap when he's giving you all of these airplanes. You're like, "Jesus." So you kind of ride each other just a little bit to keep everybody's level up. (Center, M)

A few controllers observed that the FAA managers and supervisors don't do a good job giving feedback and criticism on performance, and so it was left to controllers to do it themselves. Is it effective? With those controllers who are having an off day, yes, they get it. With those who have a basic weakness, the effect is dubious. Ironically, the training creates a Teflon controller:

> When somebody is not up to your standards, whatever your standard happens to be, there's already something in your mind about how much you'll take from them. And I'm pretty good at cutting them down. It's hard to tell if they get it though or not, because there's enough of the good natured joking, putting down, going on, that they might believe that you're doing the good natured joking, putting down, that basically means that you accept it. And it'd be hard to know the difference. Once you get to be a controller, there's this sense that you have to be right. And you have to believe that you can do it. If

> you don't believe that you can do it, you can't do it, so anyone who believes that they can do it, and gets that put down, probably is not going to realize that it's a "you can't do it" put down. (Center, F)

Gallows Humor

Gallows humor has special meaning. Realization of risk and the consequences of mistake materializes as gallows humor—making a joke in the face of death. The classic example for air traffic controllers is "they can't fire us all," which originated during the PATCO strike when Reagan threatened to fire all striking controllers if they didn't go back to work in forty-eight hours. They didn't take the threat seriously because there were more than fourteen thousand strikers and Reagan had promised to support the union in exchange for its members' votes during his presidential campaign. Later, the phrase became a running joke:

> So the supervisor wants to take someone to task and we look around and say, "Well, they can't fire us all." Knowing damn well they could. (Center, M)

Gallows humor distances people from grim realities they would rather not face or openly discuss, putting them in a more manageable light. So, in common with other professionals who do risky work, it is another way controllers handle risk and the possibility of fatal mistakes and accidents. They display irreverence:

> John's moved. He's whackin' 'em together at Providence now. (Tower, F)

> No, we don't think of it as risky. As a matter of fact, there was a tour that had come in, and they were standing behind me. I think it was mostly elderly people. And the person giving the tour was explaining what the data blocks were. For instance, he goes, "This plane here, this is a DC-10. There's probably two hundred people on that plane." Well, I jumped out of my chair and said, "People! Nobody told me there were people on these planes!" And those people just . . . they kind of backed up. You know, I was just kidding, but no, you don't think. It's you and a pilot. There's just two of you. So you don't think about anybody else. You can't. (Center, M)

> Joking helps everybody. It's just a way of letting out steam. Because if you really thought about what you were doing, I think you'd make yourself a wreck.

> Even when two airplanes get close, you just make a joke or say something silly about it. You know it's dangerous, you know it could be critical. But if you just thought like that every day "Oh my God. There are 320 people on that airplane . . ." so we just make a joke, "Oh, well, maybe next time." (Center, F)

Gallows humor is a well-known coping mechanism, occurring everywhere—after disasters or personal tragedies—and has a known association with certain professions that deal with trauma or its possibility: psychologists and medical personnel, soldiers on the battlefield. Gallows humor is an attempt to control the uncontrollable. As one controller said, "If you can make a joke of something, you own it."

Nicknames that symbolize mistakes and accidents are a form of gallows humor. Sitting at a scope with a controller at the TRACON on my first day, I introduced myself. He said that his name was Pete but everyone called him "Butch." I said, "Oh, that's an unusual nickname. How did you get it?" He said, "It's a term of endearment." "A term of endearment?" "Yes, when I was training they said I butchered everything, so they nicknamed me 'Butch.'" Many controllers received nicknames as a historical marker of their training mistakes. The TRACON roster included "Jigsaw" (everything he did was a puzzle), "Cement Head" (self-explanatory), "Rock" (same). One controller's nickname was "King." This seemed to be of a different category from the training mistake nicknames. One evening, I was in line with the group of controllers gathered around the supervisor's desk looking at restaurant menus to order dinner. One said, in a loud voice, "Diane, why do you think John is known as 'King'?" They started pushing him to tell me about his nickname. He laughed but didn't seem eager to tell me. Later that night, when John and I were both in the break room, he brought it up and explained:

> Have you ever heard of "The Flying Elvises"? There is a group in Vegas. They're called the Flying Elvises. They travel around the country basically. OK, these guys, they dress like Elvis Presley. There are like five of them, or six of them, or something like that. . . . They parachute out of airplanes for events. And their capes are lit up so you can see them at night. And they had a situation down in Quincy where they were . . . building a restaurant or something. It was opening night for the restaurant, a very, very windy night. And they had decided, the restaurant people, had hired the Flying Elvises. They flew out of Norwood. The winds were 330 at about twenty-five knots, which meant once you dropped you would probably fall in the ocean. Yeah, they should not have

> been dropping at night. But they got paid for it, were willing to do it. They had already agreed and all of that. And I was the controller, and the guy [the airplane] came out of Norwood. And we actually basically physically stopped traffic 'cause we were on 33 Left and we left a huge gap so they could drop. And I gave the guy, I said the winds are 330 at twenty-five knots. "OK, we're dropping." So I gave them the wind again, I said 330, twenty gusting to thirty. "We're dropping. OK."
>
> *Q: And where was he at this point?*
>
> Basically over Quincy. So he dropped. I think it was four thousand or five thousand feet. He dropped the four guys. Two of them landed in the water. One of them landed on the ground safely, and the other one landed in the parking lot and died.
>
> *Q. Oh, no.*
>
> So Elvis being the king, since I killed Elvis, I became the king. [*He laughs.*] Is that morbid? That's how it all started out. So you have to laugh at these things, otherwise you'd go nuts. (TRACON, M)

Like "Butch" or "Cement Head," "King" is also "a term of endearment"—a symbolic marker that also says: "We get it. We understand your experience."

Gallows humor extends to controllers who've worked major disasters, so the joke, like a nickname, simultaneously marks for posterity the controller's connection to the accident and functions for the group as a distancing mechanism. Take the case of the controller who was working TWA 800 when it suddenly and mysteriously went down. A controller in his area told me:

> Whenever there is an emergency in a sector in our area, someone will say, "Send Dave over there," because now he is disaster-proof, and what are the odds of that happening to the same controller twice? That's how we deal with it. People aren't going to say, "Oh, Dave, you poor thing." You're not going to say that because that acknowledges our vulnerability and we're not going to acknowledge our vulnerability at all ever, ever. I don't like to think that when I fuck up I could kill five hundred people. I am never going to think about that because I would find it immobilizing. So there is not "You poor dear," nothing like that. (Center, F)

Having denied that the work is stressful or risky, when talking about humor, they are, in fact, acknowledging both stress and risk when they recognize that play, joking, banter, the zingers, and gallows humor are ways that they deal with the emotional experience generated by the work. "Nothing's too cruel if it's funny enough" and "If you can make a joke of something, you own it" are both true. Humor is emotion work that indicates the social psychological strain of the job and functions to transform feelings of risk and stress into emotions that are acceptable to the group. Laughter and anger conform to group feeling rules, but tears and whining, whether metaphorical or real, are unacceptable. But the wild and hilarious backstage of air traffic control sustains those who choose this occupation in additional ways. They love to come to work because of both the challenges and the fun. For a highly standardized job, which at every facility except Bedford was described as "assembly line" that caused many to say that the main problem is "boredom, not stress," humor was a way of adding spontaneity and the unexpected.

Although air traffic controller humor itself takes on a pattern, the form of it—what will be said at a given moment, the competition for best dig of the day, or the trick that will crack everyone up—adds the element of the unexpected to the everyday routine. Banter is also a way of testing each other. The work of each controller is interdependent with the others. They need to know each other's capabilities and how they are going to work together to get the job done. So they zing each other "to keep everybody's level up." The meanings are culturally complex because the zinger is a sign of affection and respect that makes them feel part of the group, as well as a signal (whether recognized by the receiver or not) that a controller is marginal to the group: not fitting in, not doing the job, not liked, not respected. Humor is a cultural device that works not only to normalize their experience of risk and stress but also functions to reinforce social solidarity and, finally, to upgrade performance because beneath the humor, collectively they understand their work is moral work.

They openly talk about the varieties of humor as a response to risk and stress that meet the need for camaraderie and as a safety valve that allows them to diffuse emotional situations and to blow off pent-up emotional energy. But ironically, they fail to realize another crucial function: humor produces energy. I got my first clue about this effect in an interview with a controller at Bedford. She was talking about gossip—backstabbing—during breaks in traffic, when people had nothing to do but talk about one another, how they thrived on the mistakes of others, and how cruel it could get. Although she had never gossiped about people before coming to air traffic control, she described how, in the small group, she felt pulled into it. "I wanted to be a part of the group," she said, but she didn't like herself for it, didn't like what she had become. Then she said, "But sometimes it's the only rush we get all day." This statement has profound symbolic meaning.

Many controllers had talked about the emotional high of a busy session, some even describing it as addictive. "The only rush we get all day" indicated that zingers and humor also produce a high, suggesting a need for the mental and emotional energy. This struck me as making particular sense for controllers at Bedford, a facility where they had more downtime than at any of the others. Of the 188 controllers I interviewed, only one—on her own, without any leading questions from me—confirmed my interpretation. Explicitly associating humor with the production of mental and emotional energy, she explained:

> We can seem unprofessional and irreverent. But it's that we put so much emphasis on the safety part. So much of your mental and emotional energy is sucked into that, that safety vortex, that it gets sucked out of other places. So you get to be irreverent. I don't really think that there is this vacuum of this energy that you can just pull from it, that it's infinite. Energy comes from someplace. (Center, F)

Behind the statement "It's like every facility needs a football to kick around" is this very point. During my fieldwork, I saw how, after a busy session, when traffic was normal or low, humor diffused the tension, kept controllers mentally alert, and generated energy that prepared them for the next wave of airplanes. Energy comes from the play, constant banter, ragging, joking, collective signing, targeted humor, gossip, and sexual humor, so when the airplanes come, controllers do not have to go from a full stop at zero or a speed of twenty to sixty miles per hour to cope with the onslaught. Humor not only diffuses the experience of risk and stress, producing an acceptable outlet for emotion; it literally reduces the risk by engaging controllers in mental and physical activity that prepares them to handle whatever comes next. A cultural device, humor protects them from mistake and error as well as the experience of risk and stress. Mistake and error, risk and stress, are system effects. Here we see the agency of controllers who enact their work experience to make it enjoyable and at the same time meet the system goals of safe, orderly, and expeditious delivery of air traffic.

PART V

"That Little Frisson of Terror"

On that clear and beautiful September morning in 2001, terrorists hijacked four commercial airliners to use as weapons in an attack against the United States: American Airlines Flight 11, United Airlines Flight 175, American Airlines Flight 77, and United Airlines Flight 93. The two that terrorists flew into the World Trade Center—American 11 and United 175—both departed Boston Logan and were worked by controllers in three of the four facilities in this study. These three chapters examine dead reckoning on September 11 and in the days, weeks, and months of the first year after. Although some aspects of the air traffic control response to September 11 have been made public, this is a different, untold story. It is controllers' personal story. Also, it is a sociological account that demonstrates the action and agency that contributes to institutional persistence and transformation of the socio-technical system across time and social space. As such, it illuminates how this unprecedented attack precipitated system effects at the individual, local, national, and international level both on September 11 and in its aftermath, which for controllers was a second crisis, brought for the first time here to public view. A crisis opens up the inner workings of institutions, organizations, and system effects for observation, especially when people, skills, structures, cultures, routines, and technologies are put to different tests from those for which they were intended. Managers, controllers, and supervisors were making decisions under conditions of extreme uncertainty. The characteristics of the system conveyed in previous chapters are made more visible by these twin crises. Ethnocognition and boundary work are curated to meet the changing circumstances. We can see controllers' series of distributed, embodied, situated, local and expert practices of working things out moment to moment with material objects while simultaneously enacting their trained methods of thinking, problem solving, hearing and feeling.

On September 11, we see how the impact and response to the attacks varied by facility as a result of differences in space and place. We see how the boundary problems that drove controllers nuts every day were conquered in a remarkably smooth coordinated effort across local, national, and international air traffic boundaries. Controllers deployed common sense—the training and experience with institutional rules and procedures, local knowledge, tacit knowledge—as they rose to a challenge that could not be rehearsed. They identified early warning signs but in this instance were unable to correct to prevent catastrophe. Trained to control, they lost control: here was the collision of aircraft they had never anticipated and could not prevent. We see the emotional impact upon them when forced to confront the realities of what they never dared think about: fear, risk, death, and destruction. Yet they systematically worked through it. Crucially, in a standardized, rule-based system, they improvised. The result was that managers, supervisors, and controllers nationwide brought down 4,395 planes in two hours and fifteen minutes, without incident.

How did they do it? The key is in these comments, quoted in earlier chapters, that two controllers said about their work one year before September 11, when such a catastrophe was beyond imagination:

> Structure and routine, structure and routine—the rules and regulations [structure], and how we respond that becomes automatic [routine]. If we have something new, we have at least had something like it before. If we have to improvise, we improvise from the base. (Bedford, M)

> You need neat and orderly, but you also need flexibility, and you also need the ability to just pull something out of the air. And you have to because there are going to be times in your career when neat and orderly is going to go out the window and you're going to have to pull an idea out of your head. And it had better be there. (Center, F)

Structure, routine, and improvisation combined: as the tragedy unfolded, time and again these principles explain how controllers emptied the sky.

The second crisis occurred when, in another unprecedented, unrehearsed effort, they had to get the airplanes flying again in a work environment that had been completely transformed by the attacks. Initially, the military took control of the airspace. Then air traffic controllers returned to work, policing the sky. Dead reckoning dramatically changed as their job expanded to include national security strategies, escalated vigilance, the suspicion that any anomaly could be a hijacking, and the awareness that they, too, could be at risk. In getting the planes off the ground and into the air again, they faced new boundary problems. Myriad hard-to-decipher new rules and procedures for the return to flight changed airspace boundaries. Authority relations changed. New dictates were formulated and handed down from above by people who had never worked air traffic: the government's national security hierarchy. For a second time, air traffic controllers lost control. Again, structure, routine, and improvisation combined as controllers worked to figure out how to make the rules, procedures, and technology changes fit the local situation. In contrast to the improvised workarounds they created on September 11 to *overcome* standard procedures, this time improvisation was in the service of *creating* standard procedures.

Fully revealed by these two crises are dynamics of the air traffic control system as a whole. Following the actions of controllers within their separate facilities demonstrates that the resilience, reliability, and redundancy of the system is enacted by them. The exchange between the four facilities in New England and the Command Center are a microcosm showing how the combination of structure, routine, and improvisation at the individual level repeated at the system level; the dynamic quality of boundaries—their flexibility and porousness, their capacity to include or exclude, and the import of the human-technology interdependence in those accomplishments; system variation in vulnerability and experience of risk and stress; and the variation in emotional experience and emotion work over time and place. Further, the system effects of September 11 and the year after were on a grand scale: reorganization of structures, technologies, rules, and routines at both the national and local level. In combination, the three chapters in this part demonstrate how redundancy, resilience, and reliability at both the individual and organizational level work in a large-scale socio-technical system to enable a fast response when necessary and also long-term organization change. Piece by piece, controllers rebuilt the system to protect air transportation and the system itself. For security reasons, I did not have access to detailed information about the substantive changes that constituted that long-term transformation, but I did not need that detailed content. Consistent with our interest in system dynamics, these chapters reveal agency and the process of repair in response to the crises and system transformation.

The 9/11 Commission Report extensively chronicled many federal agencies for their mistakes in responding to the terrorist attacks on September 11. Whereas the New York Police Department and Fire Department, the North American Aerospace Defense Command, and FAA headquarters were "understandably unprepared," FAA headquarters was (wrongly) singled out for poor communication.[1] Further, the report was rife with examples of failures of national security agencies in the years preceding the attacks. There were missed signals: opportunities to detect and stave off such attacks were blown time after time. The mistakes that were made fulfill all the stereotypes of slow-responding bureaucracies: rules were not followed, or rules were followed in situations for which logic would dictate a different action. The result was structural secrecy: information that might have saved lives was not communicated across agency boundaries.[2] In contrast, the report praised air traffic control operations personnel:

> We do not believe that the true picture of that morning reflects discredit on the operational personnel at the Northeast Air Defense Sector [NEADS] or FAA facilities. . . . They actively sought out information, and made the best judgments they could on the basis of what they knew. Individual FAA controllers, facility managers, and Command Center managers thought outside the box in recommending a nationwide alert, in ground-stopping local traffic, and ultimately, in deciding to land all aircraft and executing that unprecedented order flawlessly.[3]

The report credits improvisation: individuals and their out-of-the-box thinking received well-deserved praise. Structure and routine were equally important. In this, the accomplishments of the system's operational personnel resulted from what Chambliss called "the mundanity of excellence."[4] Although the whole operation for getting the planes down and back up again had never been rehearsed, the multiple bases of controllers' expertise—institutional rules, local knowledge, tacit knowledge, technique, emotion work, boundary work, technologies, phraseology, room awareness, and cultural understandings about safe, orderly, and expeditious delivery of air traffic—were embodied knowledge. As stricken as the rest of us, controllers' ability to act almost as if "without thinking" was essential to getting the job done. As a system and as individuals, air traffic control operations was as prepared as it could have been for this unimagined event.

These three chapters are important both as a social history of a national and international tragedy and as a study of an occupation, its work, and how air traffic operations responded, as a system, collectively and individually in small groups in the facilities. In addition, the chapters are relevant beyond the incidents and actions of that day and era. Showing the on-the-ground, moment-to-moment intersection of institutions, organizations, technologies, and individuals allows us to consider which kind of socio-technical system this is and its larger significance for organizations, system effects, boundaries, and risk discussed in chapter 13. Chapters 9 and 10 follow the chronology of events, which I constructed from published sources and my interviews. In the aftermath, many different accounts of the timeline of the events that occurred on September 11 were published. To be as accurate as possible, I consulted multiple sources, especially official reports, then compared with sources available from the documentary records at my facilities, and verified and added detail based on my interviews. Fascinating accounts from the internet and published accounts captured the sequence of events from different perspectives, such as the airlines. To keep the focus on events, actors, and actions, I do not cite them here.[5]

In chapter 9, the events of September 11 begin at Boston Center, where the hijackings first were discovered; the chapter then follows controllers and events at the other facilities as they became involved, paying attention to coordination and interaction across the four facility boundaries. Their exchanges with the Command Center for Air Traffic Control Operations in Herndon, Virginia, show how system boundaries were crossed and pulled together between the New England Region and the Command Center, and between the Command Center and all regions to coordinate the nationwide response.

Continuing the chronology, chapter 10 examines policing the sky in the year after September 11. It reveals how the system response to the war on terror changed the system, boundaries, work practice, and perceptions of risk, and how those changes distributed across airspace space and place. Chapter 11 begins with the effects of these events on controllers. It shows how controllers overcome the objective boundaries of the system that physically divide them, instead creating symbolic boundaries that distinguish them from other occupations, binding them together in an occupational community. Moreover, they resisted the inequalities built into formal objective boundaries that stratify them by facility, instead constructing symbolic boundaries to redefine their status and their moral worth, in relation to one another, based on competence and the moral nature of their work.[6]

Security was very tight after the attacks. Even controllers' family members were barred. No visitors were allowed unless on official business and cleared by several bureaucratic layers of approval. In 2002, I received permission to return to the four New England facilities under the topic of "official business," based on an agreement made at the beginning of my project that I be allowed to return to clear up questions arising during my data analysis. Although crucial to understanding the terrorist attacks and their impact on air traffic control system, the air traffic control facilities working flights United 93 and American Airlines 77 necessarily are peripheral in this analysis. Absent a pre–September 11 connection and baseline experience with them, I could not have gotten access to the Cleveland, New York, and Washington region facilities that handled them in the taut security net drawn up after the attacks, nor would I, as a stranger, have had the trust of individual controllers to discuss these sensitive matters.

Although we know that the relationship between space and place means that each of the facilities that worked the hijacked planes was different and so controllers had different experiences, we also would expect some common patterns. We can assume that the responses of the facilities working the hijacked planes near Cleveland, New York, and Washington were similar in many ways to the New England four, given the standardization of training, architecture, structures, rules and procedures, and technologies. We also could expect that their experiences and actions during those crucial, difficult months afterward would reveal the same combination of structure, routine, and improvisation, variation in the experience of risk and stress, porousness and flexibility of boundaries, and variation in emotional experience and emotion work over time and place that will be exposed in these chapters. Nonetheless, their responses would not be identical: individual and collective problem solving would be tailored to local conditions of airspace and work.

9

September 11

Boston Center

American Airlines Flight 11, a Boeing 767 flying nonstop from Boston to Los Angeles, departed Boston Logan at 7:59 am on September 11. Boston Tower and TRACON controllers assigned it headings and climbed it, handing American 11 off to Boston Center without incident. At Boston Center, American 11 entered the airspace of Area C—"we are known mainly for TWA 800"—it passed uneventfully through the low-altitude sector, Bosox, which Greg was working. Greg continued its climb and handed it off to Peter, who was working the Boston High sector next to Greg's. Both were experienced controllers. At 8:15, Peter instructed the pilot to climb to thirty-five thousand feet, but the pilot did not respond. Peter tried several times to contact the pilot, with no success. Suspecting a radio failure, he tried using an alternative emergency frequency, again with no success. Repeatedly, he tried to contact the aircraft. At 8:21, American 11's transponder signal disappeared. With the transponder gone, the plane no longer was sending its identification signal. The data block disappeared from the scope. Peter no longer had information about its call sign or altitude. The radar still allowed him to follow its path, however, because even after the raw radar blip disappears, the radar continues to send the "primary signal"—the target—that marks the aircraft's position.

Losing radio contact with a pilot was common. It could happen for many reasons, including that the pilot set the radio to the wrong frequency. Now with both radio and transponder gone, however, Peter, Greg, and the Area C supervisor were very concerned. Other controllers in Area C heard and were paying attention. Peter followed the procedure for "no radio." He checked back with the TRACON to see if the pilot was still on the TRACON frequency. He contacted American Airlines and other airplanes in the area to try to communicate through other channels. Then American 11 deviated from its route and left Peter's sector, crossing into Lino's sector, still in Area C. Other controllers in Area C began moving their aircraft out of its way and asking pilots in their airspace to watch for American 11. Greg was working his own planes and helping Peter with his. Then, at 8:24, the following transmission came from American 11, which was still on Peter's frequency: "We have some planes. Just stay quiet, and you'll be OK. We are returning to the airport."

Peter didn't catch what was said, but he knew from the spoken rhythm it wasn't phraseology, so not a professional pilot. Also, he heard a foreign accent that he couldn't identify. He said, "And, uh, who's trying to call me here? American 11 are you trying to call?" Almost immediately came a second transmission: "Nobody move. Everything will be OK. If you try to make any moves, you'll endanger yourself and the airplane. Just stay quiet. Nobody move please we are going back to the airport. Don't try to make any stupid moves." The transmission was chilling. Peter realized it was a hijacking.[1] Greg described the moment: "Peter all of the sudden jumps up. He heard something. He starts shouting to the sup, 'John, John, something's really wrong here.' I hit the button so I could hear on my headset. Peter was trying to flip the speaker switch above his scope so everyone could hear."

Greg took over all Peter's airplanes so Peter could concentrate on American 11. The other Area C controllers, still working their own traffic, used the computer capability to pull the flight up on their radar scopes to watch the hijacked plane. The supervisor contacted the operations manager at the watch desk, located in the Traffic Management Unit at the top of the control room's center aisle. At 8:28, the operations manager, Terry Biggio, notified the facility's air traffic manager, the New England Region headquarters, and the Command Center in Herndon of the hijacking and the fact that the plane appeared to be headed toward the New York Center airspace. Next Biggio contacted Boston Center's quality assurance representative to pull the tape of the first radio transmission and report what the hijacker said. The Command Center opened up a teleconference line that brought on New York Center and New York TRACON. Other facilities and agencies would be brought in as events continued to unfold.

At 8:34 the last transmission from American 11 was broadcast on the speaker, heard by all in Area C: "Nobody move please. We are going back to the airport. Don't try to make any stupid moves."

At 8:37, Terry Biggio and the TMU manager Dan Bueno, both physically in TMU at the top of the control room, deliberately violated protocol, improvising to save precious time. Instead of requesting military assistance from up the hierarchy, the TMU manager directly called the Northeastern Air Defense Sector (NEADS), asking to scramble F-15 fighters after the hijacked American 11 heading toward New York airspace. With the transponder off, NEADS personnel spent a few minutes searching the radar for the American 11's primary radar return. At 8:46, the terrorists crashed American 11 into the north tower of the World Trade Center.

Area C controllers helplessly watched the primary target on the radar turn to "coast," showing that it was not associated with an airplane any longer. At 8:50, the NEADS personnel were still searching for the target on their radar. Area C controllers assumed American 11 had gone down somewhere in the New York airspace. Peter, badly shaken, was pulled off position immediately. The in-house procedure was much like that for an operational error or an accident investigation. Retrieve the tapes of controller-pilot conversations and the radar computer data that retraces the movements of the airplanes. Find out what happened. Have involved personnel make written statements about what they saw and heard as soon after as possible. Peter and the supervisor left the area for the investigation. Greg worked on. He recalled: "I was so angry, angry right away. I don't remember what I did but others said I was slamming my fist on the console and yelling. I felt so invaded, threatened. I was so angry and pissed that they would do this to passengers of that airplane. We were helpless. Had to watch and do nothing. I was horrified." Some controllers on break who'd been watching TV in the cafeteria hurried to the control room to report that a plane had crashed into the World Trade Center. Although the news reported that it was a small plane, Area C people immediately grasped that it was their airplane. They took it hard. They all had heard the hijacker's voice. They were all shocked, talking to each other, comforting each other. But still working airplanes, people went into "do your job" mode: "When that fight or flight response kicks in, we learn to work through it." Word traveled quickly throughout the control room.

Meanwhile, United Airlines Flight 175 had departed Boston Logan at 8:14. Also a Boeing 767, it too was going from Boston to Los Angeles. United 175 was handed off from Boston Tower to the TRACON, then through the same sectors in Area C. Greg talked to the pilot, Peter talked to him, then handed United 175 off to the Center's Area E. United 175 was heading southbound, toward New York Center's airspace. American 11 by then had gone off course, taking a sharp left-hand turn, heading south. Controllers in Area E asked United 175 if he could see that aircraft. United 175 spotted American 11 and reported its location to an Area E controller. Area E soon transferred United 175 to New York Center. A routine flight. No problems while in Boston Center's airspace. However, at 8:41, after United 175 checked in with New York Center, the pilot reported hearing a suspicious transmission on the Boston Center frequency, the one threatening any "stupid moves" that was broadcast by speaker in Area C at 8:34. Then United 175 was not heard from again. No radio. The transponder code kept switching and then disappeared from the radar.

At 9:03, the Boston Center quality assurance representative, Bob Jones, reported back, verifying that the first transmission heard from the American 11 hijackers was "We have some planes." Biggio, the operations manager at the watch desk, relayed this information to the Command Center's teleconference line. Biggio then innovated a second time, sending an unprecedented message to all Boston Center controllers to announce to all aircraft on their frequency to heighten cockpit security. He followed with a message to the Command Center, suggesting they send out a cockpit security alert nationwide. A second or two later, the terrorists flew United 175 into the south tower of the World Trade Center. In the subsequent confusion at the Command Center, the Boston Center operations manager's message never made it to its destination, the desk of Command Center's national operations manager.

Peter was sequestered in the NATCA office. Once off position, events really hit him. He broke down, started crying. He was shaking. Greg, also relieved from his position, had gone upstairs to call his wife. He felt weak, very upset. He went downstairs to the cafeteria just in time to witness, with other controllers on break, the second plane plunge into the tower on CNN. Everyone was overwhelmed, filled with horror mixed with disbelief. With the exception of the controllers who heard the hijackers, there was a noticeably different emotional response between the people who were working traffic and Boston Center controllers who were not in that morning, because they had the day off or were working a later shift. Those who were working were stunned, upset, even devastated, but they controlled their emotions to do the work. Those who were at home had an immediate emotional reaction. Tom, a controller in Area C and a member of the National Critical Incidents Team, was one of the many not on the morning shift that day. In his reaction, we see that the usual defenses against awareness of risk and human life that controllers have in place while working air traffic were not operating:

> I was at home on my day off. I was watching TV. And they broke away from the *Today Show*, about ten minutes of nine [o'clock], saying that a plane had hit the World Trade Center. I said, no. It's a beautiful day, you could see a million miles. I said, this is a deliberate act. This is some madman that is hell bent on killing himself. And I'm assuming it was some kind of smaller aircraft. And then I saw the other one go in. And all I could think of was—something that controllers never ever think about. I never think about it, we even talk about it from time to time, how you never think of the people on the airplane when you work your job. Ever. It just doesn't work that way. All I could think of was, there's people on that airplane. And somebody flew it into that building. On purpose. Another one. And so now there were two airplanes.
>
> Well, I started my own little critical incident reaction right there. I started to tremble and shake as the minutes—after about ten minutes, it just—you know, they kept showing the images over and over, the plane, different angles, the explosion, the engines coming through the building. After about ten minutes I was shaking so uncontrollably that, you know, I finally had to get up and leave. I can't watch this anymore. My cell phone is ringing. And it's from the union president of the local here, and he says, you got to get in here. He says those were both 767s out of Boston. One was American 11, the other United 175. I said, shit! And that was just—whoa, OK. I hung up my cell phone and I started to cry. It took me ten to twenty minutes before I could get out the door. Had to take some deep breaths. I was like, I could barely hold it together. I had to collect myself and quickly. OK. So I did that. Got in my car and headed out.

Like Tom, either spontaneously or because they were called, many controllers began the drive to Boston Center to help.

Responding to this second hijacking and crash, Biggio and Dan Bueno, the TMU manager, again improvised, skirting the bureaucracy a third time. Two from Logan. Would there be more? Together they decided to impose a ground stop: no more departures for the New England Region. If any more terrorists were on planes at Logan or anywhere else in the region, they would not be allowed to take off. Improvisation and standardization combined as they drew from their base. They already had the technology and procedure to do this. Ground stops routinely were used at terminals during thunderstorms or heavy congestion, and they had the ability to communicate the message quickly to all facilities in their airspace. By computer, Bueno issued a general-instruction notification to stop all departures from all airports in the region. Biggio informed the Command Center and centers bordering the Boston airspace on the teleconference line. New York Center soon followed, stopping departures from all New York airports. Then New York declared "ATC Zero," closing New York airspace. No planes could depart, arrive, or travel through their airspace. ATC Zero is a procedure for extreme situations—power outages, weather, radar, radio, or computer failures—that prevent controllers from controlling air traffic safely. It, too, had been used before. Now it was deployed for a new kind of extreme situation: national emergency.

The Command Center

Two horrific, devastating disasters and the open-endedness of the statement "We have some planes" began a series of unprecedented decisions at the Herndon Command Center that affected the entire system. At 9:08, the national operations manager, Ben Sliney, on his first day in a new job—relayed to all facilities in the nation that New York had gone to ATC Zero.[2] Then he closed all New York airports. Shutting down the multiple airports and high-traffic New York airspace immediately backed up the system nationwide. All airborne traffic destined for New York had to be diverted. Next, following Boston Center's lead, Sliney ordered ground stops for all Washington and Cleveland airports, the other two regions ringing the New York airspace. Then began a series of unprecedented actions. At 9:25, Sliney put in a national ground stop, a decision that was a natural progression of ground stops immediately before and the ongoing discussion with Linda Scheussler, the tactical operations manager, and the Command Center first-line managers and area managers gathered around them, all putting together information from their different sources. In the ten minutes between 9:30 and 9:40, a stunning sequence of events occurred. The Command Center learned that Cleveland Center had a suspicious transmission from United 93; that American Airlines Flight 77, which had disappeared from Cleveland Center airspace, was heading toward Washington; and then, that American 77 had crashed into the Pentagon. In response, the new national operations manager, independently and under his own initiative, and having advice and information in steady conversations exploring options with the Command Center managers, made two more unprecedented decisions. At 9:43, Sliney issued an order that no international flights be admitted to the US airspace. Immediately following, at 9:45 he gave the order to land every plane in the air over the United States at the time: all aircraft were to land immediately at the nearest suitable airport, regardless of destination. With this last of his series of unprecedented actions, US airspace was closed to both domestic and international traffic for the first time in history.[3] The entire system had gone ATC Zero.

Those in charge passed on the decision on the teleconference line to FAA headquarters in Washington, the Department of Transportation, and other relevant agencies.[4] Two Command Center controllers, assigned to related tasks in different parts of the Command Center control room, described what was going on behind Sliney's official announcements. Mark was working at the severe weather desk in the front of the room, directly in front of the giant, ten-by-seventeen-foot screens. The coincidence of the beautiful day across the country freed up the severe weather desk controllers to play a major role. Mark described it:

> I was the coordinator on severe weather that day. And, it was a very strange situation. It was horrific. I mean, we got in on the phone call when they first identified this one plane out of Boston, it was American 11, wasn't talking to anybody and made this big turn which was unusual. All these facilities were on [the teleconference]. I think we had Boston and New York Center, and New York TRACON. Talked with them about what this airplane was doing. And nobody could talk to him. Kept descending. And he kept—finally wound up in New York TRACON's airspace. We got Newark and JFK Tower up [on the teleconference]. And started asking, "Do you see this airplane? He's descending pretty well." And then, you just hear this girl, and I think she was from Newark Tower or, not Newark Tower, I think she was from JFK Tower, saw it, just say, "Oh my God an airplane hit the building." And she started screaming. Just lost it. And, so it was pretty traumatic for like the next thirty minutes. Everybody trying to figure out—I mean, we all knew what had happened in our minds, I think.
>
> Our primary source of information here, believe it or not, is CNN news and World National Weather System. There's a small TV up near the national operations manager's desk. We have the capability on the floor to switch that up to the big screen up in the front of the room. We immediately turned CNN on, but there weren't any reports yet. And then the reports came out that they thought it was a helicopter or something. And we all knew, again, that that wasn't the true story. And we knew what was going on. There was a lot of in-

formation coming in. And then there was another airplane that was reported that New York Center was not talking to that had gone below their radar coverage. A lot of pieces here and there of information. So the first tower was smoking, so they put it up on the big screen. So, when the second came in, everybody saw it.

And then, so, I think, for at least the first half an hour, it was just—it was shock and trauma and "Wow, what's going on here?" And then we started getting tasked with some things. Start rerouting airplanes around New York Center. And then they went into, we're going to ground stop everything going to New York Center. Well, we in severe weather started taking phone calls from Indy Center. "We got an airplane we're not talking to." Chicago: "We got an airplane we're not talking to." Denver: "We got an airplane we're not talking to." And so, we had this whiteboard out there. I was writing down the call signs of these airplanes. At one point, we had twenty-two airplanes written down there that we weren't talking to. We had kind of a spontaneous stand-up briefing. Talked about what was going on here. And now two more planes had gone down, into the second tower and the Pentagon.

And I just brought up, "Look, I got twenty-two airplanes down here on the board and I can't tell you what they're doing. Now some of these are LA to Florida flights. I mean, we don't know if they're going to get to Memphis and take off for the Pentagon. We don't have any way to keep track of it." The decision was, we just get everything that's in the air on the ground. And they'd already agreed to stop everything that was on the ground and leave it there. They took that and I don't know who they—management—talked to, but it was less than five minutes later, they came back and told me, "Just get everything that's in the air on the ground." (Command Center, M)

The whiteboard was a device for working it through to make a plan. Tom, the controller who was at the Command Center's West position, for all regions west of the Mississippi River, organized a teleconference with all the regional centers, and a second telecom with the airlines to pass on the orders for the national ground stop. Tom's unprecedented announcement drew an unprecedented response:

I had a national teleconference which brought up all of the centers, Boston Center, New York Center, Los Angeles . . . and brought them all on line. All traffic in the continental United States is now ground stopped. In other words, traffic that's sitting on the taxiway is not allowed to take off. All airborne traffic is ordered to land at the closest available airport due to a national emergency. And the strangest thing happened that day. When I've done national conferences before, you have upwards to twenty centers on there, twenty people. No matter what you broadcast, there's always somebody who says, "Say that again?" or "What's this?" or "I didn't copy that." That day, I made the announcement, there wasn't one question on there. Everybody—all you ever heard was, "Roger," "Roger," "Roger."

Then we conferenced all the airlines and gave them the same spiel. We just asked everybody in the FAA to take every airplane that you've got and put them on the ground somewhere. I don't know where they're going to wind up. I can't guarantee that it's going to be at an airport with the facilities you need. But we've got too many airplanes that we don't know where they are going. This is costly for them [the airlines], later, to get them back to where they are supposed to be. And everybody said "OK." Not one piece of grief about it. (Command Center, M)

In the national emergency, consensus on the common goal overrode the usual boundary problems that created endemic turf wars on the ground and in the sky. Controllers at the Command Center remarked afterward that they had never seen the system work so well. Everybody was doing whatever they had to do to make it work. Jim, the Command Center NATCA president explained:

It was instant understanding. I mean, I think that that spread throughout the world. We know exactly why you're doing what you're doing and we don't have a problem with it. Yeah, it's not going to make us money, it's not going to do us any good, it's going to hurt all of us, but it's perfectly legitimate. (Command Center, M)

The tools they worked with every day—the rules and procedures for moving airplanes and for coordinating traffic across airspace boundaries—were basic to accomplishing this enormous task. Ground stops, ATC Zero, closing airspace when an air sector or tower was overloaded or in emergency were routine procedures in local conditions. Now, for the first time, these same practices were applied to the entire system. Moreover, they already had a technological communications infrastructure they could build on. The Command Center and all operational units were practiced with strategic planning teleconferences between all center TMUs for regulating national traffic flows. In fact, they had practiced these same connections the year before for a new program, the Severe Weather Avoidance Program, an innovative system in the making for standardizing reroutes to reduce the chaos in the skies during thunderstorms. In addition, they had technological "bridges" that could be opened up between all FAA facilities and with the airlines. They had "bridges" with the North American Air Defense Command and all their sectors and the military, the Federal Bureau of Investigation, the Secret Service, and the Capitol Police. They already knew the key players and their contact points.

On September 11, they built on experiential knowledge and existing formal and informal practices. They patched together existing bridges as the number of places and planes and people and agencies involved expanded. Event by event, bridges were opened. When they got information on a particular flight or something suspicious going on, it would be reported on this hotline and communicated instantly across the entire system. Expanding or contracting system boundaries was routine. When the Command Center closed the national airspace to all international traffic, it was a simple matter: contact all centers with international borders. Centers on the northern and southern borders informed Canada and Mexico. West Coast facilities contacted Pacific Rim countries and told them not to send anyone. East Coast facilities contacted facilities in countries on the other side of the Atlantic oceanic tracks. In a great moment of boundary spanning, country by country, word had spread through the international air traffic system. Distributed cognition, normally associated with small groups in which coordination among members is visible, expanded to encompass nations and pull them together in a shared effort.

Getting the planes out of the sky was a monumental task never even conceived as a possibility. It proceeded incrementally and quickly. Typically, there are about five thousand planes in the air during rush hour traffic. However, the timing of the terrorist attacks had an effect. Given the time difference on the West Coast, departures had just begun when the national ground stop was imposed at 9:25, eliminating all departure traffic. So the total in air at the time of the attacks was down a little, at 4,395. Moreover, by 9:45 am, when the US airspace was closed, controllers had already brought down close to one thousand planes: between 9:00 and 9:15, first American then United independently had ground stopped their fleets and ordered all in the air to land. Then when New York went to ATC Zero, Boston, Washington, and Cleveland had begun landing the planes New York turned away. When Sliney closed the national airspace at 9:45, the computerized Traffic Situation Display (TSD) of all the airborne traffic in the United States showed 3,395 planes.

As controllers began bringing the planes down at an accelerated rate, the

Command Center's severe weather desk began sorting through the list of the twenty-two planes on the whiteboard that were out of communication with controllers. They called, facility by facility. Mark recalled:

> "Have you heard from this plane?" "Yeah, we did hear from him, he's back with us." And at that time, there was still the unknown. We didn't know what was going to happen. And when we finally got done, the one plane that was left was the one that made the U-turn and went into Cleveland Center, came back and crashed in Philadelphia, out in the field. It was just a little after 10:00 am. What would be next? Keep in mind, we didn't know if somebody was going to declare war on the United States. We thought it was a precursor to war. (Command Center, M)

Boston Center

As some airspace boundaries closed, other boundaries opened up to absorb unaccustomed traffic, changing the distribution of work across the system. As New York went to ATC Zero, Boston Center's Area E, normally feeding into New York Center airspace, was immediately swamped, putting traffic on spin and rerouting planes around the New York airspace. Traffic in the other Boston Center areas backed up because of Area E's extra load. The effect rippled back to the TRACON, Boston Tower, and Bedford. At all altitudes, planes that had just departed were returning to Logan. Others were directed to land at other towers in the region. The big problem was about 250 international jets, airborne from Europe, coming across the Atlantic to the United States on the oceanic tracks. Those planes that were midway from Europe turned around and went back. Internationally, the system expanded to accept more traffic. Canada accepted those closer to the US boundaries. This created a huge strain on the Canadian system, which was not designed to handle so much traffic. Canadian controllers directed them to land, but soon airports ran out of places to put them. Airfields got full, and planes were parked in rows on runways and alleyways, dragged there by tugs after landing. Some large planes landed in airports too small to handle them and with runways too short for a takeoff for large international aircraft. A large number of planes landed in Gander, Newfoundland. About 5,000 passengers deplaned in a town of 2,200. People put them up in their homes.

The controllers in Boston Center's TMU, the main link between the Command Center and managing the traffic flow in the region, were deluged with rerouting. The work was labor intensive in terms of figuring out where to send the airplanes, especially for the high-altitude sectors, which was where most of the rerouting was being done. Controllers who usually griped about TMU decisions because they doubled the workload never questioned; they just worked the traffic as it came to them. Terry Biggio, the operations manager, described it:

> After three planes, we weren't sure where the end was. Imagine what it was like. We had hundreds of planes in our airspace. We had no playbook. The other side of the mic [pilots] had never done this before. The model of hijacking they had before was very influential. Hijackers took over the flight, asked for something and a safe landing, they landed the planes at some destination they requested, and the passengers walked off. Air traffic controllers had to get a buy-in from pilots, and it was difficult to get the message across. You are not leaving our airspace, pick an airport, contact your company, tell us where you are going to land. We had to use our best-guess estimates as to how we are going to react to each and every one of those planes out there, decisions made under warlike conditions, with unbelievable precision. It was an incredible feat by our workforce.

Whereas previously, most controllers had told me that being busy was not stressful, they all agreed that this day was stressful, busy or not, because controllers experienced it not only as a loss of life but as a loss of control. The concept of ownership, so central to boundary issues and everything air traffic controllers do, was threatened and lost that morning to terrorists. Everyone at Boston Center was reeling from knowing it was two of their own planes that crashed into the World Trade Center. Stress combined with grief. The task was made more doable by routine procedures—reroutes, descends, handoffs, holds—and by the ground stop, which effectively cut traffic in half. It was the uncertainty. American 77 and United 93 had crashed. No one knew what was going to happen next. Carol, in Area B, described the response in the control room:

> I think that it broke everybody's heart, and I think everybody was physically just a wreck over it. However, it was amazing to see how well people held on in there. . . . But I think that we are just so programmed to—we have to do this. As much as we goof off in there and as much as everybody says it's such an easy job 90 percent of the time, everybody in there knows what we're do-

ing and how important it is. And it has to go, it has to keep going no matter what happened. You know, everybody's hearts were aching and a horrible pit in your stomach. Just like the rest of the country. But it had to keep going, you know? (Center, F)

Tom, a NATCA official, was coordinating, moving between Boston Center's areas and TMU:

> When everything was going down on 9/11, I had firsthand knowledge and understanding of how chaotic it was outside the control room. TMU was going through a bunch of things. Walk down the control room aisle—and you've been on the floor for a number of thunderstorm sessions—you would not be able to tell the difference between a thunderstorm day, heavy-traffic day, a hijacking, a terrorist attack on this country . . . You know, my worst feeling that day was when I was off position. Sitting back here, or up at traffic management, or going between areas, coordinating sector to sector, was when you could feel the terror. But when you put the headset on, you come into a position, it's business. People are still flying airplanes. You have a job, you do it. (Center, M)

Areas C and E were especially hard hit emotionally. Everyone in Area C was traumatized by having heard the hijacker's voice. Several Area E controllers had talked to United 175 before it was hijacked. But Area E was also experiencing a personal tragedy: the wife of one of the Area E controllers was a passenger on American 11. Andy's wife left for Logan Airport at dawn that morning to fly to Los Angeles for a business meeting. He was assigned the noon shift but came in early to help because of the attacks. A sitter had taken the kids to school. When Andy woke up at nine, he heard the news and saw the second plane hit the World Trade Center. He didn't think of his wife because she was going to Los Angeles from Boston, so would not pass through New York airspace. He arrived at work as controllers were clearing the skies of all aircraft. A controller working with him the night before knew she was flying American 11 next day. When it went down, word spread in Area E that she was on it. The Area E supervisor, two of his Area E coworkers, and the NATCA vice president were waiting for Andy at the door. They took him to a secluded room to break the news. Some controllers went home with him. One of them drove to Maine to pick up his parents and bring them to his house.

In quick succession came two more crises at Boston Center. Emptying the sky continued at an astonishing rate. At 9:59, when the south tower of the World Trade Center collapsed, the TSD showed 2,651 airborne planes. Nationwide, controllers had landed almost a thousand planes in the first fifteen minutes after the order was given. In the following fifteen minutes, the number flying was reduced to 1,695. By the time the second tower collapsed at 10:28, there were 901 aircraft remaining in the skies over the United States. Nearing 11:00, Boston Center's Day Care Center next door had a bomb threat. A few controllers rushed over to help, grabbing cribs and children. Almost simultaneously, Terry Biggio had a call from the regional headquarters. A small airplane, an unidentified target, had been spotted on the Manchester TRACON radar moving at a low altitude down the Merrimac River toward Boston Center. It was close. The events of the day gave them every reason to believe Boston Center was under attack. A lot of people had been clustered around the operations manager at the watch desk, listening in on the teleconference, including the facility's air traffic manager and NATCA president. When this message came, it was chaos. Some ran to tell people in the administrative offices upstairs and the technical support downstairs to leave. The air traffic manager ran down the control room center aisle shouting to each area, "Unplug and run, unplug and run." They were not told why. They were trained on ATC Zero procedure each year: instruct all aircraft to descend to eighteen thousand feet, the altitude at which pilots can operate on VFR so they can guide themselves down if necessary; announce on the frequency that the facility was going to ATC Zero; transfer planes to a frequency at Boston TRACON. But they had never had an evacuation rehearsal.

At about 11:00 am, with 367 airplanes still in the sky, Boston Center controllers evacuated the building. At this moment, Boston Center's airspace was nearly empty. After telling pilots to descend to eighteen thousand and contact the TRACON, some walked but most ran out the control room doors and into the parking lot. The NATCA president had phoned ahead to the small Holiday Inn a half mile down the road to make arrangements for them to gather there. Everyone got in their cars and drove the short distance to the hotel. Someone from security had apparently called the police, because they were already stopping traffic for them on the narrow road. FBI agents, who had driven to Boston Center in a hurry to listen to the tapes of the terrorists, fled the building with the tapes and the supervisor and NATCA president who had been with them as they all listened together. Everyone crowded into the Holiday Inn. Some were directed into the courtyard garden and pool area.

Some were sent to a large closed room, a bar called the Bounty, designed as a mock-up of a pirate ship. NATCA officials located the Area C controllers who had heard the hijackers in a room of their own. The word most frequently used when people described the evacuation and the return to the facility after was *surreal*.

Mandy, who was on break and had walked into the cafeteria just as the second plane hit, describes it:

> It was bad—I don't even know how to word it. It was surrealistic. And then Andy came in, and they ushered him through the door. That was bad. Andy's wife was on American 11. It was a bad day. . . . There weren't that many flying in the skies at that point when they said, "Get them down." So we'd already sent a lot of people back that were departing in the area. And then the other ones, you just cleared them through into an approach control. So it was kind of intense but kind of not that big a deal either, at that point, compared to everything else happening. And I was kind of spaced out at that point, too. I had a few military planes left when the ATM came running down and told us "unplug and run." I broadcast [on the frequency] what was happening and for them to contact their military base. I looked at the supervisor, and he said, "Do it." And so we both unplugged and ran out of the building as fast as we could. (Center, F)

Keith had missed the ATM's announcement in the control room:

> I had gone down to the cafeteria when I had a break—after I found out the airplanes had been hijacked, I went down to the cafeteria and I was watching it on CNN, like probably everybody else in the world. And it must have been sixty, seventy-five people there. When I went back to the aisle, the supervisor said, "Why don't you round up everybody that's here and bring them back and we're going to figure out what we're going to do." The first thing I did was I went back to the cafeteria. This was like a minute or two later. And there was nobody in the cafeteria. It had gone from like seventy-five people to zero. "Jeez, everybody knows something that I don't." And they'd begun to evacuate our building. And it was one of those almost *Twilight Zone* experiences, in that I panicked. I was here a minute ago, and I know that there were seventy-five people here. There wasn't a single person in there. I got out of there. Everyone was driving out. Then there were police out there that just stopped traffic and waved us on. It was a very surreal thing. (Center, M)

A half mile down the road at the Holiday Inn, after the hijackings, the Area E tragedy, the bomb threat, the lone plane heading down the river channel toward them, and "unplug and run," sitting around in the sun in the garden and pool area and in a bar like a pirate ship had to be a bit surreal. Carol reflected:

> People were somber. Yeah, I mean, a little overwhelmed by the news. Maybe a little shell-shocked, but nobody overtly emotional, sobbing or anything like that. It was very orderly. Kind of like, sitting there at a car accident when you're sitting around, watching it, talking about it with other people. It wasn't real. I don't think it really had affected everybody right then. People were sitting around talking. You know, we're a bunch of problem solvers. People were saying what they knew, trying to put the pieces together. Of course, being controllers, somebody made a joke about the plane heading toward the center, "Well, this will be the one pilot who's off course by a half mile." (Center, F)

After a while, everybody moved into the Bounty. They had been cleared to go back to the facility. The unidentified small aircraft that was flying down the Merrimac toward Boston Center had disappeared. They never found out what it was. At 11:30, the TSD showed twenty-six planes left in the airspace, most of them military. They asked for volunteers, three for each area, to stay. Everyone else was told to go home. Tom, the critical incidents specialist, stayed behind to talk to Peter:

> I met up with Pete who was working the hijack, so he's who I wanted to see first. And we're walking through the parking lot next door, and you know, I have a tear coming down underneath my sunglasses. Because it was all still so fresh. You know? I was still in a place where I wanted to be comforted and taken care of. But when you're a caregiver, you can't do that. And actually, you know, when you help people they help you. And so talking to Pete, I began to feel better. I spent a good hour with Pete. We were sitting on a guard rail in the parking lot. He was so keyed up, he couldn't remember exactly what happened when I was first talking to him, which is very typical of trauma. While this was going on—and he's a nineteen-and-a-half-year veteran at the time—he said, "I don't want to do this job anymore."
>
> And believe it or not, that's a pretty common reaction. Even people that go through a bad accident without any fatalities. . . . Initial reaction, you know, fight or flight, forget it. "I'm fleeing." That's fine. Take as much time as you

need, but don't make any kinds of judgments like that right now . . . Those are the kind of things you say to anybody. And that's what I said to Pete, just to get him through. Make sure that he had people when he went home, because I know he lives alone. He called his parents and his parents were coming over. And Carol is a good friend of his, spent the night over there. You want to make sure people are taken care of. So I was on the phone with him, day after day after day after day . . . There was Pete, there was Greg, but everyone in Area C heard the hijack because it was on speaker. I set up a debriefing for Thursday. Whoever wanted to talk to me. (Center, M)

The following week, Tom would go to Boston Tower and TRACON, then with other members of the National Critical Incidents Team, on to New York, Newark, and Cleveland.

Those who volunteered to return to Boston Center described the return as one more strange experience:

And then we drove back. And that was the spookiest thing. That was probably the first time that was something that was really spooky. Leaving, you know, you're in your own car and you know, granted, everybody is leaving at the same time, and they're stopping traffic, which is a bit unusual. But still, that's just you driving to the Holiday Inn. Which I've done for union meetings. And when I was younger, used to go there after shifts. But we're now driving back into work and there's a guy with an M-16 at the front gate. You know, there's a national cop there in a SWAT uniform with an M-16. So that was the first time that it was really—that was the first bizarro thing. That's something you don't see every day. And up to that point, our guards never had guns. Now there's a national cop with an M-16 standing at the front gate. (Center, F)

Then they walked into an empty facility. Three shifts around the clock. Always, controllers are on position and those coming in replace them. Now, three or four controllers walked into each empty area and opened up one or two sectors. When American 11 and United 175 slammed into the World Trade Center, 4,395 planes were aloft. In two hours and fifteen minutes, from 9:15, when American and United ordered their fleets to land, to 11:30, controllers throughout the system had brought them all down. At 12:16 came the official announcement that the airspace was clear of all civilian aircraft. With no airplanes, controllers had nothing to do. The military had taken control:

> We just basically gave them the sky. It was still our airspace and we still have sectors and responsibilities. And there were F-15s out of Syracuse and we just said, "Well, you can do whatever you want, we have no airplanes. And the only civilian airplane we worked all day was George Bush Sr. coming back to Kennebunkport from somewhere. He was the only civilian traffic. I think I was there eight hours. I probably worked the position like fifteen minutes, that's about it. It wasn't really anything. We just played cards. Got out a deck of cards and started playing cribbage. And people trickled in for the night shift. We had no interest in watching TV. Couldn't take it anymore. (Center, M)

From this beginning skeleton crew of volunteers, the numbers of controllers on duty quickly went back to normal when the next shift came in. The events of the day had changed their perception of collective risk. They talked openly about it:

> At first they told us we only needed three people for the night shift. We found three people for every area. Was able to find people that volunteered to come in, knowing what was going on, knowing that we might be a target too. And everybody else was told to stay home. And then George W. Bush decided to be tough and said, "Business as usual. Tell everyone to go to work." Even though there was no traffic. There wasn't a plane in the sky except for the military. So, you brought everybody in here to be a target. So now we're sitting in this building and all this stuff is going on and we're thinking, "Jeez, are we a target, too?" It was very stressful. Then the next day was the same. Everybody sitting in the aisles [areas]. There was no traffic, but there was like twelve, thirteen, fourteen people all sitting in the aisles. It was terrible. There were only two scopes running in our area. It was like a ghost town. There were plenty of people working, but there were no planes. (Center, M)

The unprecedented experience of an unidentified plane hurtling toward them and the hurried evacuation had an impact on everyone, including those who were not working that morning. Previously, most Boston Center controllers, as at the other facilities, had said they did not experience their work as risky. They had many reasons for this, among them that they did not personally feel at risk. They could not be hurt. They had not ever thought about the possibility of attack. Boston Center was in a rural location, which made them invisible to the public. Now, controllers there believed they could be a target. They believed that they, as individuals and as an occupation, were at risk:

> And we were scared, too. You came in September 11, there were armed guards at the entrance. You know what I mean? It was horrible! So when I saw those guys with guns at the front entrance I went, "Oh, man, they're thinking like I'm thinking. That's not good. I want to turn around and go home. I don't want to be here right now." (Center, M)

Ironically, these feelings of personal and occupational risk were reinforced daily by the new security at the center. The armed guard with the M-16 was only part of a new routine. The police were at the front entrance, checking everybody's cars. Controllers, administrators, and all staff were required to have and show new identification badges. Before, they simply flashed the badge and were waved through the front gate. Now they were asked to step out while the car was carefully searched and sniffed by dogs. And for that first week, once they were in, they were in. They couldn't leave the facility for any reason. No one was permitted to go outside on break, whether to sit, walk, or jog around the fenced-in property. No visitors.

From Tuesday to Saturday, controllers watched a sky empty of all civilian aircraft. It was clear, even at that early stage, that their work as controllers had changed:

> I worked September 11 and I worked September 12 and I can remember coming back into this building on September 11 and it was eerie. There were no con trails and there were no planes, it was just military traffic flying around and MedEvacs going in and out of New York. And basically we were passing messages to the military telling them who these planes were or else they were going to shoot them all down. (Center, M)

The first planes began flying Saturday. The US airspace was still officially closed, but waivers were issued that permitted certain flights to take off. Saturday night during the midshift, the stranded internationals in Canada began entering the United States. Those commercial airliners were the first to enter Boston Center's airspace. The protective device distancing risk by thinking of airplanes as call signs and data blocks was in abeyance. There were people on those airplanes:

> I worked the midshift that Saturday night and Sunday morning. All the planes were recovering from Canada. They were recovering back to Kennedy and stuff. And when I was talking to pilots I was getting—if I can find a word for it,

> sentimental, I guess. As I'm talking to pilots and realizing that this is just more than a just a job. There's so much humanity. And I'm just by myself in the area. And as they checked in, I'd say, "Welcome home." And the pilots are, 'Oh, it's so good to be going home.' And I'd chitchat a little bit with them. There were probably half a dozen that came into our airspace [his area]. (Center, M)

It was a sentimental moment, but it was quickly overshadowed in the new reality of post-9/11 air traffic control. Everything had changed—including dead reckoning, because now risk was at the forefront of their minds. The country was at risk, airplanes and passengers were at risk, and controllers, too, could be at risk. Controllers brought a different frame of reference to the interpretation of anomalies in the sky:

> The next day, some of the international stuff began flying out. I remember a Delta didn't call me that was coming over Kennedy, headed up towards Maine on its way over to Europe and he didn't call. And in the past it was no big deal, but he's not answering me. New York Center said they switched him and I said, "Well, I'm not talking to him." Fighters that were circling over New York peeled off and followed the guy. And when he saw the two F-15s go by him, he woke up and started looking for what frequency he should be on. And that happened quite a few times. So now, if somebody doesn't answer you, you wonder. (Center, M)

The TRACON

At Logan, the first information in the TRACON came from Boston Center when Peter was looking for American 11. "Would you see if you have American 11 on the frequency? Did he come back to you?" The TRACON controller tried calling American 11 several times with no response. About five minutes later, Boston Center's TMU called the TRACON watch desk, telling Dan, one of the two supervisors on the shift, that American 11 had been hijacked. Then they called again to say they'd lost contact with the pilot, who was heading toward New York. Dan didn't announce to the controllers because American 11 was not on their radar. Also, even though a hijacking was bad, the typical pattern for hijackers was that they made demands, landed somewhere, and when their demands were satisfactorily met, passengers disembarked unharmed. Better not to distract his people, Dan thought.

Then the controllers in the break room called him in to look at CNN. Dan remembered:

> I went there and first they said a bomb went off in the World Trade Center. And I'm looking at the World Trade Center and I'm thinking, uh-uh. That's American 11. Too coincidental. I said, "That's not a bomb, that's our airplane." At that point, the controllers had no idea there was a hijacking. Then CNN was saying eyewitnesses were reporting an airplane hit it. (TRACON, M)

The supervisor announced to the radar room that one of the towers of the World Trade Center was on fire, that there was a big gaping hole in the top of the tower, and that it could have been caused by American 11. Pulling up the flight's data block on the TSD display, they saw only "coast" in the vicinity of New York City, showing the radar was not associated with a target any longer. Boston Center called again, reporting that United 175 was being hijacked—the same thing had happened: the transponder was turned off, the airplane had quit communicating with them, and then it turned toward New York City. Another one of their airplanes. Dan went back into the break room to watch the news and saw United 175 go into the south tower. The news had no details, but controllers knew where the planes came from and which planes they were.

The teleconference line was open at the TRACON watch desk. The line was turned very low, but everyone in the room could hear what was going on. The telephones started going "nonstop crazy." Brian, another controller on duty, helped with the phones. He had been working initial departure, so had talked to pilots of both the hijacked planes. And whereas during those moments, when all was routine and uneventful, he might have thought of the exchanges in terms of call signs or pilots or airplanes, the air traffic cultural guards against acknowledging risk and harm to human life were shattered:

> I was actually thankful to not be on position because my heart was racing. Normally I feel like I like to be in the action. But that was more action than I'd ever hoped for. That was the action that none of us every really wanted to be in. The thing that I remember feeling so chilled about afterwards, I worked United 175 also, he came out off of 9, and there was no traffic in his way, I climbed him straight up to 14 and when he got out about eight miles from the airport, I turned him to [heading] 270 and shipped him to the next controller.

> I had very little communication with him. But both airplanes, I said, "Have a good day." Which I say to all airplanes pretty much, [it] is just a habit I have developed over the years doing this. And I felt very creepy. You know, I said, "Have a good day," and minutes later these men's lives ended in this horrible way and I had to believe that they were coerced out of the cockpit and they had to know what was coming. They're very low, they're going fast, they're right over downtown New York. It's just, it's still, it makes the hairs stand up in the back of my neck to even think about it.
>
> The whole thing was just unnerving. Watching it. The image of the airplane crashing into the building. United 175. It's gruesome. It's an entire nation of people that have seen something that, because it was on TV we witnessed something so horrible and horrific as—at that moment, how many people died? All the people on the airplane. And how many in the offices? Who knows. And the thing that was so sick about it was like the building looked like it absorbed the airplane. It was like the airplane; it was like flying into a waterfall. Before it blew up, it just like it absorbed into the building. And I thought, what must it be like on the plane? What must it be like sitting in the seat and all of a sudden, the plane is collapsing toward you? For an instant, you know, you realized what's going on, you must. I don't know. I mean, I can't imagine what it's like. I wouldn't have imagined what it looked like for an airplane to fly into a building at four hundred miles an hour until I saw that. (TRACON, M)

Then the controllers on duty experienced a cascade of emotional jolts in quick succession. Over the teleconference, they learned before CNN that American 77 had hit the Pentagon and United 93 crashed near Pittsburgh. The realization of some greater plot raised, for some, the possibility that they might be at risk, by virtue of their occupation and Boston location. This, in contrast to Boston Center, where the same perception was raised by an announced threat of attack on the facility:

> Thing after thing. And I was stunned, I remember, when they said that one of the towers had fallen down, that just seemed beyond belief that that could even happen. This humongous building, absolutely massive, to fall like that. It was just a sick feeling. I just couldn't believe it. When the Pentagon guy hit, man, that was like, "Oh my God!" Everyone was going, "What the hell is going on? Are we in danger here?" And nobody really knew. (TRACON, M)

In a system where dead reckoning depended on standardization, routines, and predictability, controllers were working in conditions beyond anything they had experienced. Controllers began working fast to get the planes down as quickly as possible. Initial departure was shut down. The only positions open were the feeder positions and the final approach. They had traffic coming from every direction, traffic that they had never handled before. Many pilots were unfamiliar with the airport, requiring extra coordination. Most of the rerouting was being done at Boston Center, so the TRACON controllers' main task was to talk to each airplane and to work them into the airport pattern. Controllers were coming in from the break room and telling the people on position what was happening on CNN. Al was working final, so was the busiest controller in the room. Brian, his friend and carpool buddy, kept him updated. "I went over and sat next to him. When he stopped for a second, I would go 'one of the towers just fell down,'" Brian said. "And you know, he'd shake his head and then he'd go back to turning airplanes."

They had a sky full of airplanes to work. Steve, the other supervisor on duty, walked up and down the row of controllers telling them to keep their emotions bottled up and get the job done, keep their heads into the work, because it was hard to concentrate with everything that was going on. In addition, they were told to watch for off-course airplanes heading toward the city. Dan said:

> All we did was get them in, keep getting them on the ground. And then we were watching for—we were told to watch for primary targets, coming towards us, coming towards the city. Because we knew that there were more airplanes hijacked in the air. And nobody knew where they were going. I opened every single position after it happened, because I knew it was going to get out of the ordinary in there, so I had all the controllers in there and I told them, "We got to do this. We got to get these planes down." Because everybody wanted to go watch CNN. Everybody did. Everybody was numb. Everybody was in shock. We all knew what had happened. We all knew that those two airplanes were ours. And we also thought there were up to two more that were ours. Turned out that we didn't have any more. But everybody—to every single person in there—knew that we had to do this. We had to stay there. No one said anything about getting out. Matter of fact, people at home were coming in. "How can I help? How can I help" They wanted to help, you know? That's what we train them for. (TRACON, M)

There was a huge push during the first hour after the two planes hit the towers. By 11 am, the planes in the sky were nearly all down, except for a few still coming in over the ocean. When Boston Center evacuated, controllers there instructed pilots to contact the TRACON, but they didn't tell the TRACON they were going to ATC Zero. One of the TRACON controllers announced to the room, "I've got these guys calling me, there's two or three of them, and I don't know who they are." Dan asked, "Is Boston Center gone?" And people were trying to call and nobody was answering. The planes were still in Boston Center's airspace, so not on TRACON radar. Improvising from their base, the controllers resorted to the flight progress strips that the TRACON had stopped using some years before. Resurrecting the "manual" technique they had trained on at the Oklahoma City academy, they printed out the strips and began working Boston Center's planes without radar.

By 11:30, the TRACON's planes were down. By then, all four hijacked planes had crashed and the twin towers had collapsed. The full 12:00 shift came in early. As the next shift had approached the airport, what they saw was startling:

> Something that I'll never forget, as I drove up to the Ted Williams Tunnel, were the number of people who were walking away from the city. You know, hundreds. They were not in the streets and not running. But apparently people who had been released from work and whether because there was no public transportation in that area, over in south Boston by the civic center, or whether transportation was full, whether they just didn't want to wait. You hardly ever see anybody walking over there. And this is quite a few people just walking away in business clothes, obviously just came out of an office or whatever. Leaving the city. And we were going opposite to that, going in. All the roads out of the city were full. As if it was rush hour going home. (TRACON, M)

With all the TRACON's civilian traffic down, the controllers had very little to do. Dead reckoning consisted solely of new security responsibilities:

> It was very quiet. Solemn. A lot of people were watching tv, trying to get updates on what was happening. Of course there was almost nothing for us to do. Really, all we did for that day and for a few days afterwards, was look for targets that aren't supposed to be there. And generally, that meant most everything because the only flights that were going anywhere were military and

after a while some police or state police helicopters or whatever, and medical emergency flights. (TRACON, F)

The FBI came, with guns, to listen to the tapes of the conversations between controllers and the pilots of the hijacked airplanes. The controllers who had talked to them also listened. Hearing the pilots' voices was emotionally wrenching. There was some small relief in hearing that the controllers' phraseology was perfect, no mistakes, no missed signals, nothing that indicated that controllers could have prevented it if they had acted differently. Their awareness of what post–September 11 air traffic control would be like was speeded by a memorable incident in their airspace that afternoon. The military scrambled two F-15s on a state police helicopter whose flight had not been coordinated through the air defense command. The TRACON contacted Boston Center to stop the fighters because they knew the "target of interest" was law enforcement. One controller said, "That was scary, because you don't know if they're going to pull the trigger on him or not!"

The new security procedures, instead of making controllers feel secure, had the opposite effect. In the next few days, they watched as air traffic control and air travel, as they had known it, changed. New protections—human, material, technological—were very quickly put into place. The airport was transformed. Romaine, who had only a few months before passing her final check ride to become certified, observed:

> You drove into a war zone, basically. The military and the tow trucks towing tons of cars out of the parking garages. That's all we saw. Tow trucks and us. And military and police set up everywhere. To come in and see a guard and another guard and another guard come to your work, to your place of work, that's a horrible feeling. To look at all the cones and all this other stuff, and these new cameras in the TRACON, that show the downstairs lobby, the outside parking areas and stuff—areas underneath us where they could blow us up and stuff. You know, one day they're finding this car in the garage [that belonged to the hijackers] . . . our sense of security was just gone from our work environment. We used to have pilots call up and say, "Hey, we got some time before our next flight, could we come up for a tour?" "Yeah, sure, come on up!" That was our world. (TRACON, F)

Everything was being scrutinized by the police, and the controllers were pulled into it. In addition to searching the sky for targets that didn't belong there, they were asked to go on checks of the areas to make sure everything was secure, both inside and outside the building. Some controllers on break would walk around. If they found doors that weren't supposed to be open, they'd report it. If they found a box or bags unattended, they'd have cops and dogs check it out:

> We all became suspicious—of things that you never thought of before. Of course, we were required to become much more suspicious because of the nature of the procedures that we had to change, security-wise. But things that you never would have thought much of before, people were coming up with, "Well, what about this, what about that? How come that door is always open downstairs? Isn't that supposed to be locked? Why isn't that locked?" (TRACON, M)

Saturday night, when the national airspace opened to planes that had landed in Canada, the Command Center decided to bring them to Boston, unless their original destination was New York. They didn't come all at once. They trickled in to Logan, one at a time. While the return of civilian airplanes to the sky was significant for everyone, it was especially significant for controllers:

> I remember going to the break room window when the first one came in. It was United I believe. I had to watch them land. It was weird. I had to. You know, there was a whole bunch of us. We're just watching. I mean, we see landings—a thousand a day? Or five hundred a day? We were all lined up at the window. (TRACON, M)

However, the familiar work environment was transformed into one of suspicion and vigilance. Incidents provoked moments of real fear. As civilian air traffic was resuming, the TRACON and Boston Tower had a scare equivalent to the scare at Boston Center when the unidentified target appeared flying low above the Merrimac River toward the facility. Dan and his crew were again working. The controller working final had just handed off a Continental flight to Boston Tower. Dan, a supervisor, recalled:

> It was exactly one week after, on the eleventh [of September]. My shift again. I've never been so scared in my life. 'Cause that's—you know, 9/11's in the back of your mind. We're running these planes down from 4 Right and I'm sitting at a scope watching. And I says to Bill, who was working final, "Con-

tinental's coming off the final. What's he doing?" He's making this left hand turn off the final, towards the city. Bill says, "He's talking to the Tower." So I call the tower. Tower says he's screaming, "May Day!" And he's diving toward the city. And we were watching him [on radar], going, "Oh, my God." And I told the controllers, "Break everybody out. Get everybody away from the airport. We don't know what's going on here, but get everybody away from the airport." And we went into real combat mode then. And stopped all the arrivals coming in.

And this guy went, he was heading for the Prudential building so we thought he was going to hit the Prudential. That's what we thought. He missed it. And he was as low as I've ever seen an airplane go by the Prudential. Ever. And then he missed, he went right by the city and he's still descending. Now, five hundred feet out there is awful low for an airliner to be out there. And on the radar we're watching him and he's just coming right at the city and descending like this. Then he went around the Pru, starts coming back this way. And I says to the tower, "Is he going to make the airport?" and he said, " We don't know." I told everybody, "Climb them. Get them all high, get them all high." The guy came back around, and was at about three hundred feet, which is about the height of the tower. Came back around and landed 27.

When we learned he was screaming "May Day!" it was that time where you can't do anything. I picked up the hotline and I says, "We got a situation." I told them what was going on. And they scrambled F-15s off the Cape. Never made it here on time. Guy was on the ground before the F-15s got to us. When he was down I remember looking at my hands and they were literally like this, just shaking. The whole thing happened, ten minutes, tops. More realistically, seven or eight.

That one scared me more than anything. 'Cause of the past history and how close he went to those buildings. We thought he was hijacked. We thought he was doing it. 'Cause it was one week to the day, too. And that was a scary one. That one, we had issues with people, emotions, as it was happening. One of the controllers got up off of Rockport [sector] and tried to run out of the room. I literally had to stop him. Get back in here. Get in here and sit down and work airplanes. I physically had to block the door. He thought we should have evacuated the facility. I says, "We're not evacuating anywhere. We got all these airplanes. We're not going anywhere." We all thought the plane was crashing. Everybody in the room thought he was crashing. And we still had now a full airspace full of airplanes. (TRACON, M)

On September 11, TRACON controllers had all stayed while they were getting the planes down. There had been no discussion of evacuation. As a facility, they had never felt vulnerable. Darlene talked about how it was before:

> You know, I never in my conscious rational mind felt like we were much at risk in this job, in this place. If someone wanted to crash an airplane into the tower, which is an obvious target in my mind, we could certainly become collateral damage because of the proximity, but I never felt the average person if they were trying to do damage would pick a nondescript building where most people don't even realize we exist. (TRACON, F)

But the Continental flight changed that. They felt vulnerable. And that experience changed dead reckoning: common sense now included different assumptions that governed their interpretations of what was or was not an anomaly, and the meaning of risk to them. John talked about how it was after:

> What hit me afterwards was that we've had something like that [an out-of-control airplane] happen a thousand times in the past. And our first reaction was, "What can we do to help this guy?" Try and think of something we can do to help him. What can we offer him? Who needs to be told? Except this time. This time, it was, "My God, they're trying to kill us." (TRACON, M)

Boston Tower

The call about a possible hijacking of American 11 traveled from Boston Center to the TRACON to Boston Tower, reaching Boston Tower's traffic management coordinator around 8:25. Mike, the supervisor on duty that morning, didn't pass that on to the controllers because, he said, "I didn't want it to affect them. Possible hijack, you know. Possibly just a loss of communication." Following procedure, he phoned Massport to let the airport manager know. Then about fifteen minutes later, Mike got a call from the TRACON supervisor that United 175 was another possible hijacking. Trying not to attract the other controllers' attention, Rich, Boston Tower's TMC pulled up the two flights on the TSD. Together, he and Mike saw that American was off course, going south toward New York, and United was still westbound. Then someone on break came up to the cab and reported that CNN had announced that a small plane hit the World Trade Center. Chris, the other TMC on duty, was on the floor below when coworkers called him to the break room to see.

He got there just in time to see United 175 fly into the south tower. Alarmed, Chris rushed up the stairs to the tower. Mike described what transpired in those next few minutes. Airspace and place—urban location, high visibility, easy access—had tower controllers thinking about risk to the city, the facility, and themselves:

> Chris was graphic, describing one tower smoking and big fire balls and smoke coming from the second. And that caused everyone to be concerned and I still had them on position. They were concerned about knowing what happened, but also about their safety. They were concerned about the city of Boston being hit and also about the Tower itself being hit by a small plane. There was some talk about a runway change. We were landing 4 Right and maybe we should land 33 Left, just to get them farther away from us. And then we said, "Well, that's kind of directing them right towards us." And so there was some talk about that, but not much.
>
> It's a vulnerable-type situation. We're visible, so the location, and we have elevators that go up and down, anybody can get on. Four levels of parking lot that the elevators stop on. And I could see it, so I decided—we were holding everything—to get as many people out of the Tower that wanted to get out. I had too many people up there because now everybody's talking and we still have a job at hand. We have a lot of airplanes around on the runway, they have to be brought back in. And there were planes that are still landing at Boston and there was just too much chatter going on. And somebody had turned the radio on in the back and I told them they were going to have to turn the radio off because we can't do our work.

Innovating in the moment, the supervisor Mike made a plan A and a plan B:

> So, the radio got turned off and I asked if there was anyone who was concerned about their safety, they were free to leave. That I would be staying and I needed a couple of people to stay with me and we'd run the tower. So then about half the people—I had about ten or eleven people up there—and five or six of them said that they wanted to go. And I told them to go to the fire station [near the airfield, just south of the terminal, visible from the tower cab, about five football field lengths away] and I would call them over there. Because I didn't know if we'd bring them back into work, or just let them go home. This was just a decision that we're making on our own just to get the people evacuated from the building. I was just thinking, just to buy some time.

> I can't decide "go home" then. Get them out of the tower, and then if something happened to the tower we could open up the, you know, if you have five or six people over there, we'd have—they could grab even a fire truck or something that could, we could actually run the tower from a fire truck if it came down to it. Because it's got a radio in it and you could just get a couple of portables [phones] put in, if they get a tower phone or radio, you could actually get it done. It would be slow, but you could get it done. (Tower, M)

After United 175 drove into the south tower, those controllers who wanted to go left for the firehouse, unaware of Mike's innovative plan B. Staying in Boston Tower were the two TMCs, Mike, and three controllers were working the airplanes. One of the three cracked, "Wouldn't you know it. America is attacked and two Italians and a Polack stay behind to save the city." They had a lot of distractions. The national teleconference open line was continuously broadcast into the room. The phone was ringing off the hook: Massport, the region, Washington, reporters. Mike was working with a telephone receiver at both ears. The three controllers were putting the ground stop into effect, telling pilots to return to their gates: "No one's departing. I have no other information for you. Please, contact your company. Please." A lot of pilots were asking why they couldn't depart. Telling them would create more questions. Fearing loss of control of the frequency if they told pilots what had transpired, controllers were instructed not to pass on information about the hijackings. With no explanation and anxious about meeting their departure times, pilots were not happy. One pilot especially was pushing for more information: "What do you mean? What are you talking about? Why can't I go?" Another pilot got on the frequency and said, "Hey, buddy. Call. Do what he says. Call your office. You need to know what's going on." Then, just minutes later, silence. Everybody found out what was going on. Following ground controller instructions, they began returning to their gates.

Massport had closed the airport. Crew and passengers alike were instructed to leave the airport by a specified route. Luggage was going to remain and be searched. As people deplaned, every airplane in the airport was going to be searched by security police and dogs. Massport immediately wanted to set up an emergency operations center at the firehouse. Like the Command Center, Massport needed to pull together all relevant agencies. They wanted a liaison with the FAA. After an hour of briefing, Chris, a TMC, went. He was astonished by what he saw. The emergency operations center was already done and in operation:

> Massport did an unbelievable job. I couldn't believe—it was like they had done this a hundred times before. I walked in the door, and there was a seat for me with my own telephone and my own telephone number. They had every agency that you can think of—FBI, Secret Service, State Police, every agency at the airport, a representative from all the airlines, from the parking department. FAA security. I don't know. I knew they had this giant auditorium, I had been in it before. But not set up like this, I'd never seen telephones or anything. But when I got there an hour after all this whole thing went down, it was ready! They had giant—like three big screens across the front of the room with local news station on one, national news station on another, and closed circuit to the governor's office. I mean, they immediately had it catered. I mean, they had food and coffee and everything for all these people. They had office equipment and secretaries in there, working. (Tower, M)

They had "improvised from the base." The rapid response was possible because Massport had used previous network connections, pulling them together at one time to respond to the crisis. Massport was also setting up a center at the airport Hilton Hotel for the victims' families. From experience, they knew how it would go and were organized to help. In event of accident or a missing flight, people go to the airport ticket counter. There they would be directed to the hotel. Massport basically bought out the Hilton. They had all the conference rooms. American and United would have their own rooms. They would be sending trained counselors. They reserved rooms for the families of victims and for experts. A block of rooms was reserved for Massport personnel to stay over for the next three weeks.

When the second tower collapsed, the controllers waiting at the firehouse were released to go home. The NATCA facility representative came in, and one of the three controllers in the tower—the only one with a family and also the one with the longest commute—went home. The three controllers staying were single, electing to stay so those with families could go. Except Mike, the supervisor, whose wife was a controller in the TRACON. She had been released to go home. From the time US airspace was closed, all that controllers in the tower did was take in arrivals. With all the internationals going to Canada, things went rapidly and smoothly except for one alarming incident. A Delta shuttle, a 737 that had departed Logan for New York LaGuardia, was turned back from New York airspace and returned to Boston. Like other events of that day, this incident would be forever in their memories. One of the controllers described what happened:

> The pilot sounded very shaken on the frequency. Of course he would have seen the Trade Center burning. On that particular day, I'll tell you, one thing I remember about it was that there wasn't a cloud in the sky. The visibility was, I mean literally, you only get one or two days a year like that. Ever. And it was really kind of uncanny that they picked this particular day and they must have had that in their planning. Because they didn't really have, well, they had a fair amount of flight training, but the hijackers weren't that well trained. But on that particular day over Boston at twenty thousand feet you'd have no trouble seeing New York City from Boston. You could see two, three hundred miles easily. You know, it would have been simple to navigate. You could see the rivers and everything. So the Delta came back here—the shuttle—and he wanted to make his approach. We were landing 4 departing 9, but he wanted to land on 33 because he wanted to make his approach from out over the harbor and keep his plane over the water. He thought there was a bomb on it. (Tower, M)

They called out the emergency equipment. The airplane landed uneventfully. Security people with dogs boarded. No bomb was found.

The Delta shuttle was one of the last to land. The airplanes designated to land at Logan were all down by 10:00 am. It was only then those who remained in Boston Tower had time to think:

> We were all just shocked. Then you know you realized you were one of the last people to speak with the crews that were going from here. It was almost like disbelief that this happened, that the airplanes came out of this airport. And it was such a tragedy, the magnitude of it, the loss of life. People were shocked. Shocked. You know it all unfolded very quickly so by the time we realized what had happened, where the airplanes came from, everything was basically stopped. There was not much going on. There were people that were in communication with all the different emergency centers, but it was very, very subdued. I think everybody was really tense. (Tower, M)

> It was like World War III. The only thing operating was the military. You could look on the TSD . . . at 11:00 am on September 11, there was a few hundred airplanes in the air. I'm going to say around 250 aircraft. And they were all military, command and control, mostly fighter planes and then tankers that keep the fighters fueled. And I'm looking at the display and I'm thinking, "Boy, this

must be what it would be like in the doomsday scenario. This is it. This is it." (Tower, M)

> It was really scary, even afterwards. It was quiet. You looked out and that was one thing that we remarked. It's a beautiful clear day and nobody's moving. It was very eerie. You know, to see these airplanes sitting at the gates. It was empty. Nothing on the radar scopes. Just, like, silence, almost. You know, unbelievable peace and quiet. It was shocking. Because it's very rare that you see something that. Maybe on the midnight shift. But it was middle of the day and all the airplanes sitting at the gates, nobody's on the runway. Very, very eerie. And the unknown, what else is going to happen. You know, what other aspects of this attack on the United States will unravel? Because obviously the person that made the decision to shut the system down probably saved a lot of people and saved airplanes from destruction. Who knows what else was planned that day? (Tower, M)

Controllers came in early for the 2:00 pm shift, wanting to be with coworkers and participate in the effort. Like the TRACON controllers, the scenario driving in to the airport was beyond the realm of their imagination. As they neared South Boston, people on foot were walking away from the city, walking where people were never seen walking, trying to get home. Traffic was all headed the same direction—out. Two controllers came in for the midnight shift, along with Mike, who had gone home for a while and returned. There was no civilian air traffic anywhere in the system. They watched air force fighter jets and refuelers overhead at twenty thousand feet circling the city. And they watched about twenty tow trucks towing cars out of the Terminal B parking garage, where the tower was located. Then a scare. About 2 am, Massport called to report an alarm in one of the "silos," the legs that support the control tower, on the twentieth floor. Someone had opened a door. The supervisor called the state police, who were stationed around the airport and terminal. Controllers locked the door to the tower but had no weapons. The fire extinguisher would have to do. The state troopers showed up and, with the supervisor, searched the silo. They found no one. Later that same night, Massport ordered the building evacuated because it was considered a possible bomb target. Everybody left.

Through the week, when nothing was flying, people came in for their regular shifts to staff positions. Controllers learned that the search of the parked airplanes revealed a close call. In one of them—a plane scheduled to fly from Boston to San Diego on September 11—box cutters were found hidden in the seats. What more might have happened, or would still happen? Wondering, they sat idle in the tower and the break room. The military owned the airspace. With little to do, controllers talked a lot about what had happened. What really struck the controllers was that those were their airplanes, and everybody felt a great sense of loss. They had a continuing conversation about it. The crew that had been on that morning went over their role in it, as if trying to reassure themselves that they did everything they could. That everything was done correctly.

Following FAA procedure, the Boston Tower controllers who were on duty the morning of September 11 and talked to American 11 and United 175 were required to participate in the official investigation, in the usual way. Complete a statement of what happened. Listen to the voice recordings. They had to sign, authenticating that the tape recording represented what they said. So, one at a time, the supervisor and his crew went down to the floor below with the quality assurance representative and the NATCA facility representative. Unlike Boston Center controllers in Area C, who became aware of the hijacking and heard the hijackers' voices at the time, controllers at Boston Tower had no such information. For them, these were two planes in the midst of a normal morning rush, two pilots out of many, whose voices were lost to their memory. Listening to the recordings of their own voices talking to the pilots after the tragedy changed that. They knew what the pilots sounded like and would remember them because they listened to the recording. People were somber, deeply affected.

For others who had been there that day and were actively involved, the impact came later, as the week wore on, waiting and watching with no airplanes flying. Chris, the TMC who was the liaison with Massport, talked about the exposure to risk and loss of life that controllers never allow themselves to think about. Unlike working traffic, his work during the crisis focused all his attention on the fact that people were on those airplanes. Chris was working with victims' families:

> It hit me hard, very hard. Probably the hardest of anything in my entire life. Like that day that it happened, went right through the day, no problem. It shook me, I can't believe this is going on, but no problem. Then, a couple days later, I just crashed when it hit me. I don't know why, because none of my family members were on board, I didn't talk to the pilots, I wasn't vectoring them

> when they were hijacked, I didn't see planes crash in Boston, I didn't know anybody that died. But I knew everything. I knew it all. Every little intricacy that controllers don't necessarily need to hear. You know, immediately, you know we're talking there were people on that airplane. I mean, obviously they must all be dead. And where—what are we doing for their families right now? We're setting up a crisis center here. So those are things that I think a lot of people didn't even have to hear. And it's not bad to hear it, but I think that after, you process it over and over again. (Tower, M)

Not all controllers at Boston Tower were hit equally as strong emotionally as those who were working the day shift on September 11. However, they all reported having the same feeling of vulnerability. They were aware of their personal vulnerability, given their occupation and Boston Tower's location and the vulnerability of the air traffic control system and the country. They were reminded of this daily by the new security requirements. State Police were stationed at the entrances to the tower. They watched construction of the security enhancements to the airport and to Boston Tower and TRACON. They saw workers putting concrete barriers up around the base of the Terminal B tower and parking garage to prevent any possible bomb attack from the base, like the one at Oklahoma City, when a truck loaded with explosives was used to blow up the Federal Building in 1995. Guards and dogs were stationed throughout the airport.

As flying slowly began over that first weekend, they wondered whether something else might happen. And it did. The Continental flight. Boston Tower and TRACON spotted the trouble at the same time. The emotional impact on controllers at Boston Tower was greater. The TRACON was no longer talking to the pilot and watched on the radar as the Continental peeled off final approach and headed toward the city. A controller recalled:

> Probably a week after the attacks, a Continental, an MD-80, was on four mile final. He was landing 4 Right. He had slowed to final speed. He was at an altitude of five hundred feet. Suddenly he breaks off Approach and makes a left turn toward the city. He is heading up the Harbor Channel. The pilot was shouting, "Mayday! Mayday! Mayday!" We never hear that, that urgency. We thought he's headed right into the Pru [Prudential building]. He's going to crash into the city. But he circled and came back, still descending even lower, now about the height of the tower, came in over the water towards us and landed runway 27. We learned he had a flight control problem. Normally I

> leave here, I forget everything that's happened. That one stayed with me a long, long, time. (Tower, M)

Dan, the TRACON supervisor, came up to the tower afterward to hear the tape of what the pilot had said on the frequency. No sign of trouble on the TRACON frequency. When the TMC cued up the tape for him, he said, "You gotta hear this. You gotta hear what we were hearing." When the Continental veered off of final, heading toward the city, controllers had put the pilot's voice on the speaker. Listening to the tape, the TRACON supervisor was startled to hear a Middle Eastern accent. He said, "We weren't talking to a Middle Eastern—the voice we heard was American. And the voice shouting 'May Day' was Middle Eastern." They put it together: there were two pilots.

The experience of risk varied with space and place. Recognizing the greater impact on Boston Tower given the visual spectacle and the emotion of what they heard on audio, the TRACON supervisor said:

> One was doing the radios when they were talking to us, the other one was doing the radios when they were talking to the tower. I said, "Oh my gosh. You guys were listening to that and watching him out the window!" We were watching on radar, they're watching him out the window, diving at the city! That was even worse for them I got to believe. At this point for both us, it was just another thing. We were already beat up. (TRACON, M)

Bedford Tower

Andy and Jim were the only two controllers in the Bedford tower. They were busy. Traffic was typical for the morning: departing corporate jets, private pilots, and students from the pilot schools in the pattern doing touch-and-gos. Chris had just finished training Kristine, and they were one floor down, debriefing in the kitchen. The printer sent out a message from Boston Center to all facilities in the region saying to hold all traffic. A ground stop on a bright, clear day? Their immediate thought was that something was wrong with the radar at either Boston Center or the TRACON. So they held the instrument flight traffic, but continued the visual flight traffic. Then a clue: the ATM called upstairs and told them to forward any calls about an airplane accident in New York to him because an airplane had just gone into the World Trade Center. A second printer message from Boston Center specifically said hold all IFR and VFR traffic. Andy and Jim had no idea what was going on. Then, alarmed, Chris and Kristine rushed back up to the tower cab, saying they had watched the news and saw a second plane go into a tower. The supervisor was downstairs with the manager. Neither had phoned the tower cab. When pilots started questioning them about stopped departures, Andy and Jim told questioning pilots to go into the terminal and watch the TV:

> The only thing I knew was, watch the media as I did, and then go from there. So, you have to understand that we were—Bedford was so low on the pecking order that we didn't really have an idea of what happened. I mean, we were learning everything from the media. (Bedford, M)

It was not only that Bedford was low on the pecking order. Bedford was not central to the crisis. Bedford was never advised by any of the other regional facilities that the planes were hijacked, nor did controllers there know that both had originated from Boston Logan. They knew only what TV audiences knew. When the second plane hit, they assumed hijackings but thought the two planes were inbound to LaGuardia or Kennedy when overtaken, had broken off final approach, and been flown into the towers. The Bedford airfield by then was quiet. Traffic that had departed had returned, and those visual flights with students from the training schools all came down. It all happened very quickly. Everyone was very upset, either openly emotional or quietly awestruck by the impossibility of it all. They used their cell phones to call home frequently, about what was happening and to decide what to do about children in day care and at school. Everyone wanted to be with their families:

> It was funny because we had no traffic but everyone was signed on to a position because we obviously still had to man the positions. Actually they allowed us to bring—once they shut down everything, they allowed us to bring the TV into the tower. The best source of information was off the TV. I mean we watched it live from upstairs. Every one of us, you know. And I'm pretty sure the word came from regional level or headquarters level that whoever was at the facility, keep them there, because they're not sure if other controllers will show up to work. But there was an overwhelming response of controllers that were calling us, saying, "Hey what can I do? How can I help?" And it ended up being that Bedford stayed open for twenty-four hours. (Bedford, F)

Staying open for twenty-four hours was unprecedented. For the next several days, they ran a midshift with a supervisor and two controllers. The airfield stillness was occasionally broken by a Civil Air Patrol flying blood supplies down to New York, taking off from Bedford and flying directly into LaGuardia. No military operations originated out of the adjacent Hanscom base, but Bedford controllers could see F-15s on the D-Brite , their antiquated radar, and heard them patrolling the sky over Boston. The airbase's security was upgraded. The main gate was locked. The military personnel who normally stood guard at the gate were joined by police, who were stationed also throughout the civilian terminal. A police officer was assigned to the tower three shifts a day, even after the tower reverted to its normal two shifts, from 7 am to 11 pm. Jim reflected on the changes:

> I went to Saudi Arabia during Desert Shield. And the entry point onto our airbase, it reminded me exactly like that. Two cops, maybe three cops standing at the gate. And then you'd have two more cops standing at a truck inside the gate, you know? So if you were under suspicion, they would pull you over. But that was a weird feeling, because I mean, I know what it was like being in a different country and to deal with that, but to be like that in our own land, that was weird. (Bedford, M)

Like the other facilities, Bedford tower controllers were pulled into security operations. They devised an evacuation plan for the tower. Also, they were all physically involved in searching vehicles around the tower and securing the facility. The days with no traffic passed with nothing else to do. They watched the news.

The effect of September 11 on Bedford deviated in two major ways from the pattern for the larger facilities. One was the perception of risk. No Bedford controllers talked about the tower being "at risk" or experiencing increased personal fear and risk associated with the work of controlling traffic itself. No threatening incident occurred, like those that caused controllers at Boston Tower, the TRACON, and Boston Center to fear they were under attack. By physical proximity to the military base and the increased security there and at the tower, they were certainly aware of increased risk, and the risk to the country and air traffic system was undeniable. But it was more distant from them, physically and psychologically. The terrorists had gone to major airports, taking over commercial airliners to use as weapons. It was possible that terrorists would board airplanes departing from Bedford, because they regularly had commercial airliners flying in and out. But the controllers believed it was not likely. Bedford was in a rural area, so passengers had greater visibility and less anonymity. Furthermore, Bedford's airspace was low altitude and small. The hijackers' activities had become visible only when the planes were in high-altitude airspace. So as flight resumed, Bedford controllers were not policing the sky as were the other facilities, watching with extra scrutiny for deviations from standard performance that suggested hijackings or kamikaze pilots with evil intentions against the United States or other air traffic facilities.

The second difference was the lack of information. Boundaries determined by space and place affected the flow of information. The national teleconferences were between the Command Center in Herndon and major facilities, including all center TMUs responsible for dispersing information within their own regions. Bedford received the general notices sent to all facilities in the region but was not privy to events as they unfolded on September 11. When flight resumed, Bedford again was left out of the loop. There was no announcement from the FAA or from Boston Center TMU that traffic was going to be released for flight. The Bedford traffic manager was a daily participant in the regional telephone conferences, but controllers never had definitive word until traffic had actually resumed:

> I think that came from President Bush with the advice of [Jane] Garvey, but I don't think that anyone knew until it came down. At least we didn't at our level. And you know when it happened they knew in advance that it was going to happen, but we didn't know. I remember coming in to work and us having traffic again. But it was just the IFR stuff, no VFR for the longest time. (Bedford, M)

All facilities were important in getting the planes down and getting them up again, but not all facilities were as high on the "need to know" scale given the size and location of their airspace and their centrality to the main events. Taking into account the rapid succession of incidents on September 11 and the shock and intensity of work at Boston Center, Boston Tower, and TRACON, it is not a surprise that Bedford got only bare bones messages about the terrorist attacks, relying instead on the news. Similarly, it makes sense that the centers and major terminals would be informed of the return to flight, but smaller facilities might not. However, it may not have mattered. Remember that Bedford controllers described their work as "seat of the pants" air traffic control; they dealt with surprises as they happened, which with their traffic mix, they often did. Perhaps for them, being informed was less important because they were used to improvising. Thus, Bedford—and by extension other smaller facilities in the system with similarly diverse, often unpredictable traffic—were as prepared as the larger ones who had been forewarned.

The Attacks: System Response and System Effects

On September 11, emptying the sky after the attacks was an unprecedented feat of coordination by controllers throughout the system. Unrehearsed, moment to moment each controller did what he or she did everyday: worked the traffic as it came. Recall now these principles, told to me prior to the attacks:

> Structure and routine, structure and routine—the rules and regulations [structure], and how we respond that becomes automatic [routine]. If we have something new, we have at least had something like it before. If we have to improvise, we improvise from the base. (Bedford, M)

> You need neat and orderly, but you also need flexibility, and you also need the ability to just pull something out of the air. And you have to because there are going to be times in your career when neat and orderly is going to go out the window and you're going to have to pull an idea out of your head. And it had better be there. (Center, F)

In this new situation, the old methods worked. Drawing on common sense, controllers throughout the system brought planes down using the technologies, standardized rules, procedures and methods of coordination that they had always used. As in the past, they combined standardization with improvisation. They improvised from the base, were flexible, and demonstrated "the ability to just pull something out of the air."

Trained to be decisive, they were decisive. At the Command Center, the national operations manager took unprecedented actions, closing the US airspace to domestic and foreign travel. At Boston Center, the operations manager at the watch desk and the Traffic Management Unit manager took the initiative, acting without precedent: bypassing bureaucratic procedures to get F-15s in the air fast, calling for the initial ground stop, then warning pilots in the airspace to secure cockpit doors and keep passengers seated. The supervisor at Boston Tower improvised, coming up with a plan A and a plan B that included possibly running the airfield from the back of a truck, drawing on the same background knowledge as the controllers at Bedford had the day that all their radios failed. TRACON controllers improvised from the base, resurrecting flight progress strips to manage Boston Center traffic outside their airspace. For all controllers throughout the system, improvisation was required—working planes they had never worked before, devising new routes for planes that had to land in unfamiliar places, organizing a mass exodus from the sky.

At Boston Center, Boston Tower, and the TRACON, the extreme character of the crisis exposed controllers' emotional experience, showing the variation in emotion and emotion work integral to the job. The cultural and architectural factors that normally worked to suppress emotions were not in play because of the overriding power of the experience itself. For the first time, these controllers openly acknowledged risk, fear, and stress. Their recollections of their experience were full of talk about their feelings—shock, horror, fear, and vulnerability to risk and harm—and the struggle to suppress them while working traffic, the moments when those efforts failed, the need for supervisors to remind them to control their emotions, and how emotions overcame them in off position moments. Normally distanced from personal risk and seemingly impervious to threat, they believed that they, too, could be attacked and killed. The crisis also showed how the experience of risk, stress, and the exercise of emotion work can vary over time and place. The emotional experience and impact of the crisis varied by facility: greater at Boston Center, Boston Tower, and TRACON, less at Bedford. Position in the air traffic system affected the extent of involvement in the attacks and the degree to which the architecture and technology of each facility shaped perceptions and experiences, as well as by whether people were working or off that day.

Controllers acknowledged and respected these differences. No one got hammered by their colleagues for being scared. The daily private individual experience, normally transformed by controllers' emotion work, training, organizational culture, and official language, became publicly acceptable to admit and discuss when a threatening experience was shared by the group. Scares kept happening, keeping emotions near the surface. Paradoxically, perceptions of risk and fear were reinforced by the very system responses designed to increase security: the rules and procedures for watching the sky, material technologies for emergency response communication, and the physical protections erected and inspection processes put into play to safeguard each facility. The national cop with an M-16 at the gate, dogs sniffing cars, police in the terminal, and security checks all were symbolic reminders of risk to the country, the air traffic system, and themselves.

Emotionally, all controllers felt fully the loss, the uniqueness of the event, its place in history and in their lives. But of the work itself, they insisted that they did what they always did. But this was not entirely the case. In fact, they were doing one thing that they seldom did—at least at the scale that it was undertaken that day. As unprecedented as the terrorist attacks was the smooth coordination across the boundaries of the entire system. The response of these four facilities is a microcosm showing analogies with the system response that reached from local to national and international levels. Conflict about who owned the airspace and how the separate parts of it should fit together and be used were built into the air traffic system. To keep airplanes flying under normal conditions, conflicting interests had to be negotiated daily in continuing hassles between parties: military versus civilian needs, the airlines versus the Air Traffic Organization, labor versus management, FAA headquarters versus facilities, region versus region, facility versus facility, controller versus controller. On September 11, against a common enemy and for the greater good, the system coalesced. A Command Center controller described it as follows:

> If we have a common goal and everybody understands what the goal is, we have the ability to get it done just like that [*snaps fingers*]. We're very good at that as a system. It's when we do the everyday mundane stuff that we trip. You know? There's always a better way. But like 9/11 or other situations, where a facility loses everything, everybody bellied up and just got the job done. They knew what they had to do. (Command Center, M)

The resilience, redundancy, and reliability of the system were revealed in full. Individually, each person was working airplanes, using the technologies, procedures, and cultural understandings that had always guided them. The coordination between controllers bridging airspace and organization boundaries brought the large parts of the system together—region with region, country with country—to get the planes down. The system became tightly coupled in response to the crisis. At the same time, boundaries in the air and between and within air traffic facilities on the ground proved both porous and flexible in the crisis. Cooperation not only was evident between controllers and facilities working the traffic; it occurred in other parts of the system as well. Massport improvised, expanding structure, pulling together usually embattled constituents into an emergency operations center in an hour. The Command Center, its separate desks epitomizing the boundaries in the system and its inherent conflicts, pulled together and in turn, pulled the parts of the US system together and incorporated the military. As some parts of the

US airspace closed, other parts opened up, receiving aircraft not normally in their airspace. Similarly, when US airspace closed, air traffic facilities in other countries opened to embrace the aircraft denied their US destination.

Bringing the planes down on September 11 was an unrehearsed effort successfully executed. Foundational to controllers' achievement were ethnocognition, boundary work, and "the mundanity of excellence": the repetition of basic tasks such that they become routine and embodied, and can be done "without thinking," freeing up the mind and body to respond to the unusual.[5] Also crucial to their success were components of the socio-technical system: their training, multiple technologies of coordination and control, a technological infrastructure that enabled coordination across great distances, and the resilience, reliability, and redundancy built into the organization of the system, as enacted by its people. The Command Center is properly named. By virtue of organization and technology, it can "see farther" than any of the individual facilities; also, historically, it has standing authority for system coordination. The system not only worked; it worked better than it usually did because each separate part set aside local interests to achieve a common goal. On September 11 and during the first week after the crisis, the boundary problems that always had defined and plagued the system disappeared. These patterns—the nexus of standardization and improvisation, vulnerability to risk and stress, variation in emotion and emotion work across facilities, system dynamics and the flexibility and porousness of system boundaries—remained through the following year, when air traffic controllers became key instruments in the war on terror.

10

The War on Terror

Policing the Sky

In the aftermath of September 11, the physical transformation of security at air traffic control facilities and the new constraints upon them were controllers' first introduction to the new air traffic control: national security was the priority. The goals of the safe, orderly, and expeditious delivery of traffic were joined with the additional responsibility of policing the sky. Indeed, for a while, "expeditious" was the least important goal. No one was flying. Safe and orderly took priority as controllers worked to get the system going again. The attacks themselves and the government's response affected the air traffic control system by creating new boundary problems that immediately affected controllers' dead reckoning.

Like emptying the sky, getting the airplanes in the air again also was unprecedented. For the safe and orderly return to flight, controllers were charged with implementing rules known as "restrictions" that designated areas of airspace as no-fly zones. Long standard operating procedures, these restrictions were used to create temporary altitude and distance boundaries around small and localized airspace to keep airplanes away from some activity in the sky or on the ground for safety purposes. So, for example, if *Air Force One* flew into Boston Logan, controllers created a restricted space around it. Restrictions were imposed over Fenway Park during Red Sox home games, during Fourth of July fireworks on the Esplanade, and whenever the tall ships sailed into Boston Harbor. Now, however, airspace restrictions had become a strategy in the war on terror. They blocked off large areas of sky. They were designed to protect crucial cities—twenty-eight major urban areas, with Boston, New

York, Newark, and Washington among them, and other sites deemed crucial to the safety of the country—and to transform the flow of traffic around and above them by creating boundaries for altitude and distance that were not to be breached. With the airspace clear of traffic, an intruder crossing the boundaries would be visible early. Air traffic controllers and the North American Aerospace Defense Command (NORAD) would monitor the no-fly zones.

A second type of restriction regulated the return to flight. Airspace boundaries had been closed. Categories of aircraft were ranked, with some given priority for readmission to the airspace and not others. The rankings of aircraft categories were determined on the basis of importance to air transportation and to the economy. At first, all categories were restricted. Then restrictions were lifted gradually. Waivers allowed certain categories of aircraft to return to flight before others. This maximized safety, allowing flight to begin again in an orderly way. Moreover, it maximized the potential for surveilling the sky. For controllers, the creation and implementation of these two kinds of restrictions made their work more complex, changed the flow of traffic, and redistributed the workload within and between facilities.

Implementing both types of restrictions simultaneously was made immensely difficult by a second change that altered the organizational structure and existing boundaries of authority. The rules governing the return to flight were created by a new regime. The operational component of the Air Traffic Organization—the Command Center at Herndon, the centers, towers, and TRACONs—had brought the planes down. It was orderly, smooth, and rapid. At each level of operations, the decision makers were all controllers who knew the airspace that they owned. Invention was necessary, but the existing procedures for landing planes were known and standardized, and they worked to accomplish the task. In contrast, the return to flight was masterminded at the highest levels of government, in the interest of national security: FAA headquarters, the military, the White House, Department of Defense, and the FBI. No one at the operations level was quite sure who was participating, but it was clear that the restrictions were being created from above the Command Center. The bureaucracy had taken over.

Within the FAA, the air traffic control system operated autonomously, with every level run by controllers. Now, the boundary separating the technical expertise necessary to run the system was penetrated by the actions of government decision makers external to the system who did not know the airspace or the language of those working operations who would be putting airplanes back in the sky. Moreover, the diversity of airspace, traffic, geography, and the complexity of the system required rules constructed for particular situations. Incrementally, decision makers at FAA headquarters, national security, and other government bodies were inventing rules on a case-by-case basis. Airspace boundary restrictions and restrictions governing the return to flight were changing daily. Uncertainty prevailed. Controllers were used to standardized rules. When standardization went by the wayside, the two kinds of dead reckoning—predicting the position of objects in the sky and the actions of controllers on the ground—were jeopardized. As a consequence, they continued to experience a loss of control and, with it, a months-long period of intense stress.

In the aftermath of the terrorist attacks, controllers in the facilities throughout the system struggled to create order out of the confusion and disorder that the changes had created. They solved problems with local coordination: in the absence of clear guidelines from decision makers located above in the hierarchy and geographically distant, people in positions of responsibility below them charged with local implementation were interpreting and inventing, tailoring to fit the local situation.[1] They pulled together resources and organizations to meet local conditions and needs, moment to moment, as they experienced them. Although the rules from above were dictating what to do in every specific eventuality, there was no coherence to them. The external decision makers provided no general guidelines that could produce uniform decision making that met the needs of a facility airspace. Air traffic controllers coordinated locally, creating new structures, both organizational and technological, to meet local conditions. They did not invent these guidelines out of thin air, however. Controllers built on work-related personal ties, technologies, and structures already in existence to sort out the problems created by the national security and government bodies who were micromanaging. The resilience of the system manifested over and over as controllers refashioned the system to meet the changed political situation and future threats. By September 11, 2002, when the air traffic control system had again stabilized, the system's response to the attacks had changed the system.

Another effect of the attacks on dead reckoning was that scanning the skies took on an additional dimension: boundary surveillance in the interest of national security. Vigilance, always the basis of dead reckoning, had new intensity and purpose. Controllers were looking for anything suspicious—any anomaly that could be a threat to the country, the city, or air traffic control facilities. Moreover, they brought to all their work an awakened sense of risk. Boston Center, Boston Tower, and the TRACON had all had the hijackers in their airspace and moments in which they feared for their own safety. These emotions were kept alive by additional scares in the following several months as aircraft returned to the sky. So salient were these experiences that, in the aftermath of the attacks, the social and cultural factors that normally held awareness of risk and danger in abeyance were ineffective. A Boston Center controller expressed the feelings of all: "It just felt like a totally different job all of the sudden." And it was. Controllers described the year after as months and months of the worst experience of their careers. Traffic was down, but the work had become more complex, little was rote, and risk and stress were at the forefront. Although a year out from the attacks the system and the work had stabilized, for controllers it was a fragile stability. Anything could still happen.

This chapter shows the dynamics of the large-scale socio-technical system in the midst of rapid change, produced by problem-solving controllers who responded to the mandates of external actors and factors. They were preoccupied with reinventing the system under time constraints with renewed dedication to restoring safe, orderly, and expeditious delivery of air traffic. As September 11, 2001, and the week after are splayed open to view, the themes of system effects, boundary work, interpretive work, human-technology interaction, emotional labor and emotion work, and precedent and innovation take on new form and meaning as controllers remade the system.

Changing Boundaries: Restrictions, Translation, and Local Coordination

Incremental return to flight was the plan. The FAA and the Command Center had a precedent. In preparing for the expected reduced staffing of the 1981 PATCO strike, the FAA created, then implemented, a specially designed contingency plan "Flow Control 50." The plan assigned quotas to airlines in order to reduce the number of planes flying (see chapter 2). It classified types of aircraft, assigning quotas for departures by national importance (e.g., some commercial airlines versus others and those versus general aviation) and within these types, each type was rank ordered. Historically, this quota system had been invoked from time to time, so the aviation industry expected quotas from the Command Center as routine. Now the quota system was being invoked in returning planes to the sky. Some categories would be given priority to fly first, but all categories had restrictions applicable to them.

In the first few days, FAA headquarters and the military and national security agencies controlled the sky, issuing waivers that granted special dispensation to fly. It began with only waivers for selected individual flights like those admitted from Gander, Newfoundland, and others that landed in Canada. Other flight requests would be decided on a case by case basis. Then categories of aircraft would be released for flight: first commercial aircraft, then general aviation with special permission, then all general aviation, which was the last barrier to fall before traffic was allowed to fly as normal.

As restrictions were issued from the bureaucracy above, they created enormous confusion at all facilities. First, there were so many. They were coming by e-mail and pouring out of fax machines at the Command Center and in the facilities. Some applied to the whole country. Special restrictions were created for special cities. New York, DC, and Boston were very restricted, and the restrictions were in effect longer at those cities than at other places. Sometimes the restrictions contradicted each other. Another problem was that they kept changing. A restriction one day would be replaced by another the next day, or elaborated on with new exceptions. In any given moment, nobody seemed to be sure exactly which ones were in effect and which were not.

The restrictions lacked clarity. Translation became a major activity. In the first few weeks after the terrorist attacks, traffic was practically nonexistent at every air traffic control facility. Even two months later traffic was light, which was a good thing, because controllers in all four facilities were spending most of their time debating the meaning of the restrictions. At Boston Center's watch desk, a podium was moved in to hold two eight-inch-thick binders for easier tracking of the latest restrictions to see which was current and which superseded which restriction. Supervisors would gather around the podium and try to come to a collective interpretation, then daily they would go to their areas and brief their controllers. In the absence of clarity from above, local coordination produced the guidelines for implementation:

> At the TRACON, we had two files, the "Operations Immediate" file that we're supposed to read every shift before we start, and the "Read and Initial file," which is important but less time critical. And sometimes they would be put in both. They might be thirty pages long. Single spaced. We had to search for the paragraph that applied to us. That happened daily. We'd get a reissue of the same things written in a different way. And we'd have a Boston TRACON interpretation of this rule each time it came out in an attempt to at least have all of us doing it the same way. (TRACON, F)

The language was a major obstacle. The restrictions looked like legal documents: "Pursuant to FAA regulation . . ." Controllers had to translate the language into something they could understand and interpret similarly:

> The rules they gave us to work with were written by somebody whose primary language was bureaucratic. And their secondary language was ambition. They could not put this down on paper in a way that we could understand. And we, every day—every day for weeks would say, can we do this? What does this mean? And when they would say back what they mean, every day we'd have to get interpretation and every day the person who would give us the interpretation was a different person. Not the one who wrote it. We had a lot of changes that were required in who was allowed to fly and who wasn't allowed to fly. And the general aviation load suffered a lot for a long time. Boston, New York and Washington had special rules. Even more special than everybody else did. And it restricted them [general aviation] a lot. (TRACON, M)

The new rule makers generated waivers—exceptions to the flight restrictions—for individual aircraft or for company-owned fleets. Controllers were sorting through long lists of waivers for who could do what and who could not do what: "What about blimps? What are we supposed to do with them?" Then we'd try to find out, then they'd make up something about blimps." Issues came up, plane by plane: "Can this guy go? Can this guy go?" When training, controllers always learned the rule and the logic behind it. Not this time. There was no logic to the restrictions, at least not one that they could understand:

> [The restrictions] made no sense at all. Literally no sense. They are saying no one can take off or land at Beverly. Take off or land Beverly? We've got pilots that are on the ground there that need to get their airplanes up. They've been sitting there for two weeks. They can't take off cause they're too close to Boston. So then they came up with rules. Well, if you depart on this heading, you can depart, but you can't turn around. The restrictions to the airports like Beverly, Norwood, Lawrence, and Bedford that had been in our airspace were unbelievable. It was hard to determine who could go into one of those airports and who couldn't. They were trying to make it appear that Boston was being protected, but in fact if you were being hijacked, none of those rules would apply because the hijackers are going to fly wherever they want to fly

> their airplane. So the ones you're really penalizing are the law-abiding people that are going to follow the rules that are being put into place. (TRACON, M)

The restrictions creating boundaries around no-fly zones were no clearer:

> There was a block of airspace around Boston that they said air carriers could go through, but air taxis couldn't. And we can't tell because there's a mile-long list of who's an air carrier and who's an air taxi. We don't necessarily know who's an air taxi and who's an air carrier. So as this guy approaches this restricted airspace, you don't know whether he fits the qualification to transition to this airspace, or if he needs to be turned around. OK, can he go through? Can he not go through? You ask your supervisor. He's got to ask somebody. By then the guy's either halfway through the airspace, or you've sent him halfway around the restricted area. (Center, M)

It was hard to get a ruling. Pilots would call controllers with questions. Controllers would ask their supervisor for interpretation. A lot of times, the rules seemed to depend on who the supervisor on duty was. If the supervisor wasn't sure, she or he would phone TMU at Boston Center. The waivers created an enormous workload for TMU, which was already deluged with figuring out reroutes around restricted airspace. So in addition, they were besieged with "Can this guy fly?" questions. One TMU controller complained, "Moving an airplane would take about ten calls. It was chaos every day. Every day for a month or two." If TMU didn't know the answer, they would phone the region or they would phone the Command Center to get an interpretation. The Command Center became the major translator and the arbitrator of disagreements. The whole focus of their operation was gone. The Command Center dealt with airplanes in the air and airplanes trying to land. Now they had neither. They had planes waiting to take off. In place of their usual task—regulating the national traffic flow—they became the link between the FAA's air traffic control system and the new national security hierarchy that was generating the new rules.

The return to flight changed not only the content but also the structure of Command Center operations. In response to changed demands on the system, the Command Center created two new positions. First, the crisis pulled the formerly competitive military into partnership with the air traffic control system. The ownership of US airspace had been turned over to the military, but the military quickly realized that they were in no position to take over the operation of the air traffic control system. The military had a presence in the Command Center before but worked from a top-secret room and never interacted with the controllers there. To get the planes flying again, the Command Center opened up a new security desk, staffed by military and Command Center personnel. The goal was to coordinate the return to flight to ensure that no civilian aircraft was shot down. For the time being, the competition for use of the airspace between military and civilian was set aside.

Second, to handle the enormous volume of calls about the restrictions, the Command Center opened up a room and filled it with nearly thirty telephones and people. It became another controller position, known as the emergency operations room. Controllers' time there was spent translating the restrictions issued from above in order to answer questions from air traffic facilities and pilots that were pouring in from all over the country. The controllers at the Command Center had worked at the busiest facilities in the country. They variously described the return to flight as "a disaster," "a nightmare," and "worse than any day working traffic in any environment." They said it was the most stressful part of their FAA career.

A controller at the Command Center's emergency operations desk described the confusion and intensity of the experience:

> They put out this eight-page-long advisory. "If you're a Part 91, you can fly VFR, if you're part 94, you can fly with no passengers, if you're part 101, you can operate within the eastern half of the United States." So every two minutes, somebody is saying, "What's a 101 again?" "Oh, it's an air taxi." "What's a 91?" "Oh, that's an air carrier." That whole document, I wish they'd burned it. But it was very legal and very precise, but was very confusing. And I think it generated even more phone calls for us. Because some guy would call, "Hey, I got my crop duster, I don't know if I'm a 101, 121, or 131. Can you tell me?" "I don't know. Just don't fly. You'll be fine." So that was pretty confusing for the first few days. Who knew, you know? We don't have those rules—we didn't even know what they were! The people at headquarters, the security people, the certification people, they operate on what rules different categories of planes fly under. Air traffic controllers have no idea. Are you IFR or VFR? Are you a heavy jet or are you not? What's your altitude? It's simple. But this was really whacked out.
>
> Then they said, "OK, well there are certain countries that planes could fly from," so there were some that couldn't. So they would give us the countries that planes could fly from in three-letter identifications, like you would see on

> a baggage tag. So people giving us this information, they had no idea what air traffic controllers used. It's a four-letter ICAO [International Civil Aviation Organization] identifier. Like London Heathrow. On a baggage tag it is LHR. In air traffic parlance, it's EGLL. You know, E for Europe, G for Great Britain, LL is the airport, London Heathrow. So we dealt with the four-letter international identifiers, and then we're getting these three-letter things. And then we said, "Well, can you translate it into the four-letter codes?" The disconnect between the hierarchy within the FAA and us was remarkable. It was startling! It was scary. And it became very clear to me how something like 9/11 could happen. Because we're screwed up. We are really screwed up. So if this was happening in the FAA, what was happening between the National Security Agency, the CIA, the FBI, and the military intelligence agencies? (Command Center, M)

The emergency operations room was handling incoming queries from pilots and controllers about who could fly. Every day it got more convoluted. The Command Center had books with lists of airplanes that had special waivers. People would call saying that they had dispensation to fly from so and so and then controllers in emergency operations had to call above to check. Or pilots would call to see how soon they were going to be able to fly. Controllers were dealing with a huge, and unfamiliar, audience: the flying public. Answering the questions coming in generated unexpected emotions. The experiences were new, and they were unprepared for them:

> I worked in the national emergency room for some twenty-eight, thirty, forty days and they were the hardest days of my life. Because then we had to deal with the ongoing issues day in and day out. An awful lot of general aviation people went out of business after. They depend on their ability to fly, and we were not allowing anybody to fly at that time. We had these horrendous calls from people whose lives we were destroying. And well, just thousands of calls. Thousands coming in. You know, crop dusters. This was their last two weeks, three weeks of operation to make their money for the year. They're grounded. You try and tell some farmer in Iowa that he can't fly his crop duster to save his income for the year. Off of Houston, there's 1,200 helicopter flights a day that go out to the oils rigs and back. Just taking people out and bringing people back home. Those were all grounded. A gentleman called me, he ran a banner-towing business. That's what he made his livelihood at. With three airplanes, I believe, two pilots, plus himself. And he said, "I understand why we can't

fly." He wanted to know when they might because of contracts. Because if he didn't fly pretty soon, he was going out of business.

And it tugged at my heart quite often. And my temper. Some of it was very nasty. I took a call from a gentleman out of Southern California that was not allowed to fly his Beech[craft] Baron because of the restrictions. And he said that he had summer homes down in Mexico and this had a terrible impact on his family's vacation time. That they were not able to go down and enjoy their summer homes. And I had been dealing with a lot of phone calls that day, and my patience was about that long. I said, "Sir, have you been watching TV at all?" In other words, with the towers and stuff. And he says, "I really don't care." One gentlemen, South Texas, called. He owned a huge, huge ranch with his own private airstrip. And he called, very adamantly complaining that his daughter was not getting her flying lessons. Her private flying lessons. And he—for the most part, he didn't care about whatever. Oh, her birthday was coming up, that one, and he had promised her that she could fly by then. And it was so narrow. That people were dying and people were worrying about their daughter's flying lessons. (Command Center, M)

In the facilities, controllers also were struggling to translate from above and answer questions from below. Tensions were often high. Trying to reach consensus, they argued a lot about the meaning of the restrictions. Bedford, Boston Tower, and the TRACON were the main recipients of calls from pilots still stuck on the ground. Pilots were angry at controllers, not understanding that controllers were not making the rules. Arguments were probably the most frequent at the small airports, like Bedford, where general aviation pilots were grounded the longest and where controllers personally knew many of the pilots. The no-fly restrictions were most difficult for those pilots who ran and taught at the five training schools at Bedford:

The poor person running this, he earns his money by flying. Now the government says he can't fly at all—for months. So then we're getting calls, are they going to lift this yet? So I have to talk to pilots who I know their lives depend on their companies. They understand, but they've got bills coming in, and they've got to vent on somebody. It was hard being the person being called all the time, and being yelled at, "What!" You know, I'm telling the pilots, "No, you can't fly. You can't fly." And then when they started it was so limited. And telling the guy, you have to go refile your flight plans." "Well, can't you just file

> them for me? I talk to you every day." "I can't. I know you. You fly out of here all the time, I know. But I can't. Sorry." "Can't you just do, you did it . . ." No, I can't do that anymore. (Bedford, F)

The new airspace boundaries changed traffic flows. Restricting certain airspace to protect it called for reroutes that were nonstandard ways to get to and from destinations. The most pronounced effects were felt at high altitude. At first, so much of the airspace was restricted to the military that Boston Center controllers had a hard time working civilian traffic around it. Also, as time went on, restrictions changed, affecting both routes and altitudes. For example, at first the restrictions for New York's JFK airport was no-fly within twenty-five miles and from surface to infinity. Then as time passed, the restrictions were fine-tuned so they blocked off a smaller amount of airspace. But every time they were fine-tuned, surrounding routes changed, which had an impact on traffic flows. Moreover, flight levels changed: for example, airplanes that were normally supposed to be at twenty-four thousand feet were changed to twenty-nine thousand feet. And then all the flight levels would be fine-tuned.

Even with traffic down from its normal levels, the daily changes kept controllers off balance and added stress. Moreover, the alterations in traffic flows redistributed the workload within and between facilities. A Boston Center controller from Area C said:

> [There was] the CAP [Combat Air Patrol] over New York where there were fighters flying for six straight months, constantly, day and night. That was where the Boston arrivals and the Bradley arrivals used to come in. They couldn't go that way anymore. They all came in through Area C. Yes, Area C, everything in and out of New England. That was the only way to go. And this is what killed us. New York Center, basically they went on vacation afterwards because all their traffic moved east. Area E, which was a fairly busy area before September 11, they didn't get any traffic. They fed into and received from New York. So planes didn't come in that way anymore. So everyone essentially went out on a break and we'd be working our butts off, doing all these unusual traffic flows and it was very stressful. (Center, F)

In addition, all four facilities experienced a change in traffic mix. They lost most of their VFR traffic and the number of regional jets increased, affecting the prediction essential to dead reckoning. At Boston Center:

> CEOs decided not to fly commercial, so they started flying on the little Citations and Learjets. They'd always fly bad routes and fly slow and get in the way, and that's our traffic now, a different mix of traffic. It changed everything for us. They fly high. They have different performance capability. A Citation jet will do like Mach 6.5 and carry ten people. So there's a lot more of them. For us it didn't matter if the 747 was full or empty, it was one airplane. It mattered for the airlines. Now instead of flying a 727 on a route, they'd fly a regional jet and try to fill it up. So the airlines changed up their fleets because regionals are more fuel efficient. They only take two crew instead of three. Cheaper. (Center, M)

Traffic at Bedford had been a mix of small, slow, VFR planes, IFR jets, and commercial airlines, with VFR dominating. Immediately after September 11, none could take off or land there because of Boston's restricted airspace. When they began releasing the IFR traffic on that first Saturday after September 11, the commercial jets that had landed at Bedford in the rush to get down at the nearest available airport were released first, and then no traffic for months: "Nothing but boredom, boredom, boredom." The diversity of Bedford's traffic mix was not only the challenging part of the job; it also had been their entertainment: pilots in fancy aerobatic planes, pilots in training, weekend warriors, celebrities, political figures, blimps, and sports teams flying in and out. Moreover, the strict security prohibiting visitors left them isolated in the small tower. This had additional, less visible system effects:

> Pilots couldn't even visit anymore. You used to get to know the pilots that way, they used to come and visit. Or they'd bring their students up, get the point of view. They can't get the perspective of the tower anymore. And a lot of the students don't know anything about what we do, or why. (Bedford, F)

As time passed, however, Bedford's jet traffic doubled from what it had been before. General aviation was still banned, so the choice to charter a small VFR was nonexistent. People didn't want to take off from Boston, and they didn't want to fly in an air carrier. They wanted to fly a way that was "safer." Soon the rich began chartering small executive jets. Corporate jet traffic at Bedford rose dramatically. Then, when general aviation finally began to be released:

> We had to have a different way of running traffic. Because our jet traffic increased, when we did start getting some VFRs or some little guys flying IFR,

> squeezing them in there was harder. First of all, we're rusty. And second, we have more jet traffic, so they're only leaving three miles in trail, and either you can't fit a little slow guy in three miles or the jet behind him is going to run him over, you know. So there were adjustments. Then the whole aspect of Bedford changed. Because it was more jets, not as many little guys. Some went out of business or they were just not flying. (Bedford, F)

Controllers dislike change for a good reason: they need standardization to work traffic and to be able to do it "without thinking." The extensive boundary changes in the aftermath of September 11 were one part of the new air traffic control. They presented new challenges, both cognitive and emotional, as controllers engaged in translation of new sets of rules that had them struggling to create order from disorder and maintain calm while translating across the boundaries of the system to pilots and the flying public. Simultaneously, dead reckoning was transformed further by the new surveillance responsibilities. The first year after the attacks, the job ran the emotional gamut: boredom and stress mixed with fear and a changed sensibility about risk kept alive by new scares, both major and minor.

Police Work, Emotion Work

As the return to flight began, controllers had only a few airplanes. More time for griping and less to distract them from the harsh reality of police work. "We were dying," a Boston Tower controller said. There was a renewed sense of camaraderie, based on their shared experience of the attacks and as they tried to get the system going again, together. But there was a definite edge to the job—that awakened sense of risk, fear, and extra vigilance—that hadn't been there before. It went on for months after September 11. A Boston Center controller recalled:

> There was a change of attitude here. After September 11, coming in and seeing the scopes with no airplanes and nothing but military planes flying around, it was a very eerie feeling, and I would say for a good three months afterwards. Just a knot in the pit of your stomach when you're working traffic just thinking, could there be something else that's going to happen? Thinking about what had happened, when they lost communication on September 11, we saw them both [American 11 and United 175]. So in the back of your mind, when you lose communication now, you're thinking, could this be something else?

> Is he doing something he shouldn't be doing for a less than honorable reason? (Center, M)

They coped, as usual, with humor: "When a pilot doesn't answer, you no longer think he's checking his pay scale." Anomalies took on new meaning. They let nothing slide by. When they heard an intercom at another frequency, they thought twice about it: "What was that? What exactly was said? Who said that?" When a pilot did not respond on the frequency or turned in an unusual way and was not responsive, controllers gave it extra attention. Whereas formerly their experience reassured them that most of the time, the pilot was not paying attention or had turned the frequency dial to the wrong spot, their experience had changed.

They were policing the boundaries of their airspace for intruders. There were scares at every facility. Center controllers were particularly vulnerable to them because when the September 11 hijackers struck they were in Boston Center's high-altitude airspace. The expectation was that if it happened again, it would happen there. An air carrier came out of Boston. He was slow responding to controller instructions, and when he did, he used the wrong call sign a couple of times in a row. Was this really the pilot, or had someone taken over the airplane? An Area C controller who was working during the hijackings and had heard the hijacker's voice on the speaker dreaded the possibility:

> There's always something in the back of your mind now, that it could happen again. When you try to talk to an airplane and he doesn't answer you, in the beginning, you're just thinking that he's lost. And you go to try and find him. Then you go back to another frequency, you can't find him. And he just keeps trucking along, you're like, "I hope this guy isn't the next one." Because I don't want to see that again. I'm not interested at all in being the guy that was working the guy. I hope to God it never happens to me. (Center, M)

Policing responsibilities were built into the new procedures. Controllers had been instructed to report any suspicious activity immediately. After the controller tried repeatedly and unsuccessfully to contact the pilot, the controller would tell the supervisor, who would use the hotline—now in all the major facilities—to scramble F-15s on "the target of interest." Many activities that had been normal were now suspicious activity. Pilots of small planes were often terrorized. Not realizing the airspace was shut down, a farmer someplace would go out for a Sunday fly. Controllers could see the plane on the radar but didn't know who it was. So they reported it, and the F-15s would fly to bring him in to an airport.

One of the NOTAMs (Notice to Air Men) that routinely informed pilots of hazards or temporary restrictions cautioned about loitering over high-population areas or other points of interest, like dams, nuclear power plants, and industrial areas. If a VFR airplane went within a certain distance of a prohibited area, fighters had to be scrambled to check out the airplanes. Controllers' ability to order this was a consequence of the new cooperative relationship between air traffic control operations and the military. In parallel to the change at the Command Center, airspace boundary disputes between the military and air traffic control facilities were set aside as the former competitors joined in pursuit of a common enemy. Constant and immediate contact was possible by the twenty-four-hour-a-day open teleconference line connecting all the air traffic control centers, some TRACONs, FAA security, and the Command Center with the regional Northeast, Southwest and Western Air Defense Commands, NORAD, and US Customs. The TRACON and TMU at Boston Center were both on, cementing the tie with national security. A Boston Center TMU supervisor described—in classic understatement—how the direct connection with the military had made controlling traffic "a little intense" in the aftermath of September 11:

> We've had four hundred and some odd fighter intercepts take place over the last year, nationwide. I've actually handled six of those. The FAA and the military handle the airspace. It's a very joint effort. We work the airplanes and we talk to the military. The military's talking to their airplanes. Prior to September 11, only the president could order a shoot down of domestic air carriers. And that authority has been delegated to two air force generals now. If they can't consult with the president, they can order an airplane to be shot down. We now have the capability of hooking up a telephone link between NORAD and through our equipment to actually simultaneously talk to their weapons controller on one phone and on the other phone talk to the supervisor that's actually watching the airplane and watching the controllers. So we'll talk it over and make a determination as to whether or not they should send the fighters out. And the weapons controller is asking me, "Is the pilot taking your commands?" So the process would be, our controller would say, "No, he's not taking my commands." And I tell NORAD, "No, he's not taking our commands." And then at some point in time, they could possibly decide to

> shoot the airplane down because of that information. So that's a little intense. (Center, M)

As time passed, the awareness of risk and danger was regenerated by these incidents daily. They were false alarms, but from the moment an anomaly was spotted, controllers were gripped by uncertainty and the possibility that it might be another one. Controllers' sensitivity to possible risk and danger was kept alive by orders that shifted their vigilance from threat to threat:

> Every time you came in, there were new requirements or new restrictions, or different ones from the day before. Even during the course of a shift, you'd have, "All right, today we're paying particular attention to nuclear power plants." And then five hours later, it would be like, "Forget the nuclear power plants. Watch the coast." (Center, F)

They were never told what was going on that accounted for the changes.

They worried about the uncertainties that would result if shooting down a plane were ordered. It was a continuing topic of speculation:

> Even if we knew the intentions of an aircraft taking off, just who was going to have the balls to order the shoot down of a civilian aircraft with so many innocent civilians on board, before it had done anything? If the order was given to shoot, just where and when would you give it? Would we have time to get the other traffic out from under the guy? The plane's gonna land somewhere, and if it's not over water, just who makes the decision about who on the ground gets killed? Even if you could shoot it down over water or relatively unpopulated areas, can you assure the plane will go straight down where you want it to? Can you be positive that you would obliterate it in the air? (Center, M)

Fear of more terrorist attacks was confirmed by two major incidents in Boston Center's airspace before the end of 2001. Two months after the September 11 attacks, American Airlines Flight 587 crashed in Rockaway Park, a neighborhood in Belle Harbor, Queens, New York. It was an Airbus out of New York traveling to the Dominican Republic. It was first suspected to be a terrorist incident, but investigation showed that the crash resulted from a combination of pilot error and mechanical problems.[2] When it happened, however, controllers reacted in the moment, uncertain about the causes, suspecting the worst. At the center, Lori—who in 1996 had been working TWA 800 twenty minutes before it disappeared out of the sky—was working the airspace right above the crash site:

> The military block was active, so the fighters came out of the military block and went down, and we didn't know what was going on. We were told to tell the planes on our frequency that it had happened, because we didn't know if it was another hijacking, and if it was, then the pilots needed to know to increase their cockpit security, make everybody sit down, don't let them walk around the cabin.
>
> I had to tell the planes on the frequency—everybody did in the whole center, the planes that they were working—that this had happened. Another plane had crashed in New York. We didn't know what had happened. Increase your cockpit security. But it was that quiet. And it was really tough to tell these planes. I was sitting there like, I'm going to say it, and then I started to get tears in my eyes. But I think everybody felt that way. And the people that were here on both days, on September 11 and this time, I'm sure that that just plays through with them constantly, on how they feel and stuff.
>
> I think my voice cracked right at the end. And you could hear it in their voices, the pilots were just—and the pilots that I told were right over New York City, so they could see the fire. And then the pilots started saying to us after their [airline] companies sent out the word, "Hey, we need to land right now. We need to divert." It was so intense. The fighters came out of Otis and they were over Boston in five minutes, over New York. Very tense time. It was hair trigger. (Center, F)

Then, almost one month to the day later, on December 22, 2001, was the "shoe bomber" incident. On American 63, Paris to Miami, a passenger boarded with explosives embedded in his shoe. During the flight, while attempting to light the fuse on the explosives, he was subdued by flight attendants and passengers. The flight was diverted from its Miami destination to land at Boston Logan, passing through the airspace of Boston Center, the TRACON, and Boston Tower. The center's TMU called up the fighter escort, which followed the plane all the way in and down to the ground. Controllers in each facility turned other planes in the airspace out of the way. The landing at Logan was met with police, fire equipment, emergency vehicles, and dogs. The F-15s circled above. For controllers in the New England Region, it was jarring. December 22, 2001—the significance was that anything could still happen. Uncertainty, awareness of risk, and vigilance remained high.

A Fragile Stability: 2002

In March 2002, I was cleared by security to return to Boston Tower and TRACON for a period that included the sixth-month anniversary of the attacks. Barred by the elimination of visitors post-9/11, I was cleared as "official business" for the promised follow-up to clarify questions during my data analysis of interviews and observations. I was armed with plenty of those questions but hoped also to get permission to talk to controllers about September 11. I had not been to Logan Airport since June 2001. The security differences—the concrete barriers around the tower base to keep traffic away, the police at every entrance and in the terminal, the stringent procedures for getting into both facilities—were sobering. I recalled my first visit to the Logan facilities in 2000. Having just come from months of fieldwork at Boston Center, with its barbed wire enclosure and guarded gate, I had been struck by the naked vulnerability of a tower innocently constructed in the 1960s to rise out of a multi-tiered parking structure. Despite the extensive protections, it was still vulnerable by its height and visibility.

Even with official permission to return, my admission each day was subject to rigorous scrutiny. Once in, however, my access to the facility and the controllers was as unrestricted as before. I was the first visitor admitted after September 11. Upon seeing me, the typical reaction was surprise: "How did you get in?" After a moment, one controller who said that continued, "A lot has happened." Some began talking about it on the spot as if our conversations had not stopped. Others brought it up in hallways and break rooms without my asking. Interviews were, as before, voluntary. At six months after the attacks, a semblance of order had been restored. All categories of aircraft were flying. New routes had been established to take into account the altered traffic mix: fewer large commercial jets and VFRs, and more regional jets and corporate jets. Controllers had developed techniques for handling these changes that by six months had become routine and could be done "without thinking." However, the restrictions over New York City went on for six straight months, day and night. By one year, the restrictions around Boston, New York, and Washington were no longer operating full-time, but there were increases in the amount of restricted airspace for limited periods of time for security reasons, and fighters went up. With restrictions reduced, workloads within and between facilities had resumed a distribution that was as it was before September 11, except traffic was down significantly from its former highs—as was true nationally.

The system itself—technologically, organizationally—had changed. Technological innovations overcame boundary problems of hierarchy and geography with a new flexibility that enabled the system and/or selected parts of it to come together in an instant. The new cooperative arrangement with the military was institutionalized through a national security communication system that comprised "bridges" cobbled together to share information so that innocent people were not shot down in the hunt for terrorists. Another technical system connected regional facilities directly with the military, institutionalizing the spontaneous, unscripted end run around government and national security bureaucracy that operations manager Terry Biggio and TMU manager Dan Bueno at Boston Center took on September 11 to get the F-15s in the sky as soon as possible after the hijacked planes. This new system was an active hotline, always ready in case of a national security emergency. The TMU manager explained how they improvised from the base:

> We were tremendously impacted. It's changed but to where it's now manageable because we've got the procedures. We've proceduralized. We got used to the security hotline being there all the time. One is from the Command Center, we'd had that already. There is a brand new, twenty-four-hour-a-day, seven-days-a-week national security hotline. Dedicated, totally whole new people on the other end. Military, everything. It goes all the time. We had those bridges originally, built for thunderstorms and other events needing coordination. So we had that infrastructure, and we knew what we were doing, taking the planes down. We knew how to do it and so we did it. Then, when we needed to get back to operations, we used those bridges and added new stuff. There's four new dedicated lines at the watch desk. National security hotline. (Center, M)

In addition, the restrictions being doled out from above gradually had been translated and codified in a book in each facility, as people coordinated locally " to come to the best consensus" for their airspace. The hard-to-decipher rules for the return to flight had been translated into local knowledge. Although no longer in use all the time, the routes that had been invented, seat of the pants, as no-fly zones to protect vulnerable places in the New England Region in case of attack were now in a "play book," so "when these things happen, it's stored and we know what to do." Controllers had rehearsed both getting the planes down and getting them up again, coping with changing airspace boundaries around major cities and changing restrictions to air travel. Now they had a codified policy for a rapid emergency response to a national security threat. New ties were formed between air traffic control operations and other agencies with national security responsibilities, the collaborative experience leaving them better able to interact quickly in the future. By one year after, the system had stabilized with new capabilities for policing the sky. It was prepared for a similar attack, in which airplanes were hijacked to be used as weapons, as well as other kinds of incursions in the air and on the ground. Was the system prepared for any eventuality? No one knew.

Although at six months beyond September 11, the system had changed substantially and had stabilized, controllers were still feeling the emotional effects of the attacks and the effects of the securitization of their facilities and their work. When they talked about what had transpired and their experience of it, as much as their words it was their faces, voices, and body postures that conveyed their sense of loss, sorrow, anger, consternation, fear. It was no doubt more on their minds because the six-month anniversary was imminent. Yet people talked about how they were reminded every day because their work environment reminded them. It was not just the police work: the extra vigilance, anxieties about nonresponsive pilots, ordering up the F-15s, and the changed security technologies and procedures all directly affected them. They had gotten used to those changes. It was the little things that had been part of the pre-9/11 routine that had become imbued with new meaning:

> When we are departing 2-2 Right, the planes get a 140 left heading for noise abatement. When we've got a 76[7], it's like—I flash back to the second plane turning into the World Trade Center. You know how it looked; it's the same turn. (Tower, M)

> You are constantly reminded when you go down to the terminal, and a lot of times the terminal is empty. Why? Because people aren't flying. Then you go down and you see the tremendous lines where people are waiting to get screened and checked and you see how different that is. So, you know, that is sort of a constant reminder of why we are where we are. Because of September 11. So, I think you're always thinking of this security aspect. (TRACON, M)

For many of the controllers who were working the morning of September 11, the strong emotional hit they took that day was still strong. Several at the TRACON and Boston Tower were worrying about a six-month anniversary CBS television documentary about September 11 that would be shown that week.[3] It would be a visceral close-up of what happened. It was filmed by two French brothers who coincidentally were in Manhattan making a documentary about the training of a rookie firefighter when the attacks occurred. While out on a run on a fire truck that morning, they heard and caught on film first American 11 and then United 175 slamming into the towers, and they continued filming while the fire truck and crew rushed to the World Trade Center. The documentary makers spent the day filming the rescue efforts. It was the only known footage of events inside the Towers that day. Controllers were emotionally vulnerable. One said:

> I have a hard time seeing that—the videos of those planes hitting the buildings. Today, I have a hard time seeing that. Even though there's nothing we could have done about it. I still take it very personal. Those were my airplanes. That special is on—tomorrow night or Tuesday night? I don't know if I'm going to be able to watch it. (TRACON, F)

Controllers did take it personally. They felt that both their airspace and the basic premise of their job had been violated. Brian, who talked to both American 11 and United 175, was still feeling the impact of the attacks keenly:

> You know, it still kind of sickens me. I feel a personal violation. Like I was personally violated somehow. Because these guys came to my airport where I work and they took airplanes that I was in control of, and there was nothing I could do about it. The job of the air traffic controller, number one priority, is the safe movement of aircraft. And I don't want to say I feel personally responsible, but I do feel personally responsible. I mean, I did nothing wrong and nothing happened in my airspace. But for that to happen, for them to take away two airplanes and use them as weapons against the United States, it's like someone coming to your job and taking away pens and pencils and stabbing people to death with them or something. I don't know how to explain it, it's not really your fault, but yet you feel like it in some sick way. I've got this ugly place in history. You know, where a tape recording of my voice and conversation with these airplanes will probably be sitting in the FBI offices down in Washington long after I'm dead. . . . It's one of the ugliest things

> that's ever happened to this country, and somewhere in there, I'm part of this history . . . as ugly as it is. It was just a shock to my system, or, in fact it was a bit of posttraumatic stress syndrome. I don't honestly know, just how I felt. (TRACON, M)

Not everyone who was working that morning experienced the same emotional impact. That variation carried through to the six-month anniversary. A Boston Tower control explained:

> I taxied both of them out, but I don't think I reacted in any way that people expected. I remember going home that day and people would say, "Oh, I thought about you. Were you working?" "Yeah, I was there and I talked to both those planes." "Oh, that must have been horrible!" It's like, "Well, no, because at the time, I said taxi to runway 9 or taxi to runway 4 Right. I didn't think about, 'I hope this plane, everything goes well, I hope nothing tragic happens.' I was more cognizant of what was going on after the fact, than before. And you know I played a minimal role. So I wouldn't say that that [taxiing them] really affected me at all. I was more affected by it like everyone else was. I didn't clear them for takeoff. I wasn't vectoring them when they were hijacked. You know, I suppose those folks that were doing that, it was very different for them. Or the people in New York that actually, from their tower saw the planes hit. . . . They weren't even working them but they saw it happen. (Tower, M)

His observation confirmed what I had been thinking myself. I had believed early in my research that in the case of an accident or serious incident, a facility's airspace and technology would determine the emotional impact on controllers and their awareness of risk. It was, in the beginning, a simple "airspace and place" hypothesis, based on proximity to an incident. So, for example, tower controllers who saw incidents in real time through the tower window would have greater awareness of risk than controllers in TRACONs or centers, who were more distant, experiencing it by virtue of their radar technology. Mainly, I was thinking about differences between facilities based on the obvious. Then at Boston Center, I learned through the accident experiences of controllers and their fear of being the last person to talk to a pilot that radio contact brought them near, and that visually, the disappearance of the transponder code from a scope could have an emotional impact, the controller knowing the rest without seeing it. I was not fundamentally wrong about airspace, place, and proximity, but the explanation was even more complex than I thought. The emotional impact and awareness of risk varied by proximity, which varied both within and between facilities. Proximity included nearness, in interaction and involvement. More finely drawn, proximity must take into account a person's overall exposure, determined by their position in a facility, as it shapes the intensity, frequency, and duration of exposure to the activity, incident, or event.

Not everyone had the same exposure that day. Controllers who were working and who were not working that morning had different experiences. Also, among those working, there was variation in proximity. The controller in the previous quote makes a distinction between his involvement, based on his position as ground controller, and the effect on other controllers whose job had them getting the plane airborne or guiding its trajectory while flying. He also notes what he saw in relation to what controllers in New York saw. But think also about the duration of his experience: after the terrorists flew American 11 into the World Trade Center, Mike, Boston Tower supervisor, sent almost the entire day shift out of the action, away from the control tower, to the fire station. Then they went home. They did not take part in bringing down the planes and were absent for the scare when the Delta shuttle returned with the pilot who feared a bomb was on board. That controller was one of those who missed the other events that morning. Frequency of exposure also seems to generate a cumulative effect in perpetuating—or perhaps deepening—the impact. Tower and TRACON controllers who were "on" on September 11 and also "on" when the Continental lost control the week after September 11 talked about the double exposure as making a difference.

My proximity hypothesis was confirmed when I was readmitted at Boston Center for five weeks, including September 11, 2002. Here also the heightened security from my first visit was striking. Getting past the front gate required a different identification system. And although the guard in SWAT uniform and M-16 was gone, the usual facility guard who was there was armed. Once inside, I did not have the same freedoms to roam as before. All visitors had to be escorted from the front entrance to their destination in the building. Because controllers' alertness and concentration on the sky had to be especially sharp during the period around the first anniversary, I could not be in the areas. When I wasn't interviewing, I observed in TMU at the top of the control room aisle. For interviews, I was given an office in the administrative wing. I would be there during certain hours and anyone who wanted to come by to talk could do so. I was not allowed to ask about September 11, but if controllers brought it up, talking about it was OK. I did not ask; most who came by to talk brought it up.

The first anniversary of any major life event is an emotional trigger. The first anniversary of September 11 would have been on everyone's minds anyway, but as before, Boston Center personnel's sensitivity to risk and threat was regenerated by the very procedures designed to maximize safety and security. The air traffic control system was going to be in "lockdown mode" for a week bracketing September 11. Lockdown mode meant a return to the high-security strategies deployed in the weeks immediately following the attacks: restrictions around major cities, full military readiness in the skies, enhanced security at individual facilities (including increased police presence), no controllers leaving the area during their shift, and no visitors. No one knew what might happen. Moreover, air traffic controllers would be in the media spotlight. An NBC TV crew had been at Boston Center with cameras to film the anniversary special that would air on *Dateline* with Tom Brokaw at 8 p.m. on September 11.[4] They were filming controllers at Boston Center, New York Center, New York TRACON, Newark Tower and TRACON, and Cleveland Center. The documentary was featuring controllers whose position had required them to actively respond to the hijacked airplanes. Not everyone who was closely involved volunteered, but an impressive number did. They had to relive the experience publicly, under conditions that were strange to them. There was a logic behind their participation, however. At the time of the attacks and during the weeks after, controllers received very little national attention for their unprecedented achievement of grounding planes on September 11. Controllers understood this neglect as politically motivated. The Bush administration was moving toward privatization of the air traffic control system, so had no desire for the system to be viewed as exceptional or the accomplishment as extraordinary. Now, with national attention on the anniversary, the *Dateline* program was that opportunity, and NATCA had supported the idea.

The memory of the experience for everyone was emotionally keyed by the anniversary activity. In casual conversation and interviews, controllers at Boston Center talked about still experiencing the emotional effects. Some said it took them about six months to round the corner, settle it enough that it wasn't at the front of their minds when they were working. For safety reasons, those who felt emotional were offered the anniversary day off. Like some terminal controllers, a number of controllers mentioned feeling invaded and violated, taking it personally at the time and indicating that they were still feeling that way a year later:

> Well, the first day it was just, nobody could believe it happened and how are we going to deal with this. And the next few weeks, we just wanted revenge. I mean, we wanted to get back. People took—everyone of us had spent our entire professional career ensuring the safety of airplanes in the air and somebody took that away from us. Somebody took away our ability to make sure that that happens. And we don't want to let them get away with that. I worked very hard as an air traffic controller to make sure that two airplanes don't run together. For somebody to take one of those airplanes and run it into a building became very personal. It still is. And I think it is for a lot of people. (Center, M)

No one was unaffected, and controllers talked openly and emotionally about how they felt at the time and in the present. But like controllers at the terminal, the emotional effect and awareness of risk for a controller varied depending on proximity, defined as variation in exposure and involvement, as determined by position in the air traffic control structure: whether a controller was working that day, and in which area, involvement with the hijacking, the exposure over that day, and the cumulative effect of events in the weeks following. Peter, who talked to the hijackers of American 11, did not come by for an interview. However, the week prior to my arrival, he had participated in the *Dateline* interviews. On the program, he said that after the attacks he took six weeks off, during which time he was upset and unable to sleep. Even one year later, it was still very much with him: "I can still hear their voices, that will never go away from me. Just horrific, just the feeling of it, the voices, the, you know, they had control. You know, they had control and we didn't. And that was very scary. Because as controllers you're taught to have control. And there was none that day." Greg, who was working beside him and took over his airplanes, told me he was still feeling very deeply the experience of that day. He described the physical and emotional impact:

> At first I was afraid to leave the house, then couldn't stay in it and had to get out. I couldn't watch the TV anymore, but I read the newspaper every day to see are we getting them back for what they did. I never thought I would feel this way, not able to make a decision, not able to work, down about my job. (Center, M)

These feelings had lasted quite a long time. He said that the impact was such that he would have not been able to talk to me if I had been there within a few months after September 11. He had organized his days off so that he would not have to come in on the anniversary day.

All controllers were affected systemwide. The greatest affect was surely on those controllers whose proximity and exposure included personal ties and loss. Andy, whose wife died on American 11, went home after receiving the news at work and never returned to his job. Bob, a TRACON controller, knew Captain John Ogonowski, who was flying the same flight. Controllers who had friends who were pilots, as many did, knew the sorrow and fear they felt and tried to be supportive. For those who had no personal tie, we would expect that the emotional impact and awareness of risk for a controller would vary by position in the air traffic control organizational structure. So it would be greater for controllers in the Boston, New York, Cleveland, and Washington regions than for controllers in the Midwest or the South; in the New England Region, it would be greater for Boston Tower, TRACON, and Boston Center controllers than for Bedford; within facilities, it would be greater for controllers in Areas E and C than other areas, greater for controllers who handled the airplanes at Boston Tower, TRACON, and Boston Center than for those who did not; greater for those who had more intense experiences, based on shift assignments and exposure to the frequency, duration, and cumulative effects of fear-generating incidents while on duty.

Lori talked about the impact on people, like herself, who had two experiences—being on position on September 11 in Area C and again for the Queens accident—as having an even more enduring impact, resurrecting the emotions at times when pilots were silent, the course of flight appeared to be suspiciously deviant, or something was going badly wrong. As position in the air traffic control structure decreased or increased proximity to these events, the emotional effect, awareness, risk, and duration of the system effects would vary. Not that anyone would ever forget. In particular, Peter and others who talked to the pilots—the last ones to have contact with them—would have it not only in their mind but as a permanent part of their professional identity. But when a full year had passed, many controllers came to work and could invoke the usual techniques to neutralize risk and danger while working traffic, keeping awareness on the back burner—but only until an anomaly, like a pilot's suspicious activity or the first anniversary, brought it all back. Emotion work, too, was a fragile stability. Even then, as they had done during the worst moments on September 11, controllers did as they were trained: they sat there and worked through it.

This is not to suggest that the socially acquired ability and techniques to transform risk were firmly back in place, unchanged. To the contrary, the need for them was even greater in order to keep working traffic when flights resumed after September 11. On ordinary days with no surprises, emotion work was sufficient to transform risk and stress was enough to allow them to do the work, with one change: they could not yet do emotion work without thinking. They were aware of the gap between their experience of risk and the emotion work necessary to do the job. In recognition of this, an experienced controller said to me in a hallway, with serious countenance and mocking tone of voice:

> It's a game. It's a game, right? Controllers do it because it's fun. Controlling emotions is part of the game. (Tower, M)

The War on Terror: System Response and System Effects

The chronology of events in the year after the attacks shows the dynamics of a system undergoing rapid change. It was not a change planned by the FAA, like the well-coordinated and much-tested "cascade" of the DSR system in all centers. Instead, change was initiated from outside the air traffic control system: first by the attacks, then by the response to the attacks by the FAA headquarters and national security agencies that developed rules to protect the airspace and organize the return to flight, then by new FAA security procedures to limit access and material and technological changes to physically protect every facility, and then by airport procedures and technologies to protect passengers and airports. Not only were controllers charged with new responsibilities for surveillance of the sky, their very workspace was the subject of surveillance: security cameras, new identification systems, physical barriers, warning systems, police, guards, weapons, and dogs. The effects of the war on terror on controllers were as extensive and enduring as were those of September 11.

The combination of changing boundaries due to restrictions and recurring incidents meant daily uncertainty and the perpetuating of fear, risk, and stress for months. Again, structure, routine, and improvisation were the basis of controllers' achievements of that first year. However, in contrast to emptying the sky on September 11, during the initial months afterward, "neat and orderly went out the window"—standardization was absent and improvisation was the most essential factor to their work. Whereas normally, airspace boundaries had remained stable and air traffic moved through a sky marked by known routes and boundaries, controllers were working in a system where the boundaries protecting cities and sensitive areas shifted daily, and along with them, traffic routes, traffic mix, and workloads. Further, translating new rules, finding local-relevant meanings, and conveying them to an anxious flying public was new and unfamiliar work. Uncertainty and improvisation became an everyday matter. Controllers, trained to organize and order, worked hard to convert the disorder into order. Order was a locally coordinated accomplishment of problem-solving controllers in the facilities, knitting together new ways to work that fit the peculiarities of their particular airspace. After months, improvisation resulted in the codification and institutionalization of new structures, rules and procedures, routes, work practices—to secure air traffic boundaries both in the air and on the ground against incursion by intruders bent on destruction.

As remarkable as the air traffic control system's response to September 11—but unrecognized by the 9/11 Commission and the public—was its resilience and flexibility. This capacity was built in during the system's history: the crisis showed the system to be a complex web of different parts with boundaries that could be transformed from porous to impermeable to porous again, or from tightly to loosely to tightly coupled as needed, by the actions of controllers making decisions at all levels. Yet the dynamic flexibility of the system was hard-won. One of the biggest challenges—and aggravations— for controllers was the boundary problem: moving airplanes across segments of the sky so different that it was a source of daily conflict between controllers and facilities. Recall the comment of one controller, who observed that the boundaries gave the system a built-in fragility. However, September 11 and the year after made visible controllers' role in the system's resilience and flexibility revealing their ability to negotiate the intersection of standardization and improvisation as a vast source of the system's strength.[5]

Resilience and flexibility showed in many ways. By the first anniversary of the attacks, the system had undergone rapid change, achieving a fragile stability. Local coordination produced new sets of rules and procedures that were institutionalized, ready for a quick response to future terrorist attacks and other system intrusions. New technological communications systems were cobbled together from separate parts of both organizations and technologies, creating high-speed emergency networks. These gave new meaning to the idea of distributed cognition:[6] originated to describe cooperative meaning systems in small groups that combine humans and technologies, the expanded air traffic control communication system distributed cognition across national and international boundaries, its capability selectively including some or all, as necessary. Further, the new cooperative organizational arrangements established on 9/11 and the weeks after persisted, showing the flexibility and porousness of organization boundaries, as social relations shifted.[7] Normally conflicting goals that required daily negotiation with external players were set aside as air traffic control and the military, the airlines, and regulators worked together toward the common goal of protecting the country, air traffic, and the air traffic system. Within-system combatants over ownership, use, and design of the airspace and work arrangements—labor-management, controller-controller, facility-facility, and region-region—were moved similarly to cooperative coordination. System resilience, it turns out, is highly dependent on individual workers being willing to hammer out disagreements and compromises to craft new work arrangements on a daily basis in local settings.

This fragile stability extended to emotions and emotion work. Policing the sky was still taking its toll. Lax pilot practices set off emotional alarms and a system response: deviations from standard routines by pilots had controllers initiating F-15 chases. In the aftermath of the attacks, the variability in controllers' experiences of risk and stress continued to be defined by the relation between space and place. The impact was determined by controllers' positions in the air traffic control system—their facility and the exposure, frequency, and intensity of airspace incidents on them. The procedures designed to provide security to air transportation and to air traffic facilities perpetuated feelings of vulnerability, altering controllers' interpretation of anomalies. Added to this, translating and implementing the restrictions made stress a constant, for everyone. The emotional effects of dealing daily with calls from anxious general aviation pilots on controllers at the terminal facilities and the Command Center were new to them and exhausting. Talking within the facility about experiences of risk and stress was accepted because the source was clear and experienced by all, and thus not a sign of individual weakness. Whereas changes to the air traffic control system had been institutionalized and normalized, at the first anniversary controllers' emotions were resurrected—by the lockdown procedures, the public attention, and the collective memory, indicating continuing vulnerability. No one was certain what would happen next, or when it would, but controllers expected another incident.

11

Symbolic Boundaries

Distinction, Occupational Community, and Moral Work

The effects of September 11 and the year after on controllers—their airspace, technology, procedures, organization, work, emotions—were dramatic and enduring. As the system settled into the new normal, the attacks left controllers with painful memories that could surface unexpectedly at the low turn of a 767 or a clear blue sky on a fall day. I asked if their experiences on September 11 and the year after had drawn them closer to one another. They answered that no, it had not:

> I think you're going to find that regardless of what people will admit, air traffic controllers are a pretty close knit group. Because we have to be. We rely too much on each other on a daily basis. And we have for years and years and years. What other occupation do you know, you go to the same job with the exact people every day for fifteen years of your life? It doesn't happen in many occupations. You think of even being married to the same person for fifteen years, that's what it's like. It really is. You're with the same people, forty hours—some weeks more than forty hours a week. Day in, day out, it never changes. So you have—I think you have to rely on people and that in itself makes us close knit. But I don't think September 11 made it any closer.
>
> I think one thing that it probably did bring out in us is a little bit more pride in what we do. Just the fact that we shut down the air system, got all those planes landed the way that we did in the time frame that we did, is just an unbelievable feat. I mean, if you had told me it could be done that quickly before September 11, I would have told you you were nuts. But I mean everybody—

> didn't complain, just buckled in and worked plane after plane, put them down wherever they could. Fit them in as closely as they could. And got them on the ground. So I think there is a tremendous pride knowing that, you know, we could do that. As a profession. And that it was done. (Center, M)

This controller defines September 11 as the collective achievement of air traffic controllers as a profession, regardless of facility, geographic location, or FAA region. He recognizes that closeness characterizes them as a profession, distinguishing them from others. Although he does not name system effects, he infers them when he says that closeness is built into the regularity of their work ties and their necessary interdependence daily and over years. His comments reflect his sense of belonging to an "occupational community" that takes pride in its work.[1] In spite of the formal physical and social boundaries of the system that divide them one from the other by specialization and geography, they are joined in an acknowledged and felt occupational community, bound together and at the same time distinct from other occupations.[2] Lamont and Molnàr observed that symbolic boundaries are distinctions people construct that separate them into groups and generate feelings of similarity and difference as well as group membership, and bestow status, including and defining some people or groups and excluding others.[3] These scholars contrasted symbolic boundaries with objective social boundaries that divide and separate individuals into groups that block opportunities to resources for some but not others.

In air traffic control, symbolic and social boundaries intersect and work in unexpected ways. In addition to the purely physical boundaries of the system that separate them, controllers work in a system that stratifies and classifies them by social and physical location according to the type of work they do. The physical and social boundaries within and between parts of the formal system—regions, facilities—stratify them, regularly put facilities in competition with each other for opportunities for resources, salary, and social status. Even as controllers actively overcome or override the objective boundaries of the system by constructing symbolic boundaries that join them in an occupational community, in a remarkable response to these additional system effects, they construct symbolic boundaries that distinguish members of the community, one from the other. Within their separate locations in the system, controllers transcend the formal hierarchy and inequalities integral to the formal status ascribed to them by their social location.[4] Resisting standardization and a formal classification system, they elevate their status in relation to other controllers by constructing symbolic boundaries that make distinctions based on the importance of the work requirements of their airspace and individual competence at their own facilities. This chapter explores these two collective responses to system effects, which show the agency of controllers as they construct symbolic boundaries that override the formal boundaries of the system, enacting an informal status system that works for them.[5]

This form of boundary work is different from the boundary work they do to move airplanes across boundaries in the sky and across boundaries of facilities on the ground (part III). There, the work is physical, cognitive, and material; they are moving material objects by raising and lowering the position of aircraft in airspace. Here, the work is cognitive, subjective, and discursive; they are raising and lowering their social status by constructing symbolic boundaries that redefine their standing in relation to that of other facilities and controllers within their own facility. The collective construction of symbolic boundaries include strategies of social differentiation and distinction that allow them to transcend the constraints of the formal system three ways. They construct a collective occupational identity, correct inequalities and relations of power, and monitor and reinforce the morality of their work.[6]

Formal Structure and Occupational Community

In part III, ethnographies of the four facilities demonstrated that the airspace a facility owns affects the place—the technology, architecture, and the social organization of work—in ways that produce a different culture at each facility. In addition, as an occupation they collectively share a distinctive culture across facilities.[7] Van Maanen and Barley write that the physical and social conditions of work are responsible for the particular work cultures of occupational communities.[8] The experiences themselves have symbolic meaning, giving the work a special significance, setting those who do it apart from others. One of the identifying characteristics of occupational communities is related to boundaries: the crucial dimensions for recognizing one another are not ascribed characteristics or formal organizational boundaries, but "social dimensions used by the members themselves for recognizing one another." Occupational identity is social in that it is constructed and reconstructed in daily interaction with others, but it is also cultural because the experiences that constitute it are collectively experienced and common to the occupation.

Van Maanen and Barley observe that the physical and social experiences of work are the material from which occupational communities are formed and identify several typical characteristics: the shared experience of difficult and rigorous training, an esoteric skill, service and public trust, and risk and danger. These characteristics give air traffic controllers much in common with the military, police, and firefighters. However, the work of air traffic controllers has additional physical and social conditions that differentiate it from these other occupations: task interdependence and system embeddedness, a shared occupational history, and emotions and emotion work. Together, these three combine with those identified by Van Maanen and Barley in a set of shared experiences across facilities that produce a particularly strong and enduring occupational identification and bond.

Rigorous Training

All developmentals experience their training as a mental challenge and humiliating hardship. Regardless of training variations across the years, the process of producing controllers has retained certain basics. At the FAA Academy and at their assigned facility, controllers must survive time pressure, constant criticism, and public degradation by their coworkers. They must acquire the ability to do a lot of work in a short time; survive simulators; master technologies, airspace maps, phraseology, rules, and procedures and also apply them; concentrate when bombarded with distractions; tolerate stress; and continue despite daily failure. They talk about lasting bonds between those who trained together, giving them a network of ties with controllers in other facilities. Those who survive the full training process report their pride that upon passing that final hurdle. For them, it is a status passage with symbolic meaning. As one controller put it, it is like joining an "elite club."

Moreover, training and work experience changes them, giving them characteristics in common. Acquired cognitive skills, habits, and routines necessary to their work trickle over into their everyday lives. They apply dead reckoning while driving and other predictive activities. They prioritize, organize, and have in mind plan A and plan B. Automatically, in everyday social situations they multitask and scan, and room awareness makes them acutely aware of everything going on around them. They are supremely conscious of time, schedule, efficiency, and impatient with delays. Further, they report that, collectively, they are changed fundamentally as people, becoming more like what they define as "a type A personality": take-charge, confident, assertive, decisive, "my way or the highway"—in other words, controlling. They talk about possessing these characteristics collectively, "we" as controllers. These shared physical, psychological, and social traits become part of their collective identity, adding a distinctive quality to the work culture of the occupation.

An Esoteric Skill

Controllers share a common career trajectory in that they develop cognitive, linguistic, and technological capabilities so specific to their occupation that they cannot transfer to other specializations. Although some general skills are transferable (controllers mention the ability to prioritize, to organize, to think on your feet, to be efficient), the specific skills and experiential knowledge about aircraft, airspace, and controlling airplanes are not. Moreover, the system is always changing: airspace, procedures, and technological innovations are part of the FAA's continuous attempt to improve service and safety. Consequently, controllers' knowledge and the material practices necessary to the job are always being refined and are increasingly specialized. Upward mobility is tied to moving to new facilities. Retraining through the career is an ongoing enterprise.

Service and the Public Trust

Service and the public trust are officially scripted into the job. Reagan fired striking controllers in 1981 because controllers were public service employees, making the strike illegal because it denied the public safe air transportation and hurt the national economy. Beyond that official legal designation, part of controllers' collective self-definition as an occupation is that they are responsible for life and death, public service, and public trust. This cultural belief is rarely publicly visible. When on the job, their humor downplays the importance and seriousness of their work. As one controller announced while putting on a headset and plugging in, "OK, I'm ready to serve the American public."

However, in interviews, uniformly they spoke about service as a goal and the weighty responsibility that accompanies it. In response to the September 11 terrorists attacks, they spoke passionately about their professional responsibility to keep passengers safe by separating airplanes—"The job of the air traffic controller, number-one priority, is the safe movement of air traffic," "Every one of us had spent our entire professional career ensuring the safety of airplanes in the air and somebody took that away from us," and "Those were our airplanes." "Ownership" of parts of airspace is part of their professional identity. Airspace boundaries were penetrated by people who violated every possible principle on which the system was based: respect for boundaries, rules of separation, safety of passengers and crew. For controllers, their planes were stolen and their passengers were killed. Collectively, as a profession, they took it personally. They felt helpless to respond in a preventive way that lived up to the trust placed in them.

Risk and Danger

Van Maanen and Barley observe that "recognition that one's work entails danger heightens the contrast between one's own work and the safer work of others."[9] For controllers, shared experiences of risk and danger have symbolic meaning. Whether the everyday experiences of close calls, mistakes or deals, threat of accident, or the reality of the September 11 attacks, these experiences reinforce controllers' identification with the larger community of controllers. However, their collective construction of risk and danger—and their responses to it—are complex. They acknowledge the emotional impact of the work on them when talking about mistakes they have made in their career. Every controller had an incident. No matter how much time had passed, every detail of the situation was embedded in their memory, including the strong physical and emotional effects.

Paradoxically, controllers said the work was not risky. Because of the standardization of the system, they explained, its rules (in particular about spacing between aircraft), render the possibility of airplanes colliding practically nonexistent. So routinized is it that some of them complain that boredom can be a problem. Further, they deny that theirs is a stressful occupation. They say the work becomes risky and stressful in circumstances when they feel out of control: during thunderstorms, a pilot emergency, or a technological failure. Despite their convincing statements about how the standardization of their work minimizes risk and danger, symbolically, controllers inadvertently acknowledge the risk and danger when, individually and in groups, they develop strategies that redefine the conditions of their task, mediating their experience of risk and stress.

In addition to the physical and social conditions of work that Van Maanen and Barley see as responsible for the particular work cultures of occupational communities, I found three others specific to air traffic control: task interdependence and system embeddedness, a shared occupational history, and emotional labor and emotion work. Collectively, these aspects of the work also have symbolic meanings that generate feelings of similarity, group membership, and status that binds controllers together, distinguishing them from other occupations.

Task Interdependence and System Embeddedness

Controllers never work alone. Their work consists of distributed cognition between people, technologies, and material objects occurring in small spaces. *Coordination* is a word central to their vocabulary and their moment to moment activity on the job. The work of a single controller is continuously interdependent with the work of others, within the same small workspace, at other locations in the same facility, and across facilities. Similarly, they are affected by the actions of these others and beyond. Their interdependence and system embeddedness mean that many, or even all, can have the same or similar experiences due to some change in the external environmental that affects the entire system, despite the geographic distance between controllers in different facilities.

Many external changes—national and international economies, political administrations, airline mergers, budget allocations, technologies, labor issues, wars, regulatory policies—affect the National Airspace System and therefore the everyday work of every controller. A thunderstorm in the Southwest or a power failure at Denver Center changes traffic flows regionally and throughout the system. Similarly, special events, such as the launch and reentry of a space vehicle, affects air traffic in several neighboring regions, the effect spreading beyond local. An accident has both local and national implications: typically, an accident investigation results in new nationwide rules and procedures designed to prevent a similar incident, not just for the airspace and facility where the accident occurred.

A Shared Occupational History

Controllers' careers have a distinctive temporal trajectory. Those who pass all the requirements stay in the system, working for the same employer. Because of hiring waves, they have shared experiences with many others: the training, organizational and technological changes, accidents, management and labor disputes, facility lore, negative and positive work experiences with coworkers, and historical events outside the system. Some of these are local, and others are systemwide: airline deregulation, automation, government shutdown.

Some are sudden and unexpected shocks to the system, like September 11. These shared experiences are passed on from generation to generation, creating a common cultural heritage.

A defining moment was the PATCO strike in 1981 and the firing of approximately fourteen thousand striking controllers. This historic event was diffused and carried into the present by the experiences of generations of controllers. Obviously, those most immediately affected had gone on strike and were fired. In interviews, many PATCO controllers echoed the same sentiment in nearly identical words: "It wasn't just that I lost my job. It was like I lost my identity." A second group affected were those controllers who decided not to strike and so kept their jobs and their identity—like those fired, thought, they also lost their friends. Most have retired, but some still work as supervisors, other administrators, or staff. A third group affected were the "replacement hires" who began training in the first months after the firings. Individuals in each of these affected groups have stories that trace the special situation in which all of them found themselves at the time of the strike. In 2001, thirty years after, the replacement hires who were still working acknowledged that if it weren't for the strike, they would be doing something different. As one said, "We owe our jobs to those guys."

Union membership is another common bond among controllers. Starting with its formation in 1987, NATCA engaged in a new set of shared struggles and victories that renewed and strengthened ties between controllers. Collectively experienced, binding generations of controllers to their common history, being in the union was a defining identity and a component of controllers' work culture. Gaining power during the Clinton administration, NATCA actively worked with the FAA to improve technology available to controllers. Then under the Bush administration, those opportunities as well as rights and benefits were taken away. The replacement hires passed on the legacy as they trained the next generation, including some children and relatives of PATCO controllers. In today's FAA, the cultural heritage of the past lives on in gallows humor: "Well, they can't fire us all," someone will say. Collectively, across history, the union has always been central to the work culture and the occupational community. "We are union," they say.

Emotional Labor and Emotion Work

When discussing the shared experiences of risk and danger, Van Maanen and Barley note that workers identify as a distinctive occupational community through awareness that only coworkers can truly understand the "attitudes, behavior, and self-images for coping physically and psychologically with threat."[10] For air traffic controllers, emotions and emotion work come with the job. Routine and collectively experienced, this aspect of everyday work deserves to be singled out as a factor in its own right in the work culture of the occupational community. Emotions and emotional work are integral to the rise and fall of adrenaline throughout the day, in correspondence to the rise and fall of traffic rhythms, and in those moments when air traffic controllers lose control—thunderstorms, equipment breakdowns, and aircraft emergencies—and again at the commission of an operational error or when suddenly they suddenly fear that they have missed something. Then there are the unusual events—accidents, for which the emotional impact is typically local, but that is not always the case. In contrast to other occupations, emotional labor and emotion work can affect everyone in the system simultaneously. The collective impact on controllers, physically, socially, psychically, emotionally, has symbolic meaning, their work having a special significance that sets those who do it apart from other occupations.

Status and Moral Work

In seeming contradiction to the construction of an occupational community that transcends formal physical boundaries of the system, controllers construct symbolic boundaries that distinguish members of the community from one another. In both cases, they overcome or override the objective boundaries of the system that stratify and classify them by social and physical location, according to the type of work that they do.

Within this socio-technical system is a status system: a standardized system of classification by rank in which the status of a facility in the system determines the status of the individuals who work there.[11] The relevant formal classifications of air traffic control facilities in the National Airspace System are by type of facility, as determined by task and flight stage of the aircraft, so towers, TRACONs, and centers, and then a standardized numerical system that ranks each facility by airspace complexity on the basis of a complicated formula that measures traffic count, runway layout and geography, fleet mix, weather, and other factors that vary the difficulty of the job. The complexity formula was developed to eliminate pay inequalities in the system, so that controllers would be paid according to airspace complexity.[12] In 2000, the highest possible complexity level was 12: Boston Tower, TRACON, and Boston Center were level 11; Bedford was level 7. Throughout the system, then, and now, the status of the facility, salary, and prestige of the individual controllers who work there are determined by the facility's assigned complexity level.

In response to social distinctions of rank and prestige imposed by the formal system, between facilities air traffic controllers construct their own, creating symbolic boundaries that mediate inequalities in the system. Within facilities, controllers refine the occupation's collective sense of belonging to an occupational community by strategies of inclusion and exclusion: some members are in better standing than others.[13] Both between and within facilities, they construct an informal status hierarchy.[14] The creation of informal ranking systems within and between groups is a taken-for-granted aspect of all social life. What is interesting is the meaning the rankings have for air traffic controllers and the criteria upon which the lines of distinction are drawn.

Air traffic controllers create distinctions based on cultural understandings about competence in the occupation: the technical and cognitive skills they view as necessary to effectively do the work of controlling air traffic. Not only do these distinctions contest and contrast with the status hierarchy imposed by the formal organizational system; they also affirm social identity and a code of conduct for the occupational community. Controllers' categorization schemes are cultural because they are consistent across time and social space. In response to system effects, they transform the meaning of objective social boundaries, symbolically converting the meanings of the official and normative hierarchy of the organization in ways that have an impact on the material and cognitive practices of work.

Individual competence is a preoccupation of the formal organization that is equally significant in the occupational community. The requirements for becoming a certified professional controller are standardized, extensive, and difficult to complete, so all controllers must achieve a high standard of competence. Yet as with all occupations, there are variations in competence among those who all pass similar hurdles. Whatever the "true" variation in level of individual competence, the ranking distinctions that controllers construct takes on an insiders' "us versus them" quality that contradicts the formal numerical ranking of facilities bestowed by the system. These constructed distinctions within the community are acts resisting the formal hierarchical and bureaucratic organizational system of control in which they work. Controllers adjust the formal ranking system, creating informal distinctions that corrects perceived inequalities. In doing so, they overcome the formal status assigned by the system, adjust their facility standing in relation to other facilities in the system, and enact a code of conduct to pull all controllers to a high standard of professional performance required of their moral work.

Using competence as a criterion, controllers actively correct system inequalities by distinguishing controllers from management, facility from facility, and controller from controller. We have seen how the system accomplishes the social transformation of risky work; here we see controllers responding to the system by making symbolic distinctions that accomplish the social transformation of power relations in the system. This is also boundary work, but it is invisible work that is a process of ordering and reordering with a moral dimension.

Controller-Management Distinctions

A striking feature of the air traffic control system is that all administrators and staff at the Command Center at Herndon, regional offices, and in air traffic control facilities across the United States are air traffic controllers. Management and other specialized personnel, such as in cartography and quality assurance, have all had the training, become certified professional controllers (CPCs) and worked air traffic. This is true also of personnel at FAA headquarters in Washington, the only exceptions being FAA directors and their entourage of administrators. Consequently, all managers, supervisors, and controllers in the air traffic control system self-define as members of the same occupational community.

However, controllers distinguish themselves from management, increasing their relative status as they downgrade that of management. The typical controller view is that the controllers who move into management are "failed controllers": people who couldn't work air traffic and so opted to shift to a management track, first becoming a supervisor, then moving into other administrative positions or to a staff position. Controllers believe that controller incompetence leads naturally to managerial incompetence. This cultural belief is reinforced by the fact of civil service: once people are in, they stay in. In general, all the FAA's failures and problems are, in the view of those controllers working air traffic, due to the practice of internal promotion and the principle of "incompetent controller, incompetent manager."

The us-them dichotomy by which controllers working traffic discursively distinguish themselves from managers is not surprising, given that distinctions of this sort are typical in organizations because of the distribution of power and authority that typically undergirds management-worker relations.

What is different about this occupation is that skill and competence, not seniority, level of authority, or salary, are the criteria on which management-worker distinctions are based. In each facility, the belief in managerial incompetence is mediated by experience with particular management personnel. Controllers give anecdotal evidence to back their views, citing individual manager histories, including early career events at other facilities that become part of a person's "rep," personal characteristics, training problems, errors in judgment, or mistakes to support their evaluations. These reputations and anecdotal histories are a focus of much humor that publicly constructs and reinforces status differences on a daily basis. I heard several controllers repeat what appeared to be the standard joke about one of their supervisors: "He's a nice guy, but don't let him touch anything."

Controllers acknowledge the exceptions: many supervisors, operations managers, and other administrators and staff not only do not fit the incompetence model but also were respected as controllers and are supremely competent and therefore respected in administrative positions. Facility size may have something to do with it: controllers' belief in managerial incompetence was more frequently voiced in conversations and interviews at Boston Center, which, due to its size, is more bureaucratic and has more supervisory personnel and controllers than the other facilities. The managerial incompetence belief was rarely heard at Boston Tower and TRACON. Both are small facilities that cut short the physical and social distance between administrators, staff, and controllers, thus increasing the interaction (for better or for worse). Also, though, at the time of the study both were headed by an air traffic manager who stayed in touch, knew their names and family situations, did required retraining as controllers during busy times not low-traffic times, did not micromanage, and whom most controllers respected. In contrast, at Bedford, also small, controllers held their ATM in great disdain, mainly because he displayed little interest in controllers' problems, seemed out of date, and seldom left his office to visit the tower. Despite acknowledging many exceptions, managerial incompetence stands as a widely accepted cultural belief about status within the occupational community.

Further reversing the formal authority hierarchy, controllers self-identify as the essential component of the system. Recall the comment of the controller who said, "The pyramid is upside down in this organization," or, the most important decision makers, those ultimately responsible for safety and human life and the ones the FAA absolutely cannot do without, are controllers. Controllers complain about being the grunts in the formal structure. However, salary and shift work have symbolic meaning to them that opposes their formal structural position. The salary scale contradicts the formal hierarchy: controllers make more money than their supervisors and many other administrators and staff in their facilities, due to overtime and the shift worked. Second, controllers point out the essential character of their jobs with this observation "If you want to know who's important around here, look at who goes home on weekends and holidays and who stays here to work," a reference to shift work and the public necessity that controllers work around the clock. They complain about shift work because it interferes with family and social life, which it does, but collectively they define shift work as a characteristic of their essential occupation that provides a service central to the well-being of the flying public and the economic and social well-being of the entire country.

Facility-Facility Distinctions

Controllers make between-facility distinctions based on what controllers at each facility define as "real air traffic control." Their definitions of real air traffic control claim the superior status of their own work and workplace (and thus the prestige of individuals who work there) over other regional facilities, in contradiction of the assigned numerical ranking by traffic complexity for facilities. They make these distinctions on the basis of perceived differences in the challenges of the airspace and the cognitive skills controllers that are required at each place. At each facility, controllers believe that their particular traffic situation constitutes "real air traffic control," in contrast to the traffic at other places, which, in their view, does not.

Controllers at Boston Tower and the TRACON, both level 11 and both operating at Logan International Airport, define their work as "real air traffic control" because of the amount and complexity of the traffic, the small airspace and crossing runways, and the fact that traffic is relentless. Their self-assigned status matches that assigned by the formal system. Unlike the other facilities, where the traffic rhythm has highs and lows that allow controllers some downtime, controllers at Logan rarely have time to relax unless on breaks, which are short. Their sensibility about their status as controllers over those controllers at the other two facilities rests in the speed and accuracy of the judgments that they must make and the close proximity of the airplanes in the sky as they control them. Further, at Boston Tower, ground control takes special skills and pilot behavior is less predictable, especially at high-traffic times when aircraft have to get out of the way of arrivals and departures and when traffic jams occur at airport "hot spots."

Controllers at Boston Center (level 11) and Bedford (level 7) acknowledge the time pressure and hard work that the job entails at Logan, but they also downplay the cognitive work and technical skill required by referencing Logan's greater standardization. Center controllers assert their status over Boston Tower, saying, "It's just "Cleared for takeoff, cleared to land, cleared for takeoff, cleared to land." In contrast, Boston Center controllers distinguish their work and selves as "real air traffic control" because at high speed and high altitudes, commercial jets can present problems during thunderstorms and high winds, so require controllers to improvise more. They downgrade the skills at Boston Tower with the often-heard comment, "The best place [for a controller] to be during a thunderstorm is at the airport." They assert that if the airport is in the storm zone, all traffic is grounded and therefore controllers have nothing to do. In contrast, during thunderstorms standard routes are out the window and all the rerouting is done by Boston Center controllers. In their view, their traffic is more challenging than that of Boston Tower; therefore, they do "real air traffic control" and the other two facilities do not. As I overhead the NATCA representative at the center say in a phone call to the Bedford representative who was moving to Logan Tower, "When are you going to stop being a tower flower and come up here with us to do some real air traffic control?"

Bedford controllers assert their own status as doing real air traffic control by pointing out their improvisational challenges versus the standardization at other places: "At Boston Tower, the TRACON sets everything up for them. We have to set up our own [approaches for landings]." Bedford controllers define their work as "real, seat-of-the-pants air traffic control." Controllers deal with opposites in motion in a small airspace: both VFR and IFR, military and civilian, general aviation and commercial jets, aerobatics pilots and student pilots. This, in contrast to the more predictable behavior of professional pilots and commercial jets that dominate the fleets for Boston, the TRACON, and Boston Center: at Bedford, pilot error, misunderstanding, and deviation from controllers' expectations are routine. Bedford, Boston, and the TRACON downplay the skill required for controllers at Boston Center, emphasizing instead the standardization ("that place is a factory") and that they don't work very hard ("whenever you go there, you see all these people just sitting around in the cafeteria").

Controllers at the three level 11 facilities keep Bedford low in the status system by referencing its controllers' traffic count, extreme seasonal fluctuation, small airspace, and quality of controllers' work. Many controllers begin at small towers and graduate to larger ones with more complex traffic. Unlike many small towers, however, Bedford has a complex airspace and large number of operations for the size of its controller staff. Many Bedford controllers who desire upward mobility move from there to larger facilities: Manchester Tower, Boston Tower, and occasionally Boston Center. And many train and succeed, so they do move. However, other facilities' definition of Bedford rests on the collective memory of those Bedford controllers who moved up, trained at the larger facilities, but didn't make it through the training so returned. Bedford's status is defined by its "washouts," and therefore is not real air traffic control.

Controller-Controller Distinctions

Finally, air traffic controllers construct distinctions based on competence within the same facility. Each person's work is public and visible. Working with the same people in a coordinated activity, day after day, year after year, controllers know one another's preferred techniques, levels of competence, temperament, and personal work history, for they witness one another's everyday victories, failures, and good and bad days. They gain status within their facility relative to coworkers by rankings based on the safety and efficient movement of air traffic. From levels of perceived competence, they construct three symbolic categories. The "natural" is the rare person (man or woman) for whom dead reckoning is always easy and fun. They are always cool and quiet, never "go down the tubes," and are so skillful that they can handle many, many airplanes easily and still look for more; they can predict aircraft position and possible problems unusually far in advance and so make decisions early, decreasing the number of pilot contacts they have to make. These are "the artists": they are a cut above everybody else, defined by others as "born to it" or possessing a "natural talent."

Most controllers are the "technocrats." The performance of most controllers is not quite up to the level of the natural, but these controllers are still highly qualified and respected by peers and supervisors for their consistent technical skill and efficiency. However, even technocrats can have an off day. One controller said, "All of us have days when we are just not as sharp, not as into it, as other days." Their good sessions typically do not get a public sign of approval, but every miscue is met with jokes or insulting humor that calls attention to the gaffe. Humor slices through the common cultural identity of the occupation to publicly stratify and rank those within it. Humor is used by the group as a sanction in order to maintain professional standards and to pull the deviant toward the group standard for quality professional performance. Some humor, directed toward competent controllers who struggled with a particular traffic situation or were having a bad day, is sharp but a sign of group membership and affection. Publicly degraded, controllers are respected and liked, one of the group, but the message is still "you could do better."

Then there are the controllers, few in number, that the group defines as having "below-standard" professional competency. All controllers who survived training are officially competent. However, traits that the group equates with "below standard" are those that interrupt the group effort and rhythm. One person having a bad day adds unpredictability to others' dead reckoning and, for the group, interferes with the goals of safe, orderly, and expeditious delivery of air traffic. Controllers in this category may be always competent but not as fast as they used to be, or competent in routine situations but struggling in unusual ones, overwhelmed by heavy traffic, or unable to change a technique to handoff or receive traffic from people with different techniques. Having two or more operational errors over a few years can be the subject of gossip and sometimes facility lore. Alternatively, a controller may often feel out of control and so be "overly cautious." Being overly cautious is taken as a sign of weakness. On radar at Boston Center, they might expand the distance allowed by the rules of separation from five to six miles to feel comfortable ("a sissy ring"), or at the TRACON they might allocate greater distance between planes when lining them up on final. Or, if allowed to choose, they prefer the least busy position.

Those who repeatedly interrupt the rhythm of the group effort get harsher and more frequent criticism. If a mistake is egregious, they get the worst sanction that controllers on position can give: silence. Their perceived gaffes are talked about behind their backs, in conversations that reinforce the status differences between the target of the criticism and the criticizers. Nonetheless, they are not left to flounder and err when they cannot rise to a challenge. In difficult situations, one said, "We step up to the plate." They are watched out for, worked around, and helped out by controllers working nearby, so that safe, orderly, and expeditious delivery of air traffic is maintained and the group performance is minimally affected. However, symbolically, they are assigned a lower status than others because they are perceived as less competent. Marginal to the group, their presence reinforces the higher status of the others.

Maintaining Moral Boundaries

Significantly, the three types of symbolic rankings—controller-management, facility-facility, and controller-controller—are a historical pattern in air traffic control. The controllers whom I spent time with and interviewed were from a wide variety of cohorts, and many had worked at other US facilities. These conversations confirmed that the distinctions are a consistent pattern, along with the use of humor as a mechanism of social control to remind controllers of group standards. The consistency in these informal distinctions across time is not surprising, given the cultural embeddedness of occupational identity and the nature of the formal organizational system in which controllers work.

Why do air traffic controllers construct a categorization scheme that stratifies their occupational community on the basis of competence? A possible explanation is that it is a character trait: in the public imagery, all air traffic controllers are competitive and have big egos, so it is natural that they would be preoccupied with ranking facilities and selves in relation to each other. This thesis is not as reductionist as it sounds. Air traffic control is a character-shaping occupation: the rigorousness of the training and the requirement of the job to be assertive and decisive with pilots creates people who, in common, have survived a highly competitive process to earn their job and are confident in their work. However, there are limits to how much a character trait can explain. Although some air traffic controllers describe becoming competitive and having big egos as an outcome of the training, not all controllers acquire that trait. In contrast, distinctions based on competence and the way controllers actively work to keep up a certain level of performance have become cultural understandings passed through generations.

To answer this question, we need to reconsider controllers' embeddedness in the felt occupational community and the structural array of positions in the formal system. First, skill and competency matter enormously in the everyday work of controllers, from both an occupational community and an organizational standpoint. A shared cultural belief in the occupational community is that their work is, at its core, moral. In common, they understand that as an occupation they have responsibility for life and service and the sense that their work is a public trust. To fail to live up to their responsibility for the safe, orderly, and expeditious delivery of air traffic has life-or-death consequences for others and for the well-being of the nation. Not only is this responsibility shared and felt across facilities; controllers also derive symbolic meaning and identity from it. Second, and crucially, task interdependence, system embeddedness, and the need to coordinate within and across formal parts of the system link controllers in collective action: systems within systems within systems. The competence of all management personnel and support staff whose jobs are directly tied to the air traffic operations affects how well controllers are able to do their jobs. Further, in this teamwork-oriented, coordinated socio-technical system, one facility's actions can either help or hinder those of another; similarly, one controller's actions can either help or hinder another controller. For this reason, coordination is more that communicating intentions and collectively acting; it is helping one another when struggling under traffic burdens or just having a bad day.

Abbott writes about intraprofessional status systems that arise on the basis of categories and classifications in a given cultural system of knowledge.[15] They rest on judgments about "purity of practice," observing that the accumulation of such judgments produces a social structure in which they become associated with a profession's division of labor. The result is a status hierarchy that is particular to a profession. In air traffic control, one person's ill-advised action, failure to be cooperative, lack of knowledge or technical skill, misjudgment, or slow response directly affects other controllers, impeding their abilities to make the speedy decisions and actions necessary to move traffic and keep the system operating smoothly and safely. Because of system interdependence, incompetence by anyone jeopardizes this moral work. How much other people's competence affects controllers can be truly understood only by seeing how their work multiplies when others do not do their job properly. To make intrasystem distinctions using a categorization scheme that emphasizes attention to professional standards, a code of conduct for technical skills, and humor to pull people to group standards for performance is a logical response to both their social identification with the occupational community and to the interdependent nature of their tasks in the system.

Finally, although all controllers have the moral responsibility of the occupational community and are vulnerable to system effects mentioned throughout this book, the system does not reward them equally. They are ranked in a bureaucratic system that distinguishes facilities by a ranking system that affects individual prestige and salary. The insistence at each type of facility that the airspace characteristics are particularly challenging and therefore "real air traffic control" can be understood as collective resistance to working in a standardized system that stratifies facilities on the basis of a complex formula that uses objective, universalistic standards. The formal system contradicts the subjective experience of controllers who must develop local knowledge and particular skills to handle the peculiar requirements of the airspace and traffic that are built into the variation in this so-called standardized system. Examining the processes by which controllers who are bound together in their occupational community construct differences among themselves shows stratification as an emergent feature of social relations rather than solely an external apparatus that acts upon social life. In their response to the unequal access to resources, including status, that is inherent in the formal organizational hierarchy and rankings, controllers actively resist the classification system imposed on them by the standardized system, creating symbolic boundaries that reflect their work experience in their own facility, their own sense of standing in the community, and their contribution to the system.

PART VI

System Effects, Boundary Work, and Risk

The preceding historical chronicle of controllers' response to the September 11 attacks, the remaking of the system, and the symbolic boundaries and occupation community that binds all the fieldwork chapters together might seem a natural end to this book. Finished except for conclusions—at least that's what I thought while writing it. But even as I was writing, a crisis was unfolding that pulled me back to the field, allowing me—and therefore, readers—to witness a transformed workplace in crisis.

The year 2001 began a new era, the age of automation, although the sources of change affecting the system and dead reckoning were complex and intersecting. The post–September 11 era was like those that had come before: characterized by many turning points, contingency, and unanticipated consequences of planned change.[1] However, it was strikingly different in one way. The system had twice before experienced crises precipitated by unexpected actions by political actors outside the system that increased risk: Reagan's firing of fourteen thousand controllers in 1981 and the terrorist attacks against the United States. In contrast, the crisis in the age of automation developed incrementally, growing slowly from actions taken or not taken by actors in the external environment, then responded to by problem-solving FAA officials external to the air traffic control system. Seemingly small events and actions can have large consequences.[2] Each change may have been insignificant in its own right, but when implemented, unintentionally but cumulatively they increased risk. Here's what happened.

In the year after September 11, the FAA's the air traffic control system not only had recovered from organizational shock but also had reinvented itself to suit the new order. The collective accomplishments of all controllers on that day and in the next year powerfully demonstrated that the resilience, reliability, and redundancy of the system—its safety—was dependent on controllers' expertise: their ethnocognition, their boundary work, and the interdependence of controllers and their multiple technologies of coordination and control. Once again, these embodied material practices and processes had been foundational to the system's safety. Following this major turning point, controllers believed that the system was prepared for whatever came next on the national security front, and that life and work in the system would soon resume a more normal trajectory. And it did for a while.

But as history has demonstrated, institutional persistence and change live side by side in this system. In the two decades that followed, problem-solving actors external to the air traffic control system made decisions about the system that despite its aging technologies, physical structures, and infrastructure, had just demonstrated that it was in outstanding working order, unintentionally brought it to a crisis situation. During system emergence and development, we saw how trajectories of technological and organizational innovations, originating from multiple starts in different locations, intersected with system development in positive ways to advance the system.[3] Now, however, two independent trajectories—both begun in the 1990s—intersected to produce a critical situation and increase risk. In response, FAA officials made changes intended to improve the system that had unexpected negative effects on all facilities, training, and controllers' dead reckoning.

The first trajectory began as both the aviation industry and the FAA were under pressure to modernize. Other countries—Canada and Great Britain leaders among them—were engaged in modernization and as part of it, privatization.[4] Handling one-third of the world's traffic, the United States had to maintain its position of global leadership in aviation. At that time, the FAA began a systemwide modernization program to meet the anticipated demands of the future: The Next Generation Air Transportation System. Known as "NextGen," it was a paradigm shift in air traffic control. The purpose was no less than switching from a ground-based to a satellite-based navigation system. NextGen promised to double the capacity of the skies without expanding the workforce to double productivity. The FAA plan had both technological and organizational innovations: full automation; realignment and consolidation of TRACONs in each region, relocating them in new buildings to replace deteriorating structures and technical systems from the 1960s; and finally, streamlining FAA operations of the air traffic control system by centralizing and standardizing practices that were formerly the domain of the individual facilities.

As these modernization efforts moved into the implementation stage, the predicted controller staffing shortage fell upon the system full-scale. Also begun in the 1990s and moving in parallel, the shortage was triggered by repeated congressional budget cuts and fueled by the retirements controllers had warned about. In response to the staffing crisis and NextGen demands, the FAA made systemwide adjustments, followed by official changes implemented by people below them in charge of subunits in the FAA administrative system. Delegation of authority is typical in bureaucracies, large and small, but when large complex systems go from a general goal to the particulars of implementation in specialized subunits, the ramifications cannot be predicted. The coincidence of the two trajectories affected work in unanticipated ways at all facilities.

We know some things about organizational evolution and change over time,[5] and also about how change in one part of a complex system invariably will have unanticipated consequences for the other parts.[6] However, we know less about the effects of large-scale socio-technical system modernization on the workplace and workers' embodied cognition, their cultural understandings, and the material practices of their technical work. Specifically, while many have written about the effects of automation on the labor force at large and the de-skilling of individuals, but there is little accumulated knowledge on the transitional experience of how workers adjust to automation and to the workplace changes in architecture, placement of material objects, and the reorganization of tasks that necessarily follow.[7] Moreover, we know some things about organizational persistence, but much of it focuses on larger institutional structures, conformity to normative expectations for legitimacy, and mechanistic processes that stabilize organizations.[8] Less is known about the internal actions initiated by the workforce that respond to debilitating changes.[9] In these two facilities, the workforce drew upon the past, taking actions that curated changes to suit the local situation in the present and improvising tools of repair, thus unintentionally contributing to organizational persistence.[10]

Returning to Boston Tower and Boston TRACON during the fall of 2017 to see the effects of NextGen and the staffing shortage on dead reckoning, my original research questions were even more relevant: what makes the air traffic control system so safe—or is it? And what do controllers bring to the system that technology can't replace. I was surprised to discover that more than a decade after NextGen had been implemented, this major transition was still ongoing in the midst of the staffing shortage. The effects of history were visible in the workplace as controllers tried to reconcile past with present, simultaneously improvising ("How can we make this work here?") and still perpetuating, re-creating, and hanging on to what they had before.

The liabilities of NextGen's technological and organizational innovations had been extensive. Again engaged in repairing the system, managers, supervisors, and controllers had worked hard to restore it. Much had been collectively accomplished, but it had been a hard time and continued to be. Shorthanded, everyone was working overtime. Senior controllers were training the new generation of controllers that had begun trickling in. The processes of ongoing change and struggles over culture, structure, and power within facilities had created rifts and lingering resentments. It was about dead reckoning—ethnocognition, boundary work, and human-technology interaction—but it also was an unprecedented look at conflicting cultures and the ongoing interactional negotiation of repair, the redesign of architecture, technologies, and expertise and the role of agency in local adjustments over time.[11]

Although all four regional facilities illustrate difficulties during this systemwide transformation, Boston Tower and TRACON best illustrate the juncture of the staffing crisis and NextGen because in combination they demonstrate the effects of all three NextGen goals: automation, realignment and consolidation of TRACON facilities, and centralization and standardization of FAA operations. Moreover, they show that standardized systemwide changes designed for the many do not readily fit all. In part VI, I continue as I began the book, locating this new fieldwork in its historic context. I trace the post–September 11 political, social, and economic actions and actors outside the system as they affected the FAA and how the FAA responses affected the system and the workplace. While I focus on the two central trajectories, their coincidence, and the collective impact on the work of all controllers, additional trajectories of events of varying impact also shaped the situation.[12] The disempowering of NATCA by the Bush White House, the Obama administrations's legal empowerment of the union, and the union's consequent restructuring that made it more powerful than before were essential factors in the response to NextGen and the staffing crisis at both the national level and in the facilities.

Then in chapter 12, "The Age of Automation: 2002–Present," the ethnographies of Boston Tower and Boston TRACON show how the intersection of the staffing crisis and NextGen affected the workplace and dead reckoning, increasing risk. The chapter reveals the continuities between the past and the present and how two generations of controllers grappled with the crisis. Because of differences in airspace, place, and tasks, problem solving at each facility was shaped by preexisting cultural differences that resulted in contrasting responses in the two facilities. However, the comparison affirms the agency of controllers, boundary work as power work, and the empowerment of expertise in both facilities. Once again, controllers' response to standardization was improvisation: they invented tools of repair at the local level to preserve system safety. Following my strategy in chapter 6, I first combine Boston Tower and TRACON to show the common system effects and then separate the two facilities to show the differences. Chapter 6 and chapter 12 demonstrate the transformation of the system and dead reckoning before, during, and after automation.

Chapter 13, "Continuities, Change, and Persistence," takes the long view in order to answer my beginning questions: what makes the system so safe—or is it? And what is it that controllers bring to the job that technology can't replace? To answer these questions, the book has traced system history from its emergence as a new social entity, its institutionalization as an actor in its own right, and how it incrementally acquired the characteristics it has today. Linking actions of political actors in the institutional field to the agency of problem solving actors within the system, the chapter reveals not only the continuing vulnerability of the system, but shows controllers as the source of system resilience and consequently, persistence. In the concluding section, "Dead Reckoning: Coordinating Action and Anticipating Futures in Complex Organizational Systems," I extend the concept of dead reckoning to all complex organizational systems and explore the implications of this book for modernization, change, and risk.

⁂

As air traffic controllers' post–September 11 role in policing the sky became routine, a fearful public stopped flying. Civilian air traffic operations dropped precipitously to new lows from all-time high figures in 1999–2000, where they would remain for several years before beginning to pick up again. Military surveillance flights continued. Aircraft quotas were still in use at major urban airports to assure security surveillance. Bush had created the Department of Homeland Security as a cabinet office in the fall of 2001 and the Transportation Security Administration (TSA) in December 2001. With the Bush administration occupied with the war on terror, the invasion of Afghanistan in

October 2001, and later the Iraq War, wartime headlines obscured what was going on at the FAA. The Clinton administration had provided the basics to move forward with the modernization of the National Airspace System. Also, Clinton's economic policies had produced one of the longest periods of economic growth in American history.[13] Bush inherited Clinton's historic budget surplus and ongoing modernization initiatives: the new Air Traffic Organization, with its goals of a leaner, meaner businesslike structure and systemwide automation.

The two trajectories that would later intersect to jeopardize safety were unfolding. Already being tested in a few selected facilities were several Clinton-era automated technologies, and the two most ambitious were well under way. En Route Automated Modernization (ERAM) was the planned replacement for the much-patched-over 1960s Host Computer System (HCS) that were still powering controller technologies at all high-altitude centers. The companion system upgrade for towers and TRACONs was the Standard Terminal Automation Replacement System (STARS). Consisting of state-of-the-art color displays, new computer processing and communication equipment, weather information, multiradar tracking, and redundancy in case of failure, it replaced the old ARTS system. The combined estimated value of the contracts awarded to contractors and subcontractors for the two projects was $1 billion through 2012.[14]

In contrast, the essential human resources necessary to the system were in jeopardy. The staffing shortage was grim. Insufficient controller staffing had been a chronic problem even before the Reagan administration. Beginning 1992, the FAA invoked several hiring freezes in response to congressional budget cuts. Twice, in 1995 and 1996, the government shut down. Bush policies left many small towers understaffed, and agency hiring never caught up with staffing requirements for facilities in congested urban areas. In addition, the controller retirements that NATCA repeatedly had warned Congress about were even higher than predicted. A 2002 Government Accounting Office report stated that the number of retirements during the previous five years was likely to double in the following five. The FAA would need to hire and train thousands over the coming decade to meet predicted traffic demands and compensate for expected attrition, but the FAA had no plan to address the problem.

Controller concerns about future staffing surged in June 2002. Slipping past a media and a public distracted by the Iraq weapons of mass destruction controversy, President Bush moved to ensure the continuation of privately operated contract towers, then numbering over two hundred, and clear the way to privatize the air traffic control system. The idea had been proposed by every Republican president since 1973.[15] In every instance where a privatization proposal had reached Congress (not all of them did), it never moved forward because of unresolved issues about how to pay for it.[16] NATCA had always fought hard against it, arguing that the pursuit of private profit would interfere with safety in the system, and as a government monopoly, the system was protected from such competition. This time, congressional opposition was exceptionally strong in the wake of the extraordinary September 11 performance of controllers and continuing security threats. To defer the expected congressional battle, the Bush administration relented in its efforts to include privatization as part of the FAA Reauthorization Act of 2003.

Boundary Work as Power Work

Early in 2003, people began flying again.[17] Several major airports began to experience a resurgence of traffic and, with it, congestion and delays. Anticipating a tripling of air traffic by 2020, the FAA began its own form of dead reckoning, planning changes to keep the system responsive to predicted conditions. The airlines had nearly eliminated props and gone instead to regional and larger jets that could carry more people. Crucially, the larger aircraft created traffic jams on airfields at major airports, where aircraft movement and parking space was limited. Capacity, consolidation, and efficiency became the goals. The FAA returned to strategic tools of repair from the past: boundary work on the ground and in the sky, this time to squeeze more airplanes—safely—into the existing system. In the sky, capacity was increased by creating more routes and altitudes, consequently allowing for more efficient routings, saving time and fuel.[18] In addition, for high-altitude airplanes equipped with the automated warning system TCAS and flying in controlled airspace at high altitudes, the amount of separation for safe vertical distance between airplanes was decreased. The new reduced vertical separation minima allowed a reduction from two thousand to one thousand feet for planes flying at altitudes between twenty-nine thousand and forty-one thousand feet.

Modernization efforts began early and incrementally, then grew into an official major program. Perhaps the most futuristic change to boundaries in the sky was a sophisticated application of GPS in the cockpit that connected to the satellite-enabled form of air navigation known as free flight. Automation would allow the airplane to "know its own position" and make precision landings at airports, increasing capacity by allowing runways to be closer together and thus reducing congestion. The implications for dead reckoning were vast. The US airspace system would be transformed from one running on ground-based navigation to one in which pilots would fly by a system of "waypoints," navigating to any place in the world at great savings in cost and efficiency.[19] A waypoint is a reference point in physical space used in navigation, sea, or air that marks a stage in a journey, a point at which direction is changed. Pilots would be on their own. Controllers would still have full monitoring responsibilities, but little control, intervening only in emergencies. To be implemented in three stages, the year 2020 was targeted for full operational use.[20]

FAA strategies for major changes to boundaries on the ground were already in the works. First, TRACONs within each of the nine geographic regions would be consolidated in new large facilities designed to hold several TRACONs. Second, the FAA would reorganize its own boundaries. The details of the Clinton-initiated FAA Air Traffic Organization were announced. The FAA would streamline its business structure, centralizing and standardizing certain types of decisions at FAA headquarters and elsewhere in the system that had always been the responsibility of local ATMs and NATCA officials. The goal was to reduce some of the variation in work practices, training, and hiring across facilities. As the overarching structure, in January 2004, Secretary of Transportation Norman Mineta announced that the three goals of automation, consolidation of radar facilities, and streamlining by centralizing and standardizing would combine in a single interagency plan for the Next Generation Air Transportation System, or "NextGen."

Almost at the same time, the FAA initiated boundary work as power work on the ground. Acting to further increase system efficiency and cost savings, the agency reduced the power of NATCA and accelerated the staffing shortage. NATCA's collective bargaining agreement with the FAA was up for renewal in 2005. Signed in 1998 and in effect for five years, the Bush-appointed FAA head Marion Blakey extended it for two more. Negotiations began in July 2004. In November, after months of bitter negotiations between the FAA and NATCA, Blakey called for federal mediation to help the agency reach a voluntary contract agreement with the controllers' union. A central disagreement was compensation. NATCA made a cost-savings proposal that included more than $1.4 billion in pay and benefit cuts, but the agency had more severe cuts in mind. The FAA broke off talks in April 2006, declaring an impasse. Following procedure, Blakey sent the FAA's final proposal for a NATCA contract, along with NATCA objections to it, to Congress. If the legislators did not respond within sixty days, the FAA could implement its own proposal. Legislators did not respond.

On Labor Day in 2006, the FAA contract was imposed on an angry workforce. The cruelty of the Labor Day timing was not lost on them. With billions going to organizational and technological changes, the cost of the controller workforce would be reduced. Active controllers received a freeze in base pay and elimination of their premium pay opportunities (overtime pay for weekends and holidays). Pensions were reduced. New hires would be paid 30 percent less than those already on the job, creating a two-tiered system recognized by many union members as an old "union-busting strategy" designed to force out experienced controllers and hire cheaper newcomers, effectively increasing their workloads for less pay. Equally offensive, if not more so, were the FAA's new unilaterally imposed work rules. The shiftwork of thousands of controllers included the "quick turnaround"—working from 6 am to 2 pm, then reporting back to work that night, eight or nine hours later for the midnight shift. Nonetheless, the FAA decreased time between shifts. Rest breaks after every two hours of working traffic were eliminated, canceling a longstanding practice that controllers said was a major way to fight fatigue. Supervisors could order their controllers to work overtime, and vacations could be canceled or shortened if staffing was short or traffic unexpectedly high.

Controllers viewed this as a "return to the draconian management practices" of the strike era. Due to insufficient staffing, some controllers had already been working ten-hour shifts and six-day workweeks. In fighting back, NATCA warned that the FAA proposal would accelerate the staffing shortage by deterring people from applying for the job and spurring early retirements. Indeed, the staffing situation grew worse. In the first two years after the work rules were imposed, 3,356 active controllers left the workforce, far exceeding FAA predictions. In December 2007, a Government Accountability Office report expressed concerns about controller fatigue, citing the large number of controllers working a six-day week.[21] Meanwhile, with the 2008 election approaching, the economy was tanking. Bush's tax cuts, spending programs, and the cost of two wars had eroded the budget surplus he had inherited. FAA policies had allowed existing facilities to fall into disrepair while all focus and budget went into NextGen products. The NAS was a system under stress.[22] Operational errors had increased at facilities where the staffing shortage was most severe. New hires had to wait as long as eighteen months to start work given the shortage of staff to train them. The FAA treatment of controllers, staffing shortages, failure to maintain the existing system, and forging ahead with NextGen without controllers' input energized NATCA's active support of Barack Obama for president.

The early months of the Obama administration ushered in a more congenial relationship between the FAA and NATCA. In a February 2009 statement before Congress, NATCA president Pat Forrey reported union agreement with the central points of the FAA Reauthorization Act of 2009, which aimed to correct the actions of the Bush administration: it nullified imposed work rules, involved a return to the bargaining table, and included NATCA in the review of all FAA consolidation proposals and all NextGen technologies.[23] Acting immediately, the new Secretary of Transportation Ray LaHood appointed former FAA administrator Jane Garvey to oversee a team of mediators to settle the labor dispute. By the fall of 2009, mediation had produced a draft agreement that was approved.[24] Controllers were *legally empowered* to have a say in all changes that directly affected their work.

Consequently, NATCA moved to "take back the FAA." The union initiated its own power work by remaking its organization boundaries to more effectively negotiate union positions with the FAA. During the Clinton administration, the total quality partnership agreement resulted in "two parallel structures" in the system. Then, to aid negotiation between NATCA and management, NATCA had officials at every level of the FAA hierarchy. Under NextGen, the FAA had streamlined by becoming even more bureaucratic. In contrast, NATCA maintained its relatively flat hierarchy and further decentralized by expanding its decision making into lateral structures at the facility level. To this end, the union created separate national groups that comprised facility reps for all major centers, towers, and TRACONs with quick e-mail list discussions or teleconferences for collective problem solving between the same type facilities. This strategy was spectacularly successful. The changed structure enabled rapid diffusion of knowledge and problem solving within and across the groups, and also quick responses locally between NATCA and FAA ATMs and operations managers in the facilities. Moreover, the flat structure elevated the power of NATCA regional officials, who were interacting as equals with agency officials higher up and outside the local facilities.

The combination of FAA centralization and NATCA decentralization changed local, regional, and national dynamics. The FAA ATMs and other administrations in the facilities were disempowered. Because the FAA was a bureaucracy, facility ATMs still had to consult hierarchically, which slowed them down in negotiating local situations with NATCA. Further, FAA officials above them were similarly disadvantaged, often unable to support ATMs in facilities. NATCA consistently was acting to eliminate inequalities and took leadership roles, actively working to move NextGen projects forward. However, the staffing crisis further disempowered managers. Facility supervisors were also in short supply. Local facility ATMs and other administrators (all senior controllers who had worked their way up through the system) were frustrated and discouraged, and not all were willing to move up in the system to take regional district positions. Those who might move up could not without being replaced. Given the influx of controllers to be trained, experienced controllers who would make outstanding supervisors or administrators could not be spared because of the immediate need to train.

The Intersection of Two Trajectories: Implementation, Budget Battles, Shutdowns, and Failures

FAA efforts to speed up training of new hires backfired. A $437 million contract had gone to Raytheon for technologies for controller training, both at the FAA Academy and in facilities, so controllers could be retrained to work in a system far different than the 1960s one in use.[25] However, training was undercut by a series of decisions made in different parts of the FAA. In response to the crisis, the training of controllers had been changed and streamlined, the storied three- to four-month FAA Academy "screen" was done away with in order to speed the supply of additional controllers into the facilities. Instead of the screen, after their civil service exam, applicants were given a biographical assessment to evaluate personality and background in order to assess probability of success as an air traffic controller.[26] Neither the first nor the second version of this test was validated before put into use, and many qualified candidates, some graduates of the FAA's Collegiate Training Initiative schools (CTI)[27] or who had previous experience as controllers were turned away. The first cohort of new hires showed up in facilities in 2007, their numbers beneath the target numbers of the FAA Workforce Plan. Further, those accepted had an unusually high failure rate, an early warning sign that the hiring and training changes were inadequate.

Funding for modernization came out of special NextGen funds rather than the FAA budget, so the NextGen efforts to reorganize the system on the ground to increase capacity and efficiency moved forward on several fronts. Several consolidated regional TRACONs were in full operation and others in progress. Moreover, increasing capacity and efficiency extended to airports. Following the passage of the industry-supported Airport Improvement Program (AIP), millions went to the design and construction of modernized towers to replace models from the 1960s.[28] At existing towers, runways were being lengthened and added. The airport modernization project was expected to provide 330,000 additional takeoffs and landings.[29]

As changes were being implemented, they were falling on an understaffed workforce. Scarcity and fatigue again became a public issue in July 2011, when a controller at Reno-Tahoe International Airport fell asleep on the midnight shift while a medical flight carrying an ill patient attempted to land.[30] Getting no response, the pilot contacted the nearest TRACON and so was able to land safely. The controller, who was "out of communication"—asleep—for about sixteen minutes was suspended. Finally taking preventive action on staffing, the FAA announced immediately that a second controller had to be added to the shift at twenty-seven towers around the country that staffed with only one controller on duty during the midnight shift. A few days later, a similar incident happened at Miami Center. Quickly correcting on the basis of the new information, the FAA announced a new set of work scheduling rules to apply nationwide. Among them, controllers must have a minimum of nine hours between shifts, a change from the then current eight.

Starting in 2007, because of repeated government failures to come to agreement about the federal budget, the FAA had lurched from one short-term funding extension to the next. The FAA was hit hard during Obama's second term when Congress failed to pass the twenty-first FAA reauthorization legislation in July 2011. As a result, the FAA furloughed four thousand employees (technical specialists but not controllers) and halted work on a number of NextGen projects. Finally, on February 14, 2012, Obama signed the FAA Modernization and Reform Act of 2012. This law aimed to restore funding stability, forward progress, and balance to the system. First, it provided the FAA with a four-year, $63.4 billion authorization package. Second, and equally significant for controllers, was Section 804 of the law.[31] It required the FAA to develop a plan for consolidating regional radar facilities in an effort to support the transition to NextGen while reducing maintenance costs of old facilities without affecting safety. Section 804 specified the creation of *collaborative work groups* that gave controllers an active role in planning and implementing all technical and organizational change. These work groups would play a major role in NextGen implementation, vastly improving the outcome.

But the system was hard hit again in spring 2013, when across-the-board congressional funding cuts to federal programs caused mandatory sequestration.[32] The FAA had to cut $637 million from its budget. Transportation Secretary Ray LaHood announced that most of the FAA's forty-seven thousand employees, this time including controllers, would be furloughed for one day per pay period until the end of the fiscal year. The FAA was forced to suspend hiring and close down training at its academy in Oklahoma City for the rest of 2013. Contract towers would be closed. Controllers compensated for minimal staffing by spacing planes farther apart for safety. The FAA announced that because of "staffing challenges," passengers could expect widespread delays.[33]

In response to public outrage and national safety concerns (and some reports that members of Congress realized that they would face flight delays going home for their spring recess), in late April, Congress suspended all FAA employee furloughs and canceled the closing of the contract towers. However, at the October 1 start of the new fiscal year, the lack of 2014 appropriations to run the country resulted in a partial government shutdown.[34] The FAA furloughed almost a third of its forty-six thousand employees, again including controllers. A few weeks later, Congress remedied the situation, finding the money to fund the government through January 15, 2015. Rushing to compensate, the FAA opened a ten-day hiring window.[35] Deprioritizing CTI school graduates to increase diversity, the ad was for applicants "off the streets": a four-year degree or five years of work experience, no previous air traffic experience necessary. The hurried implementation was flawed such that workforce quotas were missed by wide margins the next two years. Moreover, although diversity in hiring air traffic controllers increased after the change, systematic independent research by Kinley showed that applicants in underrepresented groups were more likely to be lost through attrition in the revised FAA Academy training program.[36] In addition, Kinley showed that underrepresented groups were more successful in the field than at the FAA Academy, arguing that change needed to happen at the academy.[37]

Experiencing external political and budget threats and under time pressure for both staffing and NextGen contracts, the FAA repeatedly had been unable to predict the ramifications of changing one part of the system on the others. Often the agency had acted unilaterally, without consulting controllers about the effects of policies on the facilities, failing to meet objectives and increasing costs beyond expected.[38]

The Liabilities of Technological and Organizational Innovation

As soon as new automated technologies were in use, there were already warning signs about their liabilities. Near the end of 2013, the FAA issued a lengthy industry and government report on cockpit automation. After studying global incident and accident reports, the FAA review found that pilots misinterpreted situations, had lapses in attention, and became confused about aspects of automation, concluding that "pilots sometimes rely too much on automated systems and may be reluctant to intervene," or make calculations and take actions that result in errors.[39] Moreover, although the number of fatal crashes of commercial airlines had been greatly reduced by advances in aircraft design, the air traffic control system was experiencing new kinds of "crashes": technical system failures.

In April 2014, a ground stop was issued at all Southern California airports because of an automation problem at Los Angeles Center. The new ERAM system mistakenly had interpreted a U-2 flight at about sixty thousand feet as a typical lower-altitude operation. The number of reroutings of other flights necessary to clear a low-altitude path for it used so much computer memory that it blew out ERAM's other flight-processing functions. Lasting for about one hour, the crash created havoc, resulting in the cancelation or delay of thousands of flights, also affecting flights from across the Pacific.[40] In May of the same year, a burning electrical motor at the Chicago TRACON, a new facility in 1991, began emitting smoke through a vent in the control room, and the facility went to ATC Zero and evacuated accordingly. Traffic in and out of O'Hare and Midway airports was stopped for about four hours, and thousands of flights were canceled.[41]

The staffing crisis gained headline attention during the 2016 presidential campaign. The situation was abysmal. In January 2016, an audit by the Office of the Inspector General showed that after ten years of the FAA's periodic workforce plans, the FAA was still drastically behind in its effort to ensure enough fully trained controllers at the busiest, most complex, most critical facilities. In June, NATCA's President Rinaldi testified before the US House Subcommittee on Aviation. He laid bare the extent of the staffing crisis with an analysis collaboratively developed by an FAA-NATCA committee.[42] Staffing was at a twenty-seven-year low, having dropped 10 percent since 2011. The drop was exacerbated as the FAA repeatedly missing its hiring goals and more controllers took early retirements due to the imposed work rules. "Stop-and-go" funding made it worse. Systemwide, the shortage hit hardest at the busiest air traffic control facilities, creating delays that sometimes left people sitting in airplanes on the tarmac for hours. In addition, when funding was interrupted, maintenance of existing facilities lagged. Controllers were working in poor conditions. The interruption and restarting of modernization projects was expensive and stalled progress. The committee statement cited flaws in the FAA workforce plans, bureaucratic delays in the hiring process, the high percentages of training failures due to changes in hiring and training, and the high percentage of inexperienced new controllers at the bottom ranks.

The FAA-NATCA committee statement made proposals for a reauthorization bill that directly targeted the systemic causes of the crisis:

- The bill must include a stable, predictable funding stream for the Air Traffic Organization. The committee sought to achieve stable funding without privatization, instead calling for a government-sponsored not-for-profit corporation.
- The FAA should aim to hire to maximum capacity every year. Full capacity meant that the annual number actually hired should use the full capacity of the FAA Academy in Oklahoma City. Since the first of the ten annual FAA workplace plans, that had never happened.
- The FAA should redesign the hiring and training process, advertising for both novices and experienced controllers; change the test-oriented screening process; correct for the high failure rate by shifting those who failed to lower-level facilities where they could do better.

The committee stressed that while the National Airspace System was still the busiest and safest system in the world, the staffing shortage was a crisis that, if not corrected, would jeopardize both safety and the modernization of the system.

During 2015 and the months preceding the 2016 election, Obama had been taking actions to secure funding for the FAA in the continuing budget battles. After several temporary extensions, on July 15, the House and Senate approved—and Obama signed—the FAA Extension, Safety and Security Act, a short-term extension in response to the FAA-NATCA committee proposal that continued the existing FAA funding through September 2017.[43] Again, debates centered on how to fund it. The extensions dragged out past September. Air traffic controllers were living with the continuing uncertainty of their funding and the certainty of the staffing shortage and NextGen when I returned to Boston Tower and Boston TRACON in fall 2017.[44]

12

The Age of Automation

2002–Present

Boston Tower and Boston TRACON

Controllers are trained to predict and control. Their daily work routine had gone on during an era in which they experienced the unpredictability of their funding, government shutdowns, staffing, sudden departures of long-time coworkers, the arrival of less experienced new hires, and changes to all their familiar technologies of coordination and control. Over many years, the system had turned flying in a piece of metal thousands of feet above the earth into a safe endeavor, based not just on controllers and the development of more sophisticated technologies but also on additional technologies of coordination and control: organizational structures, processes, and extensive protocols and procedures that were continually amended and refined to match changing air transportation conditions. It was controllers and this whole package of technologies that provided the system with its resilience, reliability, and redundancy. The system and flying were still safe; however, the intersection of the staffing crisis and the unanticipated liabilities of NextGen's program of technological and organizational innovations had increased work complexity, thereby making the system more risky.

Historically colocated, Boston Tower and Boston TRACON no longer were. Formerly a ten-minute walk from the tower cab at Logan Airport, the TRACON moved in 2004 to Merrimack, New Hampshire. Boston TRACON was the first and largest TRACON to move into the new NextGen-equipped building designed to hold Boston and three smaller regional TRACONs—a result of the FAA's goal to streamline the system by consolidating radar facilities. Already burdened by the staffing shortage, the separation and relocation of the TRACON was an additional strain on dead reckoning at both facilities. Although no longer colocated, the two facilities remained tightly coupled by Boston's small crossing-runway system and contiguous airspace. A change for one still directly affected the other.

The congestion occurring at all major airports was a factor for both. In keeping with the national trend, traffic volume at Logan was down from its all-time high in 1998–1999 of fifty-one million operations to thirty million in 2017. However, traffic complexity had increased, more so than at other major airports that had more land for the airfield layout, so a less complex runway system. The change in fleet mix to larger jets carrying more passengers resulted in crowded airspace for the TRACON and congested ground movement, and not enough parking space at the tower. With less area for maneuvering, Boston's gate areas were tight, with planes parking close to the tower in an irregular layout to fit into the limited geography that the surrounding water and city allowed, rather than in parallel spacing a safe distance from the airfield. One airline's ramp area had two gates that positioned aircraft so that departing aircraft pushed backed from the gate onto a busy taxiway, causing slowdowns in ground movement to and from the runways, with ramifications for Boston TRACON.

Moreover, the industry-supported Airport Improvement Program (AIP) had made the airfield itself more complex. To increase capacity on the ground, a new short runway for small aircraft was added and a new taxiway inserted between the two main runways so more airplanes could be crowded into the already-crowded airfield. At both Boston Tower and Boston TRACON, the changes affected how controllers directed the arrival and departure of all traffic.[1] In addition, the AIP made possible the upgrade and repair of the runways and all airfield lighting systems. During peak tourist traffic in summer 2017, one of the two major runways, 4 Left, was closed for repairs the entire summer. The closure limited available runway configurations, meaning that arrival and departure rates per hour were lower than usual. The tower was running delays all summer long. To manage, many controllers worked six days a week and overtime, deferring holidays, which further demoralized staff.

Adding to staffing problems, a positive change affected staffing *availability* at both places. In contract negotiations during the Obama administration, NATCA had won a long-sought-after victory: the right to have input into the design, development, and implementation of all new technologies and organizational designs. NATCA participation had become required by law. As a consequence, in all major facilities some controllers with specific expertise were designated as subject-matter experts assigned to special NextGen projects to collaborate in special work groups with FAA technical experts, architects, and airspace and infrastructure specialists. NextGen would not have moved forward as quickly and safely as it did without the input of controllers. Controllers eagerly embraced this responsibility, although they were not compensated for it. Moreover, the collaborative work groups generated teleconferences and travel for group meetings. The time away depleted the number of those available to work traffic on a given shift, which added to overtime hours. Fatigue was a serious problem. Moreover, the staffing crisis slowed NextGen projects. During the summer when traffic was high, subject-matter experts from Boston Tower and TRACON had no input into their work-group project assignments because staffing needs came first.

Dead reckoning between the two facilities—both boundary work and ethnocognition—became more challenging. Logan's crossing runways and small airspace had created a relationship between them that was at the same time cooperative and competitive: the tower's goal was to get departures out; the TRACON's was to get arrivals in. Nonetheless, moving aircraft across the airspace boundaries between them had always been easy because of their colocation. Controllers knew the challenges of traffic at both facilities and each other's techniques. After the TRACON move, controllers knew about, but were no longer as fully aware of, the needs and material practices at the other place. For the new generations of controllers who arrived after the facilities separated, the experiential knowledge of the other place was missing completely. Boundary work did not always go as smoothly as before, and differences were harder to negotiate.

The Liabilities of Technological Innovation

The FAA initiated two new automated procedures that affected airspace boundaries at all major terminal facilities, both with major import for dead reckoning at Boston Tower and TRACON. In the first, the agency changed boundaries in the sky by adding standardized automated routes known as the "waypoints system." Boston Tower had standard instrument departures, or SIDs, and the TRACON had standard terminal arrival routes, or STARs. Instead of controllers' controlling the flight path of ascents and descents by vectoring—crafting a route by assigning headings that a pilot must fly—with waypoints the pilot sets the route and the plane flies itself at preestablished altitudes and speeds. Rather than the former ladderlike movement of incremental climbs or descents that level out, changing speeds and burning fuel, the waypoint system provides continuous ascent or descent, saving the airlines time and money. Moreover, it standardizes and reorganizes the airspace, avoiding collision, congestion, and delays. The airlines loved it. The FAA loved it.

For controllers, the waypoint system decreased errors, which they loved, but the transition from automation to manual control created new kinds of mistakes. Controller and pilot monitor only until a change in direction or a surprise calls for either or both to resume manual control. Fast-moving traffic coming together at airports calls for a rapid cognitive adjustment and physical change of procedures for both pilot and controller. Infrequent use of manual procedures often resulted in de-skilling—slowing the response time for pilots and controllers alike. With safety depending on reacting in seconds, the ability to respond to an unexpected event by immediately shifting from automation to manual control was especially difficult for the new generations of controllers who were trained on manual skills but daily used the new waypoints procedures. Infrequently used, vectoring skills atrophied, and mistakes were made.

The second FAA automation change to airspace boundaries was in response to four "conflictions," or "near misses," that occurred in the same week at four major airports. The FAA mandated nationwide installation of a new device to regulate traffic flow that changed the airspace boundaries separating arriving and departing aircraft. The technology was the arrival-departure window (ADW), an automated decision tool developed to preserve spacing between arriving and departing flights in bad weather and prevent near misses that could result from sudden go-arounds (fig. 23).[2] A confliction occurs when an arriving aircraft comes too close (closer than five hundred feet) to a departing aircraft and so must execute a go-around (thus a missed approach) while another aircraft is departing on converging intersecting runways. A missed approach is impossible to predict. The go-around was designed to prevent a midair collision, but the four near misses at major terminals indicated that changing aircraft equipment capability called for a different solution: automation.

The arrival-departure window is a technological imaginary that projects a protected space—a "no-go box"—in the airspace around the converging runways. When two planes are approaching on perpendicular runways, the

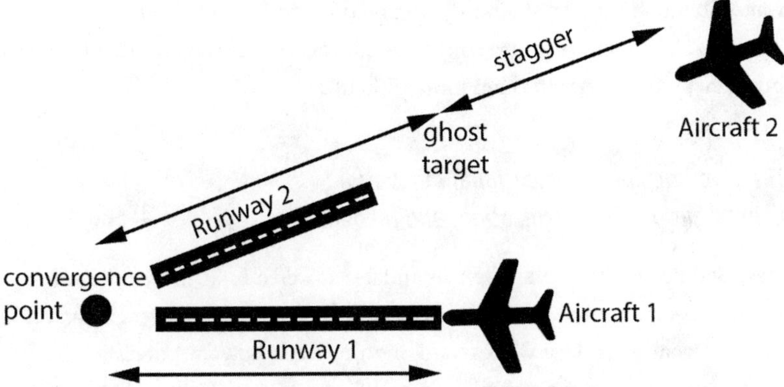

Figure 23: Arrival-departure window/Converging Runway Display Aid
Illustration adapted by Noah Arjomand

software shows the TRACON controller a ghost image of the second plane in real time, giving the controller time to stagger landings by spacing the planes before they reach the runways. On the TRACON's new STARS-powered radar monitor, a software simulation tool, the Converging Runway Display Aid (CRDA), creates a "hole" in the spacing of the TRACON's descending arrivals lined up on final approach, increasing the separation between airplanes so that Boston Tower controllers have a space to fill for a departure. The ghost target has two major advantages: it provides more space between airplanes as a defense against possible simultaneous missed approaches, and in bad weather, it keeps planes from slipping on the runway, so keeps up a traffic flow, preventing a ground stop.

The liability, though, was that in normal weather, increasing the spacing between arriving and departing aircraft increased delays. When running traffic while the winds dictated that Boston Tower use its converging runway configuration (22L, 27Arrivals; 22R Departures)—controllers' favorite configuration because with it they can land and depart the most airplanes—the ADW produced a serious drop in the arrival rate. A "48 hour" could drop to a "34 hour," and departure rates also dropped. As one tower controller put it, "We went from launch, launch, launch, to one in, one out, one in, one out." TRACON controllers were frustrated, complaining that instead of their stated mission of the safe, orderly, and expeditious delivery of air traffic, it had become "safe, orderly and wasted space." Tower controllers complained that the TRACON was no longer sending arrivals with the spacing they needed; it was "too much by at least a mile," which added to Boston Tower's limited departures and affected ground congestion. It was another change that made the boundary work between them more difficult.

The Liabilities of Organizational Innovation: Centralization, Standardization, and Inequalities

Responding to the NextGen mandate to create a more streamlined, performance-oriented administrative system, FAA headquarters initiated its own internal boundary work, changing the work flow and decision-making structures. The agency centralized several key responsibilities that had been decentralized—the domain of local management and union officials at individual facilities. Further, standardized procedures were imposed across the system. The positive benefits of these changes were offset by negative unanticipated consequences at the local level. The ramifications for dead reckoning at Boston Tower and TRACON were extensive. Designed to create system equality and efficiency, the changes increased the inequalities.

First, FAA headquarters centralized the hiring and training of new controllers to alleviate the staffing crisis. Formerly, the decision to hire was local—a facility's ATM and NATCA officials selected among applicants who applied directly to that facility, choosing those best prepared to work their airspace and who seemed a good fit with the people already there. Now, a centralized process assigned new hires to a facility according to their chronological order on a wait list, matching the top person to the staffing needs of the facility. The number of staff needed was determined by a standardized complexity index that consisted of two factors: airspace complexity and traffic complexity.

However, there was a serious mismatch between FAA headquarters staffing figures and the workplace reality.[3] Traffic volume was figured into the complexity index, but airspace complexity was not fully taken into account. Because the basis for FAA numbers was wrong, the calculation of the total staffing needed for the high complexity airspace at Boston Tower and TRACON was also wrong. As a result, the complexity index created inequalities in the distribution of staff in the region. Other towers in the region, with less traffic and less airspace complexity, were being allocated more people than Boston Tower was. The complexity index affected the tower more than the TRACON.[4] There, the number allocated would not even replace the controllers lost to retirement and resignation. True, air traffic was down, but procedures and other changes, to be discussed, had made the work more complex.

Angry about the inequalities built into the index, Boston Tower administrators and NATCA reps were working on a revised index that, if accepted, would take into account airspace complexity.

Second, the FAA plan to streamline the placement of new hires created inequalities in the ability of the new generation of controllers to meet the demands of the job. The new process did not match the experience of developmentals to the airspace complexity of their designated facility. Consequently, many of the less experienced new people were being sent to start their first on-the-job training in the largest and busiest terminal facilities. Complicating matters further, administrators in other parts of the system changed the content of the training program in order to expedite the production of new controllers. They moved through the training faster but arrived at facilities minus the cultural system of knowledge, visual, hearing, and manual skills of former generations. Most significant, the traditional FAA Academy "screen" that had produced generations of controllers was swapped for a "faster, kinder" learning that senior controllers mockingly called "train to succeed." Then, rather than developing knowledge and skills at smaller facilities and working their way up to more complex ones, as under the previous "ladder system," many developmentals were assigned to facilities with airspace complexity beyond their ability, where training programs were not designed to teach new hires how to recognize types of airplanes, airline identifications, and other fundamental knowledge and skills.[5]

Third, the FAA standardized and centralized procedures for operational errors, further eroding new hires' ability to learn on the job during and after certification. Previously the responsibility for identifying and responding to operational errors was local. Controllers always experienced having deals, the adversarial process of investigation following, and assignment of remedial learning as highly punitive, so the number of reported errors were known to be low. To produce more accurate data, a centralized, anonymous self-reporting system was instituted that was modeled on one used by the airline industry. The new procedure was successful at collecting macro-level data on trends and causes, and systemic causes were discovered that allowed systemwide changes in rules and procedures, but this, too, had effects on ethnocognition and boundary work.

The new procedure worked like this. Following an event or safety problem, controllers were asked to voluntarily file a self-report online to the Aviation Traffic Safety Action Program (ATSAP). All identifying information was removed from a voluntary report, then reviewed by an event review committee.

If the event was an operational error, a call was made from Atlanta to the facility, and an operational error would be recorded. Then the committee could take disciplinary action, either dismissing a controller for repeat offenses or recommending skill enhancement training. However, these rarely happened.[6] The need for controllers was high; the number of trainees entering the program every year was low. An investment had been made in each one. Second and third chances were given, and retraining did not occur. Mistakes were repeated. New learning did not always occur, and people were not ejected for multiple errors. Rule following and precision were not being passed on.

Managers and senior controllers lost control over the safety of their facility. They described the new ATSAP procedure as a get-out-of-jail-free card; the result was "the loss of moral accountability." Working with the new generation, they saw that it reduced the felt seriousness of an event on the person, the willingness to take responsibility for their actions, dedication to increase their competence level to that required to be a full member of the team, and the collective ability to complete safe operations. While senior controllers hated the old adversarial, punishment-oriented system, it produced a different outcome: "I didn't need Atlanta calling me to tell me I was a bad boy. I felt responsible. I had let my team down. I could beat myself up much worse than they could." Rule following and precision had been taken-for-granted parts of the job.

Emotional Labor and Moral Work

One senior controller summed up the experience: "We were decimated. It was like after the strike. New people were coming in with no experience." Training was a critical industry. To make up for these flaws in the hiring and training process, the burden fell on both the facility support staff, responsible for the classroom and simulator training classes, and on the supervisors and senior controllers responsible for on-the-job training. For some trainees, senior controllers were teaching basics they never had to teach before. For the first time since the days after the PATCO strike, a staffing crisis resulted in an insufficient number of support staff and senior controllers to train, creating a backlog. Moreover, the loss of senior personnel was a massive loss of expertise. Some senior controllers who remained—the proven most effective trainers—were spending most of their work hours training. In continuing progression, one shortage intersected with others: the retirements had also created a shortage of supervisors. Those controllers who were interested could not be spared due to training needs. The TRACON was all right, but Boston Tower was in crisis when one supervisor transferred and another was not available due to management training required for promotion.

Inside the facilities, the effect of these changes on both generations of controllers was the addition of another layer to the emotional labor that was already a normal part of the job. Although the effects were not identical at both facilities, for many senior controllers and for the new generation of controllers—those who had already qualified and those who were still in the critical apprentice phase of on-the-job training—there was a genuine sense that hard times were upon them and feelings of disappointment and loss. The generation of controllers that entered beginning in 2007—all smart, not all of the same background, some having been military controllers or pilots, others with experience at other facilities, some inexperienced from CTI schools—even those who completed training and were doing well, or still in the classroom—despaired, feeling marginal, pressured, tired of working overtime with unpredictable weekends, so were discouraged about their chosen career. Their situation seemed endless. Young and with families, they needed the job. Noting the option to retire or to move to another facility taken by some disillusioned senior controllers, one noted, "I understand why people are retiring, but we're stuck here."

A concerned supervisor at Boston Tower said:

> We are like family. I try to explain to them. They think this situation is unique, but there have been times like this in the past. We have these things to do. It's the rules. We have to do it—figure out how we can make it work here—and hang on for dear life. (Tower, M)

The senior controllers looked back at how hard they had worked to get to the top facility, the days of excruciatingly high traffic counts, teamwork with everyone pulling their weight, looking out for each other, and pride in the work. At both facilities, some senior people resigned because they "couldn't take it anymore": the lack of quality performance, the mistakes, the seeming indifference of the next generation. A few of the deeply disillusioned stayed, still effective in their jobs but cynical, casualties of the system. In the end, most did stay, working hard, dedicating themselves to figuring out how they could make it work locally and "hanging on for dear life."

Both generations blamed Congress for their discontent: "They never act until there's a crisis," "Every time there is a shutdown, all the projects shut down," and "We need a stable source of funding." However, the senior controllers at both Boston Tower and TRACON blamed the training situation on the FAA response to congressional actions: decisions made to centralize, the switch to an expedited training process, a flawed hiring scheme, facility assignment processes that placed newcomers in high-complexity facilities. Oddly, at the same time they could see their problems as system effects, these senior controllers also saw an individual-level problem: many controllers who began training in 2007 and after collectively possessed the psychological traits of "millennials": a generation that senior controllers described as having "no work ethic"; "entitled"; "don't care"; "only did it for the money"; "don't take responsibility for their actions"; "are only for themselves, not the team." Compared to earlier generations, these new controllers represented "the loss of the ownership concept"—a decrease in the personal stake in what controllers do every day: "not concerned about the next person down the road, only themselves and not the totality"; "not troubled by the increase in delays—I did my job"; "no pride in following the rules, or appreciation of advice on what to do by coworkers."

Most remarkably, many senior controllers who had been through the old process didn't recognize the loss of the FAA Academy in Oklahoma City as a factor. Only a few mentioned it. The elimination of the screen, though, accounted for many of the characteristics that experienced controllers attributed to the millennial generation. The screen, despite its abusive techniques, the many sacrifices it called for, and the high failure rate—or perhaps because of them—took people "from all walks of life" and produced in them characteristics in common. These included not only the foundational knowledge and material practices necessary to becoming a controller but also "common sense," a common way of being in the world, and a collective identity as part of an elite profession. Benefiting from the old system, the post-PATCO hires and the hires of the mid-1980s and early 1990s emerged from the screen with many embodied characteristics. In addition to the basic skills of dead reckoning, institutional knowledge, and cultural logics of the system, those who survived were dedicated to the profession, took responsibility for their actions and for looking out for one another, and—having learned to work with people from all walks of life—already had the basics of coordination within and between facilities.

No longer a product of the early hiring and training, the responsibility for this embodied transformation had shifted from the beginning of the process to the controllers, support staff, and supervisors of both facilities. As they focused on teaching cognitive and material skills, techniques, rules and procedures, and local knowledge, they were also engaged in the larger project of passing on the code of conduct of their profession: trying to correct the system's inequalities by pulling the next generation of controllers to the high standard of professional performance required for their moral work.

Boston Tower

Airspace, Place, and Work Practice

History had left visible marks, even on the airport's exterior. Logan International Airport had undergone significant reconstruction since 2001, including expanded terminals, glass walkways for passenger movement between terminals, and interior modernization. Above it all, Boston Tower retained its iconic 1970s shape, including the walkway high above the terminal traffic that was the go-between from the tower to the TRACON's former building. The protective concrete entry barriers and technology installed at street level post-9/11 were still in place.

The strict entry procedures from the airport to the tower were tightened, but once arriving at the level of offices, the administrative office locations and break room showed only small changes, the exception being the locker room. The mementos on the locker doors of the two generations of controllers mark their different histories and their merger here. For the controllers who arrived in 2007 and after, photos of their training cohorts, pilot or military careers, or photos of vacations and young families; for the senior controllers, pictures from previous facility assignments, maturing families, other interests and activities. The locker of one senior controller, a developmental in 2000 and 2001 while his father was a supervisor, displayed a single document: the original of the official letter that his father received informing him he was fired for participating in the PATCO strike ("a criminal offense for which you could be imprisoned"). Occasionally, for both generations, signs of their merger: a photo at NATCA tower or regional events, a NATCA bumper sticker or poster, a group party.

Climbing the final sets of stairs into the tower, initially all seemed the same: at a certain point, the familiar clicking of flight progress strips in their plastic holders, passing from controller to controller, the murmur of their voices, and then sunlight on the stairs and that view. Once arriving in the tower cab, however, it was a different world. Formerly, nature dominated. Controllers' dead reckoning was based on observation in real time and space, out the window. Technology had been minimal: binoculars, three ARTS radar scopes suspended from the ceiling above the "front row"—the ground control, local east, and local west positions—and mounted on the desktop at each position were ASDE monitors then in the test phase. However, most vital to operations and most physically dominant were the flight progress strips, the strip slide, and strip bays at all five positions. The strips, which originated in the 1940s, were essential to interpretive work and coordination in the room.

In 2017, the contrast was startling. History had left visible marks on the tower cab, too. It was Times Square in miniature. Bright displays of new automated technologies covered every available surface. Nature seemed peripheral. The eleven-sided, six-hundred-square-foot tower cab (compared to one thousand square feet at most major airports) was crowded not only by these devices but also by the architectural changes necessary to house the infrastructure that supported them. The layout of workstations and work flow for arrivals and departures was the same, with the "back row" of controllers—the flight data, gate, and clearance delivery positions—facing the northeast and merging into the front row, which faced east: the airfield and the ocean. Now, however, above the front row hung seven new monitors: two screens were situated above each of the three positions: the Standard Terminal Automated Replacement System replaced the old ARTS radar that showed arrivals and departures; ASDE-X was the most recent development in monitoring ground movement.

Suspended in the middle of the other screens was a novel radar that monitored marine traffic in Boston Harbor. Known as the Tall Vessel Detection System, it sounded an alarm whenever a large vessel began to move into the harbor's channel. Aircraft on final approach to land runway 4 Right flew low over the water, so a tall ship in the flight path was an accident waiting to happen. Formerly, the harbor master would have phoned the tower cab supervisor, passing on the data from ship documents and experience, and controllers used local knowledge of wind, weather, and size to assess separation and avoid collision. Now the Tall Vessel Detection System accurately showed ship size, speed, and height in relation to the flight path as the vessel moved, so controllers could see and react immediately, sending a pilot on a go-around or doing nothing, as vessel size and speed dictated. Another signal indicated when the vessel had cleared the channel. Controllers loved this and were eager to show it off. The device replaced controller's dead reckoning, but senior controllers displayed their skills anyway by looking out the window, competing with one another to guess the ship's height before the device announced it. No one remembered a collision between aircraft and ship, but many remember close calls, near misses, and go-arounds. With the new device, they didn't have to watch the channel anymore, and the supervisor could give the TRACON a heads-up earlier, so they also could take avoidance strategies with arrivals otherwise heading into the path of a tall vessel.

The workstations were as jammed in 2017 when I returned as they had been years earlier. Mounted at each position were flight information display systems monitors. Touch screens accessed the most up-to-date information on everything from weather at every facility in the New England Region, types of airline cockpit equipment, varieties of helicopters, and procedures for drones. Still amid this modern array were the flight progress strip bays. Strips had been replaced by electronic fight progress strips (e-strips) at most major towers. At Boston, the paper strips remained essential for safety because of the small, compressed airfield with its complex intersecting runways. Strips still served as memory devices, an aide to coordination, and reduced the need for conversation about flights. The immediate limiting factor, however, seemed to be that the small tower cab had no room for the ERAM infrastructure necessary to electronic strips, and neither was space available in the two floors of offices below. E-strips were in the future plans, however, that possibility being investigated by controllers assigned as subject-matter experts to an e-strips collaborative work group.

Automation also had changed Boston Tower's architecture, reducing available movement space. In the center of the tower cab, the boomerang-shaped console that had divided the room was enlarged to hold the supporting infrastructure for automation in the front row. Moreover, the back of the console had been extended toward the rear, redesigned as two facing desks that held new devices for the supervisor and TMC positions. When seated, a supervisor and TMC faced each other as they faced their own information display system. Phone and additional display systems for each were mounted at their side on the console. The workstation counter at the back of the tower cab held additional new devices, including one presenting a trend analysis that projected in advance the number of arrivals and departures per hour.

Most new technologies were "decision tools" that provided information to aid controllers' dead reckoning. However, two automated devices—the Time Based Flow Management System (TBFM) and DataLink—in combination, did much more. They streamlined the back-row process, reduced delays, and made it possible to reduce staffing. The TBFM was for sequencing aircraft and determining the expected time of departure. Formerly, the

TMC had to phone Boston Center's Traffic Management Unit to get the time, based on how an aircraft fit into the regional flow. A call could take five minutes, but it had become a slick local maneuver. The TMC could quickly search the TBFM for the assigned order of, say, all planes in the airspace going to Los Angeles and just "drag and drop" a plane into an open time slot on the screen to sequence the plane. Time spent on the phone was cut in half for the TMC.

The expected departure time was then sent to the cockpit via yet another automated device, the Controller Pilot DataLink system. DataLink sent digital messages between pilots and controllers, including departure clearances and reroutes. DataLink eliminating communication errors between controllers and pilots. This much-loved device also saved time. When weather delayed departures, clearance delivery was no longer swamped with frantic calls from anxious pilots wanting to know their departure times. In combination, DataLink and TBFM reduced work complexity. When traffic was low, one person could simultaneously handle flight data, gate, and TMC. Given the staffing shortage, being able to work three positions with one person even part of the time offered some relief.

The Liabilities of Technological Innovation: Architecture and Automation

Uniformly, controllers loved their NextGen devices. Collectively, the devices produced greater speed and efficiency as promised. They also provided redundancy, affirmation or correction to what controllers could see and hear. For many of the new controllers, automation simplified the tasks so that learning to work Boston Tower's complex traffic became possible. A supervisor observed, "Technology slowed the system down. The equation was easier to solve." However, automated devices also had unexpected liabilities for dead reckoning that were less visible and measurable. Lost with DataLink was controllers' ability to predict pilot competence by tone of voice and rhythm of speech. One controller said, "I could tell whether a pilot was going to be a problem or not by the first words out of the gate." The controller heading the subject-matter expert group on all data communication had a more pragmatic view: "I don't need them to be geniuses when they are backing out of the gate." Not all aircraft coming in and out of Logan were equipped with DataLink, so predicting pilot behavior by tone of voice was a skill they needed to retain. No one was complaining, though; DataLink reduced communication errors. However, controllers did complain about other NextGen devices that increased their workload by either creating more work or making a task more complex—or both.

In the Fire

A new safety procedure had increased the amount and complexity of workload for the ground controller. Recall that ground control is responsible for all aircraft movement areas on the ground on the way to and from the runways: gate, ramps, taxiways, alleyways. The procedural change was to prevent close calls or collisions of ground movement caused by one aircraft crossing an intersection in conflict with another aircraft. Formerly, ground control issued only one instruction: to proceed to a particular destination alleyway or taxiway, where the pilot would await further instructions. However, the new procedure called for ground control to announce every crossing on the way to the next assigned destination. Given the intersecting runways, taxiways, and alleyways on the small airfield, the talking time of ground control increased by eighteen and a half minutes an hour in busy traffic—a colossal amount of time for people who are sometimes handling two planes a minute. With attention concentrated on one aircraft, ground control was stretched to visually, cognitively, and physically work ground movement for the entire airfield during busy times.

To respond to the new procedure, Boston Tower personnel created a second ground control position for those busy times.[7] However, this change also called for an architectural redesign to accommodate technology for the new position. The tower cab became even more crowded. Responsibility for the airfield was divided between the two. Consequently, the ground 2 position was directly opposite the back row, separated from ground 1 by local west and local east. A second strip slide had to be added to the front of the tower console so that clearance delivery could smoothly shoot a strip the six feet distance to the other side of the tower cab within easy reach of ground 2. Formerly perpendicular, the front of the console was sloped toward the front row to support the waist-high track for the strip. When two trainers were standing behind their trainees, which was routine, the space in front of the console was so packed that the supervisor usually remained standing at the end of the console, rather than with the front row for a better view of the airfield and sky. Oddly, a consequence not remarked on by controllers was that the addition of this new position canceled out the reduction of a back-row position that DataLink and TBFM had made possible.

Changing Space, Changing Place, Changing Culture

The cumulative effect of these and other changes was a striking difference in the feeling of the place. Recall from part III how the variation in a facility's airspace affects a place—its architecture, technology, and organization of work—which in turn affects controllers' dead reckoning, producing a distinctive culture for each facility. At Boston Tower, these factors had resulted in a far greater preoccupation with production than at the other facilities. During the years before September 11, traffic was at an all-time high. Controllers frequently experienced "the sixty hour": sixty arrivals, sixty departures. As one senior controller recalled: "It was mayhem up there. Sometimes we would have eight hundred hours in a row. On a good day, we could handle that without a sweat." Time was always a factor. The relentless traffic was reflected in controllers' postures, their arm and body movement. The local controllers always stood; ground control always stood; supervisors and TMCs always stood.

Production and competition go together. Boston Tower controllers competed against traffic and pilot errors, working with high speed to keep up; they competed with the TRACON to get departures out and avoid delays. Everyone was keenly aware of each person's numbers when working local. The emphasis had always been more on individual performance than at other facilities. Each person was responsible for a single position but collectively responsible for the whole airspace, rather than a sector of it, so a failure to notice an anomaly and alert others, or a mistake, had consequence for the group performance. Individuals were morally accountable and accepted responsibility. The ample staffing numbers had allowed the formation of three teams assigned to the same schedule, so team members trained their own newcomers, leading to a unique teamwork style. Teams were very effective in mentoring the developmentals: "Your team always had your back." They provided a sense of identity and belonging to the team, but competition between teams was a by-product. What bound them all together was a common goal, a common history, and experience. It was a culture of production, with a moral element.

In 2017, the Boston Tower culture was not the same production-oriented culture as I had witnessed before. NextGen and changing traffic conditions were both having effects on the system, such that facilities were in the midst of a major transition of airspace, place, technologies, and work practices. Traffic was down. The forty-hour was typically the high, the forty-eight hour an occasional anomaly. Also, traffic was more dispersed throughout the day because airlines had altered schedules. No longer were controllers crushed by rush hours concentrated at three critical times of the day. The reduction of the embodied urgency in the room was visible. Routinely both local controllers sat down. Ground 1 mostly stood but no longer paced the room; ground 2 sat. The back row still sat, with one exception: often, gate was in motion, watching ramp movement on both sides of the tower. There were other exceptions: out of differences in training and experience, senior controllers usually stood in all front-row positions.

However, Boston Tower was still a production-oriented, high-pressure job. The amount of traffic was not the only variable. Additional changes had made the work more complex. When sequencing for departure, the larger number of wide-body jets called for increased spacing between aircraft to allow for wake turbulence; the addition of a major alleyway between the two major runways made colliding wingspans another factor to take into account. Planes backing out of the gate onto active alleyways and tight parking space also caused delays. With the system in transition, the culture was in transition, too. The work was made more difficult because the factors that always had aided in collective responsibility and coordination in the past no longer existed. The staffing shortage and systemwide changes to hiring and training practices had created generational divisions in people's contribution to the group effort. In addition, the low number of controllers in the tower cab resulted in not enough to divide into teams, so a source of mentoring and belonging was lost. Teams had formerly bid for a team schedule, but now individuals were competing with each other for schedules, reinforcing individual rather than collective identity. For both generations, FAA decisions to centralize and the staffing shortage combined, doing further damage to collaboration and the culture of the place. No longer did anyone in the facility—local administrators, supervisors, or controllers—have the sense that someone always had their back.

The common production concern had become the production of controllers. Becoming a controller was a long process, hard on trainers and trainees alike. Trainers were still teaching the safe, orderly, and expeditious delivery of air traffic—including setting priorities, saving seconds, and filling holes and shooting gaps. However, the result of the intersection of NextGen and the staffing crisis at Logan was that the early navigation skills of dead reckoning and other elements of the craft—technique and coordination—that had been part of controllers' embodied knowledge were eroding or missing altogether. The need for those skills remained. Trainers were striving to teach the complex airfield and airspace, and to restore the early navigational forms of dead reckoning, techniques, and coordination that had been lost in the transition.

Place, Technology, and Ethnocognition

Developmentals arrived at the tower with certain common deficiencies. They lacked the intensive FAA Academy experience, in which trainees responded to the stringent militaristic rule-bound screen by helping one another in order to survive, thereby learning how to work for the good of the group, coordinating with people from all walks of life. The cultural understandings necessary to accomplishing the conflicting system goals of safe, orderly, and expeditious delivery of air traffic were also absent. Finally, they had not worked up the ladder from simple to complex facilities, so they lacked the insights into the system that sensitized them to what happened to an airplane once it left their position: when at a top facility, they could not predict the consequences of their boundary work because they did not have the background to fully understand how the parts of the system worked as a system—what happened to an airplane after a handoff—not only handoffs between the tower and TRACON but also with the other facilities in the region.

The system effects of NextGen and the staffing crisis presented training staff and senior controllers with three training challenges: the mix of generational and experiential differences among new hires, airspace complexity, and automation. The training process began with support staff teaching intensive classes that covered all the basics, including strips, and local rules and procedures. The developmentals first learned the back-row positions in the classroom. As they advanced, they had sessions on the front row on a state-of-the-art simulator. Scenarios could be programmed for an amazing array of local airspace or airfield conditions, more or less complex, hot spots or emergencies conditions. Support staff could design programs to address individual problems. Trainers would go to the simulator room to watch their trainees work in order to make a training plan for in the tower. Their goal was to shape the training to the person, not the reverse, which had been the strategy before the staffing shortage changes to the hiring and training system. As one controller said, "We work with what we get."

When beginning on-the-job training at Boston Tower, even those in the new generation of controllers who transferred in from other FAA facilities struggled with the idiosyncrasies of Logan. Uniformly, ground control was the hardest position to learn. One trainee who transferred to Boston in 2017 had spent a few years at a small tower, then advanced to a high-volume, level 11 tower with a large open airfield and four runways with only one crossing. In our interview, he began by introducing himself using words that indicated he knew well senior controllers' comments about the millennial generation, saying: "I was not an airplane freak, never around airports, none of that. I am one of the people that took the job for the money. To me, it's just a job." After nine months at Boston, he had done well and was training on ground control, but he found it overwhelming. Feeling at the top of his game when he transferred, he expressed his fear that he was not going to make it at Boston. Comparing Boston to his previous position, he said:

> It was easier there. The biggest challenges were learning the volume and learning the next person. Here it's learning the airport. The volume is more and Logan has only one-quarter of the space and is much harder. There are great controllers here. If you can work traffic here, you can work traffic anywhere. At pushback, you've got gates right by the taxiway. You've got taxiing on an active runway. It can be a mess. The most ever times I've been going down the tubes is on ground. The tows—tug drivers–are slow on the taxiways, can't get out of the way of crossing traffic. It throws your timing off, slows everything down. This is the first time I've seen strip slides. Most airports don't have paper strips anymore. Here you need them to see because the ASDE is so small you can't see the overlap in data blocks when planes are really close together. The strip bucket even has its own little system. (Tower, M)

Then there was the weather. Nature was still an obstacle to be reckoned with, but technology had reduced controllers' vulnerability to it. The sunrise and sunset in their line of vision was eliminated by sophisticated window shades of varying darkness. Formerly, controllers' predicted a wind change (and so planned a runway change) by looking out the window: cloud formation, the movement and color of the water in Boston Harbor, the windsocks on the field, and the speed patterns of landing airplanes. They confirmed their predictions by phoning Bedford, where they knew a change in wind direction usually happened before arriving at Logan. Now the information display system supplied accurate, minute-by-minute changes in weather and wind; the timing could be verified in seconds by a quick look at the screen showing weather not only at Bedford but also movement earlier, at more distant facilities. A hazardous microburst storm—so strong that the control tower and everything in it swayed back and forth, including controllers in their chairs, when one passed directly over the tower—could be predicted earlier and safety actions taken.

For the new generation of controllers, the proliferation of automated technologies and readily available information in their immediate line of vision drew their attention like a magnet. They had grown up with computers and other technologies; they understood it. In the training before their facility assignment, they had trained extensively on NextGen equipment. For them, these material objects offered some certainty and predictability when everything else was uncertain and unpredictable. They lacked confidence. Working outside their comfort zones, they depended on the technology. No one was looking out the window.

Signals, Interpretive Work, and Automation

Looking out the window at all positions was essential. The architectural arrangement had each controller position viewing a different part of an aircraft's trajectory, so anomalies were detected by individual alerts at different phases of its movement. The obvious question about controllers' dependence on technologies is, of course, what happens when technology fails or pilots err or controllers fail to detect an anomaly? Equally important is the human contribution to ethnocognition and boundary work during routine daily activities when controllers fail to detect an early warning sign indicating a pilot deviation to come or when pilots suddenly err. When talking about technique and how they work traffic now, in the midst of NextGen era, the examples from experienced controllers, including the senior controllers and those in the new generation who arrived, have certified and been working traffic for several years beyond that, indicated that their visual, hearing, and cognitive interpretive abilities were as important in 2017 as in 2000 for identifying early warning signs and fixing them:

> We can see out the window where a plane is before the computer does. (Tower, F)

> When fog starts to roll in, we don't have to think about it or calculate. We see it holistically. Actually we can usually see it coming when moving our eyes between strips and the equipment. It varies by position and time of day. You

have to be able to recognize it. When the airport is fogged in, we can work airplanes with STARS and ASDE, but we might have to divert someone so spotting it early is safer because we can alert the TRACON. (Tower, M)

Radar does not give the same judgment as looking out the window. The ASDE only gives location and distance. We have to control for speed and variation in speed. We know where a plane is going and when too fast, we will slow it for spacing. We know with a wind shift when a plane is going to hit a head wall, so we slow the guy behind. (Tower, F)

ASDE might not go off. And it might go off when it is not supposed to. If we get CA-CA [conflict alert], we must send a plane around. It's a forced action. You can't say a controller wouldn't have caught it. Sometimes it's safer to let a plane continue, especially if it's a foreign pilot. (Tower, F)

On every position, you have to know the people in the room and listen to them. Ground and local especially have to listen to each other, so they know what's coming. If one thing goes wrong, it will all cascade. When the tower goes down, that's what happens. Somebody missed something. (Tower, M)

I'm working gate. The TMC gets a [departure] time. They don't always push back then. I watch the gate, or if I can't see the gate, I watch the ASDE and watch the gate location for when the wing-walkers come out [ground crew signaling pilot with orange batons keeping eye on the wings to avoid obstacles]. Then I know they are ready to go. (Tower, M)

It's all about experience and judgment calls. Pilots will be slow on the take-off, mess up an instruction, make a wrong turn. Pilots are on visual here. We watch and listen to the tone of voice. If something goes wrong, we've had something like it before. It's like a chess game. One move you can correct your whole issue. (Tower, M)

After mastering the fundamentals, these interpretive skills have to be taught, learned, and embodied. It takes repetition in order to be done "without thinking." Given speed and proximity of crossing aircraft, embodiment is essential for safety. It is a matter of acquiring local knowledge: "This is how we do it here." Here are two examples, from trainers. This first example requires learning a new skill.

Breaking the habit of tunneling on the equipment and teaching to look out the window was challenging at all positions, but especially ground control. As trainees strived for competence on the position, trainers began pushing them to decrease dependence on the equipment:

> When I teach the scan, it's not just about where to look, it's about where to look when. On giving an instruction, say, to advance to Bravo, as soon as you start to say Bravo, as you're giving the instruction, you look. It becomes a motor skill. (Tower, M)

In the act, the cognitive has to become associated with the physical move. Standing behind them, trainers would reach over and physically turn their heads to show something in progress that they were missing. As the developmental gained in comfort and skill, the trainer would turn the ASDE off during training sessions. These lessons had to be repeated. Unless they become embodied and could be done without thinking, in the heat of the moment looking out the window was forgotten and technology was the best option. People missed things.

This second example describes learning a maneuver that calls for a greater level of expertise from the developmental. For the position of local control, vectoring an aircraft—assigning a heading and altitude to turn an aircraft a specific direction—was an essential part of dead reckoning. Before standard instrument departure routes with required waypoints, vectoring was necessary to get airplanes out of the airport and on to directional headings, so it was routine and the skill was maintained. It was always necessary—urgently so—for an emergency go-around. Since SIDs, vectoring was used less often but was still essential, especially for a go-around. To save lives, the skill must be embodied so done without thinking. A trainer described the learning challenge:

> If they haven't had training at a radar facility they can't do it. If airplanes are on 27 and 33 [intersecting runways] and both go around, we have to vector, have to give a heading and altitude, and you have to know where the other guy is. The new people would know it's a radar facility, but what's the heading? If it's 33 Left, it's 360. Then you have to be able to execute. (Tower, M)

The new generation of controllers bring to Boston Tower different backgrounds, and so different capacities to learn skills and advance to expertise.

Some who transferred in from high-volume facilities—whether tower or radar—were able to acquire skills and move on to expertise.

On a September Tuesday around 1:00 am, the ASDE failed: controllers had no radar showing the positions of aircraft on the airfield. All positions were affected. Fortunately, the weather was perfect. The tech people fixed it temporarily, but a replacement part would have to be sent from Oklahoma City. The ASDE worked haphazardly during the night and failed again Tuesday afternoon around 1:00 pm, a very busy time. Scheduled to train was the controller who transferred in from a high volume airport—the same one who just weeks before expressed his fear of failing at ground 1. The supervisor, seeing a training opportunity with the ASDE down, put him in at ground for his regular training session. At that time of day, ground 1 was working both the ground 1 and 2 positions. He pulled it off. His trainer was exuberant: "He was outstanding, calling them out faster than I ever could." Immediately after, the supervisor certified him on both ground positions and he moved on to train on local. The ASDE continued to work haphazardly and then failed again early Wednesday, before the part arrived.

He succeeded, and many did, but not everyone could achieve the same mastery. Others could acquire the skills, use them productively, become technically proficient and competent, but never quite attained the level of expertise necessary for the volume, speed, accuracy, and complexity that Boston Tower airspace required. In addition, and historically, every facility had controllers whom others saw as competent but also a "weak link," limited in certain situations, so colleagues looked out for them and helped in those situations. Coworkers did more work to compensate. Among this new generation, a few acquired the basic skills but remained very limited, with weaknesses that create mistakes and operational errors. They never took responsibility for their actions, never learned from their mistakes, so repeated them. They put passengers and pilots in jeopardy. A dark example from the tower:

> I was working [runway] 33L. Another controller was working 32 (parallel to 33L). I said, "Your airplane is lined up with my runway." This happened twice. They pulled him off and got on him for it. But they did nothing more. It was a deal. Before it would have been a local issue. Thanks to ATSAP, it would have become a national issue. I saw this with other people, too. They file a self-report. They blow it off. Nothing can be done to you. When someone gave us a heads up like that, we would say "Thank you for your assistance." To this day,

> he has never said thank you to me. These people are not helping each other. I don't think it's a generational issue. They cut the Academy and substituted technology for the old process. (Tower, M)

The FAA Academy was no longer there to drill in rule following and accuracy or screen out those who did not master the basics and conform. Missing also was shared work ethic, perfectionism, drive to succeed, and the sense of collaboration and responsibility to the group that earlier generations had acquired in the training. Contributing to this was situation awareness and understanding how a team worked as a system: "On every position, you have to know the people in the room and listen to them." In addition, other outs had been inserted in the training program. For example, if a controller was not doing well, he or she could request a change in trainer or ask for an extension to certify.

A decrease in program rigor showed up in the degradation in the quality of work. Local managers and supervisors were very concerned about "drift": the gradual loss of attention to rules and procedures.[8] Former standards of accuracy and rule following were replaced with an attitude that rules were stupid or inefficient, or counterproductive. Deviation from certain rules had become normalized. "They are not reading the weather out," a manager stated. In the past, announcing the weather to pilots was routine, but the practice had attenuated. Then, an incident occurred at Boston Center when a pilot flew through turbulence. Passengers were injured. The Boston Center controller had not advised the pilot about the weather. Although the weather was available to pilots and controllers alike via information display systems and DataLink, litigation followed. Consistently, the FAA loses cases "when we were not doing everything we can." A contributing factor, a few senior controllers thought, were too many rules, but "it's not ours to question. Pilots expect it to be the same everywhere. We have to do it."

It *was* like after the strike. People were coming in with no experience. They were getting by because during a session senior controllers were taking on more responsibility, including teaching as things came up:

> I will do more work to compensate. If I have something I need to get done, I tell them what to do and explain afterward. They are going to do it my way, then I give feedback. It's exhausting. There is nothing left when you go home."

The system was still working because experienced controllers were passing on the necessary skills and experiential knowledge, exerting the extra effort to pass on information formerly acquired as people worked up the ladder to the complex Boston airspace. Meanwhile senior controllers were maintaining safety, filling in for the less experienced and passing on the tradition of helping a coworker out. The future promised some relief from the staffing shortage. Four people training on the front row were expected to certify before year's end. Three had certified on the back row and were waiting to train on the front row, that stage delayed because support staff was short. Within three weeks, the three would begin. In January three people were transferring from other FAA facilities and would be training with some experience. Boston Tower would still be short staffed, however. The inequalities in the arrivals' preparation would still be present given the changed system for hiring, training, and facility assignments. Boston Tower, its culture, work practices, and collective expertise were, like the rest of the large system, in transition, a work in progress.

Boston TRACON

Airspace, Place, and Work Practice

In February 2004, Boston TRACON moved from its urban location to the new, state-of-the-art consolidated Boston TRACON in Merrimack, New Hampshire, just a few miles from Boston Center in Nashua, New Hampshire. The facility was an imposing sixty-three-thousand-square-foot, three-story structure off the main highway, hidden by surrounding woods. The bucolic setting was contradicted by the alarmed barbed-wire fence surrounding the property and entry surveillance system. Visitors were cleared by security guards stationed at a guard house equipped with metal detectors and multiple monitors for surveilling all activity on the property and in the building interior. The consolidated facility combined both NextGen programs: automation and realignment and integration of radar facilities. Of grand scale, the building was designed to hold Boston and three smaller regional TRACONs, and possibly Boston Center and its more than two hundred controllers at some future date.

Manchester TRACON moved in two weeks after Boston; the Cape TRACON was scheduled for February 2018. The final TRACON was yet to be decided. All TRACONs were relocating from cozy, dark, small, low-ceilinged control rooms in deteriorating and cramped old structures—usually near their towers or in them. In contrast, the new facility exterior was large, rectangular and plain, its nondescript doors opening into a stunning three-story atrium full of light. On the left was an open-slat wood stairway that preserved the airy ambience. Above was an open walkway that crossed the atrium at the second level, joining the two main wings of the building and its sprawling interior of private spaces. Beside the stairs at ground level was a waiting area with sofa, two chairs, a large plant—a plant!—and a table with magazines. Directly opposite the entry, some ninety feet away, a dramatic glass wall ran fully to the atrium's ceiling. Abutting the wall were a kitchen and a large open area with round tables where controllers could eat, take breaks, or have small meetings with a view of the lush lawn and woods behind. The wing to the right of the atrium held offices for the New England Region officials, the TRACON's air traffic manager, staff specialists, and a large conference room. The left wing housed the electronics room with the technological infrastructure for the building, necessary equipment for repair and design, and workspace for the technical experts responsible for all TRACON related work

The architectural design and spatial diffusion of the place offered privacy and more places to be. On the second floor, the wing to the right of the walkway had a large room of offices for training support staff. Outside it were offices for the NATCA president, both operations managers, and the supervisors. The opposite wing was for controllers. In a large room with mailboxes along one wall, controllers sign in and out for shifts, hang coats, and check the schedule. Two training briefing rooms were on the side wall, but sometimes trainer and trainee debriefed at one of the two small tables in the main room. Directly opposite was the entrance to the control room, admission by ID card only. Also in this wing were state-of-the-art classrooms, and a spacious study room with carrels, a reading area, and STARS monitors for training exercises—and a great view. In another area was a break room with lots of oversized leather chairs around a TV and few high-top tables. Next to this was a small room with vending machines, TV, and small sofa for video games. Elsewhere was a workout room and a training room for a simulator designed with multiple traffic scenarios, STARS-equipped for virtual display of live traffic. As at Boston Tower, training was an industry: the simulator room was always in use.

Although five large consolidated TRACONs had preceded it, the new Boston TRACON was an experiment. The fundamentals were the same: NextGen consolidation colocated three or more facilities in a single facility and called for realignment and integration of the facilities' airspaces. Realignment was a technical process, requiring technical expertise—engineers, architects, electricians. Each facility is a small socio-technical system with its own infrastructure connecting it to the other facilities that constitute its network of contacts in the system. The infrastructures had to be realigned and integrated to operate as one.[9] For example, for Manchester and Cape TRACONs, the short-range radar for the remote towers they served had to be realigned with the Boston TRACON's STARS, and the towers they served had to have their TRACON services transferred to Boston. Radio frequencies had to be moved and Boston workstation consoles reconfigured. STARS had to be updated to include the airspace of each new facility. At midnight on the "cut-over date," the airspaces formerly known as Manchester and Cape TRACONs would disappear from the NAS and become Boston TRACON.

But integration of the airspace also was a social process, calling for knowledge work, training, and expertise so the newly joined human systems would operate as one. Controllers from the different facilities would cross-train, first learning and then working traffic for each other's airspace. The NextGen expectation was that boundary work would be easier and coordination smoother because all large TRACONs combined regional facilities that were of nearly equal airspace complexity. Experience had proved this true. However, consolidating Manchester and the Cape with Boston TRACON was based on a different idea. It was as an experiment born out of necessity: a new tower was going in at Manchester airport, so for cost savings, the Manchester TRACON moved out of Manchester Tower into Boston TRACON in 2004. Unexpectedly, it became a forerunner of a plan for the future. In 2012, the new ruling known as Section 804 would pass that authorized consolidating small TRACONs with a large TRACON, thus eliminating the cost of constant maintenance and repair of old structures and enabling small TRACONs to be equipped with NextGen equipment.[10] Consequently, Section 804 consolidated facilities would combine facilities that *varied* in airspace complexity. The new Boston TRACON was the national prototype: the airspace of Boston was level 11; Manchester, level 7; and Cape, level 8. For both Manchester and the Cape TRACON, when the time arrived, learning the Boston airspace would be a challenge. The effect on boundary work was an open question. Every facet of the novel consolidation would be closely monitored.

Three problems were expected from the beginning. First, both Boston and Manchester were relocating to a control room of an unaccustomed architectural, organizational, and technological design—extreme in the difference from their usual workspaces. Second, with the staffing crisis and altered training practices, the new TRACON would have the same internal generational differences and training challenges that Boston Tower had. Expected but beyond imagination, however, were the difficulties that flowed from joining a level 11 with a level 7 facility. These three factors combined to produce inequalities in salary, skill, workload capacity, and competency between the two facilities that created mutual resentment. Moreover, the consolidation was a clash of cultures, statuses, identities, and ways of being and doing. It was a very tough time. In 2017, thirteen years after the consolidation, and with Cape TRACON to arrive in early 2018, Boston and Manchester controllers were still in the midst of working out many differences, all of which materialized in the control room (fig. 24).

The Liabilities of Technological Innovation: Architecture and Automation

Initially, everyone felt "out of place." In contrast to the experience of Boston Center controllers, who in 2000 moved from "an airplane hangar to a shoe box," Boston and Manchester moved from shoe boxes to an airplane hangar. The control room was a large oval with a two-story-high ceiling that curved to the walls. From the entry, an unaccustomed high level of light made it possible to see the number, location, and identities of staff throughout the room. Controllers' workstations were against the oval wall, forming the "outer circle"; the "inner circle" was for operations managers, supervisors, and TMCs. In anticipation of working each other's airspace, Manchester and Boston sectors were located next to each other at the top of the outer circle: Boston sectors were from eleven to two o'clock; Manchester sectors from two to three o'clock. Many workstations were empty, awaiting future TRACON controllers. High above on the wall and evenly spaced around the room were four large flat-screen Traffic Situation Displays that showed traffic landing and departing Boston Logan and other regional towers, each paired with an information display system projecting weather patterns.

Everyone had to adjust physical, cognitive, and material practices. The high ceiling was soundproofed to deaden the sound in the room. The light level was a major adjustment. In the front of the room, high above Boston's initial departure sector, was a live feed of the Boston Tower's departure strip bay.[11] Flight progress strips had not been used at Boston TRACON, but Boston Tower's strip camera showed the rolling sequence of strips for departures, silently informing them of planes entering TRACON airspace. In the old TRACON, controllers could recognize the hands of tower controllers working departure, so knew individual technique. They could also predict not only when but how airplanes were going to come into TRACON airspace. Now

Figure 24: Boston TRACON control room, Merrimack, New Hampshire, 2004
Illustration adapted by Noah Arjomand

that extra predictive ability was lost to the staffing changes and the physical separation of Boston Tower and TRACON.

Each workstation was wider than before to accommodate NextGen equipment and to easily pull away from the wall for repairs. Consequently, controllers were no longer elbow to elbow, and that meant their ability to hear everything going on around them declined. Monitors were square, not round, and blue, not green, and the sparkling radar sweep was missing: "The sweep left a trail, so we could see direction." Moreover, the workstations were designed to be interchangeable and moved: STARS produced specific air sectors on each, so airspace could be realigned to expedite the flow of traffic simply by relocating a workstation beside a different sector. Repair was also easier. However, when moved, controllers had to adjust traffic techniques to the new airspace alignment and work styles of different people beside them. Crucially, the inner circle interfered with coordination (fig. 25). As was the outer circle, it was designed to accommodate more management personnel as new TRACONs moved in. However, the design was awkward. Supervisors from Boston and Manchester, the Boston traffic management coordinators and operations managers were located where their computers and other devices were installed, but the supervisors' positions were not located near their crews. Further, the circle was large, with few openings; apparently the breaks were designed for symmetry rather than efficiency in problem solving and moving quickly to watch the changing action. The architectural design confiscated not only the ease of combining expert assessments and coordination but also the intimate exchanges of the past. One controller said: "It used to be that someone cracked a joke, everyone laughed. That family feeling is gone."

NextGen technologies had a different reception. Relocating from facilities with outdated, often failing equipment, controllers were happy about the upgrade. Many of the automated "decision tools" in the control room replicated those from Boston Tower. Accustomed to standing in a cramped space in the old TRACON, TMCs not only had ample desk space for all their new automated devices but also were able to participate in nationwide teleconferences on traffic flows every two hours. Controllers' STARS-equipped monitors provided greater efficiency in the execution of many tasks. These advantages notwithstanding, everyone was well aware of the liabilities of the automated waypoints system and the ADW. Designed to increase safety, efficiency, and simplify tasks, in practice, these automated procedures produced new kinds of mistakes and increased task complexity, as we will see.

Once both facilities were settled in, three boundary problems had to be addressed. The central concern was working the boundaries in the sky: training Manchester controllers to master the Boston airspace. At level 7, Manchester had two crossing runways and was the fourth busiest airport in the New England Region. Manchester controllers had extensive vectoring experience; also they had trained on STARS monitors prior to relocating. Manchester and Boston had to cross-train, but the bigger challenge was Manchester's. Second, the control room architecture changed the material boundaries on the ground. An important technology of coordination and control, architecture was supposed to enable controllers' work. All controllers' would slowly adjust to the overall control room architecture, but the inner circle was an immediate obstacle to dead reckoning. The size of the circle, the placement of the supervisors' desks distant from their crews, and the limited number of breaks in the circle for coming and going handicapped communication, coordination, operations oversight, and all essential physical movement—including the need for supervisors, controllers, and technical specialists to get to a workstation quickly. The third boundary problem was the integration of two facilities, divided socially and symbolically by the inequalities in skill that the mix of complexity levels built into the relationship and the difference in each facility's cultures, statuses, identities, and ways of being and doing. Not only did these differences need to be addressed out of a sense of fairness; repairing them was essential to the smooth and safe operation of the facility.

Although repair required the participation of both Manchester and Boston, the primary responsibility fell to Boston because theirs was the original, or basic, airspace around which the facility was built. As such, Boston's airspace was known as the "legacy" airspace, and Boston controllers known as the "legacy" controllers. Boston TRACON also was the dominant airspace, due to its size and complexity, but also because the Boston TRACON management and strong NATCA unit there were responsible for facility operations, and the Boston controllers would be training the others on the more complex legacy airspace. Once again, air traffic controllers grappled with the perennial challenge of mandatory standardized changes in a system typified by local variation: How can we make this work here?

Changing Space, Changing Place, Changing Culture

Demanding immediate attention was the inner circle architecture. Once settled in, NATCA representatives, Boston management, and some controllers from both TRACONs began to work on a remedy with FAA technical experts, engineers, and architects. The solution was a new design that divided the large inner circle into two smaller ones. The circle nearest the front of the room

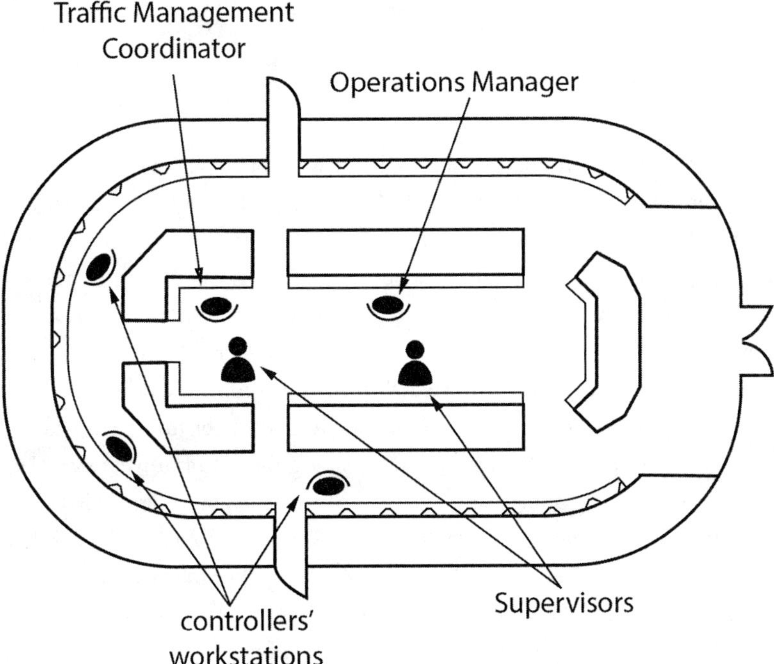

Figure 25: Boston TRACON control room architectural layout, Merrimack, New Hampshire, 2004
Illustration adapted by Noah Arjomand

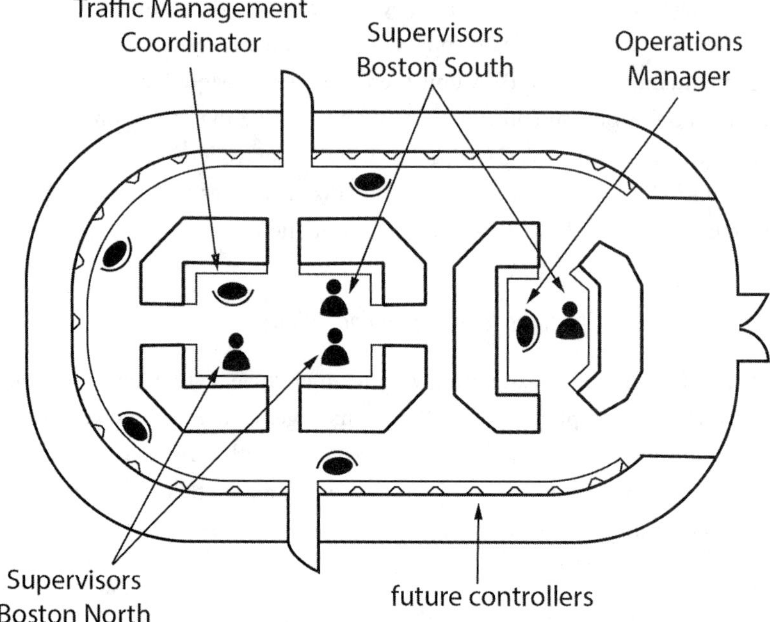

Figure 26: Boston TRACON control room inner circle redesign, Merrimack, New Hampshire, 2017
Illustration adapted by Noah Arjomand

was divided into four quarters separated by wide openings for easy access to and from controller workstations. TMU occupied top right; the other sections were for supervisors, locating them close to the workstations of their respective crews. The second circle, located between the first circle and the control room entry, had a closed area just opposite the front circle that was designed as a large desk for the operations managers and their technologies. To the right, the left, and behind were vacant desks for the supervisors of the TRACONs still to come, whose airspace sectors would be located opposite.

When completed, the new design was a significant improvement: Manchester and Boston supervisors were close to their controllers, TMU, and each other. The difference in the everyday work experience was much better for everyone. Moreover, they had worked together to improve the local situation, producing a collective feeling of accomplishment. The far more challenging boundary work was social and symbolic:[12] converting a social space riddled with inequalities into equitable and cooperative arrangements. The first sign of trouble occurred even before the move. The initial deal was that Boston TRACON would be a level 11 facility (matching legacy airspace complexity), so Manchester controllers would receive a level 11 salary. Boston controllers resented the higher pay for a lighter workload and lower skill level; moreover, Boston controllers complained that they had to relocate their families, whereas Manchester families could stay put. The expectation was that Manchester controllers would retrain and be working level 11 traffic, so the inequalities in competence and salary would be rectified. But that did not work out as expected.

The trajectory of Manchester's location in the outer circle outlines the struggle to integrate the airspace. Between 2004 and 2017, Manchester was in three locations. First, Manchester was next to Boston, at top right of the outer circle, approximately two to three o'clock. The plan was to have Manchester controllers and supervisors certify on a few Boston sectors. However, the challenge of the level 11 airspace was greater than expected. The first two trainees washed out. Everyone was devastated. A collective decision was made that, for now, it would not be a consolidated facility: Boston and Manchester would be colocated only, each working its own airspace. Boston would receive a level 11 salary; Manchester, a level 7 salary. Consequently, Manchester moved a second time, locating bottom right of the outer circle at about four or five o'clock. Manchester controllers felt like outcasts and failures: "Everything was for Boston"; "We felt like the red-haired, freckled-face step children." Manchester was in that location for about ten years.

In 2007, the first new hires began arriving to compensate for the staffing shortage. As at Boston Tower, the TRACON received a mix: military pilots, military controllers, controllers from FAA facilities with less complex airspace who had been wait-listed for Boston TRACON, and graduates from CTI schools, who had no air traffic work experience beyond school. In common, no one—even the most experienced—had worked traffic in an airspace with the complexity of Boston's. The support staff, consisting of three specially trained Manchester and Boston controllers, began classroom training for the new hires. As they moved to on-the-job training, some were failing. Early on, senior controllers began to see the "millennial generation" differences experienced at Boston Tower. The complaints were the same: they had no work ethic; they were out only for themselves and not looking out for the next guy; they had a sense of entitlement; and they were only in it for the money. Some senior controllers from the old TRACON were so disheartened and frustrated by the struggle to train and the deterioration of excellence that they began to retire or move to contract facilities. Ultimately, eleven left.

Manchester would move a third time. In 2012, the FAA Reauthorization

Act, with its novel Section 804 provision, became law.[13] Consequently, the Boston TRACON 804 Collaborative Workgroup formed, the first of its kind. The group combined national, regional, and local expertise for advance assessment, planning, implementation, and project management.[14] Locally, a core group of engineers, district regional and facility personnel—both FAA and NATCA—took the lead in implementation. They worked to realign the facility airspace in a way that best expedited the traffic flow in the region during the busiest times. Section 804 mandated that both Manchester and Cape TRACON airspace had to be integrated, so in realigning the airspace, the group had to consider both the inequalities in salary and in competency built into the consolidation. How could they fairly distribute the workload across the three facilities?

The goal was nothing less than to redesign the TRACON's conflicting cultures and structured inequalities, The solution, agreed to by all parties, had two parts. First, the local work-group members converted the TRACON into a two-area facility: the outer circle was divided into Boston North and Boston South. The air sectors of each TRACON would remain together, but new boundaries would be drawn. Boston legacy sectors would be split, part in Boston North and part in Boston South. Physically, Manchester and Cape would abut Boston sectors and would be working several sectors of Boston airspace in different parts of the room: Manchester in Boston North; Cape in Boston South. Salary would match competence level: Manchester and Cape controllers could opt to stay at their home facility's original airspace and keep that salary, or they could apply to work a predetermined number of sectors of Boston airspace, therefore raising their salary. After certifying on those, they could apply to train on all of it, reaching the level 11 salary. If they trained and failed, they would drop back to their previous level.

Accordingly, Manchester moved from the bottom right of the outer circle to the newly created Boston North, at nine to ten o'clock, adjacent to Boston's initial departure position. The move was another big adjustment for Manchester. Their airspace sectors were physical relocated in relation to one another, so in a different order than before; moreover, Manchester sectors were divided by a control room door, separating controllers. Second, the Boston North airspace was realigned so that Boston was using a shelf of Manchester airspace to expedite arrivals, provoking territorial fighting. Status differences were reinforced. As a Manchester controller later reflected, the transition back to integrating the airspace added to existing "smoldering resentments" between the two facilities.

The second part of the solution occurred when cross training resumed. The divisive undercurrents between Manchester and Boston, the experience differences in new arrivals, and the intense ongoing negotiations for the future arrival of Cape TRACON led the local work group to a decision to "fully unify" the TRACON. In contrast to the NextGen goal of "full integration" of airspace in consolidated facilities, the work group's goal to "fully unify" the facility was by passing on "the Boston Way": the old Boston TRACON's culture of collaboration and collective responsibility. The Boston Way consisted of the cultural system of knowledge, with its ways of being and doing, and its moment-to-moment enacting of bodily skills and material practices with each other and with pilots that historically had proved effective in achieving safety but also had bound the group together, despite its members' individual differences. They looked out for each other. They worked for the group accomplishment rather than individual accomplishment. But it was more. They wanted to restore what had been lost in the transition: the collective sense of belonging in a place, a pride in the work, the profession, and its craft, a felt obligation to help each other, a feeling of being bound together with people working toward a common goal, prioritizing the whole over the individual. These cultural understandings were as integral to ethnocognition and boundary work at the TRACON as mastery of the airspace, rules and procedures, technologies and material practices.

Place, Technology, and Ethnocognition

At Boston Tower and TRACON, the intersection of NextGen and the staffing crisis created the same three training challenges: the mix of generational and experiential differences among the new hires, mastering airspace complexity, and automation. To resolve them, controllers at each place drew on their history and a culture that had been shaped by their particular airspace and task. As we saw in chapter 6, in seeking the solutions to the problems particular to their airspace, controllers at each facility had created a distinctive culture.[15] Boston Tower historically had a competitive culture of production that emphasized individual identity and individual accomplishment, and as a result was responding with greater individual efforts at training that passed on not only the expertise essential to dead reckoning but also the code of professional conduct that was essential to their moral work. In contrast, the TRACON had a culture of collaboration and collective responsibility. Thus, the Section 804 Workgroup, in consultation with other controllers, responded with a collaborative, collective effort, creating a set of principals formalized as "The Five Core Values of PULSE." It, too, was a moral code of conduct: the initials stood for professionalism, unity, leadership, safety, and engagement. Each principle was defined, but in addition, implicated everyone in the facility—the ATM, operations managers, supervisors, trainers, and controllers—in its realization.

The five core values were introduced during the orientation of all newcomers, taught in the classroom, and permanently on display on a large colorful poster in the atrium eating area. Codes of conduct such as this are meaningless without accompanying directions for action that are followed. The core values were the principles and practices Boston controllers used in the old TRACON and were now using and teaching. By formalizing their practices, the hope was that the newcomers would see themselves and the actions of their coworkers as part of a collective process with a collective professional goal. This outcome would be possible, they thought, because of measures already taken to mediate the inequalities: dividing the airspace into north and south so it was "co-owned" and opening pathways to upward mobility and salary increase.

As it turned out, the three training challenges were interdependent: to be deficient in one had ramifications for the others. Mastering the airspace was the linchpin on which both facility unification and automation depended. The new people learned the Boston airspace in the classroom, but on-the-job training was one sector at a time. To coordinate with one another across airspace boundaries and facility boundaries, controllers had to know the rest of the airspace. This called for expertise and experiential knowledge that none of the trainees had because they had not worked "up the ladder" from facilities with less complex airspace to increasingly difficult ones. This senior controller described the effects on training:

> In my cohort, we worked our way up through several facilities. My whole skill set was based on learning the various jobs, and I learned how the airspace worked. The new people don't have that skill set. So here you have to pay attention to multiple sectors besides your own. So the issue is, how do you teach team work? Can you see the whole chessboard, what is happening around you? I tell them it's like being a taxi driver. You can't drive unless you know the neighborhood. It takes room awareness. You have to teach hearing and seeing to have a good sense of how busy others are. Listen to what he said to you. New people don't hear when they announce a runway change because they are not listening. (TRACON, M)

The new people struggled to learn the airspace. As trainers taught skills and technique, coworkers reinforced the lessons:

> They can't coordinate. They are all working sectors that are feeding final, but they don't know what to do because they don't know what the other Boston sectors are doing. If you hear the final guy [controller] slowing [arrivals] from 210 to 190, you react to that, slowing up your feed. Other places are not so interdependent. They don't know how to protect final. Some will never work final, so when I'm working final, I direct traffic—I need . . . I need . . . I need. (TRACON, M)

Volume is part of airspace complexity. Coworkers did more work to compensate, correcting mistakes and missed opportunities:

> We're paid to handle airplanes when things get busy. In the old TRACON, we were used to the volume. Now the volume is gone and we have to explain to people how to do things when it's busy. Everything is interrelated here. You need to see what everyone else is doing. That guy was busy. You should have been spinning but didn't see it. I say, "You need to spin that guy." Or when I'm working next to someone, I tell them, "I'm spinning, you need to do X." They should sense how things are going to go so we don't have to speak. (TRACON, M)

Boston controllers demonstrated the code of conduct by example: taking responsibility for a mistake, saying thank you for a criticism or correction, and the reverse—yelling at others in the break room about mistakes because "the mentality in the old TRACON was that you need to know when you screw up." While some newcomers resented "getting hammered" by their trainers and coworkers, one controller from the military recognized the strategy: "They need to break you down in order to build you up into the person they need you to be."

Boundary Work and Automation: Waypoints and the Arrival-Departure Window

Vectoring—assigning compass headings to turn an airplane in flight to a desired direction—had always been an essential ingredient of dead reckoning. It requires precision, speed control, room awareness, and timing. At the

TRACON, as at Boston Tower, automation had resulted in de-skilling and a drift away from traditional practices of dead reckoning. A trainer observed:

> The trainees now are all GPS trained, satellite-based, so no vectoring background. In training, we have them try it without SIDs or STARs. Automation is good, but they are so habituated that they can't take themselves off of it, so they've lost their tools. Manchester is experienced with vectoring, they like to vector, but we're on waypoints 30 percent [of the] time. They can't vector when they need it. (TRACON, M)

Making a transition from automation to vectoring was like going from a full stop, mentally and physically, to high-speed refined cognitive and material practices. Even controllers who had extensive vectoring experience had a hard time. A military controller training on Boston civilian airspace was switching from working F-15s and moving everything out of the way to the fine-tuned accuracy needed to bring airplanes close together for arrivals:

> It called for a total change of mind-set. At some point you have to intervene and start slowing the airplane, [with] speed controls. So a question is, when do you start vectoring? It might be at first call, or it could be situational. The biggest difference here is the proximity of all runways. Nothing compares to this final. (TRACON, M)

The TRACON version of the waypoint system connected the airplanes descending from Boston Center's high-altitude en route traffic into the TRACON's approach phase of guiding an arriving flight to a destination airport. Pilots more often made errors when going from waypoint automation to manual (TRACON arrivals) than from manual to automation (Boston Tower departures). Pilots' ability to quickly deactivate auto-control and assume manual control was an unpredictable operation. In addition, aircraft performance during the transition varied because computer programs differed between airlines, so an aircraft also could be unpredictable even when the pilot was skillful. Vectoring remained essential to safety. Concerned, this senior controller reflected on de-skilling and risk:

> For fifteen years of my career, nothing changed. From 1990 to 2005, we vectored all departures and all arrivals. In the last ten years, everything changed. We used to vector every day, but now we can go a full day without vectoring.

> The vectoring skill has been lost. On Final, we have to vector everyone. There are times we have to do delay vectors and sequencing. We can't control spacing because of wind and compression. Or a pilot messes up on final and we have to pull him out of line and send him around. The old guys are better at it. More subtle, less abrupt, takes three vectors; new guy, five or six. They can do the job, but not as quickly or efficiently. OK with low traffic. In heavy traffic or emergency, we need better than that. (TRACON, M)

The two incidents that follow reveal the intersection of ethnocognition, boundary work, and automation at moments of negative surprise that demanded controllers suddenly shift from automation to manual control, interpretive work, and vectoring. The first focuses on an individual controller; the second on a crew. Both occurred on the same day, and the second already was brewing when the first happened. Both show that gaining the expertise of a controller calls for learning not only how the airspace works as a system but also how all participants on a shift work as a system. Note in the two examples how the supervisor, TMCs, and controllers become a tightly coupled system, engaging in the form of dead reckoning that their position required, thus collectively supplying the resilience, reliability, and redundancy essential for safety. Note also that in both examples, supervisors made it a teaching moment, leaving the new people on position: "This is how we do it here."

Angie, who had certified on Boston airspace only six weeks before, was working final when she had back-to-back emergencies. Off of waypoints and heading to approach, the pilot of American 1326 called in reporting a flap problem and requesting time to see if he could fix it. Angie vectored the aircraft out over the ocean, but the pilot couldn't fix it. So the flight was coming in as an emergency, flaps-up landing—high speed with more brake use, and the risk of tires exploding. At that moment the supervisor for the night shift arrived early, saw what was happening, and supplying redundancy, quietly slipped into the position next to Angie to watch the monitor and provide support if necessary. Other controllers automatically took over Angie's other airplanes. The on-duty supervisor made the necessary phone calls. Then a pilot error: as American 1326 was coming in on short final, a departing aircraft moving to its takeoff point was crossing the American's arrival runway as instructed but stopped on the runway. Angie vectored a go-around for American 1326. Landing on the second try, the aircraft spent some time on the runway while the tires were checked; coworkers vectored other arrivals toward different runways until the aircraft moved clear. Seemingly calm throughout, in the break room afterward Angie explained that this was a "slow emergency," not a matter of life and death, so the passengers were probably not informed. For her, it was another step toward expertise:

> After you certify you will see things you have never seen before. There are things they can't teach you. Everyone helps you. While it's going on, the stress gathers in your body because there is no physical action. It's the neck. You learn by experience. After you see it several times, it becomes routine and you begin to be able to respond automatically. (TRACON, F)

The second example centers around a crew and two back-to-back incidents. The first concerned the automated ADW; the second, immediately following, was a rare wind direction that occurred maybe once a year that could be pulled off with old-school dead reckoning alone.

A runway change was, for Boston Tower and TRACON controllers, "a routine drama," as we saw in chapter 6.[16] Multiple consecutive runway changes occurred regularly, but this day three occurred in sequence that upgraded the routine drama to an unexpected challenge. It was a mid-November weekday, weather was cold and rainy. Visibility was two miles, with a twelve-foot ceiling and a wind shear. In the inner circle, one TMC was on a national teleconference; the other TMC was talking about a possible runway change with the supervisor. That morning they had been landing forty per hour, but the wind was expected to change. The supervisor was concerned about delays. They checked a regional weather source known for its accurate, up-to-the-minute weather, which indicated the wind change would arrive at Logan one hour later. On that basis, the supervisor and TMC were discussing a possible shift in configuration (22L, 27 Arr/22R Dep; see fig. 19). For converging and intersection runways, the configuration required that TRACON controllers engage the automated arrival-departure window and the Converging Runway Display Aid (CRDA). The supervisor wanted that configuration because they could continue putting down arrivals and Boston Tower could continue departures. It would substantially reduce the hourly rate but avoid a ground stop. The TMC thought crosswinds might be a problem for arrivals so was arguing for a more conservative, safer plan: going to one runway, although it would cause delays and maybe lead to a ground stop.

The ADW called for TRACON controllers to use the CRDA when two planes were approaching on intersecting runways. This software shows the air traffic controller a ghost image of the second plane in real time on the radar, giving the controller time to stagger landings by spacing arriving planes before they reach the runways (fig. 23). So arrivals alternate landing, first one runway, then the other, allowing time for the departure. It was a complex operation, even in good weather. The tool is automated, but execution still rested on dead reckoning: controllers' ability to precision vector planes into line for arrival, then create the precision spacing for the other airplane. A brilliant and much-needed innovation, CRDA allowed them to keep traffic moving safely during bad weather when runways are wet or if a missed approach called for a go-around. The ADW procedure calls for opening up both the final 1 and final 2 positions because arrivals are coming in on two runways. The two controllers sit side by side, coordinating and vectoring, calling out each heading and altitude to their respective pilots. The ghost target is another aircraft to think about. The two controllers need to have experiential knowledge of the entire airspace, vectoring, winds, wake turbulence, spacing technique, each other's rhythm, aircraft equipment capabilities, weather, and timing. Controller and technology are interdependent: ADW's CRDA software projects the accurate spacing, but the execution depended on both ethnocognition and boundary work.

Rarely had the supervisor gone against a TMC recommendation, but anticipating a possible wind shift in an hour, the supervisor made the call. He took an incremental approach so the crew could rehearse working together before it happened, and if it wasn't working, they could adjust. The configuration required sharp vectoring by both final 1 and final 2 controllers, as well as strong controllers at all feeder positions. To prepare, the supervisor began rotating the controllers who were flexible and most qualified to work the complex operation into key positions. For learning purposes, he kept the trainees in because they didn't get to work the configuration often enough to become proficient in it. Getting the right combination of people was key: changing one player could decrease the arrival rate or tilt the rhythm of the crew. The supervisor took full responsibility for the 22-27 configuration call, considering it a coordinated decision, having consulted with the Boston Tower supervisor, the TMCs, and the controllers working the airspace. At 12:30, the wind changed, exactly when expected. The two final controllers were doing precision vectoring to protect the CRDA ghost target, making room for Boston Tower departures. The concentration level was palpable. The procedure was working as it should. Moreover, the mix of experience level—senior Boston controllers, recently certified controllers, and the two controllers in training collectively were pulling it off.

The wind changed again, calling for a runway change to 27 Arr/33L Dep, requiring TRACON controllers to alter the direction of all planes in the sky above the airport and Boston Tower controllers to reverse the paths of airplanes on the ground. CRDA was turned off. However, it was a brief restful interlude because just as the night shift was settling in, the wind changed a third time—now out of the northwest, velocity 360 knots. The new runway configuration was known as "the 3s": 32, 33L, 33R. Controllers so rarely worked it that they called it "the trifecta." Just as in horse racing, hitting the trifecta is a long shot. The night shift supervisor now was in charge, putting in his strongest controllers, including some who had newly certified at the TRACON. So rare was the trifecta that the day shift supervisor stayed overtime to watch the operation. The operations manager stationed at the desk in the rear inner circle spread the word and moved to the front. The second operations manager rushed from his upstairs office and through the control room, taking a position at an empty workstation, leaning into the monitor to watch. Everyone in the front inner circle was standing.

Workload complexity increased, calling for old-school dead reckoning. The wind compressed the spacing on final, so controllers had to increase the spacing between arrivals from three to four miles. All controllers were vectoring and giving pilots instructions because all planes were on visual. The control room thrummed with the sound of controllers talking, unusual in the era of waypoints and silent monitoring: "Expect visual approach"; "Report that traffic in sight"; "Maintain visual separation"; "Airport nine miles; report when in sight." Also, controllers had to advise every arrival "Traffic landing on the adjacent runway." The configuration called for the utmost skill in spacing because some aircraft had to cross 33L to get to 32, the new short runway. As controllers settled into the rhythm of it, they were executing a flawless performance of their craft. The supervisor turned to me with excitement, saying, "Do you know the difference between a Puma and a Nike? I've got my Nikes in there." The new people were holding their own.

It had been thirteen years since the move to Merrimack. The system effects of the intersection of NextGen and the staffing crisis were still a factor. Implementing NextGen's three goals of automation, realignment and consolidation, and standardization and centralization of the system in the Section 804 experiment was itself like a trifecta: betting on three conditions that rarely come together. Lessons had been learned from Manchester and adjustments made that would be useful when the Cape TRACON moved in. Some of the "new people" had proved themselves to be Nikes, others skillful and technically competent, and others were "weak links." As many controllers had commented about competence in the previous era, "There will always be weak links." As at Boston Tower, experienced TRACON controllers worked harder to compensate, so the system was working but vulnerable. The staffing shortage, plus de-skilling and the surprise of shifting between automation and manual procedures created the potential for mistake, increasing risk. It was impossible to assess where they were in "fully unifying" Manchester and Boston. PULSE was not only about passing on a moral code of conduct; the program's principles were essential to safety.

By December 2017, the integration of Cape TRACON was under way. A critical facility in the New England Region with more than 146,000 operations annually, it was relocating from an old World War II era building of the 1950s. Eight controllers arrived and were working Cape airspace, two already had certified in the Boston South Area, and fourteen more were scheduled to arrive. Because Cape controllers were experienced with high volume, they were expected to train more quickly, but they would be from a different airspace, place, and culture, so it was hard to predict the outcome. The official cut-over date from the old Cape TRACON to Boston was February 11, 2018. On that date, the National 804 Collaborative Workgroup would accomplish the first radar facility consolidation that was designed to combine facilities working airspaces of different size and complexity.[17] The NATCA leaders who took the lead on the project expected a long, incremental transition. Estimated time for integrating the operations of Boston, Manchester, and the Cape was 2024–2025. Like Boston Tower, Boston TRACON was in transition, a work in progress.

13

Continuities, Change, and Persistence

This book has traced the life course of the air traffic control system and the changing nature of the organization, its technologies, and work over time.[1] A sociological history, it has explored the emergence, development, and operation of the system from its earliest beginnings through 2017. This outcome was never my plan. My research questions were located firmly in the then-present of 2000–2001: What makes the air traffic control system so safe, or is it? And, what do controllers do that technology can't replace? But once in the facilities, the visible effects of history on the present pulled me back to the past: how did this novel social form emerge and develop to acquire the characteristics it has today?[2] Locating the ethnography in its historical context has exposed the mix of old and new in the facilities and the actions and reactions that have mattered. I have argued that we can think of history as cause. To clarify, history is neither a scene setter nor a social actor in its own right, but it has a causal effect on the present only through the actions of assemblages of heterogeneous social actors—ideas, people, organizations, inventions, devices, material objects, rules—originating in different places and times that intersect with a developing system and throughout its life course in unanticipated ways, both positive and negative.[3]

The sociological history of the system's life course began with a formation story of system emergence in the Age of Innovators (1880–1920) that explained how air transportation came into being and the particular assemblage of social actors and actions—the airplane, the idea that human flight was possible, networks of aeronauts, cartographers, education institutions, manufacturers, the postal service—all those actors that contributed to its development, stabilizing it, thus providing the organization field from which the air traffic control system would grow.[4] As planes became capable of flying higher and air traffic controllers were "invented," physical structures were built, and early devices—lights, radios—connected controllers on the ground to the sky. Developing and becoming institutionalized as a system in the Age of Organizations (1920–1950), the system's shaping and reshaping continued through the remaining eras, as these various social actors—consistent in their character but varying in their content (i.e., policy a consistent actor but with varied source, content, and purpose) continued to affect it, accounting for the properties of the air traffic control system in a given moment, as an actor in its own right.[5] Every change in the system had consequences for controllers and dead reckoning. Across time, the book follows the development of a profession and its expertise.

Tracing the life course of the system exposed continuities, change, and persistence across time. Continuity refers to the things that stay the same, relatively unchanged, over long periods, even centuries. Ironically, one of the continuities was the pattern of eventfulness, contingency, and unanticipated consequences within and across all eras.[6] We would expect that a formation story would reveal a lack of stability and vulnerability during the beginnings of a novel social form, but then as the system matured, we might expect to see stability, or periods of stability interspersed with periods of change and instability. However, institutions can change their form or function or remain stable and persist even when they have outlived their usefulness. Moreover, institutions are vulnerable to political processes that interrupt the institutional status quo, mobilizing change.[7] The view from here and now is that the system has never been static but always vulnerable, always changing in response to changing conditions. Indeed, one of the continuities across eras was change itself.[8]

By situating the work of controllers in its larger social context, this book makes visible system effects: the connection between events external to the system—political, economic, technological, cultural conditions, and the actions of powerful actors in those realms, as they affected the life course of the system that became the National Airspace System, and the air traffic control system within it, changing it, and consequently changing the work place and controllers' work. The Jet Age (1950–1980) found the system in trouble, plagued by congestion, internal conflicts, technological lag, and congressionally produced budget shortages, bringing about the earliest FAA hiring freezes. Then during the Age of Conflict, Decline, and Repair (1980–2000) the 1981 firing of striking PATCO controllers was the first of three systemwide crises that undermined safety and increased risk. Repairing the system after was a slow process, marked by cycles of decline and repair due to changing political administrations and continuing cuts to FAA budgets that exacerbated staffing shortages.

FAA efforts to update aging structures and technologies to meet changing conditions were slow as a result of the liabilities of technological and organizational innovations, which not only added to internal divisions and conflicts but also left the system behind the times. To repair its own failings, in the 1990s the FAA began to build a structured system dedicated solely to the development of technical and organizational innovations. After a contractor had designed and tested an innovation, the FAA combined contractor representatives, retired controllers who after retiring worked for FAA contractors, controller representatives of the involved facilities, and FAA engineers and technical experts to conduct trials in FAA locations, then strategically implemented the innovation in air traffic control facilities one at a time—the "cascade" method—learning by mistake and correcting before moving to the next facility.

The turn of the century began with the deadly terrorist attacks of September 11 that killed thousands and revealed the system's vulnerability to intruders, taking the lives of passengers and crew members and threatening the future safety of the system itself. It was followed by a year of drawing together proven strategies of the past—the airport quota system—and improvising new rules and categories to get the planes in the air again. Then the FAA and the controller workforce began rebuilding the system technologically and organizationally to protect against additional threat. Speed being essential, it was a full year of systemwide change and emotional labor. Budget cuts, hiring freezes, and staffing shortages notwithstanding, both the system's response to the attacks and role of the workforce in the year after in rebuilding the system showed that the National Airspace System was in solid working order.

In contrast to the previous two crises that threatened system safety, the crisis in the age of automation was not a surprise action by a political actor external to the system, but the unintended consequence of sequences of actions—trajectories—begun in the 1990s that developed incrementally and culminated in the coincidence of the staffing crisis, which was then accelerated by retirements and the implementation of NextGen.[9] Perhaps better titled as the Age of Government Furloughs, Shutdowns, Sequestration, and Catch-Up Ball, Congress's failure to adequately provide funding for the FAA halted progress on both hiring and NextGen. Moreover, the situation was exacerbated by the unanticipated negative consequences of FAA NextGen efforts to streamline the system by standardizing and centralizing to decrease costs and increase operational efficiency. Streamlining included a plan for a speedy recovery from the staffing shortage by first, substituting a new but flawed test to initially screen applicants replacing a more time-consuming method, and second, by replacing the storied FAA Academy training with a new system designed to expedite the production of controllers. As a result, new people arrived at facilities unprepared, producing not only a high failure rate in the facilities, but those passing were not sufficiently competent to work traffic in the facilities to which they were assigned.

Examining the system across these five eras shows the continuities and changes that shaped the system's characteristics at the time of the study and now.[10] One pattern was the system's continuous vulnerability to the effects—system effects—of events, conditions, and the actions of actors external to the system. Although vulnerability to system effects can be said of all organizations, the FAA's National Airspace System became a government monopoly upon which the public depended. Moreover, the US government depended on it to be the leader in air transportation internationally as a material and symbolic reminder of US power. Thus, beginning after World War II, the NAS was in a competitive role internationally, which demanded innovation and cutting edge technology.

Being competitive, and an institution with mission, goals and tasks, maintaining leadership called for innovation that even under the best of circumstances was likely to be inherently fraught with both technological and organizational liabilities, failures, and learning from mistake. Recall that early in the Jet Age, there were the design difficulties of switching from calculating aircraft paths by moving shrimp boats on tabletop radar to developing a design combining computers and radar on an upright screen that would be able to follow an airplane from one scope to another in the same facility or from a scope in one facility to a scope in another facility.

Another distinguishing characteristic of the system was that unlike other large-scale socio-technical systems in other fields that were competitive for international leadership, the NAS was fully dependent on government—presidential administrations, Congress—for its funding, so resources always were scarce, and changing policies and funding variations contributed to cycles of decline and repair.[11] Oddly, and in contrast to this sequential, alternating pattern across time, the coincidence of the staffing crisis and NextGen had decline and repair occurring at the same historical moment, the two in combination escalating risk.

System Effects, Resilience, and Agency

Having considered system effects, continuities, change, and vulnerability over time, we must ask the question of persistence: what enabled this system to avoid disastrous accidents on a day-to-day basis and survive as a public agency, maintaining its original form, rather than failing to the point that it was replaced by privatization or single corporate ownership, which threatened more streamlining and cost cutting? In other words, what has made the system so safe—or is it? And what do controllers do that technology can't replace? This book so far has made it clear that controllers and technology are interdependent and that historically this has been true. In this section, I reverse the question of system effects on controllers and their work to examine the effects of controllers and their work on the system.

Throughout the changing political, economic, technological, and cultural circumstances, beginning in the Jet Age the glue that held the NAS together as a dynamic system of social action in its own right lay in several sources. Resilience, redundancy, and reliability—key characteristics that in the late 1980s scholars first identified as factors essential to a high-reliability organization—had been built into the system structures and processes in the preceding eras. In contrast to the social psychological orientation of most research on high-reliability organizations that locates safety and risk in individual and group interaction in the local situation,[12] I have taken a layered situated action approach, looking at the dynamic connections between assemblages of heterogeneous actors in the external environment, the organization—its structures, technologies, and processes—and the interpretation, meanings, and actions of controllers.

Consider, first, the sources of safety and persistence in system structure and the varying relations of its parts. *Resilience* refers to the ability of social actors—individuals, groups, organizations, states, economies—to absorb the impact of external and internal system disruptions and shocks without losing the ability to function and to cope, adapt, and recover from them.[13] Resilience developed as a reactive response to changing external demands. Once technology enabled the bridging of system boundaries, the airspace boundaries could be expanded or consolidated, fixed or flexible, permeable or shut.[14] As the airspace had changed, so had the organizational structure on the ground. Both became pliable, capable of being arranged and rearranged or eliminated to suit changing needs. The parts could be either loosely or tightly coupled, as situations demanded.[15] The airspace system was decentralized on an everyday basis, with the capability to quickly become centralized either in a facility, a region, in sections of the country, or as a whole. Its many nested socio-technical systems were interdependent, held together and responsive to the standardized routes in the sky and the airplanes that traveled them. Airline schedules, weather, technical failures, and accidents dictated the relationships and nature of the ties—strong or weak, tightly or loosely coupled—between facilities at a given moment.

Redundancy in the structure was built into the ability of one region or facility to pick up the traffic of another in case of a shutdown of one part of the system due to weather, accident, or technical failure. Reliability was tied to systemwide institutional rules and standardization originating with the FAA, but it also was embedded in specific rules generated by local facilities that were guidelines about the exchange of traffic internally and between adjacent facilities. The connections among air traffic control facilities are spelled out by letters of agreement and memoranda of understanding, documents that articulate the connection between the parts and the larger US system by multiple rules and procedures designed to create common material practices to facilitate coordinated activity across physical and social space. The International Civil Aviation Organization provided reliability for the entire system by developing and requiring standardized language, rules, and procedures that regulated not only air traffic but also accident investigation, cartography, and other specialties. These standards, like those of the member countries, were revised to meet continually changing international circumstances.

To understand persistence, however, calls for recognizing that the resilience, redundancy, and reliability built into the structure of the system were possible only because they were embedded and embodied in the coordinated cognitive, material, interpretive practices of work and the controllers who enacted them. Remember, air traffic control is systems within systems within systems: the workplace, the facility, and the multiple parts of the system on the ground and in the sky that form the NAS. The ethnography chapters focused on dead reckoning in the smallest systems—the tower cab, the control room, the area. In the workplace, resilience was visible in the number of controllers on positions—and the number of positions themselves—such that staffing could be increased or reduced as traffic rose and fell during the day and by the season. Reliability was supplied by collective enactment of standardized rules and procedures—local, national, and international—designed to meet particular situations, classes of airplanes, altitudes, and emergencies. Redundancy was built into the way that tasks and procedures were organized in the room: for example, the checklist procedure when one controller relieved another, so both had the same mental picture as the relief took over the scope, then the other watched from behind while the new person got going; the way a supervisor automatically moved to the controller with the difficult situation, providing a second pair of eyes and help if needed; and controllers positioned elbow to elbow in close quarters so they could coordinate, adjusting to each other and essentially becoming self-regulating.

As important as these actions are individually, they are examples, and they must be understood as a few of the many embodied, enacted cultural understandings and material practices that constitute controllers' work. Dead reckoning is about foresight: predicting the positions of objects in space and time by deduction, without benefit of direct observation and direct evidence. The ethnographies show that interpretive work is a key aspect of controllers' work that technology can't replace. Interpretive work is controllers' fine-tuned ability to give meaning to what they see, hear, and experience that is grounded in and grounds ethnocognition and boundary work. The three combine in dead reckoning.

The ethnography chapters revealed that ethnocognition is a cultural system of knowledge comprised of shared ways of thinking, being, and doing that transform controllers, enabling them to coordinate effectively both in the room and across physical and social space. These are systemwide and specific to their profession—an occupational *habitus*.[16] Their moment-to-moment problem solving is embodied, situated, local, distributed, and practical. At the same time, ethnocognition also is shaped by the larger system in which they work, as it responds to changes originating in the institutional environment: competition, resources, conflict, and shifts in power and control.[17] The ethnographies of the four facilities during 2000–2001 revealed how the intensive training process not only taught the fundamentals of their tasks and available devices but also began inculcating the conflicting goals of the system along with the other subtleties of the job, including the interpretive work, ethnocognition, and boundary work that constitute their cultural system of knowledge. The process of moving from skill to expertise began at the FAA Academy, was followed by training in local facilities, then increasingly refined by trainers, and then after certifying, involved seasoning based on daily experience with new situations and learning by mistake. In common across the four facilities, we saw how this cultural system of knowledge affected controllers' moment-to-moment problem solving with traffic and the interactions with coworkers and devices in the workplace day-to-day as well as how common ways of thinking, being, and doing enabled the workforce to handle safely the unprecedented tasks of September 11 and the year after.

In contrast, in 2017 the importance of this cultural system of knowledge to the work of controllers was confirmed by its *absence*. When training was changed to expedite the supply of controllers and prepare them for an automated workplace, the new people arriving in the facilities were not sufficiently competent. They had the ability and desire to be controllers. However, because of the shutdown of the FAA Academy, the streamlining of hiring and training systems, and the assignment of inexperienced people to facilities with complex airspaces without "working up the ladder," those controllers were missing the foundational cultural understandings, material practices, and experiential knowledge about the system essential to interpretive work and boundary work. They could learn the skills, but they struggled to acquire the necessary level of expertise.

Although the ethnographies graphically demonstrated what was missing, the presence, content, and cognitive, physical, and material practices of controllers' cultural system of knowledge were visibly displayed in the efforts of supervisors, support staff, trainers, and senior controllers to pass it on, and in the insights and practices of those newly certified controllers in the next generation who succeeded in acquiring expertise and whose work was shaped by it. Consequently, we gain further understanding about what controllers contribute to persistence by examining this cultural system of knowledge at Boston Tower and TRACON in 2000–2001 and controllers' efforts to reproduce it in 2017 in workplaces transformed by NextGen automation and organization changes. Doing so reveals controllers' expertise and the agency behind the system's resilience, reliability, and redundancy. In the following, I distill from the comparison the component parts of controllers' cultural system of knowledge—ethnocognition, interpretive work, and boundary work—then discuss how the three in combination enabled controllers in the workplace to become collective, collaborative, coordinated change makers, improvising tools of repair to preserve safety; to maintain the system's resilience, reliability, and redundancy; and, as a by-product, to contribute to its persistence.

Ethnocognition: Standardization and Improvisation

For this discussion, the definition of ethnocognition from chapter 1 bears repeating. The work controllers do is a process of working things out from moment to moment with coworkers, pilots, and devices, as they move material objects across time and social space. Their cognitive, physical, and material practices are situated, embodied, local, distributed, and practical. At the same time, these practices are shaped by historical actions and actors, external political, social, technological, economic, and cultural beliefs that are passed on through training methods that produce durable embodied transformations of their thinking, noticing, hearing, vision, and emotions, triggered by cues in the environment.[18] Consequently, ethnocognition is not only distributed beyond the room across boundaries of time and space; it is also layered. Sociologically defined, ethnocognition is the situated enactment of expert bodily, cognitive techniques and cultural understandings that work together in an active, interpretive thinking and doing in relation to others in the room; the multiple devices, material objects, and socially organized arrangements that surround them; the local situation; and parts of the larger system in which they work.

Ethnocognition is general to the profession and curated to suit the local situation. Recall that although the NAS is a highly standardized system, the airspace a facility owns varies depending on variations in aircraft design, fleet mix, traffic volume and complexity, geography, weather, and airplane equipment capability. And as airspace varies, so do architecture, technology, tasks, social arrangements of work, and culture of the place—or "personality," as controllers think of it. Controllers throughout the system know well the patterns of their own airspace, but even their routine daily work is never routine: the capacity for anomalies is built into every airspace.

The chronicle of these anomalies is recorded in every chapter: a pilot busts the assigned altitude, hears an instruction wrong, wanders onto a runway instead of holding, turns the wrong way, is too slow shifting from waypoints to manual, misses the approach, or mistakes an alleyway for a runway. Changing weather also can produce the unexpected: an unannounced summer microburst begins near the airport and all flights have to be diverted, a plane skids off an icy runway, a once-a-year wind calls for a rarely used runway configuration. Or technology fails: an engine stalls, landing gear fails to descend, or an automated device gives the wrong direction to a pilot, turns a plane into another rather than away from it, misreads a drone for an aircraft, or an automated system in the workplace suddenly goes out and the parts are in another city. Life-threatening emergencies and controller saves happen so frequently that annually NATCA awards the Archie League Medal of Safety to honor a controller or controllers in each of the nine regions for emergency saves:

smoke and fire in the cockpit, a VFR pilot lost in the clouds, a pilot stricken with illness.

Consequently, at all facilities, regardless of variations in traffic volume, complexity, and fleet mix, and regardless of absence, presence, or extent of automation, the contribution of controllers to the resilience, reliability, and redundancy of the system is essential. It rests in their embodied cultural system of knowledge to negotiate the intersection of standardization, interpretive work, and improvisation. Because airspace varies, the mix between the three factors varies by facility, so, for example, Bedford Tower controllers rely on standardization—rules of separation and sequencing airplanes, ground movement patterns, phraseology, the airfield layout—but the emphasis is more on interpretive work and improvisation than at Boston Tower, Boston TRACON, and Boston Center because Bedford's high-volume airspace is small and unpredictable, lacking a standardized sky and with a fleet mix of occasional corporate jets, students enrolled at airport flight schools, weekend warriors, aerobatic pilots, and a few commercial airline commuters.

At facilities with greater airspace volume and complexity, dead reckoning relies on standardization to a greater extent. Professional pilots traverse a highly organized high-altitude system of standardized routes and intersections, airlines have a constellation of standard routes between cities, and a plethora of information on airports, approach plates, equipment, maps, procedures, and automated systems are available in cockpit devices. Nonetheless, interpretive work and improvisation are equally essential because surprises are routine and happen at high speeds. Consequently, standardization, interpretive work, and improvisation are routine at all facilities. As one controller succinctly put it:

> Structure and routine, structure and routine: the rules and regulations [structure], and how we respond that becomes automatic [routine]. If we have something new, we have at least had something like it before. If we have to improvise, we improvise from the base.

Signals and Interpretive Work

Comparing the two facilities in 2000–2001 and then in 2017 shows how ethnocognition and interpretive work combine to shape moment-to-moment prediction and problem solving. To predict the trajectories of individual airplanes, whether on radar or out the window, they scan both the sky and the devices in the room. They don't watch a single airplane; they see the sky as a gestalt, with every airplane having a temporal and physical trajectory.[19] Glancing at flight progress strips, whether printed or electronic, controllers give meaning to an airplane on the basis of its position, equipment type, destination, altitude, and speed, so they are able to holistically visualize it and its forward trajectory in relation to other planes in the sky. With information displayed on devices in the room, they do "screen work": they do not read every word but recognize the meaning from the form that the information takes.[20] In the same way, they listen not to every word the pilot says but from the whole and the rhythm they can recognize a deviation from the expected. Anticipating the immediate future, they make a plan, changing it as relationships and trajectories in the sky change. The plan is not the solution, but a resource to orient them to a situation.[21] As controllers scan the airspace and their multiple devices, their situation awareness incorporates pilots' voices on the radio and coworkers' conversations, and their peripheral vision absorbs the bodily movement and physical actions of supervisors and other controllers in the vwwroom.

Across all facilities, standardization comes into play as institutional rules provide a constant base against which these embodied perceptual skills are framed. At radar facilities, the rules of separation give height, width, and depth to an aircraft, stabilizing it in relation to those around it by aiding the controller to perceive it in three dimensions. The assigned routing along standardized highways and intersections in the sky are guidelines to anticipate the amount or traffic to be expected at a given time. Similarly, tower controllers, whether in automated facilities or not, look out the window to rely on informal patterns from which they estimate distance, spacing, and time from physical objects such as highways, landmarks, shopping malls, skylines, and water formations.

Being able to move airplanes safely across the sky depends on identifying anomalies early and correcting them so little mistakes don't lead to near misses and accidents. Controllers bring to each session a background grid of information and patterns from which deviations from the expected can be identified. They don't watch everything equally; they give greater attention to those airplanes that stand out against the cognitive frame of reference that is their cultural system of knowledge. They watch for signals that something is wrong: a pilot makes a procedural error, deviates from the assigned route, violates a rule, or doesn't follow instructions. They notice variations in the pilot's tone of voice, speed of response or a silence, the wrong phraseology or none, and signs of fatigue, distraction, or incompetence.

On the basis of their experiential knowledge, controllers can anticipate what is to come, improvising a response in the moment. Take, for example, looking out the window, which before NextGen automation was the heart of the Boston Tower job and remained essential there in 2017. Both the newly certified controllers and senior controllers appreciated the advantages of automation and acknowledged its weaknesses in these excerpts from earlier examples: "We can see where a plane is before the computer does"; "We anticipate the fog coming holistically, we can see it forming between the strips and the equipment, so can divert traffic early before fog covers the airport." On using speed controls: "The ASDE only gives distance and location, but we know with a wind shift when a plane is going to hit the wall, so slow accordingly."[22]

Similarly, the reliance on interpretive work and improvisation remained consistent and essential at the old Boston TRACON and the 2017 consolidated TRACON after the addition of STARS, waypoints, and the ADW. In 2017, three consecutive wind changes in a matter of a few hours changed runway configurations, demonstrating the continuing importance of controllers even with automation. In this sequence, the balance of the interdependence of controllers and their technologies varied. In the first wind change, the configuration called for the ADW and the CRDA, impossible without controllers' supplying their embodied precision vectoring and precision spacing, sense of timing, and rhythm of the work. The last was that rare wind direction that called for the "trifecta" configuration, with pilots on visual and controllers relying on old-school dead reckoning.

Boundary Work

The system's many boundaries, both in the sky and on the ground, have always been divisive, contested, and sites of ongoing conflict where individuals, groups, and organizations compete and negotiate their differences across boundaries. Because each facility's airspace varies, the pieces of airspace do not readily mesh. Further, each facility's airspace shapes the place, its architecture, technologies, the social arrangements of work, and thus its culture. Consequently, coordinating across system boundaries is typically conflict ridden. The concept of ownership of pieces of airspace leads to turf wars, sky and ground. Controllers' boundary work holds the pieces of the system together as they negotiate standardization, interpretive work, and improvisation, working out the conflicts from moment to moment to move traffic along. The comparison of the ethnographies in 2000–2001 with 2017 reveal the variety of boundary work that controllers do: material, social, cultural, and symbolic.[23] In addition, boundary work as power work becomes a key element of strategic action in several unprecedented domains.

Historically, the system has always been vulnerable to the actions of social actors in positions of power external to it, in particular presidential elections and changing policies of the FAA that have affected labor-management power relations. Recall the Bush administration's boundary work: imposed budget cuts, harsh work rules, reduced salary and benefits that disempowered NATCA—all this spurred retirements and scared away new hires, exacerbating the staffing crisis. However, the Obama administration reversed these measures. Empowered, NATCA engaged in some boundary work of its own, applying the union's knowledge of organization systems to change its own boundaries. Decentralizing its structure and processes, NATCA gained leverage negotiating issues with FAA management at the facility, regional, and national levels. This change and the FAA's new centralization and standardization were divisive in the facilities, reducing air traffic managers' control over their facilities. At the same time, other Obama administration changes had countervailing effects.

The administration empowered workforce expertise. Controllers gained the permanent legal right to have input in the design, development, and implementation of all new technologies and organizational designs. Participation now was required by law. This change empowered workforce expertise in two ways. In all major facilities, supervisors, TMCs, and controllers with specific expertise were designated as subject-matter experts assigned to special NextGen projects to develop automated technical innovations with FAA technical experts, architects, and airspace and infrastructure specialists. Second, the Obama ruling also included organizational innovations. NextGen's Section 804 Collaborative Workgroup for the experimental consolidation of Boston TRACON pulled together operations managers, supervisors, TMCs, and controllers at the two facilities with other experts to coordinate, implement, and guide the project to completion.[24] To this task, the workforce brought its history of adjusting standardizing changes to suit the local situation, its experience of improvising tools of repair, and its cultural system of knowledge, all on full display as it reworked social, material, symbolic cultural, and generational boundaries at Boston Tower and TRACON. They collectively worked to maintain safety, but the system was still risky, as evidenced by tired controllers' errors other places and breakdowns of automated systems that shut down whole regions.

Then history repeated. Recall that in 2016 a combined NATCA-FAA report to the House Subcommittee on Aviation argued for three changes that would reduce systemic risk, eliminate the staffing crisis, and restore safety: first, a stable, predictable funding stream without privatization, instead calling for a government sponsored not-for profit organization; second, for the FAA to hire to full capacity annually; and finally, complete redesign of the hiring and training process. These proposed changes targeted the institutional, organizational, and interactional origins of actions external to the system that had increased risk. If enacted, they would have restored the stability to the system and decreased risk. But none of this happened. Instead, congressional budget battles continued. The longest government shutdown in history—thirty-five days—extended from December 22, 2018 through January 25, 2019. All new hiring and progress on NextGen halted. In fall 2019, I returned to Boston TRACON in Merrimack for a quick informal visit to see the control room with the Cape TRACON, the third to join the consolidated facility, installed and working: the final phase of realignment and integration was in progress. Before going up to the control room, I sat downstairs for a while with some controllers on break. I noticed that one of them was struggling to stay awake, something I had never seen. As we rose to go up, he said, "I'll see you up there. I forgot my coffee." A controller who forgets his coffee is one tired controller. If one is tired, more are tired. In this third and final phase, they were training Cape controllers and still engaged in boundary work to unify the facility. Fifteen years after the consolidated Boston TRACON became operational, the system effects of the intersection of the staffing crisis and NextGen were still ongoing.

In March 2020 contingency struck in the form of the COVID-19 pandemic. All passenger air transportation halted. Suddenly the staffing crisis was less of a problem: few airplanes, enough controllers. As one controller sardonically put it, "It all evened out." Essential service providers, air traffic controllers continued to work. Sitting elbow to elbow on position, they were being infected with the virus. Air traffic controllers are a profession of problem solvers. Improvising, the FAA and controllers nationwide collaborated in boundary work, reorganizing the system. The controller-powered resilience of system boundaries allowed them to shut down one NAS region at a time to clean and sterilize facilities regularly. For controllers' safety, they reorganized the boundaries of time, social, and physical space. They converted to backup crews, going to a three-team system that kept a team in waiting to come in after a cleaning. Each team came into a clean place, worked a ten-day shift with two-week breaks with the same five people, elbow to elbow, isolating from the others coming in and leaving and on breaks. By the first week of April, this was the new routine. In a two-week period in which time and timing were crucial, again the expertise of the workforce—management and labor, coordinating across national, regional, and facility boundaries—had preserved the safety of the system and its operation in the face of this new threat.

Dead Reckoning: Coordinating Action and Anticipating Futures in Complex Organizational Systems

This inquiry began with a goal to understand how the National Airspace System came to acquire the error-reducing characteristics it has today. The book has demonstrated the incremental development of the system structures and processes that provide its resilience, reliability, and redundancy and the role of controllers in the safety of the system, confirming its status as a high-reliability organization. Dead reckoning is about foresight: predicting the position of objects in space and time by deduction, without benefit of direct observation or direct evidence. Repeatedly, across time and change, this book has shown how controllers enact their expertise, individually and collectively, anticipating threats early, taking actions that protect the safety of aircraft and the system on the ground. Yet the analysis has shown that for the system, risk and safety are variables, the balance historically shifting in response to events, conditions, and the actions of powerful actors in the external environment that bestow it with an inherent vulnerability.

However, a similar internal force is at work. Complex organizational systems are dynamic, processual, and unpredictable. So it is no surprise that even in this standardized system focused on safety that when attempting to improve the system, the FAA has obstacles built into its structures and processes that can block the ability of leaders in different locations in the hierarchy to anticipate the outcomes of the FAA's own actions on the workforce when implementing change, thus also contributing to the vulnerability of the system. Although the National Airspace System is unique in many ways, these same problematic system structures and processes are general properties of all complex organizational systems, so the implications are relevant to consider.

Dead reckoning is not restricted to air traffic controllers, but the process can be understood more broadly as a social physics that applies to individuals and a variety of organizational forms, large and small, from nation-states to families. Individually, we all go about anticipating our futures and imagining our life circumstances in relation to others. Tavory and Eliasoph observe that for individuals, anticipating futures is coordinated and temporally oriented: individuals engage with anticipating the future, projecting trajectories of action in time and space and coordinating with others in order to make sense of future action together.[25] Although Tavory and Eliasoph focus on individual interaction and volition and air traffic controllers operate in a formal structure where interactions, plans, projects, trajectories, and actions are channeled and constrained, the everyday life process they describe is in many ways analogical not only to the dead reckoning that the controller workforce does on the job but also to the operation of the FAA itself. In organizations, people do "temporal work" in strategy making: they resolve their different interpretations of past, present, and future to coordinate and make sense of action together.[26] We can think of all organizational systems as engaged in dead reckoning, preoccupied with anticipating their own future position in social space and time in relation to other organizations by deduction, without benefit of direct observation or direct evidence.

Like the air traffic control system, other complex organizational systems are driven by concerns about temporality: history, the present, and the future.[27] In common, they exist in an organization field, competing for scarce resources, with a set of normative expectations for achievement. Crucially important among those expectations is that organizations innovate to keep up with changing times—seeking legitimacy in the field, moving ahead in status, staying in place or falling behind, but continuously anticipating and coordinating their future positions in relation to others.[28] And while this book emphasizes the contingencies, coincidence of disconnected trajectories, and unanticipated consequences of events and actions originating external to the air traffic control system that shaped its past, present, and future, this analysis reveals these same processes and outcomes can be *unintentionally reproduced intraorganizationally*. In contrast to the demonstrated ability of controllers to coordinate the actions of airplanes and people across the conflicting boundaries of time, physical, and social space, FAA attempts to improve the system were often impeded because of the inability of FAA leaders to fully anticipate the impact of changes on the workforce across the internal system boundaries of time, physical, and social space. These "failures of foresight," as Turner famously named them,[29] reveal the organizational complexity behind the unforeseen liabilities of technological and organizational innovation, showing how disconnected trajectories of plans, projects, and anticipated futures can result in flawed outcomes, even when every effort is being made for things to go right.[30]

Consider the following ways that intraorganizational boundaries can routinely alter the outcome of a planned innovation. “Structural secrecy” refers to how the structure of an organization—its hierarchy, division of labor, and specialization—limits the ability of people in different locations within it to know and understand what is going on in other parts.[31] Among the NextGen plans for streamlining operations by standardizing and centralizing were the goals of speeding up the hiring and training of controllers to cope with the staffing shortage. The changes were implemented by individuals positioned differently in the system who, separately, designed new programs for training and for distribution of new recruits. Consider the difficulties of going from a general project goal established at the top to the particulars of design and implementation lower in the hierarchy.[32] The new training program met the requirement for speed and aimed at preparing for automation but did not include the cognitive or manual skills of dead reckoning taught previously, so many new hires passed the training but were unprepared and failed early in the facility training program.[33]

Distributing those who did survive the training to the facilities was a different process but was similarly impaired by structural secrecy. To speed up and fairly distribute new hires, a standardized complexity index was developed based on estimated numbers of new trainees needed at each facility in a region. When distributed to the facilities, flaws in the count of the number of controllers in a facility and the number needed unfairly allocated facilities with high complexity airspace with proportionally fewer new hires than facilities with less complex airspace. In both examples, the designers’ positions in the structure and division of labor limited their understanding of what had gone before and what came next, restricting their ability to coordinate action with relevant others to make sense of action together: they could not anticipate the complexity of the work and workplace where the newcomers would begin. As a result, FAA attempts to catch up on the staffing shortage unintentionally stalled it, producing inequalities in the system as a by-product.

In addition, many technological and organizational innovations have a temporal dimension across internal boundaries. The trajectory from idea to plan to project and from design to development to implementation moves through different organizational settings operated by people with different specializations, so the time and change at each stage is hard to anticipate. The past also matters: the history of the system showed the relationship between precedent and innovation. Innovations do not grow in a vacuum but are built on what comes before. Some may begin as an informal solution to an unexpected problem and if it works, it becomes standardized, formalized and often institutionalized throughout the system. Other innovations set a precedent for further development, so will be applied to settings or purposes other than at the point of origin. Still others may be imported from other systems, the innovation being a redesign to fit a different need.

Regardless of the type of innovation or its origin, at each stage—idea, planning, project, design, development, and implementation—an innovation will be modified by different groups of specialists with different visions that have to be coordinated in order to make sense of action together, not once but many times as, learning by mistake, they make changes. Then the innovation goes to the next stage and the same will happen. Coordinating differences within specialist groups is likely to go more smoothly than crossing boundaries between groups, building the potential for disconnected trajectories into the temporality of the process. The final product will be a transformation of the original idea—an unanticipated consequence, we might say—hopefully better for the iterative processes behind it. Now standardized, the process of coordinating action and anticipating futures occurs once more, when the innovation is installed in the location of its use, which of course will vary from facility to facility, so repairs must be made to fit the local situation.

Historically, the ability to anticipate the effects of a change on the workplace and controllers work had been limited, only visible when fully installed in the workplace. Recall the unanticipated positive effects of the 1988 airspace realignment on the old TRACON, which transformed a competitive, conflict-ridden facility into an enduring cooperative, coordinated culture where individual achievement was replaced by group achievement as the priority. However, the innovations at Boston Center during the 2000 Display System Replacement initially had negative effects. The FAA did carefully prepare the technology and controllers for the new DSR monitors, but they were installed in a new redesigned control room, so controllers moved from "an airplane hangar to a shoe box." The FAA did not predict the extensive effects of the changed size, architectural design, and lighting of the room on controllers' cognitive, physical, and material practices. In neither case—the TRACON or the center—were the outcomes in the workplace anticipated. In both facilities, controllers who struggled to adjust in the short term eventually did, mastering the changes in a month or so. During transitions, when timing and mental quickness both slow, risk increases.

In contrast, NextGen was systemwide and involved multiple innovations, organizational and technological. Extensive preparation was done in advance. The FAA had become highly organized and proficient at developing technological innovations, having built separate structures for the planning, design, and development of projects. The FAA had also converted a few major facilities for use as test sites before installing NextGen at other locations. The effects of individual technologies—STARS, DataLink, TRACON workstations, the ADW—had all been researched and the necessary architectural changes for the technical infrastructures, anticipated and calculated. In 2000, work had already begun that involved the Section 804 Collaborative Workgroup, FAA and NATCA elected officials in the New England Region and facility levels in the design and development of the consolidated TRACON and also automation for Boston Tower. The controllers at both facilities had been retrained. Implementation began with great optimism.

Despite the extensive preparation, close supervision, and participation at the regional and local level, the full effects of NextGen on the workplace and on controllers' work had not been anticipated. No one, even those most closely involved in the New England Region and the two facilities, had anticipated *the combined effects* of all changes—social, architectural, spatial, technological, cultural, generational—which fully manifested only in the operational workplace. Even setting aside the effects of the staffing shortage, the effects on both facilities were multiple and long-term. The inability of leaders to anticipate the effects on controllers was not due to structural secrecy, or a result of drifting away from best practices—nor did people share a coherent set of meanings that allowed them to normalize discordant facts.[34] Instead, it was inherent in the interactive complexity of the organizational and technological innovations. Although these effects were unpredicted, incrementally the system had become prepared for postimplementation problems. As chapter 12 revealed, controllers in the two facilities—air traffic managers, operations managers, supervisors, support staff, TMCs, subject-matter experts, and controllers who were working traffic and training the new people were empowered not only by law but also by past experiential knowledge of the system and their history of improvising tools of repair. They developed and implemented plans to fix the local situation, again negotiating the intersection of standardization, interpretive work, and improvisation, collaborating to keep the system safe and, unintentionally, contributing to its persistence.

In thinking about the wider implications of the system safety during 2000–2001 and the system effects, increased risk, and controller response as modernization was still in progress in 2017, we must take into account that the New England Region Section 804 project was a NextGen experiment, a test run for future facilities. Also, the sheer number, specialized character, and extensiveness of NextGen automated devices and organizational innovations in the workplace was unusual, as was the context: a standardized system with everyone in the system except the politically appointed leaders at FAA headquarters having been trained and worked as air traffic controllers, then nationwide all people in the facilities engaged in the same kind of work make it unusual as well. Moreover, the formal concept of ownership of the airspace by a facility and its controllers is a significant part of controllers' collective identity. That sense of ownership was further reinforced by both the legal empowerment and the internal option to serve as subject-matter experts, thereby having input into the design, development, and implementation of innovations. When studying an exotic case like air traffic control, there always are limitations to what we can conclude in terms of application to complex organizational systems in general, but there are compensating benefits. The very scale, complexity, and temporal duration of this system reveals general patterns in the dynamic interaction of external environment, organization, workplace, and technical work that have not been acknowledged before.

The outpouring of books and articles predicting the effects of automation on labor markets, mass unemployment, the workplace, and the future of work is vast, encompassing a seemingly endless variety: robots, artificial intelligence, digital technologies, algorithms, machine learning. Then there is automation in factories, offices, the classroom, health care, entrepreneurship; the disappearance of low-skill jobs and increasing inequalities; and the disappearance of highly skilled professionals—doctors, teachers, accountants, lawyers—lost to lay expertise acquired on the internet.[35] Despite this extensive literature, little is known about the intraorganizational effects of modernization on complex organizational systems. We know about de-skilling but little about workers' temporal process of adjustment.[36] This historical cross-case comparison captures the changing nature of work, showing the unanticipated consequences—both positive and negative—of replacing aging technologies and physical structures with organizational and technological innovations. In particular, this book has revealed the little-known before and after of the introduction of automation in the workplace and workers' temporal process of physical and cognitive adjustment that follows, which was made more complex by the necessary redesign of the organization, its architecture, and the social arrangements of work.

Also unaddressed in this plethora of public writings is the fact that modernization projects go on in complex organizational systems made vulnerable by historical actions taken by powerful actors external to the system that have effects—system effects—thus changing organizations, the workplace, and the nature of work in unpredicted ways.[37] Even in those safety-conscious systems known as high-reliability organizations, the efforts of administrative leaders in the system can have negative workplace consequences because of organizational complexity: changing one part of a system can have unanticipated consequences for other parts.[38] Many such changes are subtle and impossible to predict. Seemingly neutral internal changes, such as redesigning a work space, merging subunits, standardizing and centralizing, or introducing a new technology can change the culture of a place, thus affecting workers' cognition, material practices, and the ways people think, act, and are.

This view from inside the workplace reveals the internal contradictions between past, present, and future that plague modern organizations. In our era of overwhelming technological benefactions, air traffic control is a warning to other organizations about the unanticipated consequences of technological and organizational innovations. The push toward increasing speed and efficiency in organizations by digitalization and automation in order to supplement or replace the contribution of human cognition and material practices is a global trend, affecting workers in many occupations.[39] The case raises new issues that change the debate about costs and benefits of advanced automation: unanticipated consequences of technological and organizational innovation on work, physical and cognitive adjustments required of employees, increased risk of mistake during transitions, and unimagined challenges in fitting new technologies into existing organizational structures and technical infrastructures.

Even as this book conveys warnings about complex systems, system effects, and the liabilities of technological and organizational innovation, it shows the kinds of problem solving solutions that worked over time and the importance of people. The FAA incrementally learned from mistake, developing procedures one at a time that the agency eventually combined in a systematic program for design, development and implementation of change. This scale of preparation would be not only impossible but also undesirable and unsuitable for most organizations and many varieties of automation. However, the idea of incrementalism, learning from mistakes and from the mistakes of others, and anticipating the future effects of change and innovations can reduce risk of unwanted effects on the workplace. Throughout, the analysis has demonstrated the interdependence of controllers and their multiple technologies of coordination and control. What has stood out both before and after automation is the importance of controllers' contributions to the resilience, reliability, and redundancy of the system and its persistence.

Across time, one of the unusual aspects of this hierarchical system is the capacity of the workforce—the people in the facilities doing the hands-on work—to have input both in system change and repair. We tend to think of expertise as the practices of individuals, but as Eyal observed and this book confirms, expertise cannot be fully understood by the practices alone but must incorporate the social, material, spatial, institutional, organizational, and conceptual arrangements that shape them.[40] In all organizations, skills and expertise are acquired through socialization into the occupation and the workplace, but expertise is developed only by being there, interacting with others, and understanding not only how the parts of a place work but also the social, cultural, and technical aspects of its tasks.[41] This includes experiential knowledge of the strengths and weaknesses of the system, the organization, and coworkers. Regardless of differences in organization size, complexity, function, and type of automation—from the automation of hospital pharmacy processing to algorithms in the classification of art—workers who know well the work and the workplace can participate in coordinating action and anticipating the arrival of technological and organizational innovations in the workplace and then coordinate action again, improvising tools of repair, making adjustments after the fact.[42]

In contrast to professionals who actively resist technological innovations,[43] controllers act as subject matter experts, rewarded by improved technologies, increased expertise, a new sense of ownership, and increased status in the workplace. This book demonstrates that controllers' expertise, their interactions with one another, and their multiple technologies of coordination and control are essential to the system's daily activities, safety, and persistence. In addition, over time the workforce became a source of system change that included repairing not only the unanticipated consequences of modernization but also the inequalities built into the system. As crucial as these workforce interventions have been to the system, however, controllers' efforts and achievements as change makers seldom fundamentally altered the institutional causes of the system effects that plague the air traffic control system. Other complex organizational systems are in a similar position, regardless of variation in size, complexity, and function. For example, professionals in health care, education, the criminal justice system, and others also rely on multiple technologies of coordination and control. Similarly, they all work in systems that have been shaped by history, contingency, and the unanticipated consequences of actions of powerful actors in the organization field that resulted in certain institutionalized vulnerabilities that put specific populations at risk. Chief among these causal factors are the conditions in the environment external to the system—technological, cultural, economic, political—and the complexity of the organizational system itself and the liabilities of technological and organization innovation.

Risk is a variable, increasing or decreasing in response to historical change and system effects. As this history attests, the systemic causes affecting the workplace and controllers' work have been unpredictable and of long duration, creating periods of decline that have increased risk. Repairing complex organizational systems calls for dead reckoning: taking into account how the past has affected the present and anticipating the future in order to coordinate actions that target the institutional, organizational, and interactional sources of problems. In the air traffic control system, this achievement is revealed in boundary work and tools of repair aimed at each level of the system: the FAA expanded its boundaries and created a second system dedicated to the design and testing of innovations; NATCA worked with the FAA on a joint proposal to Congress to stabilize funding and eliminate the staffing shortage; controllers at Boston Tower and Boston TRACON worked to mediate generational, cultural, and structural boundaries in the workplace to make the system safe. That said, history shows that the life course of organizations will remain eventful, shaped by contingency, conditions, events, and actors in the external environment that intersect in unexpected ways, changing the nature of the workplace and work. As I write this last paragraph in November 2020, cases of COVID-19 are again rising and the economy is tanking. Among the many relevant implications of this book for action, institutional change, and reducing risk, the importance of engaging in boundary work and developing tools of repair during this dark period and after looms large.

Acknowledgments

A historical ethnography of four facilities with three interventions across time, this book has taken many years to complete. Many people have helped along the way. In these few pages, I can only begin to thank them and acknowledge the importance of their contributions. My first and deepest debt of gratitude is to air traffic controllers. I did not come to this research project expecting to study the air traffic control system and its people for over two decades, and they didn't expect it either. There was the supervisor at Boston Center who, seeing me walking down the control room aisle one month after I started, said, "Aren't you done yet? We are used to finishing something and moving on," and the ATM at Boston Tower who, when I mentioned I had been there three weeks and wanted to make an appointment to catch him up on what I had been doing, exclaimed, "My God, has it been that long?" From those moments early in the research, I began to understand that as a profession, they have a different sense of time and timing from that of academics, and so their willingness to have me around periodically is remarkable.

None of this work would have been possible were it not for the initial 1999 permissions and cooperation of the Federal Aviation Administration and National Air Traffic Controllers Association elected officials in the New England Region and in the four facilities essential to this research. This includes, during my first fieldwork visit, the regional directors of the FAA and NATCA who approved the project. At each facility, my visits were approved by the FAA air traffic managers and assistant managers and the NATCA president and vice-president. Moreover, as all FAA and NATCA elected officials are air traffic controllers, they also completed surveys and gave time for interviews, as did the more than two hundred controllers working airplanes, as well as other specialists in the system. I am deeply indebted to all of them.

Collectively, controllers—ATMs, operations managers, supervisors,

NATCA reps, support staffs and all those working traffic—tolerated my presence in their work space and break rooms; taught me on a daily basis; volunteered for interviews and shared their history, experience, and personal insights; and passed on ideas in casual conversation and e-mail. During my revisit in 2017, other controllers occupied the official FAA and NATCA positions—I knew many from the first time, and a new generation of controllers had slowly materialized. They brought the project into the present: I learned from them about the hardships of the post-9/11 NextGen changes, the staffing shortage, and the advantages and disadvantages of a fully automated system and the pressing need for people, training, and patience. I was there for a short time, during a hard time. Still, they taught me. I thank all of them, not only for their generous contribution to this research but also for providing me with the field research and learning experience of a lifetime.

I want to stress, however, that I alone—not the FAA, NATCA, or controllers individually or collectively—am responsible for the content of the book: the analysis, interpretations, explanations, and mistakes are solely my own.

During the analysis and writing, I accumulated many additional debts. At Boston College, three graduate student research assistants, in sequence, cumulatively devoted four years to the data analysis and coding. Alison Doran processed the survey data and analysis and contributed to the writing of the report that was circulated to the four air traffic control facilities in February 2003. Cathy Steven entered the interview transcripts into computer files, dividing the text into the large codes I indicated. Aimee Van Waagenen did the extensive fine-tuning of the coding, within and between facilities. Analytically and theoretically, she was essential to the development of the data analysis and theorizing of the project. During this time, I benefited from intellectually enriching and supportive conversations with Bill Gamson, David Karp, Stephen Pfohl, Barbara Reskin, and Juliet Schor, and many discussions (over wine) with Michèle Lamont and Frank Dobbin.

At Columbia, where this writing began in 2006, I settled into the lonely life of the long-distance writer for the second time in my career. One of the great unacknowledged benefits of teaching is how much we learn from our students. Research assistants Daniel Fridman and Sarah Sachs provided assistance and comments in the early stages, as did continuing conversations with Pilar Opazo. Noah Arjomand commented on chapters and assisted with computer work. Chapter drafts received extensive written comments from Ryan Hagen, Olivia Nicoll, and Larissa Buchholz. Because chapter 2 was my first venture into historical-comparative sociology, I was fortunate to receive valuable insights from the XS workshop organized by department graduate students.

I am indebted to Howie Becker, who has lent his ideas, support, friendship, and typical clear-eyed feedback on many writings over the years, including extensive comments on this full manuscript. I thank Janet Vertesi, who, while finishing her most recent book, agreed to review mine. I am especially grateful to the many friends and colleagues who gave detailed comments on specific chapters related to their expertise: Harry Collins, Andrew Deener, Matt Desmond, Gil Eyal, Gary Fine, Roger Friedland, Daniel Hirschman, Isaac Reed, Adam Reich, Michael Schudson, and Stefan Timmermans. Along the way, I had invigorating in-depth conversations and e-mail exchanges with Steve Barley, Beth Bechky, Claudio Benzecry, Randy Collins, Paul DiMaggio, Neil Fligstein, Herb Gans, Todd Gitlin, Tim Hallett, Heather Haveman, Bridget Hutter, Sheila Jasanoff, Robert Jervis, Jack Katz, Michael Lynch, Harvey Molotch, Wanda Orlikowski, John Padgett, Trevor Pinch, Susan Silbey, Jorg Sydow, Harrison White, and Andreas Wimmer.

I thank Iddo Tavory and Colin Jerolmack and their students in the NYU Ethnography Workshop for their close analytic reading of the boundary work chapters. There were helpful discussions with colleagues at department colloquia and guest lectures: chronologically, Santa Barbara; Stony Brook; Northwestern; University of Connecticut; MIT–Harvard Economic Sociology Workshop; Penn; Princeton; Science Studies at Cornell; NYU Sociology and the Wagner School; Boston College Sociology and Carroll School of Management; the Macro-Organizational Behavior Society at Harvard Business School; the New School; and Department of Management, Freie Universität Berlin. More recently, my overall shaping of this work was greatly affected by my participation in a small, highly interactive conference, "Fields, Frames, Culture, and Cognition," in a unique format at Berkeley Sociology in April 2017.

At University of Chicago Press, three great editors contributed their experience and insights to this book. My special thanks go to John Tryneski, editor of my two previous books at Chicago and trusted friend, who encouraged me and patiently waited for something to read until he retired. Then, in 2018, Doug Mitchell, who, with an overload of authors and books in progress during his last year at the press, took over and guided the book through the review process. Unknown to me, he was seriously ill and passed away within months of his December 2018 retirement after an iconic career as an editor. I am fortunate to have worked with both of them. My admiration and gratitude go to Elizabeth Branch Dyson, who, having inherited Doug's vast list of sociology authors in addition to her own, immediately took charge and skillfully guided and enthusiastically supported both me and the book through the crucial final stages of writing and production. She always had time for conversations about the manuscript content and form, which was remarkable given her doubled responsibilities. Mollie McFee, her sharp and expeditious editorial assistant, advised me on the multiple details necessary for final manuscript submission.

I am grateful to Katherine Faydash, word lover and eagle-eyed detective, whose copy editing added clarity and precision to my manuscript, and Tamara Ghattas, production editor, for fine-tuning and orchestrating the rest of the editorial process. Finally, my thanks to Tobiah Waldron, whose indexing is marked by sociological understanding. In addition to my personal team at the press who worked many hours with me, I thank Elizabeth Ellingboe, who gathered all the permissions for the historical photos and put the manuscript in proper format, both for the initial review and then the final draft, and Noah Arjomand, whose creative illustrations add so much to the manuscript.

I must here finally publicly acknowledge my debt to my undergraduate professor and adviser Eva Sebo, who, in a course on social change at Ohio State in the 1970s, taught me how to theorize, a gift that has sustained me through this book and those before. My lifelong gratitude I give to my close friends through the years, especially Sandra Gradman, Jill Kneerim, and Bill Bell in Boston, Tina Packer and Dennis Krausnick in Stockbridge, Carol Roberts and Roger Travis in Gloucester, and Carol Mason and Paul DiMaggio in New York, who all successfully passed the extreme test of love and friendship by listening to details of this project that only an author could love. And to my children and grandchildren, again, always—Roger and Wendy, Lisa and Alex, Susan and Shane, Katie, Zach, Kristen, Sophie, Lindsey, Sara, Cameron, and Melissa—my enduring cheerleaders and support group, I dedicate this book to you, with much love and thanks.

Notes

Chapter One

1. Stephen R. Barley, "Technology as an Occasion for Structuring: Evidence from Observations of CT Scanners and the Social Order of Radiology Departments," *Administrative Science Quarterly* 31, no. 1 (1986): 78–108; Wanda J. Orlikowski, "The Duality of Technology: Rethinking the Concept of Technology in Organizations," *Organization Science* 3, no. 3 (1992): 398–427.

2. Robert K. Merton, "The Unanticipated Consequences of Purposive Social Action," *American Sociological Review* 1, no. 6 (1936): 894–904.

3. Robert Jervis, *System Effects: Complexity in Political and Social Life* (Princeton, NJ: Princeton University Press, 1997), 3–24.

4. Charles B. Perrow, *Normal Accidents: Living with High Risk Technologies* (New York: Basic Books, 1984).

5. See, e.g., Todd LaPorte, "The U.S. Air Traffic System: Increasing Reliability in the Midst of Growth," in *The Development of Large Scale Technical Systems*, ed. Thomas Hughes and Rene Mayntz (Boulder, CO: Westview Press, 1988), 215–44. For an insightful discussion of high-reliability theory, see Mathilde Bourrier, "The Legacy of the Theory of High Reliability Organizations: An Ethnographic Endeavor," *Journal of Contingencies and Crisis Management* 19, no. 1 (2011): 9–13.

6. Karlene H. Roberts, Denise M. Rousseau, and Todd R. LaPorte, "The Culture of High Reliability: Quantitative and Qualitative Assessment Aboard Nuclear Powered Aircraft Carriers," *Journal of High Technology Management Research* (Spring 1994): 141–61; Karl E. Weick, "The Collapse of Sensemaking in Organizations: The Mann Gulch Disaster," *Administrative Science Quarterly* 38 (1993): 628–52; Karl E. Weick and Karlene H. Roberts, "Heedful Interrelating on Flight Decks," *Administrative Science Quarterly* 38 (1993): 357–81.

7. Bourrier, "High Reliability Organizations."

8. Weick and Roberts, "Heedful Interrelating."

9. Two important early exceptions are Scott D. Sagan, *The Limits of Safety: Organizations, Accidents, and Nuclear Weapons* (Princeton, NJ: Princeton University Press, 1993), and Scott Snook, *Friendly Fire: The Accidental Shootdown of U.S. Blackhawks over Northern Iraq* (Princeton, NJ: Princeton University Press, 2002).

10. I say "surprisingly" because the foundational works on technology began appearing in different disciplines in the 1980s: in social studies of science, see Trevor Pinch and Wiebe Bijker, "The Social Construction of Facts and Artefacts; or, How the Sociology of Science and Technology Might Benefit Each Other," *Social Studies of Science* 14, no. 3 (1984): 399–441; in sociology of organizations, technology, and work, see Barley, "Technology as an Occasion"; and in ethnomethodological workplace studies, see Lucy A. Suchman, *Plans and Situated Actions: The Problems of Human-Machine Communication* (Cambridge: Cambridge University Press, 1987).

11. Thomas P. Hughes, "The Electrification of America: The System Builders," *Technology and Cultures* 20, no. 1 (1979): 124–61; Thomas P. Hughes, *Networks of Power: Electrification in Western Society, 1880–1930* (1983; repr., Baltimore: Johns Hopkins University Press, 1993).

12. Here, as in the *Challenger* book, I use ethnography and interviews to connect historical institutional patterns with decision making, interpretive work, and action in the workplace. Theoretical legitimacy for joining structure and agency in neo-institutional theory developed slowly. Key pieces are Walter W. Powell and Paul J. DiMaggio, introduction to *The New Institutionalism in Organizational Analysis*, ed. Walter W. Powell and Paul J. DiMaggio (Chicago: University of Chicago Press, 1991), 1–40; Stephen R. Barley and Pamela S. Tolbert, "Institutionalization and Structuration: Studying the Links between Action and Institution," *Organization Studies* 18, no. 1 (1997): 93–117; Walter W. Powell and Jeannette A. Colyvas, "The Microfoundations of Institutional Theory," in *Sage Handbook of Organizational Institutionalism*, ed. Royston Greenwood and Christine Oliver (Thousand Oaks, CA: Sage, 2008), 276–98. On expanding neo-institutional theory by taking into account technology and materiality, see Trevor Pinch, "Technology and Institutions: Living in a Material World," *Theory and Society* 37 (2008): 461–83.

13. Thomas F. Gieryn, "Boundary-Work and the Demarcation of Science from Non-Science: Strains and Interests in Professional Ideologies of Scientists," *American Sociological Review* 48 (1983): 781–95; Andrew Abbott, "Things of Boundaries," *Social Research* 62, no. 4 (1995): 857–82; Michèle Lamont and Viràg Molnàr, "The Study of Boundaries in the Social Sciences," *Annual Review of Sociology* 28, no. 1 (2002): 167–95.

14. The foundational work in workplace studies emerged in specialty journals during the 1990s; see, e.g., Christian Heath and Paul Luff, "Collaboration and Control: Crisis Management and Multimedia Technology in London Underground Line Control Rooms," *Computer Supported Cooperative Work* 1 (1992): 69–94; Graham Button, ed., *Technology in Working Order: Studies of Work, Interaction, and Technology* (London: Routledge, 1993), 127–44; Lucy A. Suchman, "Centers of Coordination: A Case and Some Themes," in *Discourse, Tools, and Reasoning: Essays on Situated Cognition*, ed. Lauren B. Resnick, Roger Saljo, Clotilde Pontecorvo, and Barbara Burge (Berlin: Springer-Verlag, 1997), 41–62.

15. Susan Leigh Star and Karen Ruhleder, "Steps toward an Ecology of Infrastructure: Design and Access for Large Information Spaces," *Information Systems Research* 7, no. 1 (1996): 111–34.

16. Paul Luff, Jon Hindmarsh, and Christian Heath, eds., *Workplace Studies: Recovering Work Practice and Informing System Design* (Cambridge: Cambridge University Press, 2000).

17. John Urry, *What Is the Future?* (Cambridge, UK: Polity Press, 2016); Andrew Abbott, *Time Matters: On Theory and Method* (Chicago: University of Chicago Press, 2001).

18. Arthur L. Stinchcombe, "Social Structure and Organizations," in *Handbook of Organizations*, ed. James G. March (Chicago: Rand McNally, 1965), 142–93.

19. Chronologically, *Controlling Unlawful Organizational Behavior: Social Structure and Corporate Misconduct* (Chicago: University of Chicago Press, 1983); *Uncoupling: Turning Points in Intimate Relationships* (New York: Oxford University Press, 1986); and *The Challenger Launch Decision: Risky Technology, Culture, and Deviance at NASA* (Chicago: University of Chicago Press, 1996).

20. Analogical theorizing develops general explanations by cross-case comparison, rather than same case comparison. See Diane Vaughan, "Theory Elaboration: The Heuristics of Case Analysis," in *What Is a Case? Exploring the Foundations of Social Inquiry*, ed. Charles R. Ragin and Howard S. Becker (Cambridge: Cambridge University Press, 1992), 173–202; Vaughan, "Theorizing Disaster: Analogy, Historical Ethnography and the *Challenger* Accident," *Ethnography* 5, no. 3 (2004): 313–45; Vaughan, "Theorizing: Analogy, Cases, and Comparative Social Organization," in *Theorizing in Social Science*, ed. Richard Swedberg (Stanford, CA: Stanford University Press, 2014), 61–84.

21. Rebecca Emigh, "The Power of Negative Thinking: The Use of Negative Case Methodology in the Development of Sociological Theory," *Theory and Society* 26 (1977): 649–84.

22. Suchman, "Centers of Coordination."

23. Although this may strike many readers as odd, the idea that material objects act has a long history in science and technology studies. See Susan Leigh Star and James R. Griesemer, "Institutional Ecology, 'Translations,' and Boundary Objects: Amateurs and Professionals in Berkeley's Museum of Vertebrate Zoology, 1907–39," *Social Studies of Science* 19 (1989): 387–420; Wiebe E. Bijker, Thomas P. Hughes, and Trevor Pinch, eds., *The Social Construction of Technical Systems: New Directions in the Sociology and History of Technology* (Cambridge, MA: MIT Press, 1987); Bruno Latour, *Science in Action: How to Follow Scientists and Engineers through Society* (Cambridge, MA: Harvard University Press, 1987); Suchman, *Plans and Situated Actions*; and Hughes, *Networks of Power*.

24. Pinch and Bijker, "Social Construction of Facts and Artefacts"; Bruno Latour, Michael Lynch, and Steve Woolgar, *Representation in Scientific Practice* (Cambridge, MA: MIT Press, 1990).

25. See Suchman, *Plans and Situated Actions*.

26. Wanda J. Orlikowski, "Improvising Organizational Transformation over Time: A Situated Change Perspective," *Information Systems Research* 7, no. 1 (1996): 63–92; Elisabeth S. Clemens, "Organizational Repertoires and Institutional Change: Women's Groups and the Transformation of U.S Politics, 1890–1920," *American Journal of Sociology* 98, no. 4 (1993): 755–98.

27. Paul J. DiMaggio and Walter W. Powell, "The Iron Cage Revisited: Institutional Isomorphism and Collective Rationality in Organization Fields," *American Sociological Review* 48, no. 2 (1983): 147–60; Roger Friedland and Robert R. Alford, "Bringing Society Back In: Symbols, Practices, and Institutional Contradictions," in *The New Institutionalism in Organi-*

zational Analysis, ed. Walter W. Powell and Paul J. DiMaggio (Chicago: University of Chicago Press, 1991), 232–63; Patricia H. Thornton, William Ocasio, and Michael Lounsbury, *The Institutional Logics Perspective: A New Approach to Culture, Structure, and Process* (Oxford: Oxford University Press, 2002).

28. Bijker, Hughes, and Pinch, *Social Construction of Technical Systems*; Donald MacKenzie and Judy Wajcman, eds., *The Social Shaping of Technology*, 2nd ed. (Philadelphia: Open University Press, 1999).

29. William J. Clinton, Exec. Order No.13,180 (December 7, 2000), "Air Traffic Performance-Based Organization," accessed at Gerhard Peters and John T. Wooley, *The American Presidency Project*, http://www.presidency.ucsb.edu/ws/?pid=61670. For the official definition of air traffic control system, which I have summarized in the text, see US Code, Section 40102(a)(47) of Title 49.

30. Arthur B. Shostak and David Skocik, *The Air Controllers' Controversy: Lessons from the PATCO Strike* (New York: Human Sciences Press, 1986); Katherine S. Newman, *Falling from Grace: Downward Mobility in the Age of Affluence* (New York: Free Press, 1988; repr., Berkeley: University of California Press, 1999).

31. At that time, the FAA's Civil Aeromedical Institute in Oklahoma City had a thriving research program and an extensive bibliography of research reports in their library. Most research reports available to me were of experimental design with small samples, so not immediately useful in preparing for me for an ethnographic study, although they did show the FAA's dominant interest then was in human factors.

32. See Jerolmack and Khan, who argue that what people say is often different from what they do, so therefore ethnography provides a more accurate picture. However, ethnography cannot always capture cognitive processes and actions that are not readily visible in one's action and interaction alone, as the work of air traffic controllers will show. Colin Jerolmack and Shamus Khan, "Talk Is Cheap: Ethnography and the Attitudinal Fallacy," *Sociological Methods & Research* 43, no. 2 (2014): 178–209.

33. The FAA's Host Computer System was an IBM model installed in all en route centers in the late 1960s for multiprocessing of flight and radar data.

34. Diane Vaughan, "Report: Air Traffic Control Survey Analysis," distributed to Boston Tower, Boston TRACON, Boston ARTCC, Bedford Tower, FAA New England Regional Office director, and NATCA regional president, February 2003.

35. Coincidentally, my reentry at Boston Tower and TRACON was for a period that included the six-month anniversary of September 11 (March 4–12, 2002), and at Boston Center for a month during which the one year anniversary occurred (August 21–September 22, 2002). Reentry to Bedford was later because controllers were moving into and adjusting to their long-awaited new tower, about fifty feet from the old one (March 16–23, 2004). I was admitted to the Command Center at Herndon, Virginia, for the first time in 2003 (June 20–24, 2003).

36. During my post–September 11 return, I did no observations, instead interviewing eighty-seven people during two-week periods at each of the four facilities. At the Command Center, because it was my first visit, I did observations at various regional desks, attended one

morning briefing on national traffic flows, and interviewed five people. During my return in fall 2017 to investigate the effects of automation and the staffing shortage, I did observations at Boston Tower and Boston TRACON, interviewing seventy-two people, for a total of 164 interviews in addition to the original 178.

37. Howard E. Aldrich and Martin Ruef, *Organizations Evolving,* 2nd ed. (Thousand Oaks, CA: Sage, 2006), 132–58.

38. Eviatar Zerubavel, *Time Maps: Collective Memory and the Shape of the Past* (Chicago: University of Chicago Press, 2004); Zerubavel, "Timetables and Scheduling: On the Social Organization of Time," *Sociological Inquiry* 46, no. 2 (1976): 87–94.

39. Eviatar Zerubaval, "The Standardization of Time: A Sociohistorical Perspective," *American Journal of Sociology* 88, no. 1 (1982): 1–23.

40. Diane Vaughan, "NASA Revisited: Theory, Analogy, and Public Sociology," *American Journal of Sociology* 112, no. 2 (2006): 353–93.

41. Michael Burawoy, "Revisits: An Outline of a Theory of Reflexive Ethnography," *American Sociological Review* 63, no. 5 (2003): 645–79.

42. Consider the continuing relevance of two classics of ethnography. Whyte's *Street Corner Society,* an ethnography of the North End of Boston published in 1943, had several editions (1955, 1981, 1993). He showed that areas inhabited by the urban poor were not socially disorganized, which was then social scientists accepted explanation of persistent poverty, poor school performance, crime, truancy, and violence. Instead, Whyte showed that these areas were in fact highly organized, forever canceling the social disorganization thesis and refocusing research and policy. Erikson's 1976 *Everything in Its Path* showed the destruction of community following the Buffalo Creek flood of 1972. The individual and collective trauma were reproduced when the effects of Hurricane Katrina devastated New Orleans in August 2005, then again when Sandy swept up the east coast of the United States in October 2012. William F. Whyte, *Street Corner Society: The Social Organization of a Slum* (Chicago: University of Chicago Press, 1943); Kai T. Erikson, *Everything in Its Path: Destruction of Community in the Buffalo Creek Flood* (New York: Simon and Schuster, 1976).

43. See Burawoy, who, for *Manufacturing Consent,* worked as a laborer in a Chicago factory for his dissertation, only to discover he was not the first ethnographer to labor there. Michael Burawoy, *Manufacturing Consent* (Chicago: University of Chicago Press, 1979).

44. See, e.g., the normalization of deviance, in Vaughan, Challenger *Launch Decision.*

45. Andrew Abbott, *Time Matters: On Theory and Method* (Chicago: University of Chicago Press, 2001), further elaborated in Abbott, *Processual Sociology* (Chicago: University of Chicago Press, 2016).

46. Aldrich and Ruef, *Organizations Evolving.*

47. Abbott, *Processual Sociology,* xi.

48. Ronald Aminzade, "Historical Sociology and Time," *Sociological Methods and Research* 20 (1992): 456–80; William H. Sewell Jr., "A Theory of Structure: Duality, Agency, and Transformation," *American Journal of Sociology* 98, no. 1 (1992): 1–29; Abbott, *Time Matters.*

49. On capturing the varieties of group life, we may have no such category as "typical" ethnography. For contrasting styles, see, e.g., John Van Maanen, *Tales of the Field* (Chicago:

University of Chicago Press, 1988); Gary Alan Fine, "Peopled Ethnography: Developing a Theory of Group Life," *Ethnography* 4, no. 41 (2003): 41–60; Brit Ross Winthereik, Peter Lutz, Lucy Suchman, and Helen Verran, "Attending to Screens and Screenness," *STS Encounters* 4, no. 2 (2011): 1–6.

50. Star and Ruhleder, "An Ecology of Infrastructure"; Geoffrey C. Bowker and Susan Leigh Star, *Sorting Things Out: Classification and Its Consequences* (Cambridge, MA: MIT Press, 1999).

51. Abbott, in *The System of Professions,* observed that "a profession's development is driven by wider social forces that influence the content of and control over its tasks and problem" (112). In response Eyal argued that Abbott's approach must be "supplemented by studies of expertise that combine assemblages of agents, devices, concepts, and institutional and spatial arrangements" (863). For *Dead Reckoning,* combining the two perspectives was essential. Andrew Abbott, *The System of Professions: An Essay on the Division of Expert Labor* (Chicago: University of Chicago Press, 1988); Gil Eyal, "For a Sociology of Expertise: The Social Origins of the Autism Epidemic," *American Journal of Sociology* 118, no. 4 (2013): 863–907.

52. Barley and Tolbert, "Institutionalization and Structuration."

53. See Hughes, *Networks of Power;* John F. Padgett and Walter W. Powell, eds., *The Emergence of Organizations and Markets* (Princeton, NJ: Princeton University Press, 2012); Powell and Colyvas, "Microfoundations of Institutional Theory."

54. Pierre Bourdieu, *Homo Academicus,* trans. Peter Collier (Stanford, CA: Stanford University Press, 1988).

55. DiMaggio and Powell, "Iron Cage Revisited"; Orlikowski, "Improvising Organizational Transformation."

56. Paul J. DiMaggio, "Culture and Cognition," *Annual Review of Sociology* 23 (1997): 263–87; Edwin A. Hutchins, *Cognition in the Wild* (Cambridge, MA: MIT Press, 1995); Hutchins, "How a Cockpit Remembers Its Speeds," *Cognitive Science* 19, no. 3 (1995): 265–88; Hutchins, "Cognitive Ecology," *Topics in Cognitive Science* 2, no. 4 (2010): 705–15.

57. John Van Maanen and Stephen R. Barley, "Cultural Organization: Fragments of a Theory," in *Organizational Culture,* ed. Peter J. Frost et al. (Beverly Hills, CA: Sage, 1985), 31–54; Beth Bechky, "Shared Meaning across Occupational Communities: The Transformation of Understanding on a Production Floor," *Organization Science* 14, no. 3 (2003): 312–30.

58. See Matthew Norton, "Cultural Sociology Meets the Cognitive Wild: Advantages of the Distributed Cognition Framework for Analyzing the Intersection of Culture and Cognition," *American Journal of Cultural Sociology* 8, no. 45 (2020): 45–62, https://doi.org/10.1057/s41290-019-00075-w.

59. Bowker and Star, *Sorting Things Out;* Geoffrey C. Bowker and Susan Leigh Star, "Building Information Structures for Social Worlds: The Role of Classifications and Standards," in *Community Computing and Support System: Social Interaction in Networked Communities,* ed. Toru Ishida (Berlin: Springer Verlag, 1998), 231–48; Stefan Timmermans and Steven Epstein, "A World of Standards but Not a Standard World: Toward a Sociology of Standards and Standardization," *Annual Review of Sociology* 36 (2010): 69–89.

60. Pinch and Bijker, "Social Construction of Facts and Artefacts"; Beth Bechky, "Object Lessons: Workplace Artifacts as Representations of Occupational Jurisdiction," *American Journal of Sociology* 109, no. 3 (2003): 720–52.

61. Interpretive anthropology called for ethnographers to interpret a culture by understanding how people within a culture were interpreting themselves and their own experiences, thus "the native view." Clifford Geertz, "Thick Description: Toward an Interpretive Theory of Culture," in *The Interpretation of Cultures* (New York: Basic Books, 1973), 14–17.

62. Clifford Geertz, *Local Knowledge: Further Essays in Interpretive Anthropology* (New York: Basic Books, 1983).

63. Lamont and Molnàr, "Study of Boundaries."

64. Mark A. Pachucki, Sabrina Pendergrass, and Michèle Lamont, "Boundary Processes: Recent Theoretical Developments and New Contributions," *Poetics* 35, no. 6 (2007): 331–51.

65. Daniel Hirschman and Isaac Ariail Reed, "Formation Stories and Causality in Sociology," *Sociological Theory* 32, no. 4 (2014): 259–82.

66. Hughes, "Electrification of America."

67. Hughes, "Electrification of America."

68. Powell and DiMaggio, introduction to *The New Institutionalism in Organizational Analysis*; DiMaggio, "Culture and Cognition."

69. DiMaggio, "Culture and Cognition"; Hutchins, *Cognition in the Wild.*

70. Suchman, "Centers of Coordination."

71. Thomas F. Gieryn, "A Space for Place in Sociology," *Annual Review of Sociology* (2000): 463–96.

72. Building from the classics, I use the concept to describe the work necessary to create, maintain, and cross boundaries. See Gieryn, "Boundary-Work"; Star and Griesemer, "Institutional Ecology"; Lamont and Molnàr, "Study of Boundaries."

73. Arlie Russell Hochschild, "Emotion Work, Feeling Rules, and Social Structure," *American Journal of Sociology* 85, no. 3 (1979): 551–75; Hochschild, *The Managed Heart* (Berkeley: University of California Press, 1983).

74. Van Maanen and Barley, "Cultural Organization." Van Maanen and Barley define culture as the solutions people arrive at as they interact to resolve the problems they face in common.

75. Michèle Lamont, *The Dignity of Working Men: Morality and the Boundaries of Race, Class, and Immigration* (New York: Russell Sage Foundation at Harvard University Press, 2000); Lamont and Molnàr, "Study of Boundaries."

76. Pachucki, Pendergrass, and Lamont, "Boundary Processes."

77. Abbott, "Conceptions of Time and Events in Social Science Methods," in *Time Matters,* 181–82; James Mahoney and Kathleen Thelen, "A Theory of Gradual Institutional Change," in *Explaining Institutional Change,* ed. James Mahoney and Kathleen Thelen (New York: Cambridge University Press, 2010), 1–17.

78. Evgeny Morozov, *To Save Everything, Click Here: The Folly of Technological Solutionism* (New York: Public Affairs, 2013). See also Iddo Tavory and Nina Eliasoph, "Coordinating

Futures: Toward a Theory of Anticipation," *American Journal of Sociology* 118, no. 4 (2013): 908–42.

Chapter Two

1. Andrew Abbott, "Linked Ecologies: States and Universities as Environments for Professions," *Sociological Theory* 23, no. 3 (2005): 245–74; Lynne G. Zucker, "The Role of Institutionalization in Cultural Persistence," *American Sociological Review* 42, no. 5 (1977): 726–43.

2. Elisabeth S. Clemens, "Organizational Repertoires and Institutional Change: Women's Groups and the Transformation of U.S. Politics, 1890–1920," *American Journal of Sociology* 98, no. 4 (1993): 755–98.

3. William H. Sewell Jr., "Three Temporalities: Toward an Eventful Sociology," in *The Historic Turn in the Human Sciences*, ed. T. J. McDonald (Ann Arbor: University of Michigan Press, 1996), 245–80; Sewell, "A Theory of Structure: Duality, Agency, and Transformation," *American Journal of Sociology* 98, no. 1 (1992): 1–29; Jeffrey Haydu, "Making Use of the Past: Time Periods as Cases to Compare and as Sequences of Problem Solving," *American Journal of Sociology* 104, no. 2 (1998): 339–71.

4. Ronald Aminzade, "Historical Sociology and Time," *Sociological Methods and Research* 20 (1992): 456–80; Andrew Abbott, "From Causes to Events," *Sociological Methods and Research* 20 (1992): 428–55.

5. Larry J. Griffin, "Temporality, Events, and Explanation in Historical Sociology: An Introduction," *Sociological Methods and Research* 20, no. 4 (1992): 403–27; Ivan Ermakoff, "The Structure of Contingency," *American Journal of Sociology* 121, no. 1 (2015): 64–125.

6. Andrew Abbott, *Processual Sociology* (Chicago: University of Chicago Press, 2016).

7. Daniel Hirschman and Isaac Ariail Reed, "Formation Stories and Causality in Sociology," *Sociological Theory* 32, no. 4 (2014): 259–82.

8. Hirschman and Reed, "Formation Stories," 260.

9. Andrew Abbott, *Time Matters: On Theory and Method* (Chicago: University of Chicago Press, 2001); Bruno Latour, *Reassembling the Social: An Introduction to Actor-Network Theory* (Oxford: Oxford University Press, 2005); Neil Gross, "The Structure of Causal Chains," *Sociological Theory* 36, no. 4 (2018): 343–67; Ermakoff, "The Structure of Contingency."

10. Hirschman and Reed, "Formation Stories."

11. Haydu, "Making Use of the Past"; Thomas P. Hughes, *Networks of Power: Electrification in Western Society, 1880–1930* (1983; repr., Baltimore: Johns Hopkins University Press, 1993).

12. Thomas P. Hughes, "The Electrification of America: The System Builders," *Technology and Culture* 20, no. 1 (1979): 124–61.

13. Haydu, "Making Use of the Past"; Hughes, "Electrification of America."

14. Ermakoff, "The Structure of Contingency."

15. Andrew Abbott, "Transcending General Linear Reality," *Sociological Theory* 6, no. 2 (1988): 173; Michel Callon and Bruno Latour, "Unscrewing the Big Leviathan; or, How Actors Macrostructure Reality and How Sociologists Help Them to Do So," in *Advances in Social Theory and Methodology*, ed. Aaron Cicourel and Karin Knorr (London: Routledge, 1981), 277–303.

16. In constructing this sociological history, I wanted to find an ordering of events that was inherent in the chronology of archival data, independent of and uninfluenced by what other scholars thought or did in constructing historical accounts. Working inductively, I created the narrative history and analysis on the basis of my previous research experience with historical chronologies in *Uncoupling* (1986) and *The* Challenger *Launch Decision* (1996). I followed the rhythm that events themselves seemed to dictate. Making boundaries in a continuing historical process is admittedly artificial and arbitrary. However, I could identify sequences of turning points of varying weights that formed trajectories of events that collectively either seemed to change the direction of or add a new dimension to the characteristics, activity, and life cycle of the system. I tentatively marked boundaries, and as I learned more of system history, adjusted them until they accounted for the full trajectory of events that seemed to mark the end of an era in the life cycle and the beginning of a different one. As I read more history of the system, I reworked boundaries until eventually new discoveries did not alter them but fit into them. Only when I arrived at the four stable era boundaries presented here did I turn to historical sociology and history of technology to find out the extent to which this chronology either fit or deviated from what was known in those areas of specialization.

17. Abbott, *Time Matters*.

18. Andrew Abbott, "Conceptions of Time and Events in Social Science Methods," in *Time Matters*, 181–82. Abbott refers to this as a "career model." Drawing on his work on professions and individual careers, he observes that organizations also have "careers" and therefore a traceable life course.

19. So, following Abbott, it went trajectory to turning point, trajectory to turning point, trajectory to turning point. See also Abbott on subsequences within sequences and turning points in Abbott, "On the Concept of Turning Point," in *Time Matters*, 250.

20. Wanda J. Orlikowski, "Improvising Organizational Transformation over Time: A Situated Change Perspective," *Information Systems Research* 7, no. 1 (1996): 63–92; Clemens, "Organizational Repertoires."

21. Susan Leigh Star and James R. Griesemer, "Institutional Ecology, 'Translations,' and Boundary Objects: Amateurs and Professionals in Berkeley's Museum of Vertebrate Zoology, 1907–39," *Social Studies of Science* 19 (1989): 387–420.

22. Charles B. Perrow, *Normal Accidents: Living with High-Risk Technologies* (New York: Basic Books, 1984).

23. Stefan Timmermans and Steven Epstein, "A World of Standards but Not a Standard World: Toward a Sociology of Standards and Standardization," *Annual Review of Sociology* 36 (2010): 69–89.

24. Susan Leigh Star and Karen Ruhleder, "Steps toward an Ecology of Infrastructure: Design and Access for Large Information Spaces," *Information Systems Research* 7, no. 1 (1996): 111–34; Timmermans and Epstein, "A World of Standards."

25. Arthur L. Stinchcombe, "Social Structure and Organizations," in *Handbook of Organizations*, ed. James G. March (Chicago: Rand McNally, 1965): 142–93.

26. Geoffrey C. Bowker and Susan Leigh Star, *Sorting Things Out: Classification and Its Consequences* (Cambridge, MA: MIT Press, 1999).

27. Alain Gras et al., *Faced with Automation: The Pilot, the Controller, and the Engineer*, trans. Jill Lundsten (Paris: Publications de la Sorbonne, 1994); condensed version of *Face a l'automate: Le pilote, le contrôleur et l'ingénieur* (Paris: Publications de la Sorbonne, 1991).

28. Bowker and Star, *Sorting Things Out.*

29. Andrew Abbott, "Things of Boundaries," *Social Research* 62, no. 4 (1995): 857–82; Perrow, *Normal Accidents*; Karl E. Weick, "Educational Organizations as Loosely Coupled Systems," *Administrative Science Quarterly* 21, no. 1 (1976): 1–19.

30. Paul J. DiMaggio and Walter W. Powell, "The Iron Cage Revisited: Institutional Isomorphism and Collective Rationality in Organization Fields," *American Sociological Review* 48, no. 2 (1983): 147–60; Daniel Kluttz and Neil Fligstein, "Varieties of Field Theory," in *Handbook of Contemporary Sociological Theory*, ed. Seth Abrutyn (New York: Springer, 2016), 185–204.

31. Michael Lynch and Steve Woolgar, eds., *Representation in Scientific Practice* (Cambridge, MA: MIT Press. 1990).

32. Janet Vertesi, *Seeing Like a Rover: How Robots, Teams, and Images Craft Knowledge of Mars* (Chicago: University of Chicago Press, 2015); Janet Vertesi, *Shaping Science: Organizations, Decisions, and Culture on NASA's Teams* (Chicago: University of Chicago Press, 2020).

33. See Edwin A. Hutchins, *Cognition in the Wild* (Cambridge, MA: MIT Press, 1995).

34. Michèle Lamont and Viràg Molnàr, "The Study of Boundaries in the Social Sciences," *Annual Review of Sociology* 28, no. 1 (2002): 167–95.

35. Walter W. Ristow, ed., *Aviation Cartography: A Historico-Bibliographic Study of Aeronautical Charts*, 2nd ed. (Washington, DC: Map Division, Library of Congress, 1960), 2–4; Hermann Moedebeck, "Ueber das Landen mit Ballons," *Zeitschrift fur Luftschiffahrt* 7 (1888): 272–76.

36. Karl-Heinz Meine, "Aviation Cartography," *Cartographic Journal* 3, no. 1 (1966): 31–40.

37. Thomas P. Hughes, *American Genesis: A Century of Invention and Technological Enthusiasm, 1870–1970* (Chicago: University of Chicago Press, 1989).

38. Gras and Delpech argue that the development of a technology (and in this case, a socio-technical system) depends on the fit between a national imaginary, the macrosocial, and the technical. Alain Gras (with participation of Sophie Delpech), *Grandeur et dépendance: Sociologies des macro-systèmes techniques* (Paris: Presses Universitaires de France, 1993), translated by and quoted in Geoffrey C. Bowker, "How Things Change: The History of Sociotechnical Structures," *Social Studies of Science* 28 (1996): 173–82.

39. Orville Wright, *How We Invented the Airplane: An Illustrated History*, ed. Fred C. Kelly, with additional text by Alan Weissman (New York: Dover Publications, 1988), 11.

40. Fred C. Kelly, *The Wright Brothers: A Biography Authorized by Wilbur Wright* (New York: Harcourt, Brace, 1943), 19–28.

41. Jakab found evidence that the Wrights began reading about Lilienthal other flight-related developments abut 1890 and were fairly well-informed prior to Lilienthal's death, but Lilienthal's death was a trigger point for serious engagement. Peter L. Jakab, *Visions of a Flying*

Machine: The Wright Brothers and the Process of Invention (Washington, DC: Smithsonian Institution Press, 1990), 41–42.

42. Otto Lilienthal, *Birdflight as the Basis of Aviation: A Contribution towards a System of Aviation. Compiled from the Results of Numerous Experiments made by O. and G. Lilienthal*, trans. A. W. Isenthal (New York: Longmans, Green, 1911). For how the Wrights built upon Lilienthal's work, see Wilbur Wright, "Some Aeronautical Experiments," *Journal of the Western Society of Engineers* (December 1901), repr. in "Classics," *Resonance* 8, no. 12 (2003): 99–114, and at http://invention.psychology.msstate.edu/inventors/i/Wright/library/Aeronautical.html.

43. Octave Chanute, *Progress in Flying Machines* (New York: American Engineer and Railroad Journal, 1894; repr., Long Beach, CA: Lorenz and Herwig, 1976). In it, Chanute included all research that originally appeared in his column in *The Railroad and Engineering Journal*, which analyzed every flight experiment for flaws and identified principles worth pursuing.

44. Their three-axis model consisted of altered wing design (mimicking the change of angle in birds' wings, or "wing warping," as they called it—the principle behind today's aileron) to create roll for lateral motion; moving the elevator, or horizontal rudder, forward to control pitch, or up and down; and making the rear rudder movable to control side-to-side motion. Warping refers to the ability to give the wings a spiral twist, so that the top and bottom surfaces are at different angles on the right and left sides. O. Wright, *How We Invented the Airplane*.

45. Jakab, *Visions of a Flying Machine*, 115–42; Tom D. Crouch and Peter L. Jakab, *The Wright Brothers and the Invention of the Aerial Age* (Washington, DC: National Geographic, Smithsonian National Air and Space Museum, 2003), 84–89.

46. O. Wright, *How We Invented the Airplane*, 17.

47. Wolfgang Langewiesche, *Stick and Rudder: An Explanation of the Art of Flying* (New York: McGraw-Hill, 1990), 163; Crouch and Jakab, *The Wright Brothers*, 91.

48. Tom D. Crouch, *The Bishop's Boys: A Life of Wilbur and Orville Wright* (New York: W. W. Norton, 2003), 242–43.

49. O. Wright, *How We Invented the Airplane*.

50. O. Wright, *How We Invented the Airplane*, 21.

51. The news of their achievements was believed in some quarters. Chanute had visited the Wrights at Kitty Hawk during the glider experiments in 1901, 1902, and 1903 so was aware of their gliding successes. He was active in aeronautics circles on both continents, so when members of the Aero Club of France in Paris heard the rumors, they believed them. However, many in France did not, even later, when, with patent granted, Wilbur was to demonstrate their flying machine at Le Mans. Famous French inventors and aeronauts publicly ridiculed their efforts. Two bicycle mechanics from Ohio? Impossible.

52. Scott A. Boorman, review of *The Emergence of Organizations and Markets*, by John F. Padgett and Walter W. Powell, *Acta Sociologica* 57, no. 4 (2014): 363–67.

53. Hughes, *American Genesis*, 16–17, 53–55.

54. Harry Collins, "Three Dimensions of Expertise," *Phenomenology and the Cognitive Sciences* 12, no. 2 (2013): 253–73.

55. Crouch and Jakab, *The Wright Brothers*, 89.

56. W. Wright, "Some Aeronautical Experiments."

57. "The Wright Aeroplane and Its Performances," *Scientific American* 94, no. 14 (April 7, 1906): 291.

58. David McCullough, *The Wright Brothers* (New York: Simon and Schuster, 2015), 70–71.

59. Fred C. Kelly, "After Kitty Hawk: A Brief Resume," appendix to O. Wright, *How We Invented the Airplane*, 51–52.

60. Hughes, *American Genesis*, 99, 101–2.

61. Hermann Moedebeck, *Pocket Book of Aeronautics*, trans. W. Mansergh Varley (London: Whittaker, 1907).

62. Ristow, *Aviation Cartography*, 2, translated and quoted from Moedebeck, *Pocket Book of Aeronautics.*

63. Ristow, *Aviation Cartography*, 3.

64. Bruno Latour, *The Pasteurization of France* (Cambridge, MA: Harvard University Press, 1993).

65. McCullough, *The Wright Brothers*, 133.

66. Bijker recognizes the patent as an artifact that constitutes an important part of the capabilities of actors. Patents confer power to act and affirm and extend the owner's influence. Wiebe E. Bijker, *Of Bicycles, Bakelite, and Bulbs: Toward a Theory of Technological Change* (Cambridge, MA: MIT Press, 1995), 265.

67. McCullough, *The Wright Brothers*, 162–70.

68. Crouch and Jakob, *The Wright Brothers*, 160–67.

69. Couch and Jakab, *The Wright Brothers*, 167–68.

70. "Shall America Take the Lead in Aeronautics?" *Scientific American* 98, no. 9 (February 29, 1908): 138; Wilbur Wright, "Flying as a Sport—Its Possibilities," *Scientific American* 98, no. 9 (February 29, 1908): 139.

71. "The Wright Aeroplane Tests: Wilbur Wright's Latest Flights in France," *Scientific American* 99, no. 9 (August 29, 1908): 135; "The First Flight of the Wright Aeroplane at Fort Myer," *Scientific American* 99, no. 11 (September 12, 1908): 169.

72. "Lessons of the Wright Aeroplane Disaster," *Scientific American* 99, no. 13 (September 26, 1908): 202; "The Construction of the Wright Aeroplane," *Scientific American* 99, no. 13 (September 26, 1908): 208–10.

73. Harry Collins, *Changing Order: Replication and Induction in Scientific Practice* (Chicago: University of Chicago Press, 1992); Bijker, *Of Bicycles, Bakelite, and Bulbs*, 86.

74. Gras et al., *Faced with Automation.*

75. Kelly, "After Kitty Hawk," 52.

76. On scientific discovery and the Wright's achievement, Crouch and Jakab explained: "Independent discoveries, no matter how revolutionary, rarely result in a new invention. It is only when such discoveries are integrated with other discoveries and technical knowledge that practical innovations normally come about. What made the Wright brothers special was that they defined their inventive method in these terms, and they had the talent to follow through

on an entire set of problems presented by the creation of a complex mechanical device and arrive at a sound solution." Crouch and Jakab, *The Wright Brothers,* 89.

77. See Thomas F. Gieryn, "Boundary-Work and the Demarcation of Science from Non-science: Strains and Interests in Professional Ideologies of Scientists," *American Sociological Review* 48 (1983): 781–95.

78. Roddam Narasimha, "How Two Bicycle Mechanics Achieved the World's First Powered Flight," *Resonance* 8, no. 12 (2003): 61–74.

79. Collins, *Changing Order.*

80. John Law, "Technology and Heterogeneous Engineering: The Case of Portuguese Expansion," in *The Social Construction of Technical Systems: New Directions in the Sociology and History of Technology,* ed. Wiebe E. Bijker, Thomas P. Hughes, and Trevor Pinch (Cambridge, MA: MIT Press, 1987), 111–34.

81. Emerson W. Conlon, "History of the University of Michigan Aeronautical/Aerospace Engineering," Faculty Memoir Project, University of Michigan 1817–2017, http://um2017.org/2017_Website/History_of_Aeronautical_Engineering.html.

82. Sinclair Lewis, *The Trail of the Hawk: A Comedy on the Seriousness of Life* (New York: Harper and Brothers, 1915).

83. Many thanks to my prolific and well-read colleague Paul DiMaggio, who introduced me to this little-known Lewis book.

84. W. Wright, "Flying as a Sport," 139.

85. "Airmail," Wikipedia, The Free Encyclopedia, accessed January 12, 2015, http://en.wikipedia.org/wiki/Airmail.

86. Yovanna Bieberich, "Remember a Historic Flight: Fred Wiseman Piloted the First Airmail Flight from Petaluma to Santa Rosa," *Petaluma Argus-Courier,* February 16, 2011, 1, http://www.petaluma360.com/article/20110216/COMMUNITY/110219562?template=printpicart.

87. Ruth Schwartz Cowan, *A Social History of American Technology* (New York: Oxford University Press, 1997), 250–54.

88. T. A. Heppenheimer, *Turbulent Skies: The History of Commercial Aviation* (New York: John Wiley and Sons, 1995), 8.

89. Heppenheimer, *Turbulent Skies,* 8–9.

90. Ristow, *Aviation Cartography,* 16–17.

91. Heppenheimer, *Turbulent Skies,* 8–11.

92. Heppenheimer, *Turbulent Skies,* 10.

93. Bowker and Star, *Sorting Things Out.*

94. Haydu, "Making Use of the Past"; Hughes, "Electrification of America."

95. Gras et al., *Faced with Automation.*

96. Lynch and Woolgar, *Representation in Scientific Practice.*

97. Hutchins, *Cognition in the Wild.*

98. Matt Novak, "The 'Highway of Light' That Guided Early Planes across America," *Paleofuture,* November 18, 2013, http://paleofuture.gizmodo.com/the-highway-of-light-that-guided-early-planes-across-1466696698.

99. Heppenheimer, *Turbulent Skies*, 10–12.

100. Cowan, *Social History of American Technology*, 252–53.

101. Novak, "The 'Highway of Light.'"

102. Edmund Preston, ed., *FAA Historical Chronology: Civil Aviation and the Federal Government 1926–1996* (Washington, DC: Department of Transportation, Federal Aviation Administration Office of Public Affairs, 1998), 26.

103. Cowan, *Social History of American Technology*, 253–56.

104. Chalk's Ocean Airways (1917); Delta (founded as Huff Daland Dusters, 1924); TWA (1925); Western Air Express (1925); Varney Airlines (1926, merged with National Air Transport to form United Airlines in 1933); Eastern (founded as Pitcairn Aviation in 1926, changed to Eastern Air Transport in 1929); Northwest (1926); Colonial Air Transport (1926, formed American Airlines in 1930). "List of Airlines by Foundation Date," Wikipedia, last modified February 18, 2014, http://en.wikipedia.org/wiki/List_of_airlines_by_foundation_date.

105. Heppenheimer, *Turbulent Skies*, 14.

106. Cowan, *Social History of American Technology*, 254–55.

107. Heppenheimer, *Turbulent Skies*, 20–21.

108. Charles A. Lindbergh, *The Spirit of St. Louis* (New York: Scribner, 1953).

109. Heppenheimer, *Turbulent Skies*, 22–24.

110. Franklyn E. Dailey Jr., *The Triumph of Instrument Flight: A Retrospective in the Century of Aviation* (Wilbraham, MA: Dailey International Publishers, 2004).

111. Preston, *FAA Historical Chronology*, 65.

112. Gene Nora Jessen, "1929 Air Race," International Women Pilots, Special Issue, The 99s in Aviation History, *99 News Magazine*, https://www.ninety-nines.org/the-1929-air-race.htm.

113. Paul McElroy, *Against the Wind: The History of the National Air Traffic Controllers Association* (Washington, DC: National Air Traffic Controllers Association, 2002), 220–21, 223.

114. Heppenheimer, *Turbulent Skies*, 45–52.

115. Dailey, *The Triumph of Instrument Flight*, 73

116. Gras et al., *Faced with Automation.*

117. Abbott, "Conceptions of Time and Events."

118. Glen A. Gilbert, "Historical Development of the Air Traffic Control System," *IEEE Transactions on Communications* 21, no. 5 (1973): 365.

119. "Signal Lamp," Wikipedia, accessed July 12, 2015, https://en.wikipedia.org/wiki/Signal_lamp.

120. Dailey, *The Triumph of Instrument Flight.*

121. Dailey, *The Triumph of Instrument Flight.*

122. Gilbert, "Historical Development," 365–66; Heppenheimer, *Turbulent Skies*, 125–26.

123. Ioannis Mansolas and Angelos Mansolas, "A Short History of Air Traffic Control," compiled June 2005, http://imansolas.freeservers.com/ATC/short_history_of_the_air_traffic.html.

124. Glen A. Gilbert, *Air Traffic Control: The Uncrowded Sky* (Washington, DC: Smithsonian Institution Press, 1973), 9.

125. Lucy Suchman, "Centers of Coordination: A Case and Some Themes," in *Discourse, Tools, and Reasoning: Essays on Situated Cognition*, NATO ASI Series F Computer and Systems Sciences, ed. Lauren B. Resnick, Roger Saljo, Clotilde Pontecorvo, and Barbara Burge (Berlin: Springer-Verlag, 1997), 41–62.

126. See, e.g., Ruth Schwartz Cowan's classic "How the Refrigerator Got Its Hum," in *The Social Shaping of Technology: How the Refrigerator Got Its Hum*, ed. Judy Wajcman and Donald A. MacKenzie (London: Open University Press, 1988).

127. Meine, "Aviation Cartography," 33.

128. See Hutchins, *Cognition in the Wild*, in which distributed cognition is located in the immediate environment.

129. Preston, *FAA Historical Chronology*, 10–17.

130. Gilbert, "Historical Development," 366.

131. Preston, *FAA Historical Chronology*, 19.

132. Bowker and Star, *Sorting Things Out*.

133. Heppenheimer, *Turbulent Skies*, 121–23.

134. Gilbert, "Historical Development," 367.

135. Preston, *FAA Historical Chronology*, 26.

136. Preston, *FAA Historical Chronology*, 29.

137. Gilbert, *Air Traffic Control*, 10–11; Heppenheimer, *Turbulent Skies*, 122–24; Gilbert, *Air Traffic Control*, 11.

138. "Air Traffic Control," U.S. Centennial of Flight Commission, https://www.centennialofflight.net/essay/Government_Role/Air_traffic_control/POL15.htm.

139. Gilbert, *Air Traffic Control*, 9–10.

140. Wendy E. MacKay, "Is Paper Safer: The Role of Paper Flight Strips in Air Traffic Control," *ACM Transactions on Computer-Human Interaction* 6, no. 4 (1999): 311–40; Jorg Potthast, "Ethnography of a Paper Strip: The Production of Air Safety," *Science, Technology, and Innovation Studies* 4, no. 1 (2008): 47–68.

141. Gilbert, "Historical Development," 368–69.

142. Lester Williams, iconic air traffic controller and inventor of the system that records all controller exchanges with pilots and other controllers when using the mic on position in their facility, Boston Center, tour and walking interview with the author, Boston Center, March 2000.

143. Gilbert, "Historical Development," 367.

144. Heppenheimer, *Turbulent Skies*, 134–36.

145. Heppenheimer, *Turbulent Skies*, 136.

146. Arthur L. Stinchcombe, "Social Structure and Organizations," in *Handbook of Organizations*, ed. James G. March, (Chicago: Rand McNally, 1965), 142–93.

147. Donald MacKenzie, *Inventing Accuracy: A Historical Sociology of Nuclear Missile Guidance* (Cambridge, MA: MIT Press, 1990).

148. Heppenheimer, *Turbulent Skies*, 177.

149. Preston, *FAA Historical Chronology*, 54.

150. Heppenheimer, *Turbulent Skies*, 181–83; Preston, *FAA Historical Chronology*, 56.

151. Laura M. Bolsom, "How Pan Am Extended Aura of Jet Travel to All," *New York Times*, October 27, 2019, A25.

152. Joseph A. McCartin, *Collision Course: Ronald Reagan, the Air Traffic Controllers, and the Strike That Changed America* (New York: Oxford University Press, 2011), 24–25.

153. Preston, *FAA Historical Chronology*, 125.

154. McCartin, *Collision Course*, 49–52.

155. Preston, *FAA Historical Chronology*, 77.

156. Arthur B. Shostak and David Skocik, *The Air Controllers' Controversy: Lessons from the PATCO Strike* (New York: Human Sciences Press, 1986), 49–50; Preston, *FAA Historical Chronology*, 109; McCartin, *Collision Course*, 78–87.

157. Heppenheimer, *Turbulent Skies*, 271.

158. Traffic Management Unit personnel and area supervisors, interviews with the author, Boston Air Route Traffic Control Center, Nashua, NH, May–June, 2000.

159. Preston, *FAA Historical Chronology*, 120.

160. In close succession in 1970, separate crashes of charter planes cost the lives of the Wichita State University football team and the Marshall University Football team. In 1972, baseball star Roberto Clemente and four companions traveling for an earthquake relief mission to Nicaragua died when a chartered DC-7, overloaded with recovery essentials, crashed into the ocean upon takeoff from San Juan, Puerto Rico. In 1978, a midair collision over San Diego between a Boeing 727 and a Cessna 172 caused more fatalities than any other civil aviation accident in US history. Preston, *FAA Historical Chronology*, 129, 130, 143, 172.

161. Preston, *FAA Historical Chronology*, 157–70.

162. Gilbert, *Air Traffic Control*, 37–38.

163. Preston, *FAA Historical Chronology*, 108, 113.

164. Heppenheimer, *Turbulent Skies*, 186–87, 191, 242–43, 260.

165. Preston, *FAA Historical Chronology*, 121.

166. McCartin, *Collision Course*, 110–125.

167. McCartin, *Collision Course*, 163–71.

168. McCartin, *Collision Course*, 176–201.

169. For the most thorough discussion of racial and gender inequality in the system during this period, see McCartin, *Collision Course*, chapter 7, "Turbulence," 177–97.

170. Shostak and Skocik, *The Air Controllers' Controversy*, 66; McCartin, *Collision Course*, 204–9.

171. Jan K. Breuckner and Vivek Pai, "Technological Innovation in the Airline Industry: The Impact of Regional Jets," *International Journal of Industrial Organization* 27, no. 1 (2009): 110–20.

172. Shostak and Skocik, *The Air Controllers' Controversy*, 64–74.

173. Hughes, "Electrification of America."

174. Most recently, McCartin, *Collision Course*.

175. Preston, *FAA Historical Chronology*, 221–22.

176. McCartin, *Collision Course*, 241–46.

177. McElroy, *Against the Wind*, 23.

178. McCartin, *Collision Course*, 240.

179. McCartin, *Collision Course*, 246; McElroy, *Against the Wind*, 28; Shostak and Skocik, *The Air Controllers' Controversy*, 101.

180. McCartin, *Collision Course*, 268–70.

181. Preston, *FAA Historical Chronology*, 186.

182. Although Reagan's past leniency with unions also figured strongly in the PATCO decision to go forward with the strike after Reagan's public ultimatum, Reagan never considered an alternative because the strike was illegal, against the government, by a government union. All controllers had taken an oath not to strike. Reagan quoted this oath verbatim in his public declaration. Therefore, they would no longer be government employees if they walked off the job. His position transformed what PATCO defined as a labor-management dispute into a challenge to the power of the government and the legitimacy of its laws. McCartin, *Collision Course*, 289–95.

183. McCartin, *Collision Course*, 301.

184. Preston, *FAA Historical Chronology*, 186.

185. Preston, *FAA Historical Chronology*, 186; McElroy, *Against the Wind*, 29; McCartin, *Collision Course*, 233, 235–36, 281–82; Shostak and Skocik, *The Air Controllers' Controversy*, 103–6.

186. McCartin, *Collision Course*, 301–2.

187. McCartin, *Collision Course*, 302.

188. Shostak and Skocik, *The Air Controllers' Controversy*, 105.

189. My sources were six fired PATCO controllers returned to work in my four facilities, interviews with the author; Twenty-two fired PATCO controllers located via PATCO email list, telephone interviews with the author, regarding the strike and its personal impact.

190. Personal interviews with the author.

191. McElroy, *Against the Wind*, 36.

192. McCartin, *Collision Course*, 332–34.

193. McElroy, *Against the Wind*, 222.

194. Heppenheimer, *Turbulent Skies*, 282–83.

195. McElroy, *Against the Wind*, 69.

196. McElroy, *Against the Wind*, 36–37.

197. McElroy, *Against the Wind*, 45–47.

198. McElroy, *Against the Wind*, 44–91.

199. NATCA alternate vice president, New England Region, telephone interview with the author, April 21, 2003.

200. Replacing the 1960s software was a major obstacle to modernization. Known as the "host" computer because it preserved the ability to run the original 9020 software, the host

would connect to new controller workstations, known as "Sector Suites," with new display monitors, processing and communication abilities. Invisible, it was another patch-on, a mix of the old with the new that kept the air traffic control system going.

201. Preston, *FAA Historical Chronology*, 211, 212.

202. McElroy, *Against the Wind*, 89–91.

203. NATCA alternate vice president, New England Region, telephone interview.

204. Albert E. Kahn, "The Surprises of Airline Deregulation," *American Economic Review* 78, no. 2 (1988): 316–22.

205. Preston, *FAA Historical Chronology*, 229–30.

206. Heppenheimer, *Turbulent Skies*, 346–48.

207. Preston, *FAA Historical Chronology*, 239.

208. Preston, *FAA Historical Chronology*, 237.

209. Fired PATCO controllers excelled other professions in surviving during the economic downturn of the 1980s. Katherine S. Newman, *Falling from Grace: Downward Mobility in the Age of Affluence* (New York: Free Press, 1998; Berkeley: University of California Press, 1999).

210. Personal interviews with the author.

211. Preston, *FAA Historical Chronology*, 242.

212. Bart Elias, "Air Traffic Inc.: Considerations Regarding the Corporatization of Air Traffic Control," Congressional Research Service 7-5700, R43844, January 5, 2015, 11–13.

213. McElroy, *Against the Wind*, 179–80.

214. McElroy, *Against the Wind*, 141–71.

215. McElroy, *Against the Wind*, 150–54.

216. Former NATCA president, Boston Air Route Traffic Control Center, and member of the Reclassification Committee, interview with the author, June 2000.

217. McElroy, *Against the Wind*, 168.

218. In 1995 Congress passed an annual appropriations act with sweeping changes that gave the agency the right to negotiate pay, but eliminated a section granting federal workers the right to union representation and collective bargaining. President Clinton intervened in the labor issue, and Senator John McCain wrote the FAA Reauthorization Act, restoring the union's collective bargaining rights with language including the right to negotiate salary. McElroy, *Against the Wind*, 156–60.

219. In 1991, the union and the FAA had formally agreed to implement "Quality through Partnership," an agreement that set an FAA precedent establishing that the union had a say in everything related to the facilities. In 1996, budget cuts by a Republican Congress had killed the agreement, but the idea of consensus decision making held on. Garvey activated the partnership in principle. McElroy, *Against the Wind*, 135, 141–42.

220. William J. Clinton, Exec. Order No. 13,180 (December 7, 2000), "Air Traffic Performance-Based Organization," accessed at Gerhard Peters and John T. Woolley, *The American Presidency Project*, http://www.presidency.ucsb.edu/ws/?pid=61670.

221. The total number of operations is the number of tower, TRACON, and aircraft handled by them annually. Federal Aviation Administration, *A Plan for the Future: 10 Year*

Strategy for the Air Traffic Control Workforce 2012–2021 (Washington, DC: Federal Aviation Administration, US Department of Transportation, 2012), 6.

222. NATCA president, Boston Air Route Traffic Control Center, interview with the author, September 2002.

223. Al Baker and Matthew Wald, "FAA Considers Consolidating Regional Air Traffic Control," *New York Times*, March 17, 2001, http:/www.nytimes.com/2001/03/17/nyregion/faa-considers-consolidating-regional-air-traffic-control.html?pagewanted-print&src=pm.

224. Lamont and Molnàr, "Study of Boundaries."

225. NATCA president, interview with author, September 2002.

226. NATCA president, interview with author, September 2002.

227. Clinton, Exec. Order 13,180.

Part II

1. Diane Vaughan, "Report: Air Traffic Control Survey Analysis," distributed to Boston Tower, Boston TRACON, Boston ARTCC, Bedford Tower, New England Regional FAA Office, New England Regional NATCA, and Boston Center NATCA, February 2003, 1–21.

2. Arthur B. Shostak and David Skocik, *The Air Controllers' Controversy: Lessons from the PATCO Strike* (New York: Human Sciences Press, 1986).

3. The national and regional FAA data that I cite was generated by FAA New England Human Resource Management Division, using the Integrated Personnel and Payroll System, Management Information Retrieval. Rather than giving a snapshot at one point in time, this national and regional database is updated automatically as FAA personnel join and leave the organization, so it changes daily. The numbers available are for the day on which a request is made. These data were generated in 2002, not 2000, which was the year controllers completed my survey. I was able to compare my survey data with national and regional data, thanks to Larry Piro, New England Region Human Resources, FAA New England Regional Office, Burlington, MA (May 2002).

4. In my survey for the four facilities, 94 percent reported that they were white; two people classified themselves as African American, three as Hispanic, and five (3 percent) chose "other."

5. The proportion of women is slightly higher in my study than among air traffic controllers nationally and regionally given the imbalance in participation of controllers at Boston Center, where about half the men but almost all the women participated in the survey.

6. Because the FAA imposed a hiring freeze in 1992 that lasted until 2002, fully 80 percent of controllers in my study started between 1981 and 1990. As a result, in 2000, the year of my survey, the most common background was being a "replacement hire": 41 percent began as FAA air traffic controllers between 1981 and 1985, the years immediately after the strike. Consequently, 49 percent of the men were between the ages of thirty-one and forty; 39 percent were between the ages of forty-one and fifty. Admitted later to a male occupation as a result of equal opportunity legislation, the women were younger. Sixty-seven percent began as controllers between 1986 and 1990. Fifteen percent were between twenty and thirty years old, compared to 5 percent of the men; 60 percent were between thirty-one and forty years old, in contrast to 49 percent of the men. These gendered hiring patterns mirrored the national situa-

tion at the time. Age distributions were affected nationally as the eight hundred PATCO controllers (known in facilities as "the old guys") took advantage of the 1993 Clinton offer. While I was in the field, they were either already certified as controllers or still training nationwide. Eight former PATCO controllers were in my four New England facilities.

7. Mark Granovetter, "The Strength of Weak Ties," *American Journal of Sociology* 78, no. 6 (1973): 1360–80.

8. This definition of multitasking is controllers' interpretation of tasks and how they do them. In contrast, research in neuroscience on brain functioning indicates that the human brain cannot process information this way, instead processing all tasks sequentially: that is, "task switching," not multitasking. However, other research has found a small proportion of the population are "supertaskers," who can effectively perform multiple tasks simultaneously in tests that add task after task to the workload, without loss of accuracy or fatigue. See David L. Thayer and Jason M. Watson, "Supertaskers and the Multitasking Brain," *Scientific American Mind* 23, no. 1 (2012): 22–29.

9. *Webster's Third New International Dictionary* (1971), s.v. "common sense."

10. Clifford Geertz, *Local Knowledge: Further Essays in Interpretive Anthropology* (New York: Basic Books, 1983), 73–93.

11. Paul DiMaggio, "Culture and Cognition," *Annual Review of Sociology* 23 (1997): 263–87.

12. Joseph Bensman and Robert Lilienfeld, *Craft and Consciousness: Occupational Technique and the Development of Work Images*, 2nd ed. (New York: Aldine de Gruyter, 1991).

13. Pierre Bourdieu, *Distinction: A Social Critique of the Judgment of Taste*, trans. Richard Nice (Cambridge, MA: Harvard University Press, 1984), 99; Pierre Bourdieu, *Homo Academicus*, trans. Peter Collier (Stanford, CA: Stanford University Press, 1988).

14. DiMaggio, "Culture and Cognition."

15. Matthew Desmond, *On the Fireline: Living and Dying with Wildland Firefighters* (Chicago: University of Chicago Press, 2007); Shamus Rahman Khan, *Privilege* (Princeton, NJ: Princeton University Press, 2011).

16. The section continues:

> *b* It is recognized that these services cannot be provided in cases in which the provision of services is precluded by the above factors. Consistent with the aforementioned conditions, controllers shall provide additional service procedures to the extent permitted by higher priority duties and other circumstances. The provision of additional services is not optional on the part of the controller, but rather is required when the work situation permits. U.S. Department of Transportation, Federal Aviation Administration, *Air Traffic Procedures 7110.65*. (Washington, DC: Government Printing Office, 1995)

The 1995 edition cited here was being used in the facilities during my 2000–2001 fieldwork. This excerpt was in effect at that time.

17. Bensman and Lilienfeld, *Craft and Consciousness*.

18. See also Daniel F. Chambliss, "The Mundanity of Excellence: An Ethnographic Report on Stratification and Olympic Swimmers," *Sociological Theory* 7, no. 1 (1989): 70–86.

19. Stuart E. Dreyfus and Hubert L. Dreyfus, *A Five-Stage Model of the Mental Activities Involved in Directed Skill Acquisition*, No. ORC-80-2 (Berkeley: University of California–Berkeley Operations Research Center, 1980); Harry Collins, *Tacit and Explicit Knowledge* (Chicago: University of Chicago Press, 2010).

20. Collins, *Tacit and Explicit Knowledge.*

21. Gil Eyal, "For a Sociology of Expertise: The Social Origins of the Autism Epidemic," *American Journal of Sociology* 118, no. 4 (2013): 863–907.

22. Harry Collins, "Three Dimensions of Expertise," *Phenomenology and the Cognitive Sciences* 12, no. 2 (2013): 253–73; Collins, *Tacit and Explicit Knowledge.*

23. Harry Collins, "Bicycling on the Moon: Collective Tacit Knowledge and Somatic-Limit Tacit Knowledge," *Organization Studies* 28 (2007): 257–62; Collins, *Tacit and Explicit Knowledge*, 119–38.

24. Eyal, "For a Sociology of Expertise."

25. Collins, "Bicycling on the Moon," 261.

Chapter Three

1. On the role of paper strips as a technology of coordination and control in knowledge work, see Abigail J. Sellen and Richard H. R. Harper, *The Myth of the Paperless Office* (Cambridge, MA: MIT Press, 2003), 51–74; Wendy E. Mackay, "Is Paper Safer? The Role of Paper Flight Progress Strips in Air Traffic Control," *ACM Transactions on Computer-Human Interaction (TOCHI)* 6, no. 4 (1999): 311–40; Malcolm Gladwell, "The Social Life of Paper: Looking for Method in the Mess," *New Yorker*, March 25, 2002.

2. Howard S. Becker, "Notes on the Concept of Commitment," *American Journal of Sociology* 66 (1960): 32–42; Barry M. Staw, "The Escalation of Commitment to a Course of Action," *Academy of Management Review* 6 (1981): 577–87; Patricia Benner, *From Novice to Expert: Excellence and Power in Clinical Nursing Practice* (Reading, MA: Addison-Wesley, 1984).

3. Clifford Geertz, *The Interpretation of Cultures* (New York: Basic Books, 1973), 4.

4. At a center, training takes several years because developmentals must be certified on approximately fourteen positions. Towers and TRACONs have less airspace, thus fewer airspace divisions, thus fewer positions, accounting for a shorter training period. Under the pass-fail system, developmentals had to demonstrate competency on each position in a specified number of hours, based on the difficulty for each position. If they couldn't master a position in the allotted time, they were out. Then in the mid-1990s, the FAA switched facility training from pass-fail to "train to succeed." In train to succeed, no time limit was imposed for each position, and people received second and third chances, perhaps working with different trainers, usually eventually meeting the requirements for certification. Some, discouraged, did drop out, but the practice of paying developmentals a salary, despite lack of competency to pass, keeps some developmentals keeping on in what controllers trained under the pass-fail system renamed "train to retire."

5. Benner, *From Novice to Expert.*

6. Iddo Tavory and Nina Eliasoph, "Coordinating Futures: Toward a Theory of Anticipation," *American Journal of Sociology* 118, no.4 (2013): 908–42; Ann Mische, "Projects and Possibilities: Researching Futures in Action," *Sociological Forum* 24, no. 3 (2009): 694–704.

7. Daniel F. Chambliss, “The Mundanity of Excellence: An Ethnographic Report on Stratification and Olympic Swimmers,” *Sociological Theory* 7, no. 1 (1989): 81.

Chapter Four

1. See, e.g., Atul Gawande, *The Checklist Manifesto: How to Get Things Right* (New York: Henry Holt, 2009).

2. Erving Goffman, *The Presentation of Self in Everyday Life* (New York: Anchor Books, 1959).

3. Jack Katz, “Pissed Off in L.A.,” in *How Emotions Work* (Chicago: University of Chicago Press, 1999), 18–86.

4. Discrepancies in numbers of controllers' responses to these questions and the total number interviewed reflect my new learning over the duration of my fieldwork. Consequently, I asked all controllers a set of common questions, then divided the new questions into two groups. So, at each facility I had two interview schedules: Both had the common questions, one with new questions set A, the other with new questions set B.

5. The controllers who either answered no change or else named one of the above characteristics explained that they had it before they began their training, usually attributed it to their family situation, for example, having a thick skin because “I grew up with brothers”; confidence “because anything I ever wanted to do my mother told me I could do it.” The two who said they didn't know explained that their young age when hired was a factor in their answer: it was hard to sort out the effect of the work from the effect of aging and maturity. The three controllers whose responses indicated “little change” did not name these type A traits, instead mentioning how the job and salary gave them status or made them more critical of others.

6. Michael Polanyi, *The Tacit Dimension* (London: Routledge, 1966).

7. Maurice Merleau-Ponty, *Phenomenology of Perception*, trans. C. Smith (New York: Routledge, 1962).

8. Colin Jerolmack and Shamus Khan, “Talk Is Cheap: Ethnography and the Attitudinal Fallacy,” *Sociological Methods & Research* 43, no. 2 (2014): 178–209.

9. Hubert L. Dreyfus, *What Computers Still Can't Do* (Cambridge, MA: MIT Press, 1991); Hubert L. Dreyfus and Stuart E. Dreyfus, *Mind over Machine: The Power of Human Intuition and Expertise in the Era of the Computer* (New York: Free Press, 1986).

10. Hubert L. Dreyfus and Stuart E. Dreyfus, “The Challenge of Merleau-Ponty's Phenomenology of Embodiment for Cognitive Science,” in *Perspectives on Embodiment: The Intersections of Nature and Culture*, ed. Gail Weiss and Honi Fern Haber (New York: Routledge, 1999), 103–20.

11. Patricia Benner, *From Novice to Expert: Excellence and Power in Clinical Nursing Practice* (Reading, MA: Addison-Wesley, 1984).

12. Dreyfus and Dreyfus neglected socialization in their work. Correcting in response to Collins and Evans, in a 2005 article studying professionals, they acknowledged socialization. H. M. Collins and Robert Evans, “The Third Wave of Science Studies: Studies of Expertise and Experience,” *Social Studies of Science* 32, no. 2 (2002): 235–96; Hubert L. Dreyfus and Stuart E.

Dreyfus, "Peripheral Vision: Expertise in Real World Context," *Organization Studies* 26, no. 5 (2005): 788.

13. Collins identifies types of tacit knowledge, based on how it is acquired. Collective tacit knowledge is distinctly human because it must be acquired from social groups. Harry Collins, *Tacit and Explicit Knowledge* (Chicago: University of Chicago Press, 2010), 119–38.

14. Gil Eyal, "For a Sociology of Expertise: The Social Origins of the Autism Epidemic," *American Journal of Sociology* 118, no. 4 (2013): 871.

15. Gil Eyal, *The Crisis of Expertise* (Cambridge, UK: Polity Press, 2019): 36.

16. Pierre Bourdieu, *Homo Academicus*, trans. Peter Collier (Stanford, CA: Stanford University Press, 1988); Bourdieu, *Pascalian Meditations*, trans. Richard Nice (Stanford, CA: Stanford University Press, 2000).

17. Erving Goffman, *Asylums: Essays on the Social Situation of Mental Patients and Other Inmates* (New York: Anchor Books, 1961).

18. Joseph Bensman and Robert Lilienfeld, *Craft and Consciousness: Occupational Technique and the Development of Work Images*, 2nd ed. (New York: Aldine de Gruyter, 1991).

19. Haridimos Tsoukas and Efi Vladimirou, "What Is Organizational Knowledge?" *Journal of Management Studies* 38 (2001): 974–93.

Part III

1. Geoffrey C. Bowker and Susan Leigh Star, *Sorting Things Out: Classification and Its* Consequences (Cambridge, MA: MIT Press, 1999); Stefan Timmermans and Steven Epstein, "A World of Standards but Not a Standard World: Toward a Sociology of Standards and Standardization," *Annual Review of Sociology* 36 (2010): 69–89.

2. Andrew Abbott, *The System of Professions: An Essay on the Division of Expert Labor* (Chicago: University of Chicago Press, 1988); Thomas F. Gieryn, "Boundary-Work and the Demarcation of Science from Non-Science: Strains and Interests in Professional Ideologies of Scientists," *American Sociological Review* 48 (1983): 781–95; Michèle Lamont and Viràg Molnàr, "The Study of Boundaries in the Social Sciences," *Annual Review of Sociology* 28, no. 1 (2002): 167–95.

3. On representation and the production of knowledge in science and technology, see Michael Lynch and Steve Woolgar, eds., *Representation in Scientific Practice* (Cambridge, MA: MIT Press, 1990); Catelijne Coopmans et al., eds., *Representation in Scientific Practice Revisited* (Cambridge, MA: MIT Press, 2014).

4. Clifford Geertz, *The Interpretation of Cultures* (New York: Basic Books, 1973), 14.

5. John Van Maanen and Stephen R. Barley, "Cultural Organization: Fragments of a Theory," in *Organizational Culture*, ed. Peter J. Frost et al. (Beverly Hills, CA: Sage, 1985), 31–54.

6. Controllers are all of the same specialization, same task, although in different organizations within the system. In contrast, research on coordination between teams of different scientific specializations collaborating on the same task shows how they must negotiate the different cultural ways of knowing and talking in order to accomplish the task. See, e.g., Karin Knorr Cetina, *Epistemic Cultures: How the Sciences Make Knowledge* (Cambridge, MA: Harvard

University Press, 1999); Peter Galison, "The Trading Zone: Coordinating Action and Belief," in *Image and Logic: A Material Culture of Microphysics* (Chicago: University of Chicago Press, 1997), 803–40.

7. To clarify, *personality* refers to how controllers define another airspace, facility, crew, or person, on the basis of their experience sending and receiving airplanes across official boundaries. *Culture* refers to the patterns of life inside a place and the meanings that they have for insiders. Although different, personality and culture nonetheless have the same source: a system change that affects the relation between airspace, place, and work practice.

8. Edwin A. Hutchins, *Cognition in the Wild* (Cambridge, MA: MIT Press, 1995); Edwin A. Hutchins, "How a Cockpit Remembers Its Speeds," *Cognitive Science* 19, no. 3 (1995): 265–88.

9. James Hollan, Edwin Hutchins, and David Kirsh, "Distributed Cognition: Toward a New Foundation for Human-Computer Interaction Research," *ACM Transactions on Computer-Human Interaction (TOCHI)* 7 (2000): 177–78.

10. In more recent writing, Hutchins points out that the unit of analysis for distributed cognition "should not be set *a priori*, but should be responsive to the nature of the phenomena under study." It is legitimate, then, to extend cognition beyond the small socio-technical system to see the role of system effects. Edwin A. Hutchins, "Enaction, Imagination, and Insight," in *Enaction: Towards a New Paradigm for Cognitive Science*, ed. John Stewart, Oliver Gapenne, and Ezequiel A. Di'Paolo (Cambridge, MA: MIT Press, 2010), 425–50.

11. See Lucy A. Suchman, *Plans and Situated Actions: The Problem of Human-Machine Communication* (Cambridge: Cambridge University Press, 1987).

12. Alain Gras et al., *Faced with Automation: The Pilot, the Controller, and the Engineer*, trans. Jill Lundsten (Paris: Publication de la Sorbonne, 1994).

13. Thomas F. Gieryn, "What Buildings Do," *Theory and Society* 31, no. 1 (2002): 35–74.

14. Harry Collins, *Tacit and Explicit Knowledge* (Chicago: University of Chicago Press, 2010), 119–38.

Chapter Five

1. *Modernization Efforts at the Boston En Route Center (106–47): Hearing before the Subcommittee on Aviation of the Committee on Transportation and Infrastructure*, 106th Congress (1999) (statement of Peter H. Challan, Acting Deputy Associate Administrator for Air Traffic Services).

2. See Lucy A. Suchman, *Plans and Situated Actions: The Problem of Human-Machine Communication* (Cambridge: Cambridge University Press, 1987).

3. Edwin A. Hutchins, "How a Cockpit Remembers Its Speeds," *Cognitive Science* 19, no. 3 (1995): 265–88.

4. For ethnographies showing how different tasks produce subtle varieties of coordinated actions between workers, see Christian Heath and Paul Luff, "Collaborative Activity and Technological Design: Task Coordination in the London Underground Control Rooms," in *Proceedings of ECCSCW '91, The European Conference on Computer-Supported Cooperative Work* (Amsterdam: Kluwer Press, 1991), 65–80; Wendy E. Mackay et al., "Reinventing the Familiar:

Exploring an Augmented Reality Design Space for Air Traffic Control," in *Proceedings of the ACM CHI '98 Human Factors in Computing Systems* (Los Angeles: ACM Press, 1998), 558–65.

5. Officially, a handoff is "an action taken to transfer the radar identification of an aircraft from one controller to another if the aircraft will enter the receiving controller's airspace and radio communications with the aircraft will be transferred." "Pilot/Controller Glossary," in U.S. Department of Transportation, Federal Aviation Administration, *Aeronautical Information Manual*, (Washington, DC: February 24, 2000), p. H-1.

6. The Expanded East Coast Plan initially was to help increase the capacity of the airspace to relieve congestion in the New York and DC areas, but it was expanded to cover the entire East Coast. The third and final phase, in 1988, was to decrease congestion from the New York area to the northeast, hence the transfer of airspace from New York to Boston Center. Edmund Preston, ed., *FAA Historical Chronology: Civil Aviation and the Federal Government, 1926–1996* (Washington, DC: Department of Transportation, Federal Aviation Administration Office of Public Affairs, 1998), 212.

Chapter Six

1. In 2000, some towers in the air traffic control system, such as Atlanta and Manchester, New Hampshire, were still "up-and-downs."

2. Demonstrating the flexibility of system boundaries, in case of a real fire or other emergency, controllers would exit and the facility would go to ATC Zero. Other facilities hold all airplanes outside the TRACON airspace, the TRACON turns control of its airplanes over to Boston Center, and Boston Tower lands the ones already on final.

3. Aram Khachaturian's "Sabre Dance" is from the ballet *Gayane* (1939–1941).

4. Preston, *FAA Historical Chronology*, 211, 213.

5. Wanda J. Orlikowsi, "Improvising Organizational Change over Time: A Situated Change Perspective," *Information Systems Research* 7, no. 1 (1999): 63–92; Karl E. Weick, "Improvisation as a Mindset for Organizational Analysis," *Organization Science* 9 (1998): 543–55.

Part IV

1. Arlie Russell Hochschild, "Emotion Work, Feeling Rules, and Social Structure," *American Journal of Sociology* 85, no. 3 (1979): 551–75; Arlie Russell Hochschild, *The Managed Heart* (Berkeley: University of California Press, 1983).

2. Donald J. Black and Albert J. Reiss Jr., "Patterns of Behavior in Police and Citizen Transactions," *Studies in Crime and Law Enforcement in Major Metropolitan Areas* 2 (1967): 1–139.

3. John Van Maanen and Stephen R. Barley, "Occupational Communities: Culture and Control in Organizations," in *Research in Organizational Behavior*, ed. B. M. Staw and L. L. Cummings (Greenwich, CT: JAI Press, 1984), 6:287–365.

Chapter Seven

1. Harold Garfinkel, "Conditions of Successful Degradation Ceremonies," *American Journal of Sociology* 61, no. 5 (1956): 420–24.

2. Maurizio Catino and Gerardo Patriotta, "Learning from Errors: Cognition, Emotions, and Safety Culture in the Italian Air Force," *Organization Studies* 34 (2013): 437–67; Shlomo Hareli, Noga Short, and Nicole Bigger, "The Role of Emotions in Employees' Explanations for Failure in the Workplace," *Journal of Managerial Psychology* 20 (2005): 152–62.

Chapter Eight

1. Recall that all controllers received a common set of questions, but questions I developed later were divided roughly in half, so each interview had the common questions, and half the interviews had a second set that was unique. Eighty-six controllers were asked about risk and stress.

2. These categories were identified by controllers, not suggested by me. Some named more than one. The consistency of experiences across the four facilities indicates similar system effects on their work, regardless of facility.

3. Jack Katz, *How Emotions Work* (Chicago: University of Chicago Press, 1999).

4. Arlie Russell Hochschild, *The Managed Heart* (Berkeley: University of California Press, 1983).

5. Hochschild, *Managed Heart*, 147.

6. Hochschild, *Managed Heart*, 49.

7. Diane Vaughan, *The Challenger Launch Decision: Risky Technology, Culture, and Deviance at NASA* (Chicago: University of Chicago Press, 1996).

8. See, e.g., Bridget Hutter, "Ways of Seeing: Understandings of Risk in Organizational Settings," in *Organizational Encounters with Risk*, ed. Bridget Hutter and Michael Power (Cambridge: Cambridge University Press, 2005), 66–91.

9. Vaughan, *Challenger Launch Decision*, 62–68.

10. Karl E. Weick, "The Collapse of Sensemaking in Organizations: The Mann Gulch Disaster," *Administrative Science Quarterly* 38 (1993): 628–52; Matthew Desmond, *On the Fireline: Living and Dying with Wildland Firefighters* (Chicago: University of Chicago Press, 2007).

11. Vaughan, *Challenger Launch Decision*, 252.

12. Clifford Geertz, *The Interpretation of Cultures* (New York: Basic Books, 1973), 203–13.

13. Erving Goffman, *Stigma: Notes on the Management of Spoiled Identity* (Englewood Cliffs, NJ: Prentice Hall, 1963).

14. Howard S. Becker, "Becoming a Marihuana User," *American Journal of Sociology* 59, no. 3 (1953): 235.

Part V

1. Some of the problems the commission identified were FAA failures in communication. For example, the commission stated that by 9:03 am, FAA headquarters, the Command Center, United, and American knew that the attacks were planned in advance, coordinated, and involved two and likely more airplanes, yet no one thought to tell the airlines to inform their pilots to secure cockpits against intruders. Several senior FAA officials said that doing so was not the FAA's job but the airlines' job. National Commission on Terrorist Attacks, *The 9/11*

Commission Report: Final Report of the National Commission on Terrorist Attacks upon the United States (New York: W. W. Norton, 2004), 11. However, the FAA was blamed for additional failures of communication that were clearly beyond their control. See Diane Vaughan, "The Social Shaping of Commission Reports," *Sociological Forum* 21, no. 2 (2006): 291–306.

2. *Structural secrecy* refers to how organization structure and specialization—hierarchy, division of labor, subunits—can create barriers to the flow of information, even when every effort is being put into communicating across these boundaries. Diane Vaughan, "Structural Secrecy," in *The Challenger Launch Decision: Risky Technology, Culture, and Deviance at NASA* (Chicago: University of Chicago Press, 1996), 238–77.

3. National Commission on Terrorist Attacks upon the United States, *9/11 Commission Report*, 31.

4. Daniel F. Chambliss, "The Mundanity of Excellence: An Ethnographic Report on Stratification and Olympic Swimmers," *Sociological Theory* 7, no. 1 (1989): 70–86.

5. Following official accident investigation protocol, on February 19, 2002, the National Transportation Safety Board's Office of Research and Engineering made public FAA transcripts of all exchanges between air traffic controllers at Boston Tower, Boston Center, and the TRACON and the pilots of the two hijacked planes (and at Boston Center, the terrorists) as well as images of the flight path of each hijacked aircraft both before and after the hijacking, based on recorded radar data that tracked the flight paths from each facility. See National Transportation Safety Board, *Flight Path Study and Radio Communications with FAA* (Washington, DC: Office of Research and Engineering, February 19, 2002). Using these data for timeline and my interviews, I reconstructed the details of the moment-to-moment events and actions taken by individuals at the Command Center and Boston Center, TRACON, and Boston Tower. From the interviews, I learned about aspects of the day not published elsewhere (e.g., United and American Airlines gave early orders to their respective fleets to land even before US airspace was closed). The detail of my interviews account for small discrepancies in numbers between various accounts reporting how many aircraft were in the sky at specific times, although all major accounts agree on the number of planes landed. Given the variation in facilities and position of the interviewees in the facility, collectively they present the most complete picture possible of the system effects of the terrorist attacks and the system response of many different air traffic and national security organizations involved. Each refines the others by adding details.

Other major sources expanded my analysis. The earliest published complete official chronology in the *9/11 Commission Report*, chapter 1, reported the responses of relevant agencies on that day, including but not limited to the FAA. Also important because based on different sources were Matthew Wald and Kevin Sack, "'We Have Some Planes,' Hijacker Told Controller," *New York Times*, October 16, 2001; Alan Levin, Marilyn Adams, and Blake Morrison, "Four Hours of Fear: 9/11's Untold Story," *USA Today*, August 12–15, 2002; Lynn Spencer, *Touching History* (New York: Free Press, 2008); and "Complete 9/11 Timeline," History Commons, www.historycommons.org.

6. Michèle Lamont, *The Dignity of Working Men: Morality and the Boundaries of Race, Class,*

and Immigration (New York: Russell Sage Foundation at Harvard University Press, 2000); Michèle Lamont and Viràg Molnàr, "The Study of Boundaries in the Social Sciences," *Annual Review of Sociology* 28, no. 1 (2002): 167–95.

Chapter Nine

1. Two possible explanations circulated after about how Peter was able to hear the hijacker. The cockpit cabin's yoke, or steering wheel, controls altitude, pitch, and roll but also holds other function buttons that can be controlled by the thumb. Possibly the hijacker, trying to make an announcement to passengers, had mistakenly keyed the talk-back button and talked on the air traffic control frequency rather than broadcasting to passengers. The alternative, and the one most controllers believe, was that John Ogonowski, the American 11 captain, furtively and intermittently hit the talk-back button so that the controller and other pilots on the frequency could hear the hijackers in the cabin. Reuters, "Pilot Secretly Alerted Controllers of Hijacking," *LA Times*, September 13, 2001.

2. Although it was his first day on the new job, Sliney had deep experience: beginning as a controller, he had been a first-line supervisor at several facilities, an operations manager and ATM at New York TRACON, and tactical operations manager at the Command Center, the position Linda Scheussler held on the day of the terrorist attacks.

3. Sliney's action was unprecedented, but a preexisting emergency order had been developed in the 1960s that was designed in case of confirmed warnings of an attack by the Soviet Union in order to clear the skies for a US military response. See Stephen I. Schwartz, "This Is Not a Test," *Bulletin of the Atomic Scientists* 57, no. 6 (2001): 50–52. Whether Sliney knew about the order, I was unable to discover.

4. At the time and shortly after, contradictory accounts appeared indicating that the decisions to shut down the airspace were not made in the air traffic control system but by officials with ranks high above the Command Center. However, my interviews at the Command Center, public statements made from FAA headquarters, and other official public testimony from Command Center management confirm that the decision was a collaborative effort, made locally and collaboratively by personnel at the Command Center. See "FAA Emergency Response on September 11," C-Span Video Library, aired August 12, 2002, on C-SPAN 1, https://www.c-spanvideo.org/program/FAAE. Additional confirmation comes from actions taken simultaneously, in various parts of the country, by other air traffic control facilities and the airlines (United and American Airlines) that were making the same decisions—ground stop and land the planes in flight—independently of one another. Although such a calamitous circumstance had never occurred before, taking these actions was standard operating procedure deployed for major thunderstorms and other emergencies. The simultaneous enactment across systems speaks to common local knowledge of the tools the system had to bring to bear on a problem and the expertise of the people with operational experience within the air traffic control system and the industry—thus demonstrating the cultural system of knowledge and agency that is the source of the resilience, reliability, and redundancy in the system.

5. Daniel F. Chambliss, "The Mundanity of Excellence: An Ethnographic Report on Stratification and Olympic Swimmers," *Sociological Theory* 7, no. 1 (1989): 70–86.

Chapter Ten

1. This practice of problem solving at the local level by translating and innovating ambiguous rules created by upper level governmental units has precedent throughout the history of the system, and confirmed in two additional organizational systems settings. See Frank Dobbin, *Inventing Equal Opportunity* (Princeton, NJ: Princeton University Press, 2009); Kerry B. Fosher, *Under Construction: Making Homeland Security at the Local Level* (Chicago: University of Chicago Press, 2009).

2. The Airbus took off immediately following a Japanese 747, on the same runway. AA 587 was affected by the larger planes' wake turbulence. The National Transportation Safety Board report concluded that reacting to stabilize the plane, the first officer worked the rudder very aggressively, causing loss of control and the crash.

3. *9/11*, directed by Jules Naudet, Gédéon Naudet, and James Hanlon, hosted by Robert DeNiro, aired March 10, 2002, CBS.

4. *Dateline NBC*, "America Remembers: Air Traffic Controllers Describe How Events Unfolded as They Saw Them on September 11th," season 10, episode 132, aired September 11, 2002, NBC.

5. Here I make the distinction between boundary processes in the system, and the experience of them as stable structures that act as barriers during every day work and define identities and specializations, thus building on Andrew Abbott, "Things of Boundaries," *Social Research* 62, no. 4 (1995): 857–82, and also his *Processual Sociology* (Chicago: University of Chicago Press, 2016), 198–233.

6. Edwin A. Hutchins, *Cognition in the Wild* (Cambridge, MA: MIT Press, 1995).

7. Abbott, "Things of Boundaries"; Abbott, *Processual Sociology*.

Chapter Eleven

1. John Van Maanen and Stephen R. Barley, "Occupational Communities: Culture and Control in Organizations," in *Research in Organizational Behavior*, ed. B. M. Staw and L. L. Cummings (Greenwich, CT: JAI Press, 1984), 6:287–365.

2. Thomas F. Gieryn, "Boundary-Work and the Demarcation of Science from Non-science: Strains and Interests in Professional Ideologies of Scientists," *American Sociological Review* 48 (1983): 781–95; Thomas F. Gieryn, *Cultural Boundaries of Science: Credibility on the Line* (Chicago: University of Chicago Press, 1999), 4–5.

3. Lamont and Molnàr distinguish between social and symbolic boundaries thus: "Symbolic boundaries are distinctions made by social actors to categorize objects, people, practices, and even time and space. Symbolic boundaries separate people into groups and generate feelings of similarity and group membership. They are a medium through which people acquire status and resources. Social boundaries are objectified forms of social differences, manifested in unequal access to and unequal distribution of resources and social opportunities." Michèle

Lamont and Viràg Molnàr, "The Study of Boundaries in the Social Sciences," *Annual Review of Sociology* 28, no. 1 (2002): 167–95.

4. Michael Sauder, "Symbols and Texts: An Interactionist Approach to the Study of Social Status," *Sociological Quarterly* 46, no. 2 (2005): 279–98.

5. Andrew Abbott, "Things of Boundaries," *Social Research* 62, no. 4 (1995): 857–82; John W. Meier and Brian Rowan, "Institutionalized Organizations: Formal Structure as Myth and Ceremony," *American Journal of Sociology* 83, no. 2 (1977): 340–63.

6. In Lamont's influential book, *The Dignity of Working Men*, she draws on interviews with working-class men in France and the United States to reveal that moral standards, not economic status, are the key principles of their evaluations of worth and perceptions of social hierarchy. Morality has long been the basis for how people of low economic and social status judged themselves in relation to others. There are many differences between her study and this one. That said, the strategies of social differentiation and distinction that Lamont identifies are highly relevant here but work in different ways. Setting aside differences in the organizational form, employment, gender, and controllers' collective identity as occupational community, the common finding is that intraorganizationally, controllers, like the working-class men of her study, also make distinctions among themselves that correct their status in relation to those above and below them in the hierarchy, and they do so in line with their perception of their work as moral. Michèle Lamont, *The Dignity of Working Men: Morality and the Boundaries of Race, Class, and Immigration* (New York: Russell Sage Foundation at Harvard University Press, 2000).

7. Each facility is a subunit of the larger organization, thus has a subculture within the larger culture. On subcultures, see John Van Maanen and Stephen R. Barley, "Cultural Organization: Fragments of a Theory," in *Organizational Culture*, ed. Peter J. Frost et al. (Beverly Hills, CA: Sage, 1985), 31–54.

8. Van Maanen and Barley, "Occupational Communities."

9. Van Maanen and Barley, "Occupational Communities," 301.

10. Van Maanen and Barley, "Occupational Communities," 309.

11. Michael Sauder, "Third Parties and Status Position: How the Characteristics of Status Systems Matter," *Theoretical Sociology* 35 (2006): 299–321.

12. For the national political scene and NATCA development of the complexity formula to remedy inequalities, see chapter 2: pages xxx.

13. Lamont and Molnàr, "Study of Boundaries."

14. Michael Sauder, Freda Lynn, and Joel M. Podolny, "Status Insights from Organizational Sociology," *Annual Review of Sociology* 38 (2012): 267–83.

15. Andrew Abbott, "Status and Status Strain in the Professions," *American Journal of Sociology* 86, no. 4 (1981): 824.

Part VI

1. Andrew Abbott, "Sequences of Social Events: Concepts and Methods for the Analysis of Order," *Historical Methods* 16, no. 4 (1983): 129–47.

2. Andrew Abbott, "Transcending General Linear Reality," *Sociological Theory* 6, no. 2 (1988): 173.

3. Andrew Abbott, "Conceptions of Time and Events in Social Science Methods," in *Time Matters: On Theory and Method* (Chicago: University of Chicago Press, 2001), 181–82.

4. Bart Elias, "Air Traffic Inc.: Considerations Regarding the Corporatization of Air Traffic Control," Congressional Research Service 7-5700, May 16, 2017, 13–19.

5. The four leading perspectives vary. See, e.g., John F. Padgett and Walter W. Powell, eds., *The Emergence of Organizations and Markets* (Princeton, NJ: Princeton University Press, 2012); Frank Dobbin, *Inventing Equal Opportunity* (Princeton, NJ: Princeton University Press, 2009); James Mahoney and Kathleen Thelen, eds., *Explaining Institutional Change* (New York: Cambridge University Press, 2010); Abbott, "Conceptions of Time and Events."

6. Robert K. Merton, "The Unanticipated Consequences of Purposive Social Action," *American Sociological Review* 1, no. 6 (1936): 894–904; Robert Jervis, *System Effects: Complexity in Political and Social Life* (Princeton, NJ: Princeton University Press, 1997); Charles B. Perrow, *Normal Accidents: Living with High Risk Technologies*, 2nd ed. (Princeton, NJ: Princeton University Press, 1999).

7. Stephen R. Barley, "Technology as an Occasion for Structuring: Evidence from Observations of CT Scanners and the Social Order of Radiology Departments," *Administrative Science Quarterly* 31, no. 1 (1986): 78–108.

8. Paul J. DiMaggio and Walter W. Powell, "The Iron Cage Revisited: Institutional Isomorphism and Collective Rationality in Organizational Fields," *American Sociological Review* 48, no. 2 (1983): 147–60; James Mahoney, "Path Dependence in Historical Sociology," *Theory and Society* 29, no. 4 (2000): 507–48.

9. Theorists have long since laid the groundwork, but research on the internal interactional dynamics in the workplace has not been as extensive. See Paul J. DiMaggio, "Interest and Agency in Institutional Theory," in *Institutional Patterns and Organizations: Culture and Environment*, ed. Lynne G. Zucker (Cambridge, MA: Ballinger Publishing, 1988), 3–21; Walter W. Powell and Jeannette A. Colyvas, "The Microfoundations of Institutional Theory," in *Sage Handbook of Organizational Institutionalism*, ed. Royston Greenwood and Christine Oliver (Thousand Oaks, CA: Sage, 2008), 276–98.

10. Wanda J. Orlikowski, "Improvising Organizational Transformation over Time: A Situated Change Perspective," *Information Systems Research* 7, no. 1 (1996): 63–92.

11. Orlikowski, "Improvising Organizational Transformation."

12. Abbott, "Sequences of Social Events," 131–33.

13. David Greenberg, "Memo to Obama Fans: Clinton's Presidency Was Not a Failure," *Slate*, February 12, 2008, https://www.slate.com/id/2183941/pagenum/all/#page_start/; "Economic Policy of the Bill Clinton Administration," Wikipedia, accessed July 25, 2020, https://en.wikipedia.org/wiki/Economic_policy_of_the_Bill_Clinton_administration.

14. Federal Aviation Administration, 1997–2014 Update to *FAA Historical Chronology: Civil Aviation and the Federal Government* (Washington, DC: Federal Aviation Administration, 1998), 63.

15. Elias, "Air Traffic Inc.," 2–7.

16. Elias, "Air Traffic Inc.," 8.

17. Research shows that power and boundaries also are tightly connected in conflicts between nation states in historical sociology as well as research confirming the power struggles of social actors in institutional fields in organizational sociology. My research is consistent with these macro-level perspectives, the difference being that I conceptualize the actions as boundary work and power work because the data reveal the agency and the professional work of agents acting in their official roles at both the macro and micro levels of a system. See DiMaggio, "Interest and Agency"; W. Richard Scott, *Institutions and Organizations: Ideas, Interests, and Identities* (Thousand Oaks, CA: Sage, 2013); Andreas Wimmer, *Ethnic Boundaries in the Making: Institutions, Power, and Networks* (New York Oxford University Press, 2014).

18. 1997–2014 *FAA Historical Chronology*, 77.

19. 1997–2014 *FAA Historical Chronology*, 71.

20. 1997–2014 *FAA Historical Chronology*, 75–76.

21. 1997–2014 *FAA Historical Chronology*, 108–9.

22. Patrick Forrey, "The View from the Air Traffic Control Tower and Scopes: How to Turn the Problems of NowGen into the Hope of NextGen," address to the Aero Club of Washington, March 26, 2008, http://aireform.com/resources/how-to-search-for-downloadable-reference-materials/news-archive/news-clips-folder/newsclip-2008–03–26-natca-president-patrick-forrey-delivers-speech-to-aero-club-of-washington/.

23. *The FAA Reauthorization Act of 2009: Testimony Before the Senate Committee on Commerce, Science, and Transportation, Subcommittee on Aviation Operations, Safety, and Security*, 111th Cong. 1–11 (2009) (testimony of Patrick Forrey, president of National Air Traffic Controllers Association).

24. 1997–2014 *FAA Historical Chronology*, 130.

25. 1997–2014 *FAA Historical Chronology*, 119.

26. "A Review of the Federal Aviation Administration's Air Traffic Controller Hiring, Staffing, and Training Plans," *Testimony Before the United States House of Representatives, Committee of Transportation and Infrastructure, Subcommittee on Aviation*, 114th Cong. 4–5 (2016) (written testimony of Paul M. Rinaldi, president of National Air Traffic Controller Association).

27. The FAA created a network of partnerships with education institutions that offer two- and four-year nonengineering degrees that teach basic courses in air traffic control and aviation administration. Graduates must still apply and be accepted by the FAA.

28. The Airport Improvement Program (AIP) was established by the Airport and Airway Improvement Act of 1982 (Pub. L. No. 97-248), https://www.faa.gov/airports/aip/overview/, last modified November 15, 2017.

29. 1997–2014 *FAA Historical Chronology*, 118.

30. 1997–2014 *FAA Historical Chronology*, 157.

31. FAA Modernization and Reform Act of 2012, Pub. L. No. 112-95, 126 Stat. 11 (2012).

32. Bart Elias, Clinton T. Brass, and Robert S. Kirk, "Sequestration at the Federal Aviation Administrations: Air Traffic Controller Furloughs and Congressional Response," Congressional Research Service 7–5700, May 7, 2013.

33. 1997–2014 *FAA Historical Chronology*, 180–82.

34. 1997–2014 *FAA Historical Chronology*, 194.

35. Office of Inspector General, Audit Report, "While FAA Took Steps Intended to Improve Its Controller Hiring Process, the Agency Did Not Effectively Implement Its New Policies," Federal Aviation Administration, Department of Transportation, Report No. AV2017028, February 15, 2017, 1–16.

36. Austin Kinley, "Race and Gender Changes in Air Traffic Controller Selection, Hiring, Attrition & Success 1940–2015" (master's thesis, Southern Illinois University, 2016).

37. Kinley, "Race and Gender Changes 1940–2015," 62–66.

38. "A Review of FAA's Efforts to Reduce Costs and Ensure Safety and Efficiency through Realignment and Consolidation," *Testimony Before the United States House of Representatives, Committee on Transportation and Infrastructure, Subcommittee on Aviation*, 112th Cong. 1–12 (2012) (testimony of Paul M. Rinaldi, president of National Air Traffic Controllers Association).

39. Final Report of the Performance-Based Operations Aviation Rulemaking Committee/ Commercial Aviation Safety Team, Flight Deck Automation Working Group, "Operational Use of Flight Path Management Systems," September 9, 2013.

40. 1997–2014 *FAA Historical Chronology*, 210.

41. 1997–2014 *FAA Historical Chronology*, 210.

42. "Hiring, Staffing, and Training Plans," 1–10.

43. 1997–2014 *FAA Historical Chronology*, 242.

44. In April 2018, after 25 extensions between 2007 and 2018, the House and Senate passed the 2017 FAA Reauthorization Act, providing a stable source of funding until 2023 and retaining air traffic control as a government operation.

Chapter Twelve

1. The funding for these improvements came from passenger revenues, not the FAA annual appropriations budget.

2. Dependent Converging Instrument Approaches (DCIA) with Converging Runway Display Aid (CRDA). U.S. Department of Tranportation, Air Traffic Organization Policy, Order JO 7110.65.110B, November 17. 2017.

3. *The FAA Reauthorization Act of 2009: Testimony Before the Senate Committee on Commerce, Science, and Transportation, Subcommittee on Aviation Operations, Safety, and Security*, 111th Cong. 4–5 (2009) (testimony of Patrick Forrey, president of National Air Traffic Controllers Association).

4. Staffing needs had been established in advance of the move to Boston TRACON in Merrimack, based on expected retirements due to the move. The TRACON also was understaffed, however.

5. *FAA Reauthorization Act of 2009* (testimony of Forrey), 4.

6. Office of Inspector General, "Long Term Success of ATSAP Will Require Improvements in Oversight, Accountability and Transparency," Federal Aviation Administration, Department of Transportation, Report Number AV-2012–152, July 19, 2012.

7. In 2000, the tower had experimented with a second Ground Controller, but had not adopted it. See chapter 6.

8. See Scott Snook, *Friendly Fire: The Accidental Shootdown of U.S. Blackhawks over Northern Iraq* (Princeton, NJ: Princeton University Press, 2002); Sidney Dekker, *Drift into Failure: From Hunting Broken Components to Understanding Complex Systems* (London: CRC Press, 2016); Jez Pigden, "Just Culture" (paper presented at the International Federation of Air Traffic Controllers' Associations 50th Annual Conference, Amman, Jordan, April 11–15, 2011), 1–16.

9. "NATCA Members Honored for Successful Team Effort with FAA to Realign K90 with A90," NATCA, June 7, 2018, https://www.natca.org/2018/06/07/several-natca-members-honored-for-successful-team-effort-with-faa-to-realign-k90-into-a90/.

10. U.S. Department of Transportation, Federal Aviation Administration, Section 804 Collaborative Workgroup, *FAA National Facilities Realignment and Consolidation Report, Year 1, Part 1 Recommendations*, Response to U. S. Congress FAA Reauthorization Bill, Pub. L. 112-095, Section 804 (March 11, 2015).

11. The strip camera conformed to the rule that mandated rolling nonverbal information on departures for TRACONs.

12. Michèle Lamont and Viràg Molnàr, "The Study of Boundaries in the Social Sciences," *Annual Review of Sociology* 28, no. 1 (2002): 167–95.

13. FAA Modernization and Reform Act of 2012, Pub. L. No. 112-95, 126 Stat. 11 (2012), section 804.

14. The Section 804 Collaborative Workgroup was comprised of representatives from the FAA, NATCA, and the Professional Aviation Safety Specialists (PASS) labor union, who initially analyzed TRACON facilities that would potentially be aligned, organized, and planned and implemented project management. New England regional and district officials, both FAA and NATCA, were included at both the national and local levels. Local decisions, project management, and implementation involved many regional and local NATCA and FAA officials representing Boston Tower, Boston Center, and the TRACON. Electronics engineers from Engineers New England, were an essential part of this core group. Equally part of the core group were the TRACON ATM, Ops Managers, supervisors, and other NATCA controllers—both Manchester, Boston, and Cape—were part of implementation and transition. "NATCA Members Honored."

15. John Van Maanen and Stephen R. Barley, "Cultural Organization: Fragments of a Theory," in *Organization Culture*, ed. Peter J. Frost et al. (Beverly Hills, CA: Sage, 1985), 31–54.

16. See chapter 6, "Boundary Work: The Routine Drama of the Runway Change."

17. "NATCA Members Honored."

Chapter Thirteen

1. Andrew Abbott, *Time Matters: On Theory and Method* (Chicago: University of Chicago Press, 2001), 181–82.

2. Daniel Hirschman and Isaac Ariail Reed, "Formation Stories and Causality in Sociology," *Sociological Theory* 32, no. 4 (2014): 259–82.

3. Hirschman and Reed, "Formation Stories"; Thomas P. Hughes, "The Electrification of America: The System Builders," *Technology and Cultures* 20, no. 1 (1979): 124–61; Bruno Latour, *Reassembling the Social: An Introduction to Actor-Network Theory* (Oxford: Oxford University Press, 2005).

4. Hirschman and Reed, "Formation Stories."

5. Hirschman and Reed, "Formation Stories," 274.

6. Andrew Abbott, "Conceptions of Time and Events in Social Science Methods," in *Time Matters: On Theory and Method* (Chicago: University of Chicago Press, 2001), 181–82.

7. Paul J. DiMaggio, "Interest and Agency in Institutional Theory," in *Institutional Patterns and Organizations: Culture and Environment*, ed. Lynn G. Zucker (Cambridge, MA: Ballinger Publishing, 1988), 3–21; Neil Fligstein and Doug McAdam, "Toward a General Theory of Strategic Action Fields," *Sociological Theory* 29, no. 1 (2011): 1–26.

8. Andrew Abbott, *Processual Sociology* (Chicago: University of Chicago Press, 2016).

9. This pattern conforms to the incremental process of change described by Abbott in "On the Concept of Turning Point," in *Time Matters*, 250.

10. Like Haydu, I use time periods as cases but go farther by comparing facilities within and across these periods. Jeffrey Haydu, "Making Use of the Past: Time Periods as Cases to Compare and as Sequences of Problem Solving," *American Journal of Sociology* 104, no. 2 (1998): 339–71.

11. On institutional interventions that disrupt the status quo of organizations in their field, see DiMaggio, "Interest and Agency."

12. Karl E. Weick, "The Collapse of Sensemaking in Organizations: The Main Gulch Disaster," *Administrative Science Quarterly* 38 (1993): 628–52; Karl E. Weick and Karlene H Roberts, "Heedful Interrelating on Flight Decks," *Administrative Science Quarterly* 38 (1993): 357–81.

13. Kathleen Tierney, *The Social Roots of Risk: Producing Disasters, Promoting Resilience* (Stanford, CA: Stanford University Press, 2014), 6. I have added "disruption" to her definition.

14. Andrew Abbott, "Things of Boundaries," *Social Research* 62, no. 4 (1995): 857–82.

15. Karl E. Weick, "Education Organizations as Loosely Coupled Systems," *Administrative Science Quarterly* 21, no. 1 (1976): 1–19.

16. Pierre Bourdieu, *Homo Academicus*, trans. Peter Collier (Stanford, CA: Stanford University Press, 1988).

17. On using case study data to reveal the connection between macro-, meso-, and micro-level factors in the formation and shaping of *habitus*, see Diane Vaughan, "Bourdieu and Organizations: The Empirical Challenge," *Theory and Society* 7, no. 1 (2008): 37–65.

18. Paul J. DiMaggio, "Culture and Cognition," *Annual Review of Sociology* 23 (1997): 263–87. Dreyfus can be a read as a micro-level supplement to DiMaggio's perspective. adding three important elements in human intelligence and action: the role of the body, the role of the situation, and the role of human purposes and needs. Hubert L. Dreyfus, *What Computers Still Can't Do: A Critique of Artificial Reason* (Cambridge: MIT Press, 1992), 235.

19. See also Phaedra Diapha, *Masters of Uncertainty: Weather Forecasters and the Quest for Ground Truth* (Chicago: University of Chicago Press, 2015).

20. Phaedra Daipha, "Visual Perception at Work: Lessons from the World of Meteorology," *Poetics* 38 (2010): 150–64.

21. Lucy A. Suchman, "Representing Practice in Cognitive Science," *Human Studies* 11 (1988): 305–25.

22. For complete list and full quotes, see page 529.

23. Michèle Lamont and Viràg Molnàr, "The Study of Boundaries in the Social Sciences," *Annual Review of Sociology* 28, no. 1 (2002): 167–95.

24. For list of experts, see chapter 12, note 15.

25. Iddo Tavory and Nina Eliasoph, "Coordinating Futures: Toward a Theory of Anticipation," *American Journal of Sociology* 118, no. 4 (2013): 908–42.

26. Sarah Kaplan and Wanda J. Orlikowski, "Temporal Work in Strategy Making," *Organization Science* 24, no. 4 (2013): 965–95.

27. Wanda J. Orlikowski and JoAnne Yates, "It's about Time: Temporal Structuring in Organizations," *Organization Science* 13, no. 6 (2002): 684–700.

28. Paul J. DiMaggio and Walter W. Powell, "The Iron Cage Revisited: Institutional Isomorphism and Collective Rationality in Organization Fields," *American Sociological Review* 48, no. 2 (1983): 147–60.

29. Barry Turner, *Man-Made Disasters* (London: Wykeham, 1978).

30. On how mismatched trajectories and plans are put into play at the individual level, see Tavory and Eliasoph, "Coordinating Futures," 937.

31. Diane Vaughan, "Structural Secrecy," in *The Challenger Launch Decision: Risky Technology, Culture, and Deviance at NASA* (Chicago: University of Chicago Press, 1996), 238–77.

32. Jeffrey L. Pressman and Aaron Wildavsky, *Implementation: How Great Expectations in Washington Are Dashed in Oakland, Or, Why It's Amazing That Federal Programs Work at All* (Berkeley: University of California Press, 1984).

33. Austin Kinley, "Race and Gender Changes in Air Traffic Controller Selection, Hiring, Attrition & Success 1940–2015" (master's thesis, Southern Illinois University, 2016), 62–66.

34. Sidney Dekker, *Drift into Failure: From Hunting Broken Components to Understanding Complex Systems* (London: CRC Press, 2016); Neil Fligstein, Jonah Stuart Brundage, and Michael Schultz, "Seeing Like the Fed: Culture, Cognition, and Framing in the Failure to Anticipate the Financial Crisis of 2008," *American Sociological Review* 82, no. 5 (2017): 879–909.

35. For an exceptional review essay on four recent books, see Judith Wajcman, "Automation: Is It Really Different This Time?," *British Journal of Sociology* 68, no. 3 (2003): 119–27.

36. For an exception, see Karin Knorr Cetina and Urs Bruegger, "Global Microstructures: The Virtual Societies of Financial Markets," *American Journal of Sociology* 107, no.4 (2002): 905–950.

37. See Vaughan, *The* Challenger *Launch Decision*; John Urry, *What Is the Future?* (Cambridge: Polity Press, 2016).

38. Nick Pidgeon and Mike O'Leary, "Man-Made Disasters: Why Technology and Organizations (Sometimes) Fail," *Safety Science* 34 (2000): 15–30; Bridget Hutter, ed., "Wrestling with Complexity," special issue, *Risk and Regulation* 18 (2001): 1–3.

39. Judith Wajcman, *Pressed for Time: The Acceleration of Life in Digital Capitalism* (Chi-

cago: University of Chicago Press, 2015); Evgeny Morozov, *To Save Everything, Click Here: The Folly of Technological Solutionism* (New York: Public Affairs, 2013).

40. Gil Eyal, "For a Sociology of Expertise: The Social Origins of the Autism Epidemic," *American Journal of Sociology* 118, no. 4 (2013): 871.

41. Harry Collins and Robert Evans, *Rethinking Expertise* (Chicago IL: University of Chicago Press, 2007).

42. Joshua Hurwitz, "Ghosts and the Machine: Dynamic Adjustments to Automation, the Reconstruction of Worker Meaning, and Its Differential Impacts," unpublished paper, Stanford Graduate School of Business, 2019; S. E. Sachs, "The Algorithm at Work? Explanation and Repair in the Enactment of Similarity in Art Data," *Information, Communication & Society* (2019): https://www.tandfonline.com/doi/full/10.1080/1369118X.2019.1612933.

43. Michael Barrett, Eivor Obom, Wanda Orlikowski, and JoAnne Yates, "Refiguring Boundary Relations: Robotic Innovations in Pharmacy Work," *Organization Science* 23, no. 3 (2012): 448–66; Katherine C. Kellogg, Melissa A. Valentine, and Angele Christin, "Algorithms at Work: The New Contested Terrain of Control," *Academy of Management Annals* 14, no. 1 (2020): 366–410; Ari H. Galper and Katherine C. Kellogg, "Accomodating-through-Bypassing: Implementing Algorithmic Technology in the Face of Professional Resistance," unpublished paper, MIT Sloan School of Management, 2020.

Bibliography

Abbott, Andrew. "From Causes to Events." *Sociological Methods and Research* 20 (1992): 428–55.

———. "Linked Ecologies: States and Universities as Environments for Professions." *Sociological Theory* 23, no. 3 (2005): 245–74.

———. *Processual Sociology*. Chicago: University of Chicago Press, 2016.

———. "Sequences of Social Events: Concepts and Methods for the Analysis of Order." *Historical Methods* 16, no. 4 (1983): 129–47.

———. "Status and Status Strain in the Professions." *American Journal of Sociology* 86, no. 4 (1981): 819–35.

———. *The System of Professions: An Essay on the Division of Expert Labor*. Chicago: University of Chicago Press, 1988.

———. "Things of Boundaries." *Social Research* 62, no. 4 (1995): 857–82.

———. *Time Matters: On Theory and Method*. Chicago: University of Chicago Press, 2001.

———. "Transcending General Linear Reality." *Sociological Theory* 6, no. 2 (1988): 169–86.

Aldrich, Howard, and Martin Ruef. *Organizations Evolving*. 2nd ed. Thousand Oaks, CA: Sage Publications, 2006.

Aminzade, Ronald. "Historical Sociology and Time." *Sociological Methods and Research* 20 (1992): 456–80.

Baker, Al, and Matthew Wald. "FAA Considers Consolidating Regional Air Traffic Control." *New York Times*, March 17, 2001. http:/www.nytimes.com/2001/03/17/nyregion/faa-considers-consolidating-regional-air-traffic-control.html?pagewanted-print&src=pm.

Barley, Stephen R. "Technology as an Occasion for Structuring: Evidence from Observations of CT Scanners and the Social Order of Radiology Departments." *Administrative Science Quarterly* 31, no. 1 (1986): 78–108.

Barley, Stephen R., and Pamela S. Tolbert. "Institutionalization and Structuration: Studying the Links between Action and Institution." *Organization Studies* 18, no. 1 (1997): 93–117.

Bechky, Beth. "Object Lessons: Workplace Artifacts as Representations of Occupational Jurisdiction." *American Journal of Sociology* 109, no. 3 (2003): 720–52.

———. "Shared Meaning across Occupational Communities: The Transformation of Understanding on a Production Floor." *Organization Science* 14, no. 3 (2003): 312–30.

Becker, Howard S. "Becoming a Marihuana User." *American Journal of Sociology* 59, no. 3 (1953): 235–42.

———. "Notes on the Concept of Commitment." *American Journal of Sociology* 66 (1960): 32–42.

Benner, Patricia. *From Novice to Expert: Excellence and Power in Clinical Nursing Practice*. Reading, MA: Addison-Wesley, 1984.

Bensman, Joseph, and Robert Lilienfeld. *Craft and Consciousness: Occupational Technique and the Development of World Images*. 2nd ed. New York: Aldine de Gruyter, 1991.

Bieberich, Yovanna. "Remember a Historic Flight: Fred Wiseman Piloted the First Airmail Flight from Petaluma to Santa Rosa." *Petaluma Argus-Courier*, February 16, 2011. http://www.petaluma360.com/article/20110216/COMMUNITY/110219562?template=printpicart.

Bijker, Wiebe E. *Of Bicycles, Bakelite, and Bulbs: Toward a Theory of Technological Change*. Cambridge, MA: MIT Press, 1995.

Bijker, Wiebe E., Thomas P. Hughes, and Trevor Pinch, eds. *The Social Construction of Technical Systems: New Directions in the Sociology and History of Technology*. Cambridge, MA: MIT Press, 1987.

Bolsom, Laura M. "How Pan Am Extended Aura of Jet Travel to All." *New York Times*, National, October 27, 2019, A25.

Black, Donald J., and Albert J. Reiss Jr. "Patterns of Behavior in Police and Citizen Transactions." *Studies in Crime and Law Enforcement in Major Metropolitan Areas* 2 (1967): 1–139.

Boorman, Scott A. Review of *The Emergence of Organizations and Markets*, by John F. Padgett and Walter W. Powell. *Acta Sociologica* 57, no. 4 (2014): 363–67.

Bourdieu, Pierre. *Distinction: A Social Critique of the Judgment of Taste*. Translated by Richard Nice. Cambridge, MA: Harvard University Press, 1984.

———. *Homo academicus*. Translated by Peter Collier. Stanford, CA: Stanford University Press, 1988.

———. *Pascalian Meditations*. Translated by Richard Nice. Stanford, CA: Stanford University Press, 2000.

Bourrier, Mathilde. "The Legacy of the Theory of High Reliability Organizations: An Ethnographic Endeavor." *Journal of Contingencies and Crisis Management* 19, no. 1 (2011): 9–13.

Bowker, Geoffrey C. "How Things Change: The History of Sociotechnical Structures." *Social Studies of Science* 28 (1996): 173–82.

Bowker, Geoffrey C., and Susan Leigh Star. "Building Information Structures for

Social Worlds: The Role of Classifications and Standards." In *Community Computing and Support System: Social Interaction in Networked Communities*, edited by Toru Ishida, 231–48. Berlin: Springer Verlag, 1998.

———. *Sorting Things Out: Classification and Its Consequences.* Cambridge, MA: MIT Press, 1999.

Breuckner, Jan K., and Vivek Pai. "Technological Innovation in the Airline Industry: The Impact of Regional Jets." *International Journal of Industrial Organization* 27, no. 1 (2009): 110–20.

Burawoy, Michael. *Manufacturing Consent.* Chicago: University of Chicago Press, 1979.

———. "Revisits: An Outline of a Theory of Reflexive Ethnography." *American Sociological Review* 63, no. 5 (2003): 645–79.

Button, Graham, ed. *Technology in Working Order: Studies of Work, Interaction, and Technology.* London: Routledge, 1993.

Callon, Michel, and Bruno Latour. "Unscrewing the Big Leviathan; or, How Actors Macrostructure Reality and How Sociologists Help Them to Do So." In *Advances in Social Theory and Methodology*, edited by Aaron Cicourel and Karin Knorr, 277–303. London: Routledge, 1981.

Catino, Maurizio, and Gerardo Patriotta. "Learning from Errors: Cognition, Emotions, and Safety Culture in the Italian Air Force." *Organization Studies* 34 (2013): 437–67.

Cetina, Karin Knorr. *Epistemic Cultures: How the Sciences Make Knowledge.* Cambridge, MA: Harvard University Press, 1999.

Chambliss, Daniel F. "The Mundanity of Excellence: An Ethnographic Report on Stratification and Olympic Swimmers." *Sociological Theory* 7, no. 1 (1989): 70–86.

Chanute, Octave. *Progress in Flying Machines.* New York: American Engineer and Railroad Journal, 1894. Reprint, Long Beach, CA: Lorenz and Herwig, 1976.

Clemens, Elisabeth S. "Organizational Repertoires and Institutional Change: Women's Groups and the Transformation of U.S. Politics, 1890–1920." *American Journal of Sociology* 98, no. 4 (1993): 755–98.

Clinton, William J. Executive Order No. 13,180 (Dec. 7, 2000). "Air Traffic Performance-Based Organization." Accessed at Gerhard Peters and John T. Woolley, The American Presidency Project, https://www.presidency.ucsb.edu/documents/executive-order-13180-air-traffic-performance-based-organization.

Collins, Harry. "Bicycling on the Moon: Collective Tacit Knowledge and Somatic-Limit Tacit Knowledge." *Organization Studies* 28 (2007): 257–62.

———. *Changing Order: Replication and Induction in Scientific Practice.* Chicago: University of Chicago Press, 1992.

———. *Tacit and Explicit Knowledge.* Chicago: University of Chicago Press, 2010.

———. "Three Dimensions of Expertise." *Phenomenology and the Cognitive Sciences* 12, no. 2 (2013): 253–73.

Collins, H. M., and Robert Evans. "The Third Wave of Science Studies: Studies of Expertise and Experience." *Social Studies of Science* 32, no. 2 (2002): 235–96.

———. *Rethinking Expertise*. Chicago: University of Chicago Press, 2007.

Conlon, Emerson W. "History of the University of Michigan Aeronautical/ Aerospace Engineering." Faculty Memoir Project, University of Michigan, 1817–2017. Accessed May 6, 2013. http://um2017.org/2017_Website/History_of_Aeronautical_Engineering.html.

"The Construction of the Wright Aeroplane." *Scientific American* 99, no. 13 (September 26, 1908): 208–10.

Coopmans, Catelijne, Janet Vertesi, Michael Lynch, and Steve Woolgar, eds. *Representation in Scientific Practice Revisited*. Cambridge, MA: MIT Press, 2014.

Cowan, Ruth Schwartz. "How the Refrigerator Got Its Hum." In *The Social Shaping of Technology: How the Refrigerator Got Its Hum*, edited by Judy Wajcman and Donald A. MacKenzie. London: Open University Press, 1988.

———. *A Social History of American Technology*. New York: Oxford University Press, 1997.

Crouch, Tom D. *The Bishop's Boys: A Life of Wilbur and Orville Wright*. New York: W. W. Norton, 2003.

Crouch, Tom D., and Peter L. Jakab. *The Wright Brothers and the Invention of the Aerial Age*. Washington, DC: National Geographic, Smithsonian National Air and Space Museum, 2003.

C-SPAN. "FAA Emergency Response on September 11." Aired August 12, 2002, on C-SPAN 1. www.c-spanvideo.org/program/FAAE.

Dailey, Franklyn E., Jr. *The Triumph of Instrument Flight: A Retrospective in the Century of Aviation*. Wilbraham, MA: Dailey International Publishers, 2004.

Dateline NBC. "America Remembers: Air Traffic Controllers Describe How Events Unfolded as They Saw Them on September 11th." Season 10, episode 132. Aired September 11, 2002, on NBC.

Dekker, Sidney. *Drift into Failure: From Hunting Broken Components to Understanding Complex Systems*. London: CRC Press, 2016.

Desmond, Matthew. *On the Fireline: Living and Dying with Wildland Firefighters*. Chicago: University of Chicago Press, 2007.

Diapha, Phaedra. *Masters of Uncertainty: Weather Forecasters and the Quest for Ground Truth*. Chicago: University of Chicago Press, 2015.

———. "Visual Perception at Work: Lessons from the World of Meteorology." *Poetics* 38 (2010): 150–64.

DiMaggio, Paul J. "Culture and Cognition." *Annual Review of Sociology* 23 (1997): 263–87.

———. "Interest and Agency in Institutional Theory." In *Institutional Patterns and Organizations: Culture and Environment*, edited by Lynne G. Zucker, 3–21. Cambridge, MA: Ballinger Publishing, 1988.

DiMaggio, Paul J., and Walter W. Powell. "The Iron Cage Revisited: Institutional

Isomorphism and Collective Rationality in Organizational Fields." *American Sociological Review* 48, no. 2 (1983): 147–60.

Dobbin, Frank. *Forging Industrial Policy: The United States, Britain, and France in the Railway Age*. Cambridge: Cambridge University Press, 1994

———. *Inventing Equal Opportunity*. Princeton, NJ: Princeton University Press, 2009.

Dreyfus, Hubert L. *What Computers Still Can't Do*. Cambridge, MA: MIT Press, 1991.

Dreyfus, Hubert L., and Stuart E. Dreyfus. "The Challenge of Merleau-Ponty's Phenomenology of Embodiment for Cognitive Science." In *Perspectives on Embodiment: The Intersections of Nature and Culture*, edited by Gail Weiss and Honi Fern Haber, 103–20. New York: Routledge, 1999.

———. *Mind over Machine: The Power of Human Intuition and Expertise in the Era of the Computer*. New York: Free Press, 1986.

———. "Peripheral Vision: Expertise in Real World Contexts." *Organization Studies* 26, no. 5 (2005): 779–92.

Dreyfus, Stuart E., and Hubert L. Dreyfus. *A Five-Stage Model of the Mental Activities Involved in Directed Skill Acquisition*. No. ORC-80-2. University of California–Berkeley Operations Research Center, 1980.

Elias, Bart. "Air Traffic Inc.: Considerations Regarding the Corporatization of Air Traffic Control." Congressional Research Service 7-5700, R43844, January 5, 2015; May 16, 2017.

Elias, Bart, Clinton T. Brass, and Robert S. Kirk. "Sequestration at the Federal Aviation Administrations: Air Traffic Controller Furloughs and Congressional Response." Congressional Research Service 7-5700, May 7, 2013.

Emigh, Rebecca. "The Power of Negative Thinking: The Use of Negative Case Methodology in the Development of Sociological Theory." *Theory and Society* 26 (1977): 649–84.

Erikson, Kai T. *Everything in Its Path: Destruction of Community in the Buffalo Creek Flood*. New York: Simon and Schuster, 1976.

Ermakoff, Ivan. "The Structure of Contingency." *American Journal of Sociology* 121, no. 1 (2015): 64–125.

Eyal, Gil. "For a Sociology of Expertise: The Social Origins of the Autism Epidemic." *American Journal of Sociology* 118, no. 4 (2013): 863–907.

———. *The Crisis of Expertise*. Cambridge, UK: Polity Press, 2019.

The FAA Reauthorization Act of 2009: Testimony before the Senate Committee on Commerce, Science, and Transportation, Subcommittee on Aviation Operations, Safety, and Security. 111th Cong. 1–11 (2009). Testimony of Patrick Forrey, President of National Air Traffic Controllers Association.

Federal Aviation Administration. *A Plan for the Future: 10 Year Strategy for the Air Traffic Control Workforce 2012–2021*. Washington, DC: Federal Aviation Administration, US Department of Transportation, 2012.

———. 1926–1996. *FAA Historical Chronology: Civil Aviation and the Federal Government.* Washington, DC: Federal Aviation Administration, U.S. Department of Transportation, 1998.

———. 1997–2014 Update to *FAA Historical Chronology: Civil Aviation and the Federal Government,* Washington, DC: Federal Aviation Administration, 2016.

Final Report of the Performance-Based Operations Aviation Rulemaking Committee/Commercial Aviation Safety Team, Flight Deck Automation Working Group. "Operational Use of Flight Path Management Systems." September 9, 2013.

Fine, Gary Alan. "Peopled Ethnography: Developing a Theory of Group Life." *Ethnography* 4, no. 41 (2003): 41–60.

"The First Flight of the Wright Aeroplane at Fort Myers." *Scientific American* 99, no. 11 (September 12, 1908): 169.

Fligstein, Neil, and Doug McAdam. "Toward a General Theory of Strategic Action Fields." *Sociological Theory* 29, no. 1 (2011): 1–26.

Fligstein, Neil, Jonah Stuart Brundage, and Michael Schultz. "Seeing Like the Fed: Culture, Cognition, and Framing in the Failure to Anticipate the Financial Crisis of 2008." *American Sociological Review* 82, no. 5 (2017): 879–909.

Forrey, Patrick. "The View from the Air Traffic Control Tower and Scopes: How to Turn the Problems of NowGen into the Hope of NextGen." Address given to the Aero Club of Washington, March 26, 2008. http://aireform.com/resources/how-to-search-for-downloadable-reference-materials/news-archive/news-clips-folder/newsclip-2008-03-26-natca-president-patrick-forrey-delivers-speech-to-aero-club-of-washington/.

Fosher, Kerry B. *Under Construction: Making Homeland Security at the Local Level.* Chicago: University of Chicago Press, 2009.

Friedland, Roger, and Robert R. Alford. "Bringing Society Back In: Symbols, Practices, and Institutional Contradictions." In *The New Institutionalism in Organizational Analysis,* edited by Walter W. Powell and Paul J. DiMaggio, 232–63. Chicago: University of Chicago Press, 1991.

Galison, Peter. *Image and Logic: A Material Culture of Microphysics.* Chicago: University of Chicago Press, 1997.

Garfinkel, Harold. "Conditions of Successful Degradation Ceremonies." *American Journal of Sociology* 61, no. 5 (1956): 420–24.

Gawande, Atul. *The Checklist Manifesto: How to Get Things Right.* New York: Henry Holt, 2009.

Geertz, Clifford. *The Interpretation of Cultures.* New York: Basic Books, 1973.

———. *Local Knowledge: Further Essays in Interpretive Anthropology.* New York: Basic Books, 1983.

Gieryn, Thomas F. "Boundary-Work and the Demarcation of Science from Non-science: Strains and Interests in Professional Ideologies of Scientists." *American Sociological Review* 48 (1983): 781–95.

———. *Cultural Boundaries of Science: Credibility on the Line*. Chicago: University of Chicago Press, 1999.

———. "A Space for Place in Sociology." *Annual Review of Sociology* (2000): 463–96.

———. "What Buildings Do." *Theory and Society* 31, no. 1 (2002): 35–74.

Gilbert, Glen A. *Air Traffic Control: The Uncrowded Sky*. Washington, DC: Smithsonian Institution Press, 1973.

———. "Historical Development of the Air Traffic Control System." *IEEE Transactions on Communications* 21, no. 5 (1973): 364–75.

Gladwell, Malcolm. "The Social Life of Paper: Looking for Method in the Mess." *New Yorker*, March 25, 2002.

Goffman, Erving. *Asylums: Essays on the Social Situation of Mental Patients and Other Inmates*. New York: Anchor Books, 1961.

———. *The Presentation of Self in Everyday Life*. New York: Anchor Books, 1959.

———. *Stigma: Notes on the Management of Spoiled Identity*. Englewood Cliffs, NJ: Prentice Hall, 1963.

Granovetter, Mark. "The Strength of Weak Ties." *American Journal of Sociology* 78, no. 6 (1973): 1360–80.

Gras, Alain. *Grandeur et dépendance: Sociologies des macro-systèmes techniques*. With the participation of Sophie Delpech. Paris: Presses universitaires de France, 1993.

Gras, Alain, Caroline Moricot, Sophie L. Poirot-Delpech, and Victor Scardigli. *Faced with Automation: The Pilot, the Controller, and the Engineer*. Translated by Jill Lundsten. Paris: Publications de la Sorbonne, 1994.

Greenberg, David. "Memo to Obama Fans: Clinton's Presidency Was Not a Failure." *Slate*, February 12, 2008. https://www.slate.com/id/2183941/pagenum/all/#page_start/.

Griffin, Larry J. "Temporality, Events, and Explanation in Historical Sociology: An Introduction." *Sociological Methods and Research* 20, no. 4 (1992): 403–27.

Gross, Neil. "The Structure of Causal Chains." *Sociological Theory* 36, no. 4 (2018): 343–67.

Hareli, Shlomo, Noga Short, and Nicole Bigger. "The Role of Emotions in Employees' Explanations for Failure in the Workplace." *Journal of Managerial Psychology* 20 (2005): 152–62.

Haydu, Jeffrey. "Making Use of the Past: Time Periods as Cases to Compare and as Sequences of Problem Solving." *American Journal of Sociology* 104, no. 2 (1998): 339–71.

Heath, Christian, and Paul Luff. "Collaborative Activity and Technological Design: Task Coordination in the London Underground Control Rooms." In *Proceedings of ECCSCW'91, The European Conference on Computer-Supported Cooperative Work*, 65–80. Amsterdam: Kluwer Academic Press, 1991.

———. "Collaboration and Control: Crisis Management and Multimedia Technology in London Underground Line Control Rooms." *Computer Supported Cooperative Work (CSCW)* 1 (1992): 69–94.

Heppenheimer, T. A. *Turbulent Skies: The History of Commercial Aviation*. New York: John Wiley and Sons, 1995.

Hirschman, Daniel, and Isaac Ariail Reed. "Formation Stories and Causality in Sociology." *Sociological Theory* 32, no. 4 (2014): 259–82.

History Commons. "Complete 9/11 Timeline." http://www.historycommons.org/project.jsp?project=911_project.

Hochschild, Arlie Russell. "Emotion Work, Feeling Rules, and Social Structure." *American Journal of Sociology* 85, no. 3 (1979): 551–75.

———. *The Managed Heart*. Berkeley: University of California Press, 1983.

Hollan, James, Edwin Hutchins, and David Kirsh. "Distributed Cognition: Toward a New Foundation for Human-Computer Interaction Research." *ACM Transactions on Computer-Human Interaction (TOCHI)* 7 (2000): 174–96.

Hughes, Thomas P. *American Genesis: A Century of Invention and Technological Enthusiasm, 1870–1970*. Chicago: University of Chicago Press, 1989.

———. "The Electrification of America: The System Builders." *Technology and Cultures* 20, no. 1 (1979): 124–61.

———. *Networks of Power: Electrification in Western Society, 1880–1930*. 1983. Reprint, Baltimore: Johns Hopkins University Press, 1993.

Hurwitz, Joshua. "Ghosts in the Machine: Dynamic Adjustments to Automation, the Reconstruction of Worker Meaning, and its Differential Impacts." Unpublished paper, Stanford Graduate School of Business, Stanford, CA, 2019.

Hutchins, Edwin A. *Cognition in the Wild*. Cambridge, MA: MIT Press, 1995.

———. "Cognitive Ecology." *Topics in Cognitive Science* 2, no. 4 (2010): 705–15.

———. "Enaction, Imagination, and Insight." In *Enaction: Towards a New Paradigm for Cognitive Science*, edited by John Stewart, Oliver Gapenne, and Ezequiel A. Di'Paolo, 425–50. Cambridge, MA: MIT Press, 2010.

———. "How a Cockpit Remembers Its Speeds." *Cognitive Science* 19, no. 3 (1995): 265–88.

Hutter, Bridget. "Ways of Seeing: Understandings of Risk in Organizational Settings." In *Organizational Encounters with Risk*, edited by Bridget Hutter and Michael Power, 66–91. Cambridge: Cambridge University Press, 2005.

Hutter, Bridget. "Wrestling with Complexity." Special issue, *Risk and Regulation* 18 (2001): 1–3.

Jakab, Peter L. *Visions of a Flying Machine: The Wright Brothers and the Process of Invention*. Washington, DC: Smithsonian Institution Press, 1990.

Jerolmack, Colin, and Shamus Khan. "Talk Is Cheap: Ethnography and the Attitudinal Fallacy." *Sociological Methods and Research* 43, no. 2 (2014): 178–209.

Jervis, Robert. *System Effects: Complexity in Political and Social Life*. Princeton, NJ: Princeton University Press, 1997.

Jessen, Gene Nora. "1929 Air Race." In "The 99s in Aviation History: International

Women Pilots," special issue, *99 News Magazine*. https://www.ninety-nines.org/the-1929-air-race.htm.

Kaplan, Sarah, and Wanda J. Orlikowsk. "Temporal Work in Strategy Making." *Organization Science* 24, no. 4 (2013): 965–95.

Katz, Jack. *How Emotions Work*. Chicago: University of Chicago Press, 1999.

Kelly, Fred C. "After Kitty Hawk: A Brief Resume." Appendix to *How We Invented the Airplane: An Illustrated History*, by Orville Wright, edited by Fred C. Kelly, 51–56. New York: Dover Publications, 1988.

———. *The Wright Brothers: A Biography Authorized by Wilbur Wright*. New York: Harcourt, Brace, 1943.

Kahn, Albert E. "The Surprises of Airline Deregulation." *American Economic Review* 78, no. 2 (1988): 316–22.

Khan, Shamus Rahman. *Privilege*. Princeton, NJ: Princeton University Press, 2011.

Kinley, Austin. "Race and Gender Changes in Air Traffic Controller Selection, Hiring, Attrition & Success, 1940–2015." Master's thesis, Southern Illinois University, 2016.

Kluttz, Daniel, and Neil Fligstein. "Varieties of Field Theory." In *Handbook of Contemporary Sociological Theory*, ed. Seth Abrutyn, 185–204. New York: Springer, 2016.

Knorr Cetina, Karin, and Urs Bruegger, "Global Microstructures: The Virtual Societies of Financial Markets." *American Journal of Sociology* 107, no. 4 (2002): 905–50.

Lamont, Michèle. *The Dignity of Working Men: Morality and the Boundaries of Race, Class, and Immigration*. New York: Russell Sage Foundation at Harvard University Press, 2000.

Lamont, Michèle, and Marcel Fournier, eds. *Cultivating Differences: Symbolic Boundaries and the Making of Inequality*. Chicago: University of Chicago Press, 1992.

Lamont, Michèle, and Viràg Molnàr. "The Study of Boundaries in the Social Sciences." *Annual Review of Sociology* 28, no. 1 (2002): 167–95.

Langewiesche, Wolfgang. *Stick and Rudder: An Explanation of the Art of Flying*. New York: McGraw-Hill, 1990.

LaPorte, Todd. "The U.S. Air Traffic System: Increasing Reliability in the Midst of Growth." In *The Development of Large Scale Technical Systems*, edited by Thomas Hughes and Rene Mayntz, 215–44. Boulder, CO: Westview Press, 1988.

Latour, Bruno. *The Pasteurization of France*. Cambridge, MA: Harvard University Press, 1993.

———. *Reassembling the Social: An Introduction to Actor-Network Theory*. Oxford: Oxford University Press, 2005.

———. *Science in Action: How to Follow Scientists and Engineers through Society*. Cambridge, MA: Harvard University Press, 1987.

Latour, Bruno, Michael Lynch, and Steve Woolgar. *Representation in Scientific Practice*. Cambridge, MA: MIT Press, 1990.

Law, John. "Technology and Heterogeneous Engineering: The Case of the Portuguese Expansion." In *The Social Construction of Technical Systems: New Directions in the Sociology and History of Technology*, edited by Wiebe E. Bijker, Thomas P. Hughes, and Trevor Pinch, 111–34. Cambridge, MA: MIT Press, 1987.

"Lessons of the Wright Aeroplane Disaster." *Scientific American* 99, no. 13 (September 26, 1908): 202.

Levin, Alan, Marilyn Adams, and Blake Morrison. "Four Hours of Fear: 9/11's Untold Story." *USA Today*, August 12–15, 2002.

Lewis, Sinclair. *The Trail of the Hawk: A Comedy on the Seriousness of Life*. New York: Harper and Brothers, 1915.

Lilienthal, Otto. *Birdflight as the Basis of Aviation: A Contribution towards a System of Aviation. Compiled from the Results of Numerous Experiments made by O. and G. Lilienthal*. Translated by A. W. Isenthal. New York: Longmans, Green, 1911.

Lindbergh, Charles A. *The Spirit of St. Louis*. New York: Scribner, 1953.

Luff, Paul, Jon Hindmarsh, and Christian Heath, eds. *Workplace Studies: Recovering Work Practice and Informing System Design*. Cambridge: Cambridge University Press, 2000.

Lynch, Michael, and Steve Woolgar, eds. *Representation in Scientific Practice*. Cambridge, MA: MIT Press, 1990.

Mackay, Wendy E. "Is Paper Safer? The Role of Paper Flight Strips in Air Traffic Control." *ACM Transactions on Computer-Human Interaction (TOCHI)* 6, no. 4 (1999): 311–40.

Mackay, Wendy E., Anne-Laure Fayard, Laurent Frobert, and Lionel Médini. "Reinventing the Familiar: Exploring an Augmented Reality Design Space for Air Traffic Control." In *Proceedings of ACM CHI '98 Human Factors in Computing Systems*. Los Angeles: ACM Press, 1998.

MacKenzie, Donald. *Inventing Accuracy: A Historical Sociology of Nuclear Missile Guidance*. Cambridge, MA: MIT Press, 1990.

MacKenzie, Donald, and Judy Wajcman, eds. *The Social Shaping of Technology*. 2nd ed. Philadelphia: Open University Press, 1999.

Mahoney, James. "Path Dependence in Historical Sociology." *Theory and Society* 29, no.4 (2000): 507–48.

Mahoney, James, and Kathleen Thelen, eds. *Explaining Institutional Change*. New York: Cambridge University Press, 2010.

Mansolas, Ioannis, and Angelos Mansolas. "A Short History of Air Traffic Control." Compiled in June 2005. http://imansolas.freeservers.com/ATC/short_history_of_the_air_traffic.html.

McCartin, Joseph A. *Collision Course: Ronald Reagan, the Air Traffic Controllers, and the Strike That Changed America*. New York: Oxford University Press, 2011.

McCullough, David. *The Wright Brothers*. New York: Simon and Schuster, 2015.

McElroy, Paul. *Against the Wind: The History of the National Air Traffic Controllers Association*. Washington, DC: National Air Traffic Controllers Association, 2002.

Mechanic, David. "Sources of Power of Lower Participants in Complex Organizations." *Administrative Science Quarterly* (1962): 349–64.

Meier, John W., and Brian Rowan. "Institutionalized Organizations: Formal Structure as Myth and Ceremony." *American Journal of Sociology* 83, no. 2 (1977): 340–63.

Meine, Karl-Heinz. "Aviation Cartography." *Cartographic Journal* 3, no.1 (1966): 31–40.

Merleau-Ponty, Maurice. *Phenomenology of Perception*. Translated by C. Smith. New York: Routledge, 1962.

Merton, Robert K. "The Unanticipated Consequences of Purposive Social Action." *American Sociological Review* 1, no. 6 (1936): 894–904.

Mische, Ann. "Projects and Possibilities: Researching Futures in Action." *Sociological Forum* 24, no. 3 (2009): 694–704.

Moedebeck, Hermann. *Pocket Book of Aeronautics*. Translated by W. Mansergh Varley. London: Whittaker, 1907.

———. "Ueber das Landen mit Ballons." *Zeitschrift fur Luftschiffahrt* 7 (1888): 272–76.

Morozov, Evgeny. *To Save Everything, Click Here: The Folly of Technological Solutionism*. New York: Public Affairs, 2013.

Narasimha, Roddam. "How Two Bicycle Mechanics Achieved the World's First Powered Flight." *Resonance* 8, no. 12 (2003): 61–74.

National Air Traffic Controllers Association (NATCA). "NATCA Members Honored for Successful Team Effort with FAA to Realign K90 with A90." June 7, 2018. https://www.natca.org/2018/06/07/several-natca-members-honored-for-successful-team-effort-with-faa-to-realign-k90-into-a90/.

National Commission on the Terrorist Attacks. *The 9/11 Commission Report: Final Report of the National Commission on the Terrorist Attacks upon the United States*. New York: W. W. Norton, 2004.

National Transportation Safety Board. *Flight Path Study and Radio Communications with FAA*. Washington, DC: Office of Research and Engineering, February 19, 2002.

Naudet, Jules, Gédéon Naudet, and James Hanlon, dirs. Hosted by Robert DeNiro. *9/11*. Aired March 10, 2002, on CBS.

Newman, Katherine S. *Falling from Grace: Downward Mobility in the Age of Affluence*. New York: Free Press, 1988. Reprint, Berkeley: University of California Press, 1999.

Norton, Matthew. "Cultural Sociology Meets the Cognitive Wild: Advantages of the Distributed Cognition Framework for Analyzing the Intersection of Culture and Cognition." *American Journal of Cultural Sociology* 8, no. 45 (2020): 45–62. https://doi.org/10.1057/s41290-019-00075-w.

Novak, Matt. "The 'Highway of Light' That Guided Early Planes across America." *Paleofuture*, November 18, 2013. http://paleofuture.gizmodo.com/the-highway-of-light-that-guided-early-planes-across-1466696698.

Office of Inspector General, Office of the Secretary of Transportation, US Department of Transportation. "Long Term Success of ATSAP Will Require Improvements in Oversight, Accountability and Transparency." Federal Aviation Administration, Department of Transportation, Report Number AV-2012–152, July 19, 2012.

———. Audit Report. "While FAA Took Steps Intended to Improve Its Controller Hiring Process, the Agency Did Not Effectively Implement Its New Policies." Federal Aviation Administration, Department of Transportation, Report No. AV2017028, February 15, 2017, 1–16.

Orlikowski, Wanda J. "The Duality of Technology: Rethinking the Concept of Technology in Organizations." *Organization Science* 3, no. 3 (1992): 398–427.

———. "Improvising Organizational Transformation over Time: A Situated Change Perspective." *Information Systems Research* 7, no. 1 (1996): 63–92.

Orlikowski, Wanda J., and JoAnne Yates. "It's about Time: Temporal Structuring in Organizations." *Organization Science* 13, no. 6 (2002): 684–700.

Pachucki, Mark A., Sabrina Pendergrass, and Michèle Lamont. "Boundary Processes: Recent Theoretical Developments and New Contributions." *Poetics* 35, no. 6 (2007): 331–51.

Padgett, John F., and Walter W. Powell, eds. *The Emergence of Organizations and Markets*. Princeton, NJ: Princeton University Press, 2012.

Perrow, Charles B. *Normal Accidents: Living with High-Risk Technologies*. New York: Basic Books, 1984.

———. *Normal Accidents: Living with High Risk Technologies*. 2nd ed. Princeton, NJ: Princeton University Press, 1999.

Pidgeon, Nick, and M. O'Leary. "Man-Made Disasters: Why Technology and Organizations (Sometimes) Fail." *Safety Science* 34 (2000): 15–30.

Pigden, Jez, "Just Culture." Paper presented at the International Federation of Air Traffic Controllers' Associations 50th Annual Conference, Amman, Jordan, April 11–15, 2011.

Pinch, Trevor. "Technology and Institutions: Living in a Material World." *Theory and Society* 37 (2008): 461–83.

Pinch, Trevor, and Wiebe Bijker. "The Social Construction of Facts and Artefacts: Or How the Sociology of Science and Technology Might Benefit Each Other." *Social Studies of Science* 14, no. 3 (1984): 399–441.

Polanyi, Michael. *The Tacit Dimension*. London: Routledge, 1966.

Potthast, Jorg. "Ethnography of a Paper Strip: The Production of Air Safety." *Science, Technology, and Innovation Studies* 4, no. 1 (2008): 47–68.

Powell, Walter W., and Jeannette A. Colyvas. "The Microfoundations of Institutional Theory." In *Sage Handbook of Organizational Institutionalism*, edited by Royston Greenwood and Christine Oliver, 276–98. Thousand Oaks, CA: Sage, 2008.

Powell, Walter W., and Paul J. DiMaggio, eds. *The New Institutionalism in Organizational Analysis*. Chicago: University of Chicago Press, 1991.

Pressman, Jeffrey L., and Aaron Wildavsky. *Implementation: How Great Expectations in Washington Are Dashed in Oakland, or, Why It's Amazing That Federal Programs Work at All.* Berkeley: University of California Press, 1984.

Preston, Edmund, ed. *FAA Historical Chronology: Civil Aviation and the Federal Government, 1926–1996.* Washington, DC: Department of Transportation, Federal Aviation Administration Office of Public Affairs, 1998.

Reuters. "Pilot Secretly Alerted Controllers of Hijacking." *LA Times*, September 13, 2001.

"A Review of FAA's Efforts to Reduce Costs and Ensure Safety and Efficiency through Realignment and Consolidation." *Testimony before the United States House of Representatives, Committee on Transportation and Infrastructure, Subcommittee on Aviation.* 112th Cong. 1–12 (2012). Testimony of Paul M. Rinaldi, president of National Air Traffic Controllers Association.

"A Review of the Federal Aviation Administration's Air Traffic Controller Hiring, Staffing, and Training Plans." *Testimony before the United States House of Representatives, Committee of Transportation and Infrastructure, Subcommittee on Aviation.* 114th Cong. 4–5 (2016). Written testimony of Paul M. Rinaldi, president of National Air Traffic Controller Association.

Ristow, Walter W., ed. *Aviation Cartography: A Historico-Bibliographic Study of Aeronautical Charts.* 2nd ed. Washington, DC: Map Division, Library of Congress, 1960.

Roberts, Karlene H., Denise M. Rousseau, and Todd R. LaPorte. "The Culture of High Reliability: Quantitative and Qualitative Assessment Aboard Nuclear-Powered Aircraft Carriers." *Journal of High Technology Management Research* (Spring 1994): 141–61.

Sachs, S. E. "The Algorithm at Work? Explanation and Repair in the Enactment of Similarity in Art Data." *Information, Communication, & Society*, 2019. https://www.iandfonline.com/doi/full/10.1080/1369118X.20191612913.

Sagan, Scott D. *The Limits of Safety: Organizations, Accidents, and Nuclear Weapons.* Princeton, NJ: Princeton University Press, 1993.

Sauder, Michael. "Symbols and Texts: An Interactionist Approach to the Study of Social Status." *Sociological Quarterly* 46, no. 2 (2005): 279–98.

———. "Third Parties and Status Position: How the Characteristics of Status Systems Matter." *Theoretical Sociology* 35 (2006): 299–321.

Sauder, Michael, Freda Lynn, and Joel M. Podolny. "Status Insights from Organizational Sociology." *Annual Review of Sociology* 38 (2012): 267–83.

Schwartz, Stephen I. "This Is Not a Test." *Bulletin of the Atomic Scientists* 57, no. 6 (2001): 50–52.

Scott, W. Richard. *Institutions and Organizations: Ideas, Interests, and Identities.* Thousand Oaks, CA: Sage, 2014.

Sellen, Abigail J., and Richard H. R. Harper. *The Myth of the Paperless Office.* Cambridge, MA: MIT Press, 2003.

Sewell, William H., Jr. "A Theory of Structure: Duality, Agency, and Transformation." *American Journal of Sociology* 98, no. 1 (1992): 1–29.

———. "Three Temporalities: Toward an Eventful Sociology." In *The Historic Turn in the Human Sciences*, edited by T. J. McDonald, 245–80. Ann Arbor: University of Michigan Press, 1996.

"Shall America Take the Lead in Aeronautics?" *Scientific American* 98, no. 9 (February 29, 1908): 138.

Shostak, Arthur B., and David Skocik. *The Air Controllers' Controversy: Lessons from the PATCO Strike*. New York: Human Sciences Press, 1986.

Snook, Scott. *Friendly Fire: The Accidental Shootdown of U.S. Blackhawks over Northern Iraq*. Princeton, NJ: Princeton University Press, 2002.

Spencer, Lynn. *Touching History*. New York: Free Press, 2008.

Star, Susan Leigh, and James R. Griesemer. "Institutional Ecology, 'Translations,' and Boundary Objects: Amateurs and Professionals in Berkeley's Museum of Vertebrate Zoology, 1907–39." *Social Studies of Science* 19 (1989): 387–420.

Star, Susan Leigh, and Karen Ruhleder. "Steps toward an Ecology of Infrastructure: Design and Access for Large Information Spaces." *Information Systems Research* 7, no. 1 (1996): 111–34.

Staw, Barry M. "The Escalation of Commitment to a Course of Action." *Academy of Management Review* 6 (1981): 577–87.

Stinchcombe, Arthur L. "Social Structure and Organizations." In *Handbook of Organizations*, edited by James G. March, 142–93. Chicago: Rand McNally, 1965.

Suchman, Lucy A. "Centers of Coordination: A Case and Some Themes." In *Discourse, Tools, and Reasoning: Essays on Situated Cognition*, NATO ASI Series F Computer and Systems Sciences, edited by Lauren B. Resnick, Roger Saljo, Clotilde Pontecorvo, and Barbara Burge, 41–62. Berlin: Springer-Verlag, 1997.

———. *Plans and Situated Actions: The Problem of Human-Machine Communication*. Cambridge: Cambridge University Press, 1987.

———. "Representing Practice in Cognitive Science." *Human Studies* 11 (1988): 305–25.

Tavory, Iddo, and Nina Eliasoph. "Coordinating Futures: Toward a Theory of Anticipation." *American Journal of Sociology* 118, no. 4 (2013): 908–42.

Thayer, David L., and Jason M. Watson. "Supertaskers and the Multitasking Brain." *Scientific American Mind* 23, no. 1 (2012): 22–29.

Thornton, Patricia H., William Ocasio, and Michael Lounsbury. *The Institutional Logics Perspective: A New Approach to Culture, Structure, and Process*. Oxford: Oxford University Press, 2002.

Tierney, Kathleen. *The Social Roots of Risk: Producing Disasters, Promoting Resilience*. Stanford, CA: Stanford University Press, 2014.

Timmermans, Stefan, and Steven Epstein. "A World of Standards but Not a Standard World: Toward a Sociology of Standards and Standardization." *Annual Review of Sociology* 36 (2010): 69–89.

Tsoukas, Haridimos, and Efi Vladimirou. "What Is Organizational Knowledge?" *Journal of Management Studies* 38 (2001): 974–93.

Turner, Barry. *Man-Made Disasters*. London: Wykeham, 1978.

US Centennial of Flight Commission. "Air Traffic Control." https://www.centennialofflight.net/essay/Government_Role/Air_traffic_control/POL15.htm.

US Department of Transportation. Air Traffic Organization Policy, Order JO 7110.65.110B. November 17, 2017.

US Department of Transportation, Federal Aviation Administration. *Aeronautical Information Manual*. Washington, DC: February 24, 2000.

———. *Air Traffic Procedures 7110.65*. Washington, DC: Government Printing Office, 1995.

US Department of Transportation, Federal Aviation Administration, Section 804 Collaborative Workgroup. *FAA National Facilities Realignment and Consolidation Report, Year 1, Part 1 Recommendations*. Response to US Congress FAA Reauthorization Bill, Pub. L. No. 112-095, Section 804, March 11, 2015.

Urry, John. *What Is the Future*? Cambridge, UK: Polity Press, 2016

Van Maanen, John. *Tales of the Field*. Chicago: University of Chicago Press, 1988.

Van Maanen, John, and Stephen R. Barley. "Cultural Organization: Fragments of a Theory." In *Organizational Culture*, edited by Peter J. Frost, Larry F. Moore, Meryl Reis Louis, Craig C. Lundberg, and Joanne Martin, 31–54. Beverly Hills, CA: Sage, 1985.

———. "Occupational Communities: Culture and Control in Organizations." In *Research in Organizational Behavior*, edited by B. M. Staw and L. L. Cummings, 6:287–365. Greenwich, CT: Jai Press, 1984.

Vaughan, Diane. "Bourdieu and Organizations: The Empirical Challenge." *Theory and Society* 7, no. 1 (2008): 37–65.

———. *The Challenger Launch Decision: Risky Technology, Culture, and Deviance at NASA*. Chicago: University of Chicago Press, 1996.

———. *Controlling Unlawful Organizational Behavior: Social Structure and Corporate Misconduct*. Chicago: University of Chicago Press, 1983.

———. "NASA Revisited: Theory, Analogy, and Public Sociology." *American Journal of Sociology* 112, no. 2 (2006): 353–93.

———. "Report: Air Traffic Control Survey Analysis." Unpublished, Department of Sociology, Boston College, Chestnut Hill, MA. Distributed to Boston Tower, Boston TRACON, Boston ARTCC, Bedford Tower, New England Regional Office director, NATCA regional president, Boston Center NATCA, February 2003.

———. "The Social Shaping of Commission Reports." *Sociological Forum* 21, no. 2 (2006): 291–306.

———. "Theorizing: Analogy, Cases, and Comparative Social Organization." In *Theorizing in Social Science*, edited by Richard Swedberg, 61–84. Stanford, CA: Stanford University Press, 2014.

———. "Theorizing Disaster: Analogy, Historical Ethnography and the *Challenger* Accident." *Ethnography* 5, no. 3 (2004): 313–45.

———. "Theory Elaboration: The Heuristics of Case Analysis." In *What Is a Case? Exploring the Foundations of Social Inquiry*, edited by Charles C. Ragin and Howard S. Becker, 173–202. Cambridge: Cambridge University Press, 1992.

———. *Uncoupling: Turning Points in Intimate Relations*. New York: Oxford University Press, 1986.

Vertesi, Janet. *Seeing Like a Rover: How Robots, Teams, and Images Craft Knowledge of Mars*. Chicago: University of Chicago Press, 2015.

———. *Shaping Science: Organizations, Decisions, and Culture on NASA's Teams*. Chicago: University of Chicago Press, 2020.

Wajcman, Judith. "Automation: Is It Really Different This Time?" *British Journal of Sociology* 68, no. 3 (2003): 119–27.

———. *Pressed for Time: The Acceleration of Life in Digital Capitalism*. Chicago: University of Chicago Press. 2015.

Wald, Matthew, and Kevin Sack. "'We Have Some Planes,' Hijacker Told Controller." *New York Times*, October 16, 2001.

Weick, Karl E. "The Collapse of Sensemaking in Organizations: The Mann Gulch Disaster." *Administrative Science Quarterly* 38 (1993): 628–52.

———. "Educational Organizations as Loosely Coupled Systems." *Administrative Science Quarterly* 21, no. 1 (1976): 1–19.

———. "Improvisation as a Mindset for Organizational Analysis." *Organization Science* 9 (1998): 543–55.

Weick, Karl E., and Karlene H. Roberts. "Heedful Interrelating on Flight Decks." *Administrative Science Quarterly* 38 (1993): 357–81.

Whyte, William F. *Street Corner Society: The Social Organization of a Slum*. Chicago: University of Chicago Press, 1943.

Wimmer, Andreas. *Ethnic Boundary Making: Institutions, Power, and Networks*. New York: Oxford University Press, 2013.

Winthereik, Brit Ross, Peter Lutz, Lucy Suchman, and Helen Verran. "Attending to Screens and Screenness." *STS Encounters* 4, no. 2 (2011): 1–6.

Wright, Orville. *How We Invented the Airplane: An Illustrated History*. Edited and with an introduction and commentary by Fred C. Kelly and with additional text by Alan Weissman. New York: Dover Publications, 1988.

Wright, Wilbur. "Flying as a Sport—Its Possibilities." *Scientific American* 98, no. 9 (February 29, 1908): 139.

———. "Some Aeronautical Experiments." *Journal of the Western Society of Engineers* (December 1901). Reprinted in "Classics," *Resonance* 8, no. 12 (2003): 99–114.

"The Wright Aeroplane and Its Performances." *Scientific American* 94, no. 14 (April 7, 1906): 291.

"The Wright Aeroplane Tests: Wilbur Wright's Latest Flights in France." *Scientific American* 99, no. 9 (August 29, 1908): 135.

Zerubaval, Eviatar. "The Standardization of Time: A Sociohistorical Perspective." *American Journal of Sociology* 88, no. 1 (1982): 1–23.

———. *Time Maps: Collective Memory and the Shape of the Past*. Chicago: University of Chicago Press, 2004.

———. "Timetables and Scheduling: On the Social Organization of Time." *Sociological Inquiry* 46, no. 2 (1976): 87–94.

Zucker, Lynne G. "The Role of Institutionalization in Cultural Persistence." *American Sociological Review* 42, no. 5 (1977): 726–43.

Index

Page numbers in italics refer to figures.